HONEY BEE COLONY HEALTH

Challenges and Sustainable Solutions

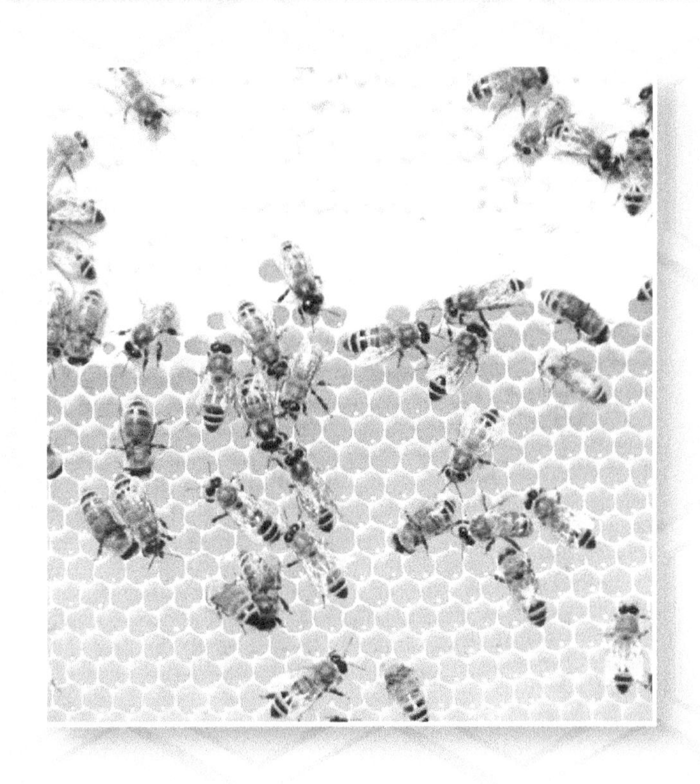

CONTEMPORARY TOPICS in ENTOMOLOGY SERIES

THOMAS A. MILLER EDITOR

Insect Symbiosis
Edited by Kostas Bourtzis and Thomas A. Miller

Insect Sounds and Communication: Physiology, Behaviour, Ecology, and Evolution
Edited by Sakis Drosopoulos and Michael F. Claridge

Insect Symbiosis, Volume 2
Edited by Kostas Bourtzis and Thomas A. Miller

Insect Symbiosis, Volume 3
Edited by Kostas Bourtzis and Thomas A. Miller

Food Exploitation by Social Insects: Ecological, Behavioral, and Theoretical Approaches
Edited by Stefan Jarau and Michael Hrncir

Molecular Biology and Genetics of the Lepidoptera
Edited by Marian R. Goldsmith and František Marec

Honey Bee Colony Health: Challenges and Sustainable Solutions
Edited by Diana Sammataro and Jay A. Yoder

HONEY BEE COLONY HEALTH

Challenges and Sustainable Solutions

Edited by
Diana Sammataro & Jay A. Yoder

CRC Press
Taylor & Francis Group
6000 Broken Sound Parkway NW, Suite 300
Boca Raton, FL 33487-2742

Version Date: 20111109

International Standard Book Number: 978-1-4398-7940-5 (Hardback)

Visit the Taylor & Francis Web site at
http://www.taylorandfrancis.com

and the CRC Press Web site at
http://www.crcpress.com

Table of Contents

Photo by G. DeGrandi-Hoffman.

Photo by D. Sammataro.

Preface

Photo by D. Sammataro.

There is no doubt that honey bees have given human beings great gifts—in the honey that sweetens our lives and in the wax candles that light our way. This highly social insect has inspired generations of writers, philosophers, artists, poets, and scientists with its industry and complex organization. Of all the insects that exist in our world, honey bees are far and away the most studied.

But all of the gifts that honey bees have bestowed do not compare with the role these industrious insects play every day—in the interaction of bees and flowers. It cannot be stressed enough that without this action of bees pollinating the numerous flowering plants that bloom every day, we poor humans would be restricted to a diet of grains and precious few fruits and vegetables. Bees were so important that the early American colonists made sure they were included in the perilous journey to the New World. This migration of bees continues to this day.

The pollinating power of bees secures for us our modern-day diet of fruits and vegetables for human consumption as well as the forage crops our animals eat. Bees also allow us to beautify our gardens by setting seeds of the many flowers we plant. Ironic, then, that honey bees have been taken for granted so long and were under-valued by many, reflected by the low prices paid to beekeepers for their hive products and pollinating activities.

This complacency is now a thing of the past. Realization of the vital importance of honey bees has been the focus of the many orchardists who grow and manage hundreds of acres of single-species crops, such as almonds, apples, and berries. The commercial monocultures needed to feed a hungry population that demands cheap food and year-round exotic fruits were the first to appreciate how dependent they are on bee pollination. Commercial orchards need commercial bees, so beekeeping became a large-scale business to meet the pollination demands.

As always when any animals or plants are forced to live together in crowded conditions, problems arise. Actions must be taken to keep them healthy and to protect them from the deleterious effects of mass feeding or chemicals to control pests and pathogens. While crop yield is undeniably boosted by such measures, it comes at the expense of depriving foraging bees of necessary nutrients and exposing them to toxins that interfere with their natural processes.

Photo by M. Frazier.

All of these factors could have been balanced had it not been for the accidental introduction of two parasitic mites. One of them, *Varroa destructor*, has proven to be the most debilitating to the Western honey bee (*Apis mellifera*) throughout its global distribution. Varroa weakens bees through their feeding, reproduction, and the transmission of new pathogens.

To make matters worse, new invaders have come, including a new beetle, which made its way out of Africa, as well as a new virus and a new fungus. All of these factors together create a perfect storm that the beleaguered bees cannot handle. Anyone who works with this wonderful insect is amazed at their resistance to pests, their tolerance to human meddling, and their ability to survive. The loss of bees and all pollinators is now the focus of intense research and speculation by many. We who work with bees are grateful that at last the importance of bees is now realized and this hard-working insect is getting the attention it needs. The downside, of course, is that bees are in trouble and, like other at-risk animals, could be the "canary in the coal mine" of the ecological changes that are currently happening in our world. The many authors who graciously agreed to contribute their expertise to this volume are united in their concern for bees during this current crisis.

The purpose of this book is to provide collective knowledge from the many scientists who work with bees, to share their research, and to inspire future generations of researchers, beekeepers, and students to continue to study bees and keep them healthy and pollinating.

Acknowledgments

Photo by M. Frazier.

The editors wish to express their thanks and appreciation to the numerous authors for their hard work in contributing to this volume. Your enthusiasm for this project demonstrates your commitment to sharing your knowledge with others so that all of us who work with bees may gain in understanding the myriad factors that influence the complex world of the bee. Your chapters made this book both important and inspirational. Thank you.

We also greatly appreciate Zach Hedges, who organized and set up the text of all the original documents. This was done under great pressure while Zach was a student and studying for exams. Thank you for your hard work, both in learning a new page-layout program and in organizing the chapters into a cohesive volume.

Judy Walker took firm control of those chapters and created this original and beautifully crafted volume. Her professional acumen and creative insights are inspiring. Not only does she have a fine eye for what is aesthetically pleasing, but also those techniques that help convey the information in a unique and forthright manner. Thank you for your diligent work; it was a great pleasure to work with you on this enterprise.

Ann Harman, as usual, was invaluable as our proofreader. Your quick eye and sharp mind are valuable assets to any author. Thank you.

We also wish to extend our gratitude to John Sulzycki, Senior Editor at CRC Press, Taylor & Francis Group, and his staff for encouraging us to publish with them. Thank you for your help and for suggesting the title.

And last but not least, we wish to dedicate this book to all the hard-working honey bees, without which we would have nothing to write about. Keep up the good work.

Diana Sammataro, Ph.D.
Research Entomologist,
USDA-ARS Carl Hayden Bee Research Center
Tucson, Arizona

Jay Yoder, Ph.D.
Wittenberg University
Springfield, Ohio

Contributors

Cédric Alaux
INRA, UMR 406
Abeilles et Environnement
Site Agroparc
Avignon, France
Chapter 18

Fabrice Allier
ITSAP-Institut de l'abeille
UMT PrADE
Site Agroparc
Avignon, France
Chapter 20

Kirk E. Anderson
c/o Gloria DeGrandi-Hoffman
USDA-ARS Carl Hayden Bee Research Center
Tucson, Arizona, U.S.
Chapter 1

Katherine Aronstein
USDA-ARS Honey Bee Research Unit
Kika de la Garza Subtropical Agricultural Research Center
Weslaco, Texas, U.S.
Chapters 10, 11

Kathy Baylis
Department of Agricultural and Consumer Economics
University of Illinois
Urbana, Illinois, U.S.
Chapter 20

Robert Brodschneider
Department of Zoology
Karl Franzens University
Graz, Austria
Chapter 20

Yves Brostaux
Gembloux Agro-Bio Tech
Department of Applied Statistics, Computer Science and Mathematics
University of Liege
Gembloux, Belgium
Chapter 20

Jean-Luc Brunet
INRA, UMR 406
Abeilles et Environnement
Site Agroparc
Avignon, France
Chapter 18

Giles Budge
National Bee Unit
The Food and Environment Research Agency
Sand Hutton, York, United Kingdom
Chapter 20

Humberto E. Cabanillas
USDA-ARS Honey Bee Research Unit
Kika de la Garza Subtropical Agricultural Research Center
Weslaco, Texas, U.S.
Chapter 11

Dewey Caron
Horticulture Department
Oregon State University
Corvallis, Oregon, U.S.
Chapter 20

Yan Ping Chen
USDA-ARS Bee Research Laboratory
Beltsville, Maryland, U.S.
Chapters 7, 8

Brady S. Christensen
Department of Biology
Wittenberg University
Springfield, Ohio, U.S.
Chapter 16

Susan W. Cobey
University of California
Davis, California, U.S.
Chapter 4

Travis J. Croxall
Department of Biology
Wittenberg University
Springfield, Ohio, U.S.
Chapter 16

Joachim R. de Miranda
Department of Ecology
Swedish University of Agricultural Sciences
Uppsala, Sweden
Chapter 8

Gloria DeGrandi-Hoffman
USDA-ARS Carl Hayden Bee Research Center
Tucson, Arizona, U.S.
Introduction, Chapters 1, 12, 16, 17

Keith S. Delaplane
Department of Entomology
University of Georgia
Athens, Georgia, U.S.
Chapter 19

Claudia Dussaubat
INRA, UMR 406
Abeilles et Environnement
Site Agroparc
Avignon, France
Chapter 18

Bruce Eckholm
c/o Gloria DeGrandi-Hoffman
USDA-ARS Carl Hayden Bee Research Center
Tucson, Arizona, U.S.
Chapter 1

James D. Ellis
Honey Bee Research and Extension Laboratory
Department of Entomology and Nematology
University of Florida
Gainesville, Florida, U.S.
Chapter 13

Marion D. Ellis
Department of Entomology
University of Nebraska
Lincoln, Nebraska, U.S.
Chapter 14

Jay D. Evans
USDA-ARS Bee Research Laboratory
Beltsville, Maryland, U.S.
Chapter 7

Richard D. Fell
Department of Entomology
Virginia Polytechnic Institute and State University
Blacksburg, Virginia, U.S.
Chapter 9

Jennifer Finley
USDA-ARS Carl Hayden Bee Research Center
Tucson, Arizona, U.S.
Chapter 17

Maryann Frazier
Department of Entomology
The Pennsylvania State University
University Park, Pennsylvania, U.S.
Chapter 14

Ingemar Fries
Department of Ecology
Swedish University of Agricultural Sciences
Uppsala, Sweden
Chapter 3

Laurent Gauthier
Swiss Bee Research Centre
Agroscope Liebefeld-Posieux Research Station ALP
Bern, Switzerland
Chapter 8

Aleš Gregorc
Agricultural Institute of Slovenia
Slovenia University of Maribor
Ljubljana, Slovenia
Chapter 15

Eric Haubruge
Gembloux Agro-Bio Tech
Department of Functional and Evolutionary Entomology
University of Liege
Gembloux, Belgium
Chapter 20

Brian Z. Hedges
Department of Biology
Wittenberg University
Springfield, Ohio, U.S.
Chapters 12, 16, 17

Derrick J. Heydinger
Department of Biology
Wittenberg University
Springfield, Ohio, U.S.
Chapters 12, 16, 17

Hannelie Human
Department of Zoology and Entomology
University of Pretoria
Pretoria, South Africa
Chapter 20

David W. Inouye
Department of Biology
University of Maryland
College Park, Maryland, U.S.
Chapter 21

Reed M. Johnson
Department of Entomology
University of Nebraska
Lincoln, Nebraska, U.S.
Chapter 14

Yves Le Conte
INRA, UMR 406
Abeilles et Environnement
Site Agroparc
Avignon, France
Chapters 18, 20

Eugene Lengerich
Professor of Health Evaluation Sciences
College of Medicine
The Pennsylvania University
Hershey, Pennsylvania, U.S.
Chapter 20

Dawn Lopez
USDA-ARS Bee Research Laboratory
Beltsville, Maryland, U.S.
Chapter 7

Cynthia McDonnell
INRA, UMR 406
Abeilles et Environnement
Site Agroparc
Avignon, France
Chapter 18

William G. Meikle
USDA-ARS Honey Bee Research Unit
Kika de la Garza Subtropical Agricultural Research Center
Weslaco, Texas, U.S.
Chapter 6

Andony Melathopoulos
Beaverlodge Research Farm
Agriculture & Agri-Food Canada
Beaverlodge, Alberta, Canada
Chapter 20

Guy Mercadier
USDA-ARS European Biological Control Laboratory
Gely du Fes, France
Chapter 6

Christopher A. Mullin
Department of Entomology
The Pennsylvania State University
University Park, Pennsylvania, U.S.
Chapter 14

Mustafa Muz
Department of Parasitology
Tayfur Sokmen Campus
Mustafa Kemal University
Hatay, Turkey
Chapter 20

Peter Neumann
Swiss Bee Research Centre
Agroscope Liebefeld-Posieux
Research Station ALP
Bern, Switzerland
Chapter 20

Bach Kim Nguyen
Department of Functional and Evolutionary Entomology
University of Liege
Gembloux, Belgium
Chapter 20

Roberta C. F. Nocelli
Center of Agrary Science
Federal University of São Carlos
São Carlos Campus
Araras, SP, Brazil
Chapter 15

Aslı Özkırım
Department of Biology
Bee Health Lab
Hacettepe University
Beytepe, Ankara, Turkey
Chapter 2

Stephen Pernal
Beaverlodge Research Farm
Agriculture and Agri-Food Canada
Beaverlodge, Alberta, Canada
Chapter 20

Jeffery Pettis
USDA-ARS Bee Research Laboratory
Beltsville, Maryland, U.S.
Chapters 7, 20

Stéphane Pietravalle
Food and Environment Research Agency
Central Science Laboratory
York, United Kingdom
Chapter 20

Lennard Pisa
Netherlands Centre Bee Research (NCB)
Utrecht, Netherlands
Chapter 20

Magali Ribière
Bee Disease Unit
French Food Safety Agency (AFSSA)
Sophia-Antipolis, France
Chapter 8

Claude Saegerman
Faculty of Veterinary Medicine
Department of Infectious and Parasitic Diseases
Research Unit in Epidemiology and Risk Analysis Applied to the Veterinary Sciences (UREAR)
University of Liége
Liége, Belgium
Chapter 20

Diana Sammataro
USDA-ARS Carl Hayden Bee Research Center
Tucson, Arizona, U.S.
Chapters 5, 6, 12, 16, 17

Walter S. Sheppard
Washington State University
Pullman, Washington, U.S.
Chapter 4

Elaine C. M. Silva-Zacarin
Laboratory of Functional and Structural Biology
Federal University of São Carlos
Sorocaba Campus
Sorocaba, SP, Brazil
Chapter 15

I. Barton Smith
USDA-ARS Bee Research Laboratory
Beltsville, Maryland, U.S.
Chapter 7

Angela Spleen
Department of Public Health Sciences
College of Medicine
The Pennsylvania State University
College Park, Pennsylvania, U.S.
Chapter 20

David R. Tarpy
North Carolina State University
Raleigh, North Carolina, U.S.
Chapter 4

Brenna E. Traver
Department of Entomology
Virginia Polytechnic Institute and State University
Blacksburg, Virginia, U.S.
Chapter 9

Robyn Underwood
Department of Entomology
The Pennsylvania State University
University Park, Pennsylvania, U.S.
Chapter 20

Julien Vallon
ITSAP-Institut de l'abeille
UMT PrADE
Site Agroparc
Avignon, France
Chapter 20

Romee van der Zee
Netherlands Centre Bee Research (NCB)
Tersoal, Netherlands
Chapter 20

Dennis vanEngelsdorp
Department of Entomology
The Pennsylvania State University
University Park, Pennsylvania, U.S.
Chapter 20

Thomas C. Webster
Land Grant Program
Kentucky State University
Frankfort, Kentucky, U.S.
Chapter 10

Selwyn Wilkins
National Bee Unit
Food and Environment Research Agency
York, United Kingdom
Chapter 20

Jay A. Yoder
Department of Biology
Wittenberg University
Springfield, Ohio, U.S.
Chapters 12, 16, 17

Introduction

Gloria DeGrandi-Hoffman

Take care of the small things
and the big things take care of themselves.

Pollinators play an important role in terrestrial ecosystems because they provide a service that is vital to the diversity and maintenance of wild plant communities (Ashman et al., 2004; Aguilar et al., 2006; Potts et al., 2010) and to agricultural production (McGregor, 1976; Klein et al., 2007; Ricketts et al., 2008). Western honey bees (*Apis mellifera*) are arguably the single-most valuable insect pollinator to agriculture because their hives can be easily maintained and transported to pollinator-dependent crops. Though most of the human diet does not depend upon honey bee-pollinated crops (e.g., grains are wind-pollinated) (Ghazoul, 2005), the production of 39 of the world's 57 most important monoculture crops benefit from honey bee pollination (Klein et al., 2007).

There has been an almost 50% decrease in world honey bee stocks over the last century. Simultaneously there has been a >300% increase in pollinator-dependent crops (Aizen and Harder, 2009). Much of this increase in production is due to the heightened awareness of the importance of fruits and vegetables in the human diet, and their role in the prevention of cancer, heart disease and obesity. Because honey bees are essential for the production of foods that are critical for maintaining human health, there is great concern about the health of honey bee colonies and the recent large-scale die-offs of honey bees around the world.

Colonies fail for many reasons. Some die from starvation and others from queen failure. Diseases, parasites, and pests can weaken colonies so that they cannot defend themselves or survive overwintering. Pesticides also can cause colony losses. Most recently, a syndrome of unknown etiology called Colony Collapse Disorder (CCD) has caused massive colony losses especially to commercial beekeeping operations. The following chapters provide the latest information on factors that undermine the health and survival of colonies. The role of beneficial microbes and bee breeding in maintaining the health and vigor of colonies also is included as these areas have the potential to create sustainable solutions to problems of colony health.

The strength and survival of honey bee colonies depend on a laying queen and sufficient numbers of workers to collect food, rear brood, and defend the hive. Key factors affecting the population size of a colony are the rate at which the worker brood is reared to adulthood and the longevity of workers as adults. Slight changes in either brood-rearing or worker longevity can have profound effects on colony survival. This is because the number of brood that can be reared depends on the size of the current adult population. If the adult population is large, more brood can be reared, and the adult population continues to grow. Large colony size in social insects is usually associated with higher reproductive output (i.e., swarming), competitiveness, and colony longevity (Wilson, 1971; Hölldobler and Wilson, 1990; Kaspari and Vargo, 1995; Karsai and Wenzel, 1998). Conversely, small colonies rear limited numbers of brood and are more vulnerable than larger colonies to death from starvation, disease, parasites, or overwintering. Colonies can show great resilience to stress factors that might perturb their populations for short periods of time. However, any factors that affect the rate that brood is reared or the longevity of workers ultimately strike at the heart of colony health and survival.

The foundation for healthy colonies is good nutrition. Honey bees obtain all their nutritional needs from nectar and pollen. Nectar supplies carbohydrates and pollen provides protein, vitamins, and minerals. The carbohydrates in nectar provide energy for hive maintenance, comb building, and foraging. The nutrients in pollen are used for brood rearing. The amount of protein in colonies affects its levels in worker bees and the age at which they begin to forage. Colonies with limited protein reserves have workers that convert from nest tasks to foraging at earlier ages (Schulz et al., 1998; Toth et al., 2005; Toth and Robinson, 2005). Workers that delay the onset of foraging have longer lives than those that begin foraging earlier (Rueppell et al., 2007). Thus food shortages set up colony declines because brood rearing is limited and worker longevity is reduced.

In addition to affecting population growth, nutritional stress affects the ability of colonies to mount immune responses to pathogens. If bees are deprived of protein, virus levels in individuals increase at greater rates compared with those fed pollen or even a protein supplement (DeGrandi-Hoffman et al., 2010). The importance of pollen diversity and immune response also has been documented. Bees fed pollen from multiple plant sources have enhanced immune function compared with those fed on pollen from a single plant source (Alaux et al., 2010a).

Honey bees obtain nutrients from their food due to the action of microbes housed in their digestive tract and in food stores. We are just beginning to fully understand the role of beneficial microbes in the health of honey bee colonies. New molecular tools such as metagenomic analysis circumvent the restrictions of studying only those microbes that can be cultured. By combining information from the honey bee genome with gene sequences from microbial DNA, we will be able to determine the role of microbial communities in the preservation, digestion, and metabolism of food. We also will be able to evaluate the full impact of environmental contaminants in nectar and pollen especially those compounds with antimicrobial activity (e.g., fungicides). These compounds might be contributing to colony losses by impacting the beneficial microbes in hives such that bees cannot obtain adequate nutrition even though there are sufficient nectar and pollen stores in the colony. The compounds also could be impacting immune function and the presence of symbiotic microbes that inhibit the growth of pathogens such as bacteria associated with foulbrood.

Colony losses and the occurrence of CCD have been associated with parasites (primarily the ectoparasitic mite *Varroa destructor*) and pathogens (virus and Nosema). Colonies that are infested with Varroa are severely weakened because the mite feeds on developing pupae. Individuals that are parasitized as pupae have lower weight at emergence, underdeveloped hypopharyngeal glands (Schneider and Drescher, 1987) and a range of other developmental maladies (DeJong et al., 1982; Marcangeli et al., 1992). There also are reduced levels of protein in the hemolymph and carbohydrate levels in the abdomen of workers that were parasitized by Varroa (DeJong et al., 1982; Kovak and Crailsheim, 1988; Bailey and Ball, 1991; Bowen-Walker and Gunn, 2001) and these workers have shorter lifespans compared with nonparasitized bees (DeJong et al., 1982; Schneider and Drescher, 1987). Varroa also vectors several viruses that contribute to morphological deformities (small body size, shortened abdomen, deformed wings) that further reduce vigor and longevity of workers, and negatively affect their homing ability and duration of foraging flights (Schneider and Drescher, 1987; Koch and Ritter, 1991; Romero-Vera and Otero-Colina, 2002; Garedew et al., 2004; Kralj and Fuchs, 2006). Varroa also weakens the bee's immune system, suppressing the expression of immune-related genes and increasing virus titers (e.g., Deformed Wing Virus) (Yang and Cox-Foster, 2005, 2007). A number of viruses including Deformed Wing Virus, Acute Bee Paralysis Virus, Chronic Bee Paralysis Virus, Slow Bee Paralysis Virus, Black Queen Cell Virus, Kashmir Bee Virus, Cloudy Wing Virus, and Sacbrood Virus are associated to varying degrees with *V. destructor* infestation (Ball and Allen, 1988; Allen and Ball, 1996; Martin, 1998, 2001; Tentcheva et al., 2004; Carreck et al., 2010a,b; Martin et al., 2010).

Colonies can be lost due to virus infections. CCD is highly correlated with Israeli Acute Paralysis Virus (Cox-Foster et al., 2007). Most recently, a significant correlation was found between CCD and the presence of two previously unreported RNA viruses, *Varroa destructor*-1 virus, Kakugo virus, and an invertebrate iridescent virus (IIV) (Iridoviridae) (Bromenshenk et al., 2010). In addition, *Nosema ceranae* also was consistently detected in bees from the failing colonies.

Nosema belong to a group called Microsporidia that are obligatory intracellular parasites with a wide host range in invertebrate and vertebrate organisms (Larsson, 1986; Wasson and Peper, 2000; Tsai et al., 2003). Two species of pathogenic Nosema occur in honey bees; *Nosema apis* and *Nosema ceranae*. Both Nosema species affect the digestive function, and lead to malnutrition, physiological aging and a reduction in bee longevity (Fries, 1997). *Nosema apis* has long been known as a parasite of the European honey bee (Matheson, 1993), and *N. ceranae* was known to infect the Asian honey bee, *Apis cerana* (Fries et al., 1996). Recently *N. ceranae* was detected in *A. mellifera* in Europe (Higes et al., 2006) and in the Americas, and Asia (Chauzat et al., 2007; Cox-Foster et al., 2007; Huang et al., 2007; Klee et al., 2007; Paxton et al., 2007; Chen et al., 2008; Williams et al., 2008; Invernizzi et al., 2009; Chen and Huang, 2010). Detection of *Nosema ceranae* and fear of colony losses from it have caused beekeepers to increase their use of antibiotics especially Fumagillin in colonies.

Though high titers of pathogenic organisms are consistently detected in collapsing colonies, whether they are the primary cause of the colony loss or a consequence of factors that have compromised the immune system remains to be determined. Strong colonies can respond to pathogens and parasites and mitigate their effects before they manifest as serious infections. Honey bees have evolved mechanical, physiological and immunological defenses against disease agents (Evans et al., 2006; Schmid et al., 2008; Wilson-Rich et al., 2008; Evans and Spivak, 2010). Behaviors that reduce the risk of disease can be expressed by individuals or as groups of workers (Starks et al., 2000; Spivak and Reuter, 2001). Selection for genetic lines of bees with greater tendencies for these behaviors (e.g., hygienic lines) can increase the ability of colonies to control pathogens and pests without the use of antibiotics or miticides. Honey bees also can mount a physiological immune response to wounding or pathogen exposure via the synthesis of proteins that recognize signals from invading pathogens and metabolites capable of halting their growth (Evans et al., 2006; Theopold and Dushay, 2007; Lemaitre and Hoffmann, 2007). Responses by insect circulating cells (hemocytes) also can be mounted to reduce pathogen loads by phagocytic activity.

In addition to parasites and pathogens, the health of colonies is being compromised by environmental toxins. Contamination of wax comb and food stores by pesticides and fungicides is common especially in commercial honey bee colonies used for pollination (Mullin et al., 2010). A survey of pesticide residues in samples from beekeeping operations across the U.S. found 121 different pesticides and their metabolites within samples of wax, pollen, bee, and hive components. More than half of the wax and pollen samples had at least one systemic pesticide and almost all the comb samples had degradates of miticides (fluvalinate, coumaphos and amitraz). The fungicide chlorothalonil also was commonly detected. Individually, many of these neurotoxins cause acute and sublethal reductions in bee health. What remains to be determined are the effects that combinations of these compounds have on colony health and if there is an association with colony losses. There is evidence indicating an interaction between Nosema and a neonicotinoid (imidacloprid) that increases colony mortality rates when both are present (Alaux et al., 2010b).

Honey bee colonies are in greater demand and are renting for higher fees than ever before. Commercial beekeeping has changed from primarily honey production to crop pollination. With this change has come extraordinary stress from moving colonies multiple times a year and increased exposure to diseases, parasites, and hive pests. Antibiotics and acaracides are being applied several times a year causing resistance and comb contamination. To continue using colonies as mobile pollinator populations will require new management methods that require fresh perspectives on nutrition, breeding practices, and the role of microbes in sustaining colony health. Finding ways to prevent outbreaks of disease and control Varroa also will be essential for reducing colony losses. The information in the following chapters could provide the basis for establishing management methods that maintain the health of colonies and secure their availability for pollination.

Honey Bee Health: The Potential Role of Microbes

CHAPTER

1

Gloria DeGrandi-Hoffman, Bruce Eckholm, and Kirk E. Anderson

Abstract Bees carry a diverse assemblage of microbes, mostly bacteria and fungi. Most microbes are commensals or even beneficial to the colony and few are pathogenic. Their role is to inhibit the growth of pathogenic bacteria and fungi, aid in food digestion and food store preservation, facilitate gene expression, and affect colony-level traits such as social immunity. This chapter summarizes the role of beneficial microbes, showing their importance for maintaining colony health, and discusses current molecular and metagenomic techniques (cf. sequence-based analysis of DNA and 16S rRNA) and findings that have expanded our knowledge about the complexity of the microbiome related to honey bees, their colonies, and community function. Of interest is that the current Human Microbiome Project (HMP) and questions related to host–microbe interactions could benefit microbiome analysis of honey bees. These studies reveal the significance of microbes in the health of all organisms and clearly indicate that optimum health depends on maintaining conditions that encourage microbial growth.

Introduction

Microorganisms, or microbes, are a diverse group of unicellular organisms, including bacteria, fungi, archaea, protists, and sometimes viruses. Microbial activity in honey bee colonies can be broadly divided into three areas: microbes that are pathogenic, causing disease within the colony; and those that are either benign to the honey bee (commensal) or beneficial. Bees carry a diverse assemblage of microbes (mostly bacteria and fungi), very few of which are pathogenic; most are likely commensal or even beneficial to the colony (Evans and Armstrong, 2006). Some beneficial microbes inhibit the growth of pathogenic bacteria and fungi; others aid in the digestion of food and the preservation of food stores (e.g., pollen). Microbes may even play a role in honey bee gene expression and affect colony-level traits such as social immunity.

While symbiotic microbes are found in many organisms, they may play a special role for honey bees. This is because bees obtain a majority of their nutrients from pollen, a food source resistant to enzymatic digestion. Furthermore, bees store pollen in the warm, moist environment of the hive that can provide ideal growth conditions for many pathogenic microorganisms. Honey bees rely on their symbiotic microbes to protect their food stores from spoilage. Indeed, it is in part through the action of microbes that the mixture of pollen and nectar is converted to a fermented food called bee bread that supplies the protein diet of honey bees. When honey bees feed on bee bread, some of the microbes are transferred to the bees where they might continue to aid in digestion and supply necessary components to the host metabolism.

Interactions between honey bees and microbes for the processing and preservation of food begin during foraging (Figure 1). Foragers inoculate the pollen they collect with nectar from their honey stomach and thus seed the pollen in their corbiculae with bacteria and fungi. Additional microbes are added to the pollen when it is packed into cells. These microbes cause the pH of the pollen mixture to be lowered and thus begin a fermentation process that converts the pollen into bee bread. Worker bees consume bee bread, metabolize the nutrients, and use them

Figure 1. An overview of the role of symbiotic microbes in the conversion of pollen and nectar to bee bread and its use in the bee colony. (*Top photo by P. Grebs; remaining photos by G. D. Hoffman.*)

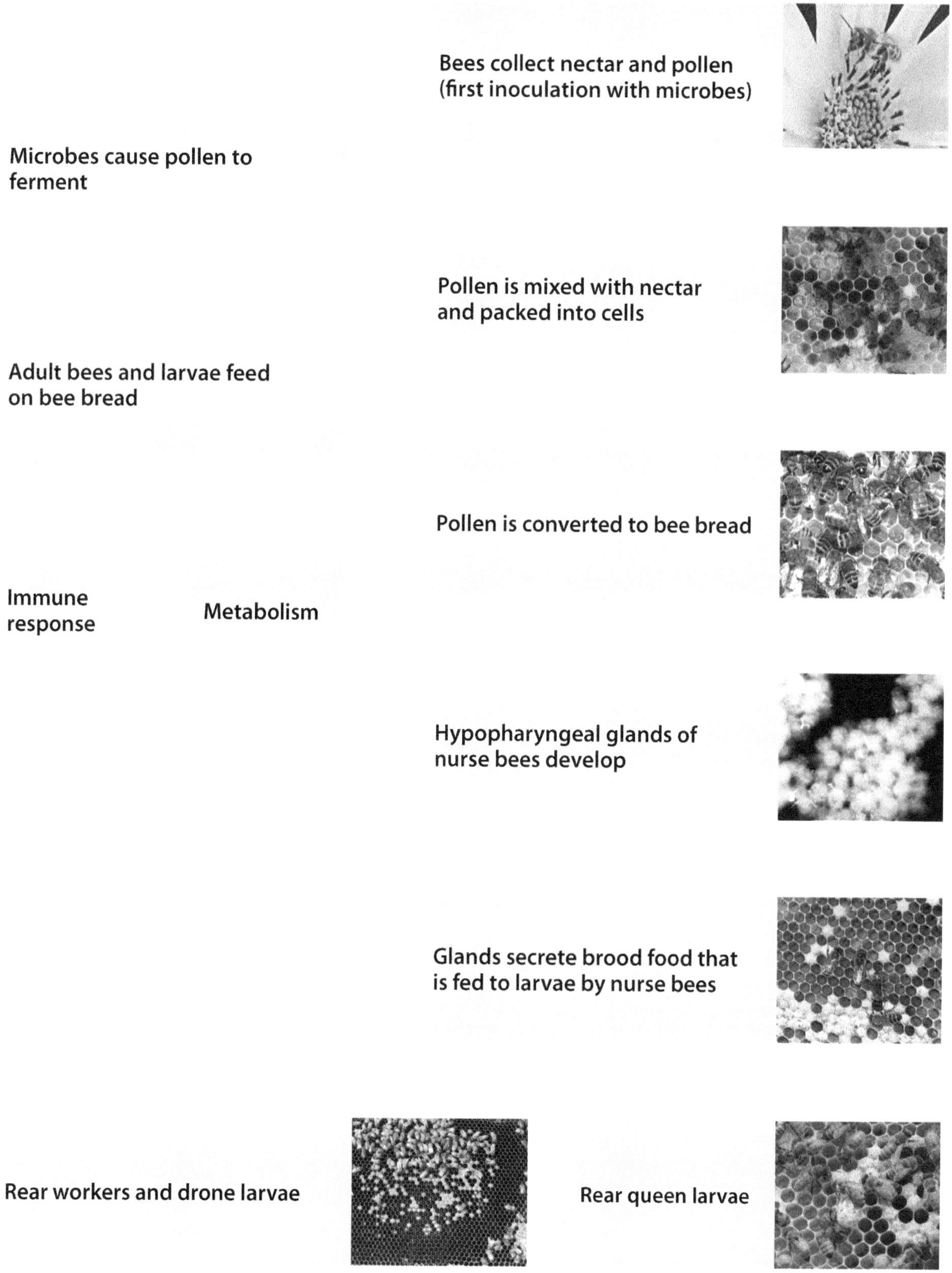

in anabolic and catabolic reactions and to generate immune responses. Young workers serving as nurse bees convert the bee bread to brood food. The brood food is used to rear larvae of all castes and also to feed adult nestmates via trophallactic interactions.

In this chapter, we will discuss beneficial microbes and their role in the nutrition and health of honey bee colonies. We will provide a historical summary of work on beneficial microbes and discuss new molecular and metagenomic tools and findings that can enable scientists to more deeply investigate the microbiome of honey bees and their colonies.

History of Research on Microbes in Honey Bees

Early studies on identification of microbes and descriptions of their function were done solely on species that could be cultured. Through the efforts of diligent and highly skilled microbiologists, hundreds of species of bacteria and fungi were isolated and identified from pollen, bee bread, and all stages of bees. The description of microbes in this section includes only those that could be cultured using the techniques available at the time. We now know that microbes that can be cultured make up only a small fraction of all those that actually reside in an organism. Therefore, when the assertion is made that certain life stages are devoid of microbes, it is based on only those that could be cultured.

Studies on the establishment of the microflora in worker bees revealed that worker-destined eggs were free internally of microorganisms. After eclosion, most young larvae have microbially sterile digestive tracts, presumably because of the antibiotic properties of the worker jelly that is their sole food source during the first 36 hours of life (Burri, 1947; Kluge, 1963). However, as larvae age and begin consuming an admixture of worker jelly, pollen, and nectar, their intestine becomes inoculated with bacteria, molds, and yeasts (Gilliam and Prest, 1987).

Most (although not all) larvae continue to harbor a diverse microbial community within the alimentary canal until they approach pupation (Gilliam and Prest, 1987). After larvae eat their last meal, they discharge their gut contents into the comb cell (Snodgrass, 1925). Fecal analysis showed that the larval gut contains numerous *Bacillus* spp., Gram-variable pleomorphic bacteria, molds (primarily *Penicillia*), Actinomycetes, Gram-negative bacterial rods, and yeasts (Gilliam and Prest, 1987). Because gut contents are discharged prior to pupation, the digestive systems of pupae and newly emerged adults are free of microbes. Upon emergence as an adult, the gut is reinoculated via pollen consumption and trophallactic food exchange with older nestmates (Burri, 1947; Kluge, 1963; Gilliam et al., 1983). As the bee ages, the gut microflora will vary. Season and geographical location also affect the species composition of the microflora; however, certain microbes are prevalent (Cherepov, 1966; Gilliam and Valentine, 1974; Gilliam et al., 1988). These include Gram-variable pleomorphic bacteria of uncertain taxonomic status, *Bacillus* spp., Enterobacteriaceae, molds, and yeasts (Gilliam et al., 1988).

In addition to inoculating newly emerged adult bees with microbes, adult bees also inoculate the pollen they collect and thus begin the fermentation and conversion processes that lead to bee bread. Floral pollen begins to change microbiologically and biochemically as soon as a honey bee collects it (Gilliam, 1979a,b; Loper et al., 1980; Standifer et al., 1980; Gilliam et al., 1989). At first glance, it might seem that foragers are adding nectar to the pollen to pack it more efficiently in their corbiculae.

While this might be so, the bees are also inoculating the pollen with microbes that reside in their honey stomach. Many studies have demonstrated that pollen, corbicular loads, and bee bread differ in pH and biochemical composition, likely due to the fermentative action of lactic acid bacteria and yeasts acting on carbohydrates present in honey and nectar (Foote, 1957; Haydak, 1958; Gilliam et al., 1989; Human and Nicolson, 2006). For instance, bee bread contains more reducing sugars than pollen from the same plant species (Casteel, 1912). There is Vitamin K in bee bread but not in the pollen used to create it (Haydak and Vivino, 1950).

The fermentation and chemical changes of pollen stored by honey bees is similar to the microbial action occurring in green plant food materials that are ensiled to improve palatability, digestibility, and nutritional value (Gilliam, 1997). Indeed, there are analogies in the microbiology and biochemistry of silage and the production of bee bread. These include microbial succession; fermentation; increased availability of amino acids; enhanced stability of silage by *Bacillus* spp., yeasts, and molds; and the production of organic acids to act as preservatives (Woolford, 1978; Gilliam, 1979a,b; Standifer et al., 1980; Nouts and Rombouts, 1982; Gilliam et al., 1989; Gilliam, 1997).

The conversion of pollen to bee bread is a multi-phase process. The first phase in the microbial succession lasts about 12 hours and is characterized by the development of a heterogeneous group of microorganisms including yeasts. During the second phase, anaerobic lactic acid bacteria (*Streptococcus*) utilize growth factors produced by the yeasts and putrefactive bacteria and lower the pH of the pollen. The growth of Streptococcus ceases and the microbe eventually disappears during the third phase, but there is an increase in *Lactobacillus* (Chevtchik, 1950, cited in Gilliam, 1979). *Pseudomonas* also is present and possibly contributes to anaerobiosis required by *Lactobacillus* and to the degradation of the walls of pollen grains (Klungness and Peng, 1984). After pollen is packed in cells for 2 to 3 days, *Pseudomonas* can no longer be cultured from it. The first three phases of pollen conversion to bee bread last about 7 days. The lactic acid fermentation process in the fourth phase is complete in about 15 days, though the microbes remain for several months. The yeasts are present in small numbers initially, but then increase after fermentation and remain in the stored pollen longer than other microbes (Gilliam, 1979). The yeasts apparently play a significant role in the fermentation process. When pollen was sterilized using gamma radiation and then seeded with *Lactobacillus*, the lactic acid fermentation process yielded an unpalatable product of poor nutritive value for the bees (Pain and Maugenet, 1966).

Floral pollen, corbicular pollen, and bee bread also differ in the predominant molds that can be cultured (*Mucor* sp. in floral pollen, *Penicillia* in corbicular pollen and in bee bread stored for 1 week, *Aspergilli* and *Penicillia* in bee bread stored for 3 weeks, and *Aspergilli* in bee bread stored for 6 weeks). The molds produce enzymes that are involved in lipid, protein, and carbohydrate metabolism (Gilliam et al., 1989).

The early work on microbes isolated from bee bread provided the first glimpse of how bees use a suite of microflora to preserve and possibly even pre-digest their food. The studies also defined the functions of the microbes with the tools available at the time. Based on these early studies, the importance of microbes for maintaining colony health became evident.

With current molecular tools we can expand on this early work, detailing the molecular mechanisms of known microbes and illuminating the functions of microbial communities that resist culturing.

New Tools for Studying Microbes: Metagenomics

Though numerous microbes were isolated, cultured, and identified from samples of honey bees, pollen, and bee bread, there are many more that could not be cultured. Indeed, most microorganisms (about 99%) cannot be grown in pure culture because microbes rarely live in isolation but rather in communities where they either enhance or constrain each other's growth. Recently, metagenomics has enabled the limitations of culturing to be circumvented by applying genomic analysis to entire microbial communities. Metagenomics bypasses the need to isolate and culture individual microbial species, thus enabling the study of microbes that cannot be isolated and cultured (Kowalchuk et al., 2007).

Metagenomic analysis begins with collecting a sample from a particular environment such as bee bread, and doing a mass extraction of the DNA from all the microbes in the sample (Figure 2). Protein or RNA also can be extracted from the microbes. Most metagenomics studies currently focus on microbes with relatively small genomes, such as bacteria or archaea (a group of microorganisms that differ from bacteria in cellular structure and function). Once the sample DNA is extracted, it can be analyzed directly or inserted into laboratory bacterial cultures for long-term storage as a metagenomic "library"—a living warehouse representing all of the DNA fragments from the sample's microbial community. Newer sequencing technologies generate gigantic volumes of high-quality sequence data for relatively little cost.

Information about the microbial community and its functions can be obtained in several ways. A sequence-based metagenomic analysis can be conducted to define the entire genetic sequence (i.e., the pattern of the four different nucleotide bases in the DNA strands). The sequence can then be analyzed to determine the complete metagenome of the community or the genome of an individual microbial species. Sequenced-based analysis can utilize evolutionarily conserved phylogenetic markers or "anchors" such as 16S rRNA, to indicate the taxonomic group that is the probable source of the DNA fragment. Such information can be used to determine the relative amounts and proportions of target microbes in a given sample.

The alternative to a phylogenetic marker-driven approach is to sequence random clones. Although relatively time-consuming, this method has produced dramatic insights, especially when conducted on a massive scale. Primary inference drawn from sequence-based analysis includes the uniqueness and redundancy of community function, and the way that microbial function may benefit or harm host function (see Handelsman, 2004). Metagenomic sequence data can be examined for genes that encode critical symbiont functions, genes involved in basic metabolism virulence, or antibiotic resistance genes. Another approach involves the direct extraction and identification of expressed proteins and metabolites (the products of cellular processes) from a microbial community. This approach has already been used to identify novel antibiotics, antibiotic resistance genes, membrane transporters, and secondary metabolites (see references in Handelsman, 2004). Nucleic acid probes labeled with fluorescent tags can also be used to quantify the relative abundance of

Figure 2. Flow diagram of a metagenomic analysis of a microbial community.
(Source: *http://books .nap. edu/ openbook.php?record_id=11902&page=20. Used with permission.*)

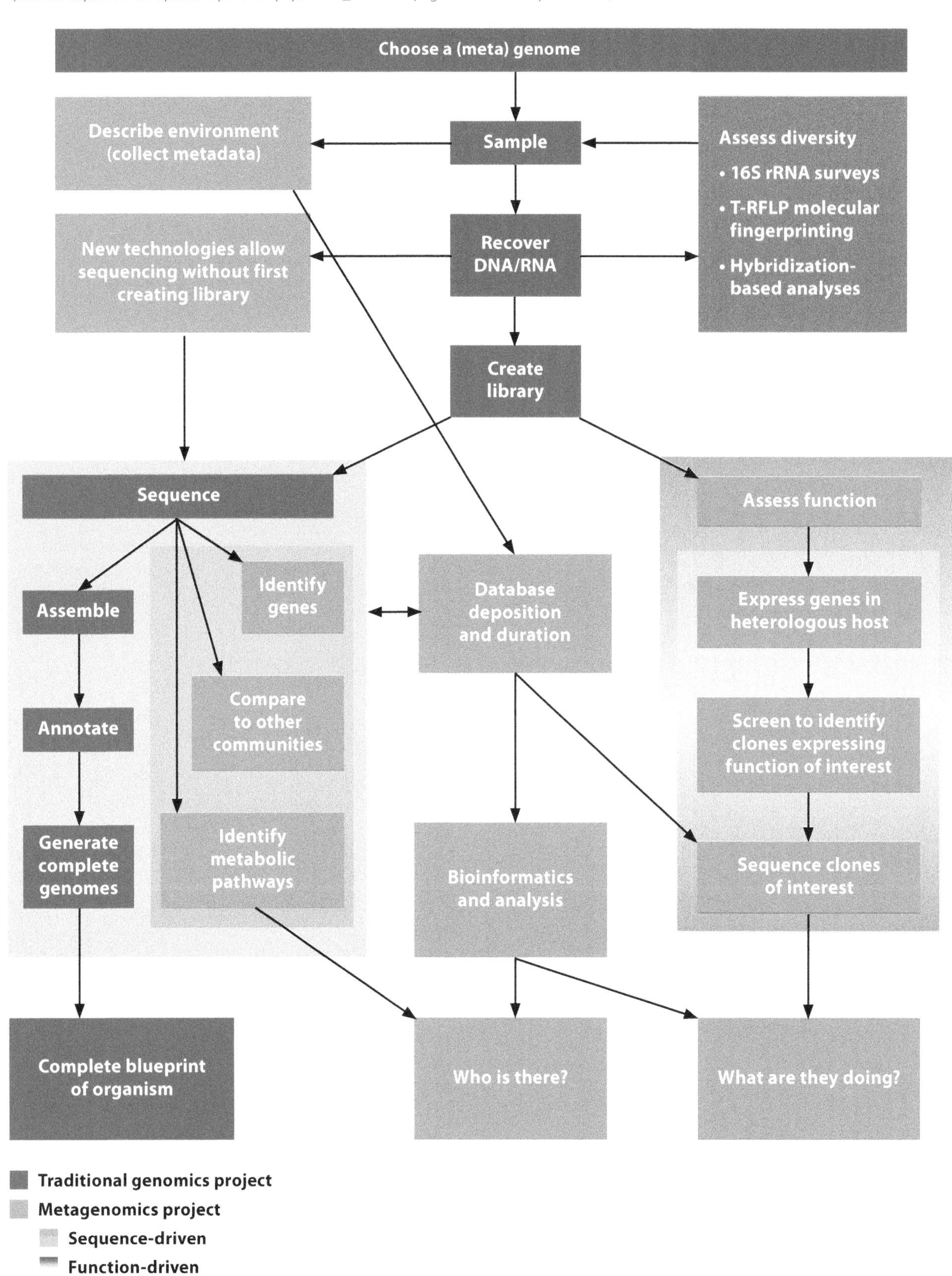

each organism within the community. This method can elucidate conditions that enhance or constrain the growth of particular microbes and can be particularly useful for understanding the effects of nutrition, antibiotics, or various agricultural chemicals.

Application of Metagenomics: Insights from the Human Microbiome Project

Research in the study of the microbiome in honey bees will be greatly advanced because of the work being done in the Human Microbiome Project (HMP). Many of the questions posed in studies of the human microbiome are similar to those of honey bee colonies. For instance, how stable and resilient is an individual's microbiota throughout the day and during his or her lifespan? How similar are the microbiomes between members of a family or members of a community or across communities in different environments? Do all humans have an identifiable "core" microbiome; and if so, how is it acquired and transmitted? What affects the genetic diversity of the microbiome, and how does this diversity affect adaptation by the microorganisms and the host to markedly different lifestyles and to various physiological or pathophysiological states? Such questions are directly transferable to honey bees, their colonies, and even beekeeping practices. The many similarities in conceptual framework and molecular techniques between human and honey bee systems will enable us to leverage findings in the HMP to advance our knowledge of the role of microbes in honey bee colony health.

The HMP is an extension of an earlier project to map the entire human genome (i.e., the Human Genome Project). At the onset of the Human Genome Project, it was thought that about 100,000 human genes would be found. Many were surprised by the finding that the human genome contains only about 20,000 protein-coding genes, not much different from the fruit fly genome. However, microorganisms that live on and within humans—the human microbiota—are estimated to outnumber human somatic and germ cells by a factor of ten. Together, the genomes of some of these microbial symbionts encode traits important to human health that humans either did not derive on their own, or have subsequently lost. Thus, the human genetic landscape is an aggregate of the genes in the human genome and in the microbiome, such that the metabolic phenotype that arises is a blend of human and microbial traits. The same is potentially true for honey bee colonies.

To understand the range of human genetic and physiological diversity, the microbiome and the factors that influence the distribution and evolution of the constituent microorganisms must be characterized. This is one of the main goals of the HMP (Turnbaugh et al., 2007). The characterization of the microbiome will provide perspective on contemporary human evolution, and on whether and how human lifestyles and environmental factors influence health and predisposition to various diseases. A similar paradigm can be applied to the honey bee microbiome and the beekeeping practices that shape it. Additionally, studies from the HMP that define the parameters needed to design, implement, and monitor strategies for intentionally manipulating an individual's microbiota to optimize physiological performance can provide a framework for manipulating the honey bee microbiota to improve colony health.

Research for the HMP has focused on bacterial communities that reside on or in the human body, including the skin and in the mouth, esophagus, stomach, colon, and urogenital region. The largest reported data sets are for the gut, although sample sizes using these culture-independent surveys are limited. Based on the studies though, it appears that more than 90% of all phylogenetic types (phylotypes) of colonic bacteria belong to just two of the 70 known divisions (phyla) in the domain Bacteria: the Firmicutes and the Bacteroidetes.

Though the phylotypes of bacteria in the human colon are limited and stable over time, there appear to be large differences in microbial-community structure among individuals (Zoetendal et al., 2001). Such high interindividual diversity in the gut microbiota might be explained by the neutral theory of community assembly, which states that most species share the same general niche or the largest niche, and therefore are likely to be functionally redundant (Hubbell, 2006). This theory predicts that highly variable communities (as defined by 16S rRNA gene lineages) will have elevated levels of functional redundancy between community members. The same might be true for honey bee colonies in that the microbiota might differ among colonies, depending upon their geographic location and diet. There also appear to be significant effects of genotype on the microbiota of humans, and the same may apply to bees and their colonies. In humans, the overall species composition of gut microbiota is significantly different between individuals of different ethnic groups (Li et al., 2008). Epidemiological observations of metabolic phenotypes also indicate differences based on ethnicity (Dumas et al., 2006); this supports the concept that host metabolic phenotype is strongly influenced by the gut microbiome (Nicholson et al., 2005). This is expected because gene sets annotated in the intestinal microbiomes identified significant numbers of bacterial genes that are not encoded in the human genome. Similarly in honey bees, the colony phenotype relative to population growth or overwintering success might also be at least partially influenced by the microbiome residing in the bees and food stores.

Another area where the HMP parallels the interests of honey bee health is in the role of nutrition in immune response. Microflora play important roles in the defense of the host by limiting the colonization of potential pathogens (Ashara et al., 2001, 2004; Kelly et al., 2007). Additionally, the interaction between a human host and microflora increases the concentrations of immunoglobulins, and the production of specific antibodies modifies the subpopulations of mucosal lymphocytes and boosts overall immunological function (Corthésy et al., 2007). In insects, humoral immunity involves the synthesis of antimicrobial peptides in response to infection by bacteria, fungi, or parasites (Hetru et al., 1998; Lamberty et al., 1999; Yamauchi, 2001; Klaudiny et al., 2005). Useable energy and the structural components required to synthesize the antimicrobial elements of the immune system are obtained from food. Without adequate nutrition and the microflora required for its thorough digestion, the immune system is clearly deprived of the components needed to generate an effective immune response (Gershwin et al., 2000).

In addition to their role in supplying the necessary nutrients for mounting an immune response, there are direct interactions between microflora and host defense. These interactions are modulated by a class of proteins in the immune system called Toll-like Receptors (TLRs) that recognize molecules produced by pathogens. Studies in humans and animal model systems indicate that macrophages play a central role in the regulation of inflammatory and immune responses. The functions associated with macrophages are triggered by the stimulation of TLR expressed on the surface of macrophages (Rakoff-Nahoum et al., 2004; Akira et al., 2006; Mitchell et al., 2007). The eradication of the intestinal microflora by antimicrobials plays a role in the downregulation of the surface expression of TLRs on the peritoneal macrophages and impaired immunological responses (Umenai et al., 2010). Whether microbes in the gut of the honey bee have similar effects on immune response requires further investigation, especially because colonies are commonly exposed to antibiotics and fungicides that might eliminate or severely curtail the growth of intestinal symbionts.

Applying the techniques and information generated by the HMP to studies on the honey bee microbiome can provide a framework for generating hypotheses and interpreting results from metagenomic and physiological studies. In turn, the HMP could benefit from studies on the honey bee microbiome because this insect can serve as an ideal model system that can be easily manipulated to answer basic questions on host-microbe interactions. Worker bees emerge as adults microbe-free and as such can be used for studies detailing the mechanistic basis for the molecular interactions that occur at the interface between microbes and their host. Studies on the role of genotype, diet, and environment on the composition of host microbiota also can be conducted using honey bees because it is possible to make specific genetic crosses using instrumental insemination and controlling diet through feeding specific protein and carbohydrates either in colony or small cage studies (DeGrandi-Hoffman et al., 2007; DeGrandi-Hoffman et al., 2010).

Recent Studies on the Honey Bee Microbiome

Recent application of metagenomic techniques indicates that the great number of microorganisms uncovered using classic culturing techniques represents only a small portion of the actual microbiota in bees and their food. Current investigations of the microflora are conducted by sequencing the 16S rRNA genes (16S rDNA). These genes encode RNA of the small subunit of the bacterial ribosomes (Egert et al., 2003; Hongoh et al., 2003; Jeyaprakash et al., 2003; Reeson et al., 2003). Genetic profiling techniques such as denaturing gradient gel electrophoresis (DGGE) (Reeson et al., 2003; Schabereiter-Gurtner et al., 2003), terminal restriction fragment length polymorphism (TRFLP) (Broderick et al., 2004), or single-strand conformation polymorphism (SSCP) (Czarnetzki and Tebbe, 2004) based on the 16S rDNA allow a straightforward comparison of the bacterial communities from a relatively large number of samples.

Using molecular techniques, qualitative and quantitative differences were found in the bacterial community structure of larvae and adult worker bees (Mohr and Tebbe, 2006). A metagenomic survey of adult worker bees from four widely separated beekeeping operations across the United States, including some affected by CCD, revealed a bacterial community that included an abundance of Gammaproteobacteria and several less frequent but widespread organisms from the Betaproteobacteria, Alphaproteobacteria, Firmicutes, and Actinobacteria groups (Cox-Foster et al., 2007). The bacterial analysis also indicated community composition similar to that in samples collected in Africa, Switzerland, and Germany (Babendreier et al., 2007; Jeyaprakash et al., 2003; Mohr and Tebbe, 2006), suggesting that *Apis mellifera* have similar bacterial flora worldwide

(Table 1). In addition to bacteria, 81 distinct fungal 18S rRNA sequences were recovered from the pooled samples. The fungal sequences are primarily from four distinct lineages: Saccharomycotina, which includes a variety of presumed commensal yeasts; Microsporidia, including the important bee pathogens *Nosema apis* and *Nosema ceranae* (Higes et al., 2006); Entomophtorales/Entomphthoromycotina, a diverse group of insect pathogens; and Mucorales. Mucoromycotina includes *Mucor hiemalis*, a species known to kill honey bees under certain conditions (Burnside, 1935).

Bacteria from larvae were isolated, cultured, and described using 16S rRNA and protein-coding gene sequences. Sixty-one bacterial isolates reflecting a total of 43 distinct bacterial taxa were identified (Evans and Armstrong, 2006). There was a high frequency of bees harboring bacteria from the *Bacillus cereus* group suggesting a stable symbiosis between bees and this taxon. Twenty-three bacterial isolates consistently inhibited *Paenibacillus larvae larvae*, the causative agent of American foulbrood. Isolates that inhibited *P. larvae* were evenly distributed across the sampled taxa. The survey indicated that older larvae (7 days old) have more bacteria and bacterial species per individual than younger larvae (1 day old). Diversity of bacteria species did not differ between younger and older larvae. The frequency of bacteria species varied among colonies and sites where they were sampled. This suggests that genotypic and environmental factors, especially as they affect the types of nectar and

Table 1. Microbes in worker honey bees as determined by a metagenomic survey of samples from four widely separated beekeeping operations across the United States. (*From Cox-Foster et al., 2007.*)

Kingdom	Taxon (rank)	Organism
Bacteria	Firmicutes (phylum)	*Lactobacillus* sp.* Uncultured Firmicutes
Bacteria	Actinobacteria (class)	*Bifidobacterium* sp.*
Bacteria	Alphaproteobacteria (class)	*Bartonella* sp.* Gluconacetobacter sp.*
Bacteria	Betaproteobacteria (class)	*Simonsiella* sp.*
Bacteria	Gammaproteobacteria (class)	Two uncultured species*
Fungus	Entomophthorales (order)	*Pandora delphacis*
Fungus	Mucorales (order)	*Mucor* spp.
Fungus/microsporidian	Nosematidae (family)	*Nosema ceranae*
Fungus/microsporidian	Nosematidae (family)	*Nosema apis*
Eukaryota	Trypanosomatidae (family)	*Leishmania/Leptomonas* sp.
Metazoan	Varroidae (family)	*Varroa destructor*
Virus	(Unclassified)	Chronic Bee Paralysis Virus
Virus	Iflavirus (genus)	Sac Brood Virus
Virus	Iflavirus (genus)	Deformed Wing Virus
Virus	Dicistroviridae (family)	Black Queen Cell Virus
Virus	Dicistroviridae (family)	Kashmir Bee Virus
Virus	Dicistroviridae (family)	Acute Bee Paralysis Virus
Virus	Dicistroviridae (family)	Israeli Acute Paralysis Virus

pollen bees collect, might influence the diversity of microbes individuals harbor and in their ability to inhibit the growth of pathogenic bacteria.

Recent investigations of the microflora in the honey stomach of adult workers have revealed the presence of novel lactic acid bacteria (LAB) of the genera *Lactobacillus* and *Bifidobacterium* (Olofsson and Vásquez, 2008). The species vary with the sources of nectar and the presence of other bacterial genera within the honey bee. Similar species to those found in the honey stomach were not found in nectar from flowers but were isolated from nectar stores, indicating that the LAB originated from the honey stomach. The nectar sugars probably act as inducers for the resident honey stomach flora and, depending on the types of flowers that the honey bees visited, enhance their numbers. When the collected nectar is regurgitated from the honey stomach into comb cells, the bacteria are transferred into the stored nectar and might play a role in its conversion to honey and conferring its antibiotic properties. The majority of the honey stomach LAB flora also is present in a viable state from both the pollen and 2-week-old bee bread. Older bee bread (i.e., > 2 months) does not contain LAB. Bee bread is fermented at least in part by the honey stomach LAB that are added to the pollen via regurgitated nectar. The presence of the honey stomach LAB and its antimicrobial substances in bee bread indicate that these bacteria might play a role in the defense against honey bee diseases for all colony members as the bee bread is consumed by both the larvae and the adult bees.

Areas for Future Research

We are just beginning to learn about the role of microbes in the health of all organisms, including honey bees. We know that many metabolic processes are not directed by host genes but rather those of symbiotic microbes. This implies that optimum health depends on maintaining conditions that encourage the growth of those microbes needed to process food, and either convert it to energy or use it to synthesize more complex molecules via anabolic pathways. Though most of what we know about microbes in honey bees is their role in food processing and preservation, there are many other areas yet to be explored. For example, the role of microbes in the conversion of bee bread to brood food is unknown. The microbial communities of queens and drones that might have roles in fertilization and reproduction also have not been explored. Whether microbial communities in queens and drones are different from those in workers or are similar but differ in structure needs further investigation. The role of microbes in the synthesis of pheromones also should be studied. There are greater supersedure rates now than in the past, and whether nutritional factors, environmental contaminants, or antibiotics cause the microbial community in the queen to be compromised—thus affecting the synthesis of pheromone signals that communicate her presence—is worth further study.

Queen fertility is another area where microbes might play a role, but a link between microbial communities and fertility has not been investigated in honey bees. However, this connection might exist based on studies of human fertility. These studies have uncovered a putative link between microbial communication systems and host infertility (Rennemeier et al., 2009). The presence of the opportunistic pathogenic yeast *Candida albicans* and the Gram-negative bacterium *Pseudomonas aeruginosa* caused multiple damage to spermatozoa, including reduced motility, premature loss of the acrosome (a cap-like structure of the sperm head that is essential for fertilization), apoptosis, and necrosis. The damage was caused by the quorum-sensing molecules that the microbes produced. This previously unknown interaction between microorganisms and human gametes might also occur in queens or drones, especially if the growth of beneficial microbes in the reproductive organs is impaired due to antibiotics or fungicide exposure.

Another substantially unexplored area involves possible changes in microbial communities particularly in workers throughout the year. In the spring, colonies are converting food to brood, but in the fall, brood rearing declines and food is stored so the colony can survive the winter. It seems plausible that the changes in the metabolic processing of food might have underlying microbial drivers.

Differences in food processing also seem to be based on genotype. For instance, the conversion of pollen to hemolymph protein and to brood differs among lines of bees, particularly between African and European genotypes (see references in Schneider et al., 2004; Cappelari et al., 2009). An in-depth comparative study of microflora of European and African bees could determine if differences in the composition of microbial communities are a contributing factor in the success of the Africanized honey bee throughout the tropics and subtropics of the Americas.

Determining the role of microbes in the health of honey bee colonies is essential for evaluating the impact that exposure to sublethal doses of fungicides and other antibiotics might have on colonies. Currently, fungicides are considered safe for bees so they are applied while crops are in bloom. Fungicides negatively affect the growth of culturable molds commonly found in bee bread (Yoder et al., 2011—this volume), and might disrupt the structure of other microbial communities that cannot be cultured but are necessary for colony health. Without knowing the roles that specific microbial communities play in maintaining the health of a colony and the effects that fungicides have on them, it is not possible to evaluate the safety of these compounds to pollinators.

A deeper understanding of the microbiota of honey bee colonies cannot help but cause a rethinking of beekeeping practices. For example, the full effect of using antibiotics—especially prophylactically—becomes clear when the effects on symbionts bees need for food preservation and colony-level metabolic functions are considered. The broader effects of carbohydrate and protein supplements on colony health will be revealed if they are studied within the context of whether the supplements enhance the growth of honey stomach and gut microbes or have deleterious effects on them. The need for clean pollen sources that can be converted to bee bread becomes especially urgent as it may represent the best solution for reestablishing the microflora in colonies and preventing the growth of pathogenic microbes. The development of probiotics to stimulate the growth of beneficial microbes especially in colonies under stress might become a common beekeeping practice.

The microbiota in honey bee colonies is critical in sustaining their health and vigor. Currently, most beekeeping practices are therapeutic rather than preventative. However, honey bee colonies have evolved strategies to process food and combat pathogens. Beekeeping practices that augment these strategies might help prevent the occurrence of numerous diseases and mitigate the impact of others. Preventing problems is always easier then solving them. In the case of developing beekeeping practices to prevent colony losses, perhaps the best solutions reside in the bees themselves.

Seasonal Microflora, Especially Winter and Spring

CHAPTER

2

Asli Özkirim

Abstract Growth of microorganisms inside honey bees, and particularly by commensal organisms that compete with pathogens preventing their colonization, is an important factor in bee health, suggesting that keeping bees healthy may be more important than treating colonies. It is critical to diagnose and treat the honey bee diseases and prevent colony losses, which necessitates routine monitoring of the microflora and its components. In this chapter, standard methods for isolation, cultivation, and identification of microflora are given along with a summary and classification of commensal and pathogenic organisms and their dynamic. Especially relevant for stabilization of the microflora for proper health are the changes in microflora composition over the seasons (winter, spring, summer, and autumn) with regard to climate change, bacterial interactions for developing new approaches to disease control to avoid commercial antibiotics, and an understanding of microflora ecology as it pertains to the health of the bee colony. After determining the standard dynamic of microflora, probiotics and prebiotics can be considered.

Introduction

Apicultural economic development strongly relies on the health status of honey bee colonies. For evident reasons, including historical ones, the normal physiology of honey bees receives much less attention than medicine and veterinary science. In order to diagnose honey bee diseases and prevent colony losses, several research projects must be supported and some are underway. Whereas it is very important to diagnose and treat the diseases, keeping honey bees healthy is as important as diagnosing their diseases. The growth of microorganisms inside honey bees is an important factor in bee health. These organisms, called commensals, grow in the honey bee and constitute normal flora that can compete with pathogens, preventing the colonization of these pathogens. By this measure, keeping bees healthy may be more important than treating colonies.

The Importance of Healthy Bees

Organisms that live in or on honey bees (or other organisms) but do not cause disease are referred to collectively as *normal microflora* or normal microbiota. Many such organisms have well-established associations with honey bees. Most organisms among the normal microflora are commensals, in that they obtain nutrients from the host. Two categories of organisms can be distinguished: *resident microflora* and *transient microflora*. The resident microflora comprise microorganisms that are always present on or in honey bees; transient microflora can be present under certain conditions in any of the locations where resident microflora are found. They persist for hours to months, but only as long as the necessary conditions prevail (e.g., winter or spring time) (Black, 2008; Madigan et al., 2009).

Among the resident and transient microflora are some species that do not usually cause disease, however they can do so under certain conditions. These organisms are called *opportunists* because they take advantage of particular opportunities to cause disease. Conditions that create opportunities include

1. Failure of the host's normal defenses
2. Introduction of the organisms into unusual body sites
3. Disruption of the normal microflora

Microflora of Honey Bees

In honey bees, the normal microflora is concentrated mainly in the posterior part of their digestive tract, namely in the middle (ventriculus) and posterior (anterior) intestines including the rectum. The most detailed information on bee microflora was published before the 1980s and required correction of the taxonomic affiliations of bacterial cultures according to presently accepted nomenclature (Kacaniova et al., 2004).

Enterobacteria of the genera *Escherichia, Enterobacter, Proteus, Hafnia, Klebsiella,* and *Erwinia* are most commonly isolated from the bee intestine (Lyapunov et al., 2008). Gut symbionts of the European honey bee, *Apis mellifera mellifera*, has been reported to contain about 1% yeast-like microbes; 29% Gram-positive bacteria (such as *Bacillus, Lactobacillus, Bifidobacterium, Corynebacterium, Streptococcus,* and *Clostridium*) and 70% Gram-negative or Gram-variable bacteria (such as *Achromobacter, Citrobacter, Flavobacterium, Pseudomonas* (Jeyaprakash et al., 2003). The description of intestinal microflora is usually limited to one of the following characteristics:

1. Occurrence of specific species in individual bees; bacteriological techniques reveal only the dominant culturable species, while molecular genetic methods yield a broader spectrum of detected species at the expense of a loss of qualitative information.
2. Total microbial numbers for specific physiological groups (mesophilic aerobes and facultative anaerobes, aerobic and anaerobic bacteria, coliforms) in individual bees, in combined samples from different sections, or from complete intestines of a number of individuals.

Another important part of honey bees that contains microflora is the honey stomach or crop. The honey stomach is an enlargement of the esophagus that can expand to a large volume (Sammataro and Cicero, 2010). It ends with a structure called the proventriculus, which ensures that the nectar is never contaminated by the contents of the ventriculus (midgut), the functional stomach of honey bees.

A novel bacterial flora composed of lactic acid bacteria (LAB) of the genera *Lactobacillus* and *Bifidobacterium,* which originated from honey stomach, was recently discovered by Vásquez and Olofsson (2009) who suggested that honey be considered a fermented food product because of the LAB involved in honey formation. They speculated that LAB flora probably evolved in mutual dependence with honey bees—the LAB obtaining a niche in which nutrients were available, and the honey bee in turn being protected by the LAB from harmful microorganisms. The findings will have clear implications for future research and will provide a better understanding of the health of honey bees and of their production and storage of honey and bee bread. It will also have relevance for future honey bee and human probiotics (Olofsson and Vásquez, 2008).

Seasonal Differences in Microflora

Bees, by nature, are vegetarian and consume only flower nectars and pollen; these are preserved and stored during the winter season in the form of honey and bee bread. In overwintering bees, the rectum (posterior intestine) is greatly distended and can occupy a large part of the abdominal cavity before defecation occurs; it becomes a storage chamber for retained feces during winter months. Around 10^6 microorganisms are present in the rectum of wintering bees and help to digest and detoxify undigested food. It has been estimated that these microflora species have been present in bees for several million years, according to the presence of microbial DNA in the oldest fossilized bee preserved in amber (Gilliam et al., 1988).

Trophallaxis (mutual feeding) of winter feed and close contact between bees in the winter cluster seem to promote the uniformity of intestinal microflora, often referred as the *social stomach*. However, the research results confirm there are differences in the microflora of individual bees. In the beginning of winter, the numbers of enterobacteria in bees may vary by six orders of magnitude. The microflora became more uniform at the time of the first cleansing flight, with the enterobacterial content decreasing to four orders of magnitude (10^4–10^7 CFU/bee).

According to Lyapunov et al. (2008), during the winter prior to the first flight, *Klebsiella* were the most frequent bacteria. In addition to *K. oxytoca, K. planticola* and *K. pneumonia* were also detected by Vassart et al. (1988), and Gilliam et al. (1988) reported the presence of *K. pneumonia* in healthy bees, as well as *K. oxytoca* in intestines of bees kept in hoarding cages, but not in the colonies. Other researchers have also reported *Klebsiella* of an undefined species and was affiliated with healthy bees.

For *Providencia rettgery*, which was previously detected in the intestines of healthy bees (Tysset and Rousseau, 1967) and in the hemolymph in septicemia cases (Fritzch and Bremer, 1975), the frequency of occurrence increased during winter in 25% to 83% of bees.

After the first spring flight, apart from a significant decrease in the number of enterobacteria, the dominant species also changed; *K. oxytoca* and *P. rettgery* were found in 43% and 54% of the bees, respectively (Lyapunov et al., 2008). This explains why the first spring flight of bees is so important.

The frequency of occurrence of *Hafnia alvei* and *Citrobacter* sp., agents of infectious diseases in bees, decreased from 12% and 10% at the beginning of winter to 3% or less, respectively. The occurrence of *H. alvei* in normal bee microflora has been reported in its original description as a bee pathogen (Toumanoff, 1951).

Interaction Between Microfloral Bacteria

A balanced association of microbial species with symbiotic and competitive interactions (referred to as an indigenous gastrointestinal microflora) forms an integral part of any well-functioning healthy organism. A regular occurrence of the typical intestinal groups of aerobic and anaerobic microorganisms was recorded in the honey stomach, the ventriculus, and the rectum of bees (Machova et al., 1997). For instance, Gilliam (1979) found that *Bacillus subtilis*, which produces antibiotics, are active against some bacterial species. Therefore, the role of the microflora in honey bee health has two sides: to stay in place and to produce antibiotics to inhibit other bacterial species in honey bees. It is reported that the presence of antimicrobial subtances in pollen and honey originated from the antibacterial activity of some *Bacillus* spp. as well as some fatty acids, especially linoleic, linolenic, myristic, and lauric acids, which have inhibitory properties against *Paenibacillus larvae* and *Bacillus cereus* (Manning, 2001). Thus, information about the interactions that occur between different bacterial species inside bees and the dynamics of the bacterial community could be important in developing new approaches for disease control and avoiding the use of commercial antibiotics. On the other hand, in order to gain insight into the microbial ecology of honey bees, we should consider the entire colony's bacterial flora as much as the intestinal microflora (Piccini et al., 2004).

Disruption of Honey Bee Microflora Using Antibiotics

Antimicrobial agents, especially broad-spectrum antibiotics, may have an adverse effect not only on pathogens, but also on indigenous microflora. When these microflora are disturbed, other organisms not susceptible to the antibiotic, such as *Candida* yeast, can invade the unoccupied areas and multiply rapidly. Invasion by replacement microflora is called superinfection and is difficult to treat because they are susceptible to only a few antibiotics (Black, 2008; Madigan et al., 2009).

In the beekeeping industry, antibiotic treatments are forbidden in Europe against bacterial diseases, especially American foulbrood (AFB) and European foulbrood (EFB). On the other hand, beekeepers use antibiotics such as tetracycline in the United States (or in some other countries illegally) in order to treat and/or prevent these diseases in their colonies. In contrast, antibiotics disrupt the normal microflora. If used too frequently, this can allow the formation of drug-resistant pathogens. Normal microfloral bacteria will be eliminated, will not be present to produce antimicrobial substances, and will not help the bees digest their food. In addition to pathogenesis of the disease, the normal physiology of the honey bees can be disrupted. This is a topic of current reseach.

The Opportunistic Microflora and Their Behavior During Seasonal Changes

In the absence of the full complement of normal microflora, either by using antibiotics or during extreme conditions (too cold or too hot, or too acidic or too basic media), opportunistic microorganisms such as *Providencia rettgery* can become established in the bee's intestinal system. The retention of opportunistic pathogens can lead to harmful alteration in the digestive function or can even lead to disease, such as septicemia.

Climate change, especially during winter and spring months, directly affects the dynamics of microflora. Some antimicrobial subtances, such as botanical compounds, can aid in this disruption and are advantageous for opportunistic microorganisms already present in the normal microflora. They grow rapidly and occupy the empty places. If the population is high enough, it is very easy for opportunistic bacteria to pass into the hemolymph and cause Septicemia (poisoning of the hemolymph) and causes bee death in a short time (Black, 2008). On the other hand, the disruption of normal microflora is also a good opportunity for viruses to become virulent; virus can fill a niche devoid of bacteria or fungi.

Dynamics of Microflora of Honey Bees

Although all bacteria seem to share the same habitat in the intestinal system of honey bees, they do not use the same kind of food and use different temperatures to grow. Moreover, one group of bacteria can convert the food to another form, and this second version of food can be used by another group of bacteria. This is a typical food chain that occurs in microflora and makes it possible for different types of flora to grow and live together. Whereas bacteria support their growth, they can also inhibit their reproduction by producing some antimicrobial subtances or other products that can change the pH or other features of the medium (Özkırım and Keskin, 2002). This well-balanced system enables microflora to change slowly according to climatic conditions.

Especially in winter and spring, the acute changes in temperature cause changes in the dynamics of the microflora and several events may occur (Lyapunov et al., 2008). This includes changing rates of bacterial growth, the sudden loss of some bacterial species, an abnormal bacterial growth (opportunistic bacteria), or a rapid viral explosion. Climate changes over the years can lead to instability in the seasonal dynamics of the microflora, which can cause the alteration of the food chain among bacteria and their metabolites and may reduce the ability of bees to digest their food.

The Conservation of Honey Bee Health by the Stabilization of the Dynamics of Microflora

We can easily understand that microflora are a very important subject for honey bee health. In fact, the stabilization of microflora may prove to be more important for keeping bees healthy. For the stabilization, we need to determine the dynamics of the microflora over the seasons (winter, spring, summer, autumn). If we knew the standard bee microflora, we would be able to detect abnormalities and how they relate to honey bee health. All microfloral bacterial growth directly depends on the conditions of seasons, because the temperature, pH, and contents of the intestinal system parallel weather conditions. After determinning the standard dynamic of microflora, probiotics and prebiotics can be also considered. According to the latest literature, there is no exact determination of honey bee microfloral members and no research on the dynamics. More research on this subject should be conducted.

Colony Collapse Disorder (CCD) and Microflora

The phenomenon of CCD was first reported in 2006 (Cox-Foster et al., 2007); however, beekeepers noted colony declines consistent with CCD as early as 2004. These colony deaths are marked by dead bees outside the hive, an incremental decline in worker population and robbing, as well as pest and pathogen invasion. One hypothesis of the cause of CCD was the introduction of a previously unrecognized infectious agent. This idea is supported by preliminary evidence that CCD is transmissible through the reuse of equipment from CCD colonies, and that such transmission can be broken by irradiation of the equipment before use (Cox-Foster et al., 2007).

Bacterial analysis indicated that *Apis mellifera* has similar bacterial flora worldwide (Jeyaprakash et al., 2003). Parts of the honey bee genome-sequence database revealed sequences corresponding to several of these bacterial groups, indicating that they are probably part of the normal flora (Jeyaprakash et al., 2003).

Although there is some evidence that explains the susceptible relationship between a microfloral opportunistic agent and CCD, there is a correlation between IAPV (Israeli Acute Paralysis Virus) infection and CCD (Cox-Foster et al., 2007). However, it is not clear what the causes of CCD are, and we still need further investigation to not only determine if it is just one pathogen, but also how the changes in microflora over the seasons affect bees.

Methods for Detection and Identification of Microflora and Its Dynamics

To determine bacteria from bee samples, generally, the abdomens are cut off and collected and then put into proper solutions according to the methods used for microflora identification.

Determination of CFU (Colony Forming Unit) Counts

The plate diluting method is used for the determination of CFU counts of respective groups of microorganisms in 1 g of substrate. The nutrient substrates in Petri dishes are inoculated with 1 mL of semi-digested chyme (the pulpy mass of semi-digested food in the ventriculus or small intestines just after its passage from the honey stomach) by flushing the surface. Homogenized samples of intestine chyme are prepared in advance by serial dilutions (10, 100, 1,000, 10,000, etc.) and at least three replicates (Özkırım and Keskin, 2002).

Dilution of the Samples

The basic dilution (10^{-2}) is prepared: 1 g of cecum content is added to a container with 99 mL distilled water. The cells are separated from the substrate in a shaking machine (20 min); basic solution is diluted to the level < 300 CFU/mL (Kacaniova et al., 2004). Nutrient substrates for the cultivation and characterization of respective groups of microorganisms that are shown in Table 1.

Enterobacteria are also classified using biochemical tests according to their fermentation system or the type of substrate they use, such as

Microorganism Group	Nutrient Substrate
Meat peptone agar	Total anaerobes
Rogosa agar with cystein	Gram-positive anaerobic acid-resistant rods
Rogosa agar	*Lactobacilli*
Yeast-extract agar with 1%(w/v) glucose	Total aerobes
MacConkey agar	Coliforms
Baird-Parker medium	Total aerobes
Slanetz-Bartley agar	*Staphylococci*
Nutrient agar	*Enterococci*
Brain-Heart Infusion agar	*Bacillus* spp.
Sabouraud-Dextrose agar	Molds
Potato-Dextrose agar	Yeasts
Endo agar and Simmons-Citrate medium	*Pseudomonas* spp.

Table 1. Nutrient substrates for the cultivation and characterization of respective groups of microorganisms. (*From Black, 2008.*)

carbohydrate, citrate, mannitol, etc. For all biochemical tests, *Bergey's Manual of Systematic Bacteriology (CM) and Micromethods (MM)* is the standard used. The following characteristics are used as markers:

1. Gas production via glucose fermentation
2. Indol and acetoin production
3. Methyl Red reaction
4. Fermentation of carbohydrates (D-adonitol, L-arabinose, dulcitol, *m*-inositol, D-xylose, lactose, maltose, D-mannitol, D-mannose, melibiose, alpha-methyl D-glucoside, L-rhamnose, D-raffinose, salicin, sucrose, D-sorbitol, trehalose, and cellobiose)
5. Utilization of acetate, malonate, and citrate, urea and gelatin hydrolysis, esculin hydrolysis, lysine decarboxylase, arginine dehydrolase, ornithine decarboxylase and phenylalanine deaminase and lipase

The metabolic pattern of the bacterial strains is also determined with API® kits (Biomereiux, France). The API50CH-system allows the study of the carbohydrate metabolism of bacteria and consists of 50 microtubes. The first tube contains no substrate and is used as negative control. The remaining 49 tubes contain a defined amount of dehydrated carbohydrate substrate. Fermentation is shown by a color change in the tube due to acid production. The system enables the biochemical identification and typing of bacilli and related genera. *Pseudomonas* spp., *Enterobacteria, Morganella morganii, Proteus* spp., *Bacillus* spp., *Staphyllococcus* spp., and *Streptococcus* spp. can be identified using API20-E, API50-CHE, API-Staph, and API-20Strep (Kilwinski et al., 2004).

Molecular Methods

The detection and identification of microflora could be made by PCR-based (polymerized chain reaction–) methods and sequencing 16S rDNA. Identification based on species is more reliable with microbiological and molecular methods together. All bacterial contents should be isolated from the intestine by microbiological methods and the concentration determined by growing a pure culture (bacteria/mL). In order to identfy each bacteria species, Real-time PCR is the most convenient method to multiply genetic materials and make semi-quantative analysis. After diluting all cultures by serial dilution, the threshold cycle (Ct) value is calculated for each sample. In conclusion, the results are compared with Ct values using a standard curve. As the final step, the specificity of the amplicons can only be verified by sequencing. In this way, all species can be determined individually and the seasonal dynamics of the bacteria can be observed semi-quantitively.

Evaluation of Varroa Mite Tolerance in Honey Bees

CHAPTER 3

Ingemar Fries

Abstract Breeding honey bees for specific defined characteristics to obtain Varroa mite-tolerant bees appears to be difficult. Instead, it is suggested that the daily rate of mite population growth during optimal mite breeding conditions be used to determine the breeding value for mite tolerance in evaluated colonies. The precision needed to establish mite population growth will determine if samples of adult bees at different occasions will suffice, or if more detailed measurements of the mite population is required. Threshold levels for mite population densities before mite control is required should be adjusted for different geographical regions and foraging conditions.

Introduction

As part of an effort to promote the development of Varroa mite-tolerant stocks of bees, the BEE DOC (Bees in EuropE & the Decline Of honeybee Colonies) project has surveyed breeding programs and available literature on the subject to formulate recommendations suitable for practical breeding purposes. The initiative has been a collaborative effort with Swedish honey bee breeders (Svensk Biavel AB).

In a Nordic climate, Varroa mites (*Varroa destructor*), and those viral diseases where the mite acts as a vector, the vast majority of colonies that are infested in collapse 3 to 4 years after the parasite first becomes established. If the mite population growth is not limited, up to 10,000 mites can be present in some colonies (Korpela et al., 1993). In more southern climates in Europe, the collapse is likely even faster. When the parasite has been established in a population of bees, the extent of viral infections is likely to increase and colony collapse may occur even at lower infestation levels. For beekeeping not to be eliminated, Varroa mites must be controlled. A recent review of the parasite's biology and how it can be combatted by different methods can be found in Rosenkranz et al. (2010).

The Varroa mite on our European bees, *Apis mellifera*, comes from the Asian honey bee, *Apis cerana*. Asia is home to more species of Varroa that have not made the host change we have seen by *V. destructor* (Anderson and Trueman, 2000). The Asian bee is not considerably damaged by the attacks because it has developed certain characteristics that makes it tolerant. The most important feature is that reproduction of the parasite, in practice, only occurs in drone brood because the infested worker brood is quickly removed (Rath and Drescher, 1990). Infested drone brood takes a long time to remove because of the strong cocoon. As the drone pupae are sensitive and often die from multiple Varroa females in the cell, the mites will die with the host (Boecking, 1993). An effective grooming behavior in which bees are helping each other to attack the mites has also been documented in *A. cerana* (Peng et al., 1987), but it has not been shown that the fallen mites are more damaged than in European bees (Fries et al., 1996). Simulations show that with reproduction in drone brood only, the Varroa population is unlikely to increase to harmful levels (Fries et al., 1994), especially if the mites are buried in the cells when multiple attacks occur that kill the pupae.

The problem with the Varroa mite of European bees is that reproduction works well in both worker and drone brood, although the latter is preferred (Fuchs, 1990) and produces more offspring per mite-cell attack (Ifantidis, 1984; Martin, 1995). It seems unlikely that European bees will develop the characteristics (solid cocoons of drones) that allow the parasite to become buried in heavily infested drone cells. It seems more likely that the ability to detect and remove infested larvae could be improved, because the property already exists to varying degrees in European bees (Arathi and Spivak, 2001). Bees selected for the removal of dead brood, known as removal or hygienic behavior, result in lower average infestation of Varroa mites in field colonies (Spivak and Reuter, 2001) but the ability to specifically detect and eliminate cells with the reproduction of Varroa (Harbo and Harris, 2009) has a greater impact on the mite population development (Ibrahim et al., 2007).

In recent years it has been shown that *A. mellifera* can survive attack by Varroa mites. Early on it appeared that the Africanized bees in South America did not succumb to attacks, at least partly because a large proportion of mite females were infertile in worker brood (Camazine, 1986, 1988). Later, the reproductive potential on worker brood improved, but still without damaging effects by the mite infestations, as mite population growth appears to slow down when the density increases (Vandam et al., 1995; Medina et al., 2002). An important reason for mite tolerance in Africanized bees, despite fertility of the mites being similar to infestations of European bees, appears to be a higher mortality of the mite offspring, also in the males, which suppresses mite population growth (Mondragon et al., 2006). When the Varroa mite came to the African continent in the late 1990s (Allsopp et al., 1997), it became clear very soon that the parasite did not have to be controlled for the bees to survive, although the reproductive potential of the parasite initially suggested damages to be likely (Allsopp, 2006). A more developed removal behavior of African bees (Fries and Raina, 2003) may have been part of the greater mite tolerance (Frazier et al., 2010), but the absence of pesticides against mites may also contribute to the evolution of mite tolerance (Frazier et al., 2010). Populations of European honey bees also appear to have developed different levels of mite tolerance in Europe (Fries et al., 2006; Le Conte et al., 2007), as well as in the United States (Seeley, 2007) through natural selection.

Selection for specific characteristics may improve the tolerance to Varroa mites in honey bees. However, because of the difficulties in recording the specifics needed, this is probably not the best way forward. Ultimately, mite tolerance will likely be a combination of qualities, a combination that may vary in different geographic areas and among different bees. In most places the infestation of Varroa mites will probably not eradicate the species *A. mellifera*. Nevertheless, the vast majority of bees are likely to die if no mite control is practiced, at least in Europe; but after a decade or so, populations may recover to build up new viable populations despite attacks by the mites. Unfortunately, the bees that survive through natural selection may have lost desirable properties for profitable beekeeping. Exposing the European bee population for natural selection in this context is not acceptable, with the implications for pollination and beekeeping in general that would result. And, efficient mite control masks any differences in tolerance leading to a continuous need for mite control. The solution must be to find a strategy that makes it possible to distinguish between colonies of greater and lesser resistance to Varroa mites. Taking into account what has been said above regarding the selection of individual parameters, probably the only realistic alternative is to study and compare the growth rate of mites in colonies of different genetic backgrounds, while allowing the mite population in all colonies to develop. In short, the success of beekeeping with Varroa mites in most of Europe is about producing healthy winter bees that have not been heavily parasitized by mites. It is fully compatible with both good honey harvest and good wintering to have a relatively high mite population in bee colonies in spring and early summer. Therefore, it must be during the time of optimal growth of the mite population that the growth rate is monitored, employing mite control at the most effective time for producing healthy winter bees if certain thresholds (in debris counts of mites or infestation rates of the bees) are exceeded for a predetermined part of the summer. What such thresholds should be, when they should be acted on, and for what geographic location need further investigation.

Recommendations

Selection Criteria

In light of what has been reported, it may be realistic to limit the selection for increased mite tolerance to two characteristics that are relatively easy to measure:

1. Hygienic behavior
2. Mite population growth rate

The hygienic behavior issue was discussed. As previously indicated, selection for hygienic behavior has only limited effect on Varroa mites, but because there are also positive, albeit limited effects on resistance to Varroa, this characteristic should also be included in this context.

Measuring the mite population growth rate in different colonies should give the best measure for Varroa tolerance. The methods used must be simple for practical beekeeping, but have as high an information value as possible. In addition, the method must allow for measurements in colonies with different mite levels, because it is unrealistic to standardize infestation levels. We can assume that the mite population growth rate is exponential (Fries et al., 1994) with a growth rate of approximately 2.5% per day if there is free access to brood, and that the mite infestations are not large enough to affect colony development (Calatayud and Verdú, 1993, 1995). By estimating the mite population size between two dates with free reproduction of the mites, a growth rate can be obtained that is comparable between different colonies regardless of infestation level and at least in part independent of the number of days the measurements include. With this information the growth rate can be calculated thus:

$$\chi = e^{r * d}, \qquad \textit{(Formula 1)}$$

where χ = number of multiples by which the population has grown

e = natural logarithm

r = growth rate per day

d = number of days during which the measurement occurred

Example: The measurement took place during 65 days (d = 65). The mite population is estimated to have increased from 100 to 580 (= 5.8). Formula 2 can now be written as:

$$r = \ln \chi / d \qquad \text{(Formula 2)}$$

$$\text{hence, } r = \ln (5.8) / 65 = 0.027$$

The growth rate is 2.7% per day in this case. This measurement should provide a basis for assessing Varroa tolerance. Measurements of mite population growth should be undertaken only in full, strong colonies (a lower limit defined) and with fully functional queens. It is proposed to begin measuring a few days after the bees fly out for the first time in spring (when willow, *Salix* spp., is blooming). This is because brood rearing takes off only with proper access to fresh pollen. A sample of approximately 300 live bees is taken in the brood area and the mites are washed off, giving a measure of the number of mites per bee. Investigations in the field show that such sampling gives a surprisingly good prediction of the overall infestation rate in a hive (Lee et al., 2010), with a precision that may be sufficient in this context. If greater precision is needed, samples from both brood and bees should be measured (Lee et al., 2010).

A second sample of bees (or both bees and brood with estimates of colony bee and brood numbers) is taken sometime in early July or mid-July in the same way. The number of times the mite numbers have doubled is calculated:

$$\frac{\text{Number of mites per bee in test 2 (or in a colony)}}{\text{Number of mites per bee in sample 1 (or in a colony)}}$$

Thereafter, the growth rate, as described in Formula 2, is used to compare colonies of different genetic backgrounds for their relative resistance to mites. Varroa populations must be allowed to grow to levels that make measurements meaningful. Only if certain predetermined thresholds, as previously discussed, are reached, should mite control be practiced. In Germany, the threshold for mite control is set to 10% infestation of adult bees in July (Büchler et al., 2010).

Concluding Remarks

The aim of this BEE DOC milestone has been to determine how to evaluate the colony's relative tolerance to Varroa mites and to be able to use this information for breeding purposes. There are many indications that the main characteristics of bees that resist Varroa are specific mite-directed hygienic behavior (such as Varroa Sensitive Hygiene or VSH) and/or a decrease in fertility and maternal fecundity of mother mites. These characteristics are very laborious to measure, so the most practical solution seems to be to monitor mite population growth, regardless of the underlying characteristics. What is proposed is mainly based on a German approach for evaluating mite tolerance (see Büchler et al., 2010, for details), but here we based our estimates solely on samples of bees, or on samples of brood and bees combined with population estimates if greater precision is needed. This latter determination must be made in the field.

Status of Breeding Practices and Genetic Diversity in Domestic U.S. Honey Bees

CHAPTER 4

Susan W. Cobey, Walter S. Sheppard, and David R. Tarpy

Abstract The many problems that currently face the U.S. honey bee population has underscored the need for sufficient genetic diversity at the colony, breeding, and population levels. Genetic diversity has been reduced by three distinct bottleneck events, namely the limited historical importation of subspecies and queens, the selection pressure of parasites and pathogens (particularly parasitic mites), and the consolidated commercial queen-production practices that have reduced the number of queen mothers in the breeding population. We explore the history and potential consequences of reduced population-wide genetic diversity, and we review the past and current status of the reproductive quality of commercially produced queens. We conclude that while queen quality is not drastically diminished from historical levels, the current perceived problems of "poor queens" can be significantly improved by addressing the ongoing genetic bottlenecks in our breeding systems and increasing the overall genetic diversity of the honey bee population.

Introduction

The decline of honey bee colonies (*Apis mellifera*) in the United States (U.S.) and Europe, in both managed and feral populations, is of significant concern. Historically, there have been periodic high losses of managed European honey bees (vanEnglesdorp and Meixner, 2009). Although colonies are challenged by numerous interacting factors, parasitic Varroa mites (*Varroa destructor*) and associated diseases play a major role. The current impact of Varroa is augmented by the worldwide spread of honey bee pathogens, the accumulation of miticide and pesticide residues in beeswax, and malnutrition (vanEngelsdorp et al., 2009). Colony losses have neither been significant in Australia (where Varroa is absent) nor in Africa and South America (where African and Africanized bees, respectively, exhibit high survival without treatment for Varroa mites).

The selection, development, maintenance, and adoption of highly productive European honey bee stocks that can both tolerate Varroa and resist diseases offer a sustainable, long-term solution to these ongoing problems. However, developing such a suite of traits has been elusive. Bee breeding is subject to unique challenges, including high labor costs, often slow progress toward breeding goals, and little economic profit. The primary importations of honey bees into North America took place between the early 17th and 20th centuries, with severely curtailed importations of additional genetic stock over the past 90 years. The limited importation of honey bees from areas of endemism, coupled with a queen production system that annually produces the majority of U.S. commercial queens from a relatively small number of queen mothers, represent potential genetic bottlenecks (Figure 1). Such bottlenecks could reduce genetic diversity and may limit our ability to select strains of bees that can both tolerate Varroa mites and be commercially productive. Here, we examine the historical importation of bees, the genetic effects of founder events and queen breeding practices, and the assessment of queen quality in U.S. populations in an effort to provide insights into current reports of declining honey bee populations.

Figure 1. The honey bee population in the U.S. has undergone three distinct genetic bottlenecks that have reduced genetic diversity. (*Modified from T. Lawrence.*)

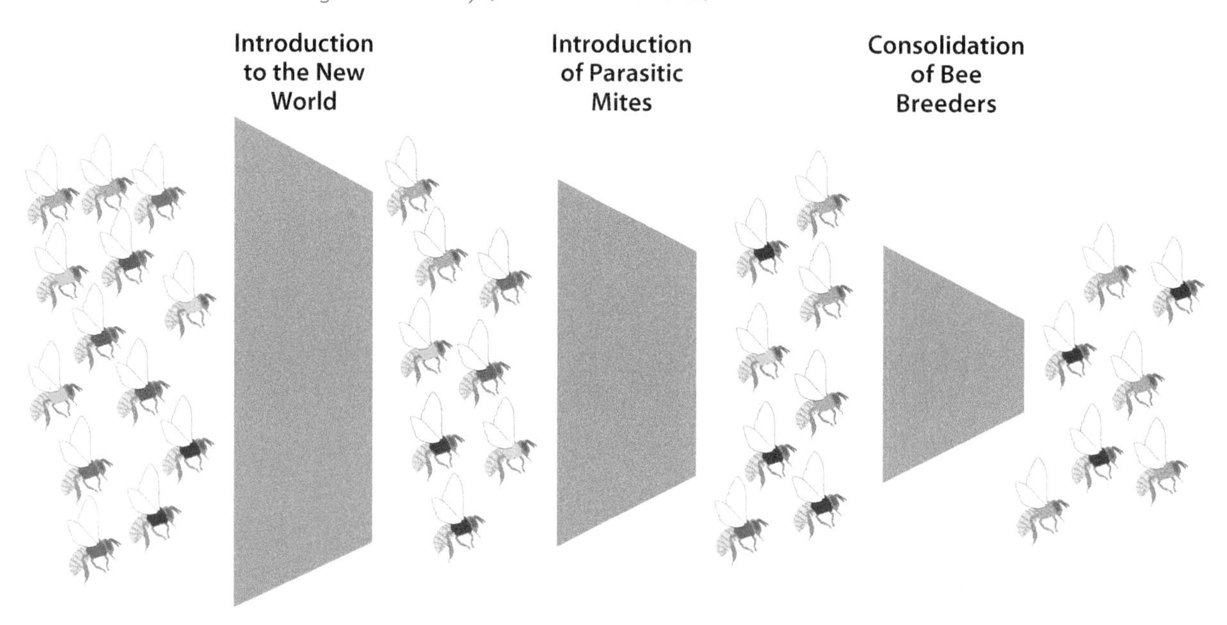

A Brief History of *Apis mellifera* in the U.S.

As with many animals of agricultural importance, the honey bee (*Apis mellifera*) is not native to the U.S. or Australia. The original distribution of the genus *Apis* was restricted to Africa, Asia, and Europe. The colonization of North America and Australia by European immigrants led to the introduction of the European native honey bee. In those early times, the honey bee was primarily of importance as a major source of sweetener and wax.

With the development of "modern" agriculture, the primary importance of honey bees from a human perspective shifted to their role in providing pollination services. In the U.S. perhaps the most striking example of the current role of honey bees as pollinators in modern cropping systems is the magnitude of the population of honey bees required to pollinate the almond crop. California alone produces about 80% of the almonds consumed worldwide on more than 300,000 hectares of almond orchards. In 2011, an estimated 1.3 to 1.5 million hives of bees will be needed to pollinate the almond crop (Flottum, 2010). This single agricultural crop, therefore, requires physical placement into California almond orchards of almost 60% of the 2.5 million colonies of bees currently managed in the United States. Similar stories, although involving fewer colonies, could be told of the pollination requirements for apples, cranberries, cucumbers, and numerous other crops (Delaplane and Mayer, 2000).

Although the pollination service of U.S. agricultural crops is by managed honey bees of commercial origin, it is important to realize that substantial "within-species" variation occurs in the honey bee across its original range of Europe, Africa, and western/central Asia. This variation reflects the adaptation of populations of honey bees throughout this vast geographic range to a large set of climatic differences. One might expect, for example, that honey bees of sub-Saharan Africa and those of Scandinavia might have evolved somewhat different behaviors related to foraging, overwintering, and swarming. Given that the honey bee across this range is only a single species (*Apis mellifera*), bees can and will interbreed when placed together in common apiaries. However, geographic differences among honey bee populations led scientists to further differentiate *Apis mellifera* into a number of "sub-species," sometimes referred to as *geographic races* (Ruttner 1975, 1988). Subspecies are designated by adding a third name to the species name and are referred to by a "trinomial." Examples include *Apis mellifera mellifera*, *Apis mellifera ligustica*, and *Apis mellifera carnica*, which refer to the Dark Bee of Northern Europe, the Italian honey bee, and the Carniolan honey bee, respectively.

The importance of having subspecies designations is primarily one of convenience in having a common vernacular to describe or refer to specific groups or populations of honey bees. Within the world of beekeeping, various subspecies have been reported to express specific tendencies or traits of apicultural interest, such as the tendency for defensiveness and high swarming rate in a subspecies from sub-Saharan Africa (*Apis mellifera scutellata*) or the tendency toward gentleness and high propolis use in *Apis mellifera caucasica* (a subspecies endemic to the Caucasus Mountains).

The history of honey bee introductions into the United States is a fascinating story in itself and, in part, is also reflective of the personalities of leading beekeepers and bee scientists of the time. The first records of honey bees existing in what is now the U.S. was in 1622, at the Jamestown colony (present-day Virginia; Oertel, 1976). These bees were imported from within the range of *A. m. mellifera* in England and became well-established in the eastern part of the country. Interestingly, these bees quickly spread out in front of the advancing European settlers in subsequent decades. By 1788, Thomas Jefferson wrote that the native people referred to these insects as "White Man's flies," indicating their role as foretelling the impending arrival of European settlers wherever they were found (Jefferson, 1788). *A. m. mellifera* (the Dark Bee of Northern Europe) was apparently well-suited for North America and was, in fact, the only honey bee present in the U.S. for the next 239 years (1622 until 1861).

By the 1850s, steamship service was established between the U.S. and Europe, reducing the time required to cross the Atlantic Ocean and opening the door to affordable and secure shipments of bees. In 1859, Italian honey bees were introduced to the U.S. from Dzierzon's apiary in Germany and by 1860 a shipment of Italian queens was received directly from Italy (Anonymous, 1859; Langstroth, 1860 reprinted in Naile, 1942). The American beekeeping public was enamored with the newly available yellow and relatively gentle bees. As a result, Italian-type honey bees form the basis for most present-day commercial beekeeping stocks in the U.S. Following the arrival and success of honey bees from Italy, U.S. beekeepers developed an interest to try other honey bee subspecies.

From 1859 until 1922, efforts were made to import and introduce a number of additional subspecies. When surveys indicated that the newly discovered tracheal mite (*Acarapis woodi* Rennie) was not present in the U.S., the U.S. Honeybee Act was passed in 1922 to restrict further honey bee importations. However, prior to the passage of the Act, a number of subspecies of European, Middle Eastern, and African origin were imported.

Following the Italian honey bee introductions of the early 1860s, the Egyptian honey bee (*Apis m. lamarackii*) was introduced in 1869. This honey bee was soon dismissed by beekeepers, although remnant genetic markers of this subspecies could still be detected in feral honey bee populations sampled in the U.S. in the 1990s (Schiff et al., 1994; Magnus and Szalanski, 2010). In 1877, the initial importations of the Carniolan honey bee (*A. m. carnica*) were made by Charles Dadant (Dadant, 1877). Larger and more sustained importations of Carniolan honey bees were made by Frank Benton, who imported substantial numbers of queens into Canada and the U.S., starting in 1883 (Norris, 1884). Interestingly, Benton later became the first apiculturist of the forerunner of the U.S. Department of Agriculture, a post he held until 1905.

Benton was also most likely the initial importer of the Cyprian and Syrian honey bee subspecies (*A. m. cypria* and *A. m. syriaca*, respectively) in the early 1880s. These subspecies were imported into both the U.S. and Canada during this time and one particularly large importation consisted of 150 Cyprian queens made in 1880 (Jones, 1880). Neither of these Middle Eastern subspecies found favor with U.S. beekeepers, although genetic markers indicating some relictual influence on the U.S. feral population have been reported (Magnus and Szalanski, 2010). In addition to the Italian and Carniolan honey bees, the Caucasian honey bee (*A. m. caucasica*) was a subspecies that came to have enduring interest to U.S. beekeepers. The initial importation details of this subspecies are less certain than for some of the other subspecies, but there were reports of importation in the 1880s and the subspecies was clearly present by 1890 (Hoffman, in York 1906; Tefft, 1890). Direct importations of Caucasian honey bees were made into Colorado in 1903 (Benton, 1905). Another African subspecies, *A. m. intermissa*, from northern Africa was introduced and established by 1891, although details of the importation are lacking. In any case, this bee was very quickly dismissed by U.S. beekeepers, especially following the publication of an article by Benton noting that the bees were "small, very black and spiteful stingers" (Shepherd, 1892).

Based on publications of the time, historical evidence points to the introduction of at least eight recognized subspecies of honey bee into the U.S. by 1922. In 1990, the descendants of a sub-Saharan African subspecies, *A. m. scutellata*, (introduced into Brazil in 1956) traversed the U.S.-Mexico border and became established in Texas. Twenty years later, these "Africanized" honey bees are found in a number of southern states, ranging from California through Texas and Florida (USDA, 2007). The imported populations that were derived from these nine subspecies therefore represent the "starting material" available for genetic selection and breeding within the U.S. Some additional importations of particular selected stocks have been made through the efforts of the USDA or queen package resellers (Bourgeois et al., 2010). In recent years, the USDA has imported "Russian" bees to increase tolerance to Varroa. Large numbers of Australian package bees and queens (primarily the Italian type) have been imported into the U.S., to assure an uninterrupted supply of pollinators for the almond crop. However, this Australian importation will be stopped in 2011 due to concerns of importing Asian honey bees and a new parasitic mite.

Genetic Bottlenecks and Diversity of U.S. Honey Bee Populations

The honey bee originated in the Old World (Whitfield et al., 2007), where it diverged into more than two dozen recognized subspecies (Ruttner, 1987; Sheppard et al., 1997; Sheppard and Meixner, 2003). Initial introduction of the honey bee (subspecies *A. m. mellifera*) to North America occurred in the 17th century and records indicate that another seven subspecies were introduced by 1922, when further importations were restricted (Sheppard, 1989). With the notable exception of the introduction of African *A. m. scutellata* into Brazil in 1957 (Michener, 1975) and subsequent expansion of descendant Africanized populations into parts of the southern U.S., no additional subspecies have been introduced into these existing New World honey bee populations.

In considering genetic diversity, it is instructive to realize that the honey bee populations originally introduced into North America were filtered through two structural genetic "bottlenecks" (Figure 1). First, the initial "sampling" of each subspecies chosen for importation consisted of a few tens to hundreds of queens, representing only a small fraction of the genetic diversity within each subspecies. Second, only nine of the more than two dozen named Old World subspecies found within the species *Apis mellifera* were ever introduced into the Americas. Thus, overall "sampling" of the within-species diversity was only partial, with two-thirds of the named subspecies never having been introduced into the Americas. Subsequent to the initial importations, additional losses of genetic diversity could have been expected due to "genetic drift." Genetic drift can be thought of as changes in gene frequencies across generations due to chance or as the effect of inbreeding in small populations, both of which can lead to loss of allelic diversity.

Prior to the establishment of parasitic mites in the U.S., there was a rather robust population of feral honey bees containing genetic markers that reflected their diverse origins from some of the original importations (Schiff and Sheppard, 1993; Schiff et al., 1994). Further, comparison with existing commercial honey bee stocks showed that this feral population contained genetic diversity that might be useful to supplement existing commercial honey bee stocks. However, the feral population of honey bees was largely decimated by Varroa mites (Kraus and Page, 1995) and only limited examples suggest recovery of feral populations (with the exception of Africanized honey bees in the southern U.S.; Seeley, 2007). Consequently, the potential for feral honey bees to contribute substantial additional genetic variation to U.S. commercial stocks for selection and breeding proposes may be limited.

Currently available U.S. honey bees are primarily derived from two European subspecies, *A. m. carnica* and *A. m ligustica*. Aside from the aforementioned bottlenecks attributable to sampling, importation, and subsequent genetic consequences in small populations, breeding practices within the U.S. are also relevant to the question of genetic diversity within U.S. honey bee populations. In studies conducted in 1993–1994 and in 2004–2005, U.S. commercial queen producers self-reported the production of close to 1 million queens for sale from around 600 and 500 queen "mothers," respectively (Schiff and Sheppard, 1995, 1996; Delaney et al., 2009). Whether this apparent decline in the number of queen mothers being used for annual queen production is a trend that will continue is unknown. However, coupled with high losses of colonies that have been reported in recent years (averaging 30% annually), declining breeding population sizes would be an additional concern. Studies of genetic markers suggest that while commercial honey bee populations have relatively limited amounts of genetic diversity, there were genetic differences between the eastern and western U.S. breeding populations that could potentially be used by bee breeders to contribute to overall sex allele diversity.

Bee Breeding Practices in the U.S.

Most economically valued livestock species are not native to the U.S. and are derived from selected strains as a result of well-designed, scientific stock improvement programs. These programs are dependent upon long-range breeding programs, as well as the routine and systematic importation and evaluation of additional resources (mostly germplasm) from within the original ranges of the species under consideration. The beekeeping industry does not have access to stocks of origin or standardized evaluation and stock improvement programs. Consequently, the beekeeping industry does not share the increased productivity that results from such programs, and which have served the poultry, dairy, and swine industries.

Bee-breeding programs must be based upon selection of behavioral traits at the colony level. Consideration must include the high mating frequency of queens, the complex and dynamic social structure of the colony, sensitivity to inbreeding, and environmental influences. Of the selection criteria and methods used, there is a lack of standardization for measuring traits and selection is often limited to too few traits at the expense of productivity. Inter- and intra-colony genetic diversity has clearly been demonstrated to increase colony fitness and survival, and lessen the impact of pests and diseases (Olroyd et al., 1991; Fuchs and Schade, 1994; Tarpy, 2003; Jones et al., 2004; Mattila et al., 2007; Richard et al., 2007; and Seeley and Tarpy, 2007). Maintaining a high level of genetic diversity is critical and challenging in any stock improvement program, which especially applies to honey bees.

Scientific bee-breeding programs have largely been dependent upon institutional and government support. Frequently subject to short-term funding, programs that have been turned over to the industry have historically lacked oversight and soon become unrecognizable. Without a long-term commitment and supporting resources, selection efforts are relaxed and gains are quickly lost.

The U.S. beekeeping industry is built upon the development of large-scale queen and package bee production, in which it today excels. Historically, private-sector breeding efforts have been largely limited to choosing a few top-performing colonies with little regard for control of mating or performance over generations. Traditionally, the terms "queen rearing" (the propagation of queens) and "bee breeding" (the evaluation and selection of breeding stock) have been used interchangeably in the beekeeping community. These are two very different aspects and require different skills, knowledge, and practices.

Queen producers represent a small, specialized, yet critical aspect of the beekeeping industry, many of which have been built upon family businesses. Production requires high overhead, is labor intensive, and the high demand for queens and package bees does not provide incentive for expensive breeding programs. The applied nature and lack of publications targeted at bee-breeding programs do not adequately foster an environment for researchers. While this situation is changing, industry support will determine the viability of such programs.

The U.S. queen and package bee industry provides about one million queens annually to replace and restock the estimated 2.4 million colonies nationwide. Of commercially managed colonies, some honey producers and pollinators rear their own queens in addition to those purchased. Some requeen their colonies annually, others every other year or as needed, depending upon queen performance. Queen-rearing operations range in size and production, mostly between 5,000 and 150,000 queens produced annually. Often, only a few queen mothers are used for propagation and with little control of mating areas. Concentrated in northern California and the Southwest, completely isolated mating is not an option and producers depend upon each other to supply adequate drone sources for mating yards.

Traditionally, commercial breeding stocks in the U.S. are based upon selection of a few queen mothers from among thousands of colonies within commercial operations. Potential breeder colonies are often followed throughout the year; others are selected during early spring. Colonies that stand out for several valued traits are selected and notations are generally recorded on hive lids. The criteria selected vary among producers and generally include large populations, good laying patterns, fast spring buildup, temperament, consistent color, weight gain, and overwintering ability. Low prevalence of pests and disease symptoms may also be noted. Increasingly and more recently, selection also includes monitoring for pest and disease levels, especially Nosema and Varroa. Testing for hygienic behavior, one known mechanism of resistance to pests and diseases, is also becoming more common. The beekeeper methods used, choosing the few best colonies within their apiaries for breeding stock, has been successful in maintaining distinct lines, yet progress in selection for resistance to pests and diseases has not been realized. Lack of controlled mating and record-keeping remains a handicap.

Most producers in the U.S. augment their programs with the purchase of breeder queens from the limited available specialty breeding stocks, including USDA "Varroa Sensitive Hygiene," "Minnesota Hygienic," "New World Carniolan," and USDA "Russian." Queen producers generally use a combination of breeders selected from their own colonies and those that they purchased. Some producers prefer not to use progeny of specialty breeding stocks in their own hives, limiting these as drone sources for mating yards.

An interest in selection of locally adapted stocks is increasing. Among hobbyist beekeepers and local beekeeping organizations, the desire to move away from the use of miticides and antibiotics and the frustration in the tightening availability and increased expense of queens promotes this. These "microbreeder" programs (*sensu* D. Tarpy, in Connor, 2008) are often based upon collection of "survivor stock" (the collection of swarms). Results are often unpredictable and disappointing due to a lack of rigorous selection criteria and controlled mating. Hopefully this will change as these programs develop.

The U.S. queen and package bee industry—concentrated in northern California, the Southeast, and Hawaii—is at full capacity with growing demand. Maintaining healthy rigorous colonies, controlling Varroa, and avoiding sublethal chemical residues in colonies is required for queen and drone production and is increasingly demanding of labor and costs. The impact of small hive beetle, SHB (*Aethina tumida*), is reducing queen production in the southeastern U.S. Mating nucleus colonies are vulnerable to beetles that are highly attracted to the small colonies and intermittent state of queenlessness between rounds. Hawaii, once a haven for Varroa-free queen production, must now deal with the recent introduction and impact of both Varroa and SHB. The SHB is also expected to spread throughout northern California queen mating areas, despite control and monitoring efforts.

The high colony demand for pollination of almonds in February and concern over the high winter loss of colonies resulted in an amendment to the Honeybee Act of 1922 to allow the importation of hundreds of thousands of package bees from Australia and New Zealand from 2005 to 2010. Due to increasing industry concern over the risk of introducing pests and diseases, as well as the potential impact of quick spread across the U.S. as colonies are trucked nationwide, the border has since been closed.

Quality of Commercially Produced Queens

The primary perceived problem for beekeepers is a diminished quality of queens, and recent survey results from beekeeping operations in the U.S. confirm this view. VanEngelsdorp et al. (2008) surveyed 305 beekeeping operations in the U.S. accounting for a total of 324,571 beehives. According to the interviewed beekeepers, their primary perceived problem was "poor queens," with 31% of the dead colonies as a result of one or more issues with the mother queen. By contrast, starvation (28%), varroa mites (24%), and CCD (9%) were significant but less prevalent causes of mortality. The "poor queens" category encompasses many different problems but most of these reports document premature supersedure (queen replacement), inconsistent brood patterns, early drone-laying (indicative of sperm depletion), and failed requeening as indicative of low queen quality. It is helpful, therefore, to place into an historical context the current quality of the commercial queen population.

Several studies have surveyed commercially produced queens, either directly or by sampling queens shipped in packages (Table 1). Farrar (1947) studied queens deriving from packaged bees over several years. Furgala (1962) collected queens from beekeepers in Canada that were ordered from either California or Mississippi, but these queens might represent a biased sample as they were either dead on arrival or queens that were lost in the first month. Jay and Dixon (1984) sampled a very large number of U.S. queens shipped in packages to Western Canada in the mid-1960s and early 1980s. Liu et al. (1987) also measured queens sent to Canada from the U.S. for various infections, as did Burgett, and Kitprasert (1992) directly obtained queens from a commercial beekeeping operation. Camazine et al. (1998) purchased sets of 15 naturally mated "Italian" queens from 13 different commercial sources across the U.S. Similarly, Delaney et al. (2011) purchased 12 "Italian" queens from each of 12 different breeders either in the Southeast or Western U.S., two in each set being temporarily introduced to colonies while the others were banked before processing.

Physical Quality

There are many measures that can serve as proxies for queen reproductive potential, or "quality." The most intuitive perhaps are standard morphological measures of individual adult insects, such as wet or dry weight, thorax width, head width, and wing lengths (Weaver, 1957; Fischer and Maul, 1991; Dedej et al., 1998; Hatch et al., 1999; Gilley et al., 2003; Dodologlu et al., 2004; Kahya et al., 2008), several of which are significantly correlated with queen reproductive success or fecundity (Eckert, 1934; Avetisyan, 1961; Woyke, 1971; Nelson and Gary, 1983).

Weight is often used as a proxy for overall queen quality. In fact, beekeepers often use size and weight as a rough indicator of the relative quality of a queen. This association may be predicated on the fact that virgin queens are smaller and weigh less than mated queens because mated queens have larger, fully developed ovaries. Nelson and Gary (1983) showed that honey productivity of colonies increased with heavier queens, although other studies have failed to show a significant relationship (e.g., Eckert, 1934). Other measures of queen size, such as thorax width, have also been used and show less variation due to environmental or colony conditions. Delaney et al. (2011) found that thorax width, but not queen weight, was significantly positively correlated with both stored sperm and effective paternity frequency.

Potential Fecundity

Queen ovaries are highly developed compared to workers, with each queen containing approximately 300 or more individual ovarioles (Eckert, 1934). Ovary development occurs soon after mating and is associated with profound genomic, physiological, and behavioral changes in the queen (Richard et al., 2007; Kocher et al., 2008). Hoopingarner and Farrar (1959) found a very strong correlation between queen weight and ovariole number, but others have not shown this same relationship (Eckert, 1934; Hatch et al., 1999; Jackson et al., in press); thus it is unclear if weight is a good proxy for potential fecundity. The important glycolipoprotein vitellogenin (Vg) is also a potential indicator of fecundity because it is the yolk precursor associated with egg production (Tanaka and Hartfelder, 2004). Transcript levels of Vg appear to be associated with queen weight independently of active egg-laying by queens (Delaney et al., 2010).

Parasites

While many parasites and pathogens either cannot or do not infect honey bee queens (in large part because of their faster development time, infrequent availability, or both), there are some notable exceptions that have been shown to diminish queen bee health and productivity. There have been several efforts to monitor and measure the gut microsporidian *Nosema apis* in commercial queens (Farrar, 1947; Furgala, 1962; Jay and Dixon, 1984; Liu et al., 1987; Camazine et al., 1998), with some studies showing as many as 38% being infested (Table 1). Most recently, however, Delaney et al. (2011) did not detect any newly mated queens infested with nosema, suggesting significant changes in management practices within the industry. With the introduction and apparent selection sweep of *Nosema ceranae* (Higes et al., 2006; Chen et al., 2008), it is unclear how

Table 1. Historical evaluations of queen reproductive potential, or "quality," from various commercial sources.

	1947	1962	1984	1987	1992	1998	2010
No. queens	835	465	777	53	200	325	136
Under-developed ovaries	—	17 %	—	—	—	12 %	7.5 %*
Nosema	14 %	11 %	8 %	38 %	—	7 %	0 %
Tracheal mites	NA	NA	—	—	21 %	20 %	2 %
Sperm counts (< 3 million)	—	29 %	11 %	—	—	19 %	19 %
Mating number	NA	NA	—	—	—	—	16

* = data may not be comparable to previous studies because of different measurement methods (Jackson et al., in press). Data from Farrar, 1947; Furgala, 1962; Jay and Dixon, 1984; Liu et al., 1987; Burgett and Kitprasert, 1992; Camazine et al., 1998; Delaney et al., 2010.

this sister taxa may affect the quality of commercial queens. Since the mid-1980s, queens have also been subject to infestation from tracheal mites *Acarapis woodi* (Burgett and Kitprasert, 1992; Camazine et al., 1998; Villa and Danka, 2005). However, this parasite also seems to have diminished in frequency among commercial stock as well as commercial queens (Table 1; Delaney et al., 2010). Finally, queens may be infected with any number of viruses (Chen et al., 2005; Yang and Cox-Foster, 2005), some of which have been shown to be transmitted vertically from queen to worker offspring (Chen et al., 2006). These pathogens, however, do not seem to have any direct association with queen quality (Delaney et al., 2010), although they may have some indirect effects that have yet to be quantified.

Mating Success

Another important characteristic that determines a queen's quality is the degree to which she is inseminated as queens with greater sperm stores can live longer and fertilize more eggs. Queens take mating flights early in life when they are approximately one week old (Koeniger, 1988), mating with multiple males on one or several flights away from their natal hive. Sperm is temporarily deposited in the median and lateral oviducts, then a small proportion migrate and are stored in the spermatheca (Woyke, 1983; Collins, 2000). Many researchers have assessed the number of stored sperm in a queen's spermatheca (Mackensen, 1947; Koeniger et al., 1990; Lodesani et al., 2004), and a fully mated queen stores approximately 5 to 7 million sperm (Woyke, 1962) and the spermathecae show a tan, marbled coloration (as opposed to whitish and opaque for partially mated queens and totally clear for unmated queens; Cobey, 2003).

Early studies of the commercial queen population assessed queen spermathecae but did not perform sperm counts. Furgala (1962) reported that 24 out of 229 (10.5%) queens that had died upon arrival had few or no sperm in their spermathecae based on visual observation, as did 34 out of 236 (12.9%) queens that died within their first month. Jay and Dixon (1984) found similar results, with 11% of the queens having fewer than 3 million sperm (termed "poorly mated"). Using the same arbitrary cutoff as defined by Woyke (1962), Camazine et al. (1998) found that 19% of the queens were poorly mated. Most recently, Delaney et al. (2011) also found 19% of commercially produced queens had fewer than 3 million sperm ("poorly inseminated"), but they also found that 80% of the queens had fewer than 5 million sperm ("inadequately inseminated"), with an average of 4 million sperm. These numbers are consistent with commercially tested queens in California in the mid-1980s (Harizanis and Gary, 1984).

Insemination is one measure of a queen's mating success, but emerging evidence suggests that mating diversity is also important for queen and colony productivity. The genetic diversity within a colony is a direct reflection of the number of drones that sire worker offspring (Tarpy et al., 2004), and several empirical studies have demonstrated that genetically diverse colonies increase the behavioral function of the worker force, reduce the likelihood for detrimental levels of inviable brood due to the csd locus, and lower the prevalence of various parasites and pathogens (reviewed by Palmer and Oldroyd, 2000). A meta-analysis of studies using molecular techniques to quantify effective paternity frequency of queens concludes that open-mated queens mate with approximately 12 drones (Tarpy and Nielsen, 2002). However, only one recent study fully quantified a cross section of mated queens. Delaney et al. (2011) found that queens mated with an average of 25 drones, with an effective average paternity frequency of 16.0 ± 9.48.

Overall, the current status of commercial U.S. queens seems to be of high quality when viewed from a historical perspective. It is clear, therefore, that the current perception of diminished queen productivity stems from alternate factors. Future research should investigate potential mechanisms that affect queen quality both prior to mating (e.g., reproductive capacity of drones) and after queen introduction (i.e., hive environment). In doing so, it will also be important to determine the genetic diversity of commercial queen and drone populations.

Future Directions

Worldwide, the apiculture community is focused on finding sustainable solutions to the multifaceted factors contributing to the current honey bee decline. The crisis has stimulated collaborative efforts on a global scale. These collaborations include major efforts to compare and document changes in honey bee populations from many geographic areas. Programs designed to select stocks for increased resistance to pests and diseases are increasingly gaining support as well, given the near unanimity among honey bee scientists that a genetic approach is necessary to ensure a long-term, sustainable managed population.

There have been several national and international collaborative efforts to help implement such goals. Perhaps the most notable international effort was the formation of COLOSS (Prevention of honeybee COLony LOSSes), a collaborative network of 17 European countries. Collaborators are working to identify honey bee populations, track changes for conservation and selection purposes, and develop certification programs for local strains and ecotypes of honey bees. While numerous studies of morphometric characteristics, behavioral traits, and molecular analysis have been conducted, there is a strategic need to establish a standard protocol to identify, record change, and preserve diverse honey bee populations in their native ranges.

In the U.S., the recognized need to increase genetic diversity and strengthen selection programs of commercial breeding stocks has resulted in collaborative efforts among universities, government researchers, and the queen industry. Honey bee semen of several subspecies of European honey bees has recently been imported and inseminated to virgin queens of domestic breeding stocks. Diagnostic programs to assist beekeepers to assess colony health are being established, as are technology transfer programs to provide hands-on assistance and instruction in evaluation and selection of commercial breeding stocks.

The current challenges facing the beekeeping industry and new technologies being developed are pushing beekeeping into a new era. With the sequencing of the honey bee genome and advancements in molecular techniques, powerful markers for evolutionary and population genetics studies are increasingly available. Discrimination of honey bee populations and subspecies may contribute to selection programs through the utilization of these technologies. Furthermore, the use of molecular techniques can assist in the identification and selection of specific traits of resistance to pests and disease in breeding stocks. Technologies to perfect the cryopreservation of honey bee semen and facilitate the safe international exchange of honey bee germplasm are priorities for development.

Finally, there is a great need to document and track the genetic diversity within and among honey bee populations in the U.S., particularly both the managed and feral (nonmanaged) populations. Determining how such genetic diversity impacts colony phenotype and productivity is affected by gene flow among populations (especially the Africanized population in the southern tier of the country), and is manifest by management and other breeding techniques; this must be prioritized by future research. In addition, new investigations into the mating behavior, pedigree relatedness within and among breeding populations, and population genetic structure of current populations will greatly inform such research. By doing so, these approaches will together enable genetic solutions to the many problems currently facing the apiculture industry.

Global Status of Honey Bee Mites

CHAPTER

5

Diana Sammataro

Abstract Parasitic bee mites have become a major problem for both beekeepers and honey bees. This chapter updates the latest information on the three parasitic bee mite genera (*Acarapis, Varroa,* and *Tropilaelaps*) as well as newly identified species that are currently infesting bee colonies throughout the world. Monitoring and treatment options are discussed as well as mite behavior and future research directions.

Introduction

Mites have numerous unique roles in the world's ecosystems and their incredibly rich body forms complement many amazing life histories. These invertebrates are in the Superorder Acari (from Greek for mite: *akari*), which include animals as small as 100 µm and as big as 10 mm (Red Velvet Mites). Mite fossils are found in the Devonian, around 400 million years ago, making them one of the earliest terrestrial lifeforms. Acarologists propose there are around 50,000 named species of mites, but suggest that number is a small fraction of those estimated to be discovered, identified, and named. This is an exceptionally large number of organisms about which we know very little; acarines are usually unnoticed or overlooked even by other scientists, mostly because of their limited economic importance and their small size.

Mites utilize any environment in which animals or plants are present and may equal or even surpass insects in the diversity of habitats and life cycles. Colonization includes soil, water (fresh and salt), in and on plants, arthropods, vertebrates, and invertebrates. Mites are found in polar extremes and desert environments and are even found 33 feet (10 m) deep in the ground (Walter et al., 1996). Some unexpected habitats include, for example, monkey lungs (Leonovich, 2010), sea snake tracheae (Nadchatram, 2006), special pouches on female carpenter bees (Scaife, 1952), and moth ears (Treat, 1957). It is only when mites cause injuries or affect the health of plants and animals (including humans) that they attract the interest of scientists. Mites that have received worldwide attention for these reasons include (but are not limited to) the spider mites (family Tetranychidae), the mange mites (family Sarcoptidae), and the allergen-producing house dust mites (family Pyroglyphidae). The tick vectors (ticks are large mites in the family Ixodidae) of Lyme disease are very well studied because of their deleterious effects on humans and other mammals (Steere, 2001); and hair follicle mites (family Demodecidae) have even been used in forensics (Desch, 2009).

Table 1. Mesostigmatic mites parasitizing honey bees, arranged according to host bee species. *(Compiled by D. Sammataro and D. L. Anderson; from Anderson and Morgan, 2007 and Navajas et al., 2010.)*

	Mite Species				
Bee Host	*V. destructor*	*V. jacobsoni*	*V. underwoodi*	*V. rindereri*	*Euvarroa sinha*
Apis florea			X Nepal, S. Korea		X
A. andreniformis					
A. cerana	X	X	X		
A. koschenikovi				X Sumatra	
A. nuluensis, Borneo		X	X ?		
A. nigrocincta, Sulawasi		X ?	X		
A. dorsata dorsata Asia, Indonesia, Palawan	## Korea				
A. d. breviligula					
A. d. binghami Sulawesi					
A. laborisoa Nepal					
A. mellifera	X Japan & Korean haplotypes	X Papua New Guinea, Irian jaya	##		
A. m. scutellata Africa	X				

X —Positive identification, ** — Currently unresolved; ## — Incidental visitor.

Mites are also one of the largest and most diverse of bee associates. And eusocial bees especially have many mites; most of these mites benefit from the conditions of the hive environment, that is, it is protected, thermoregulated, and contains an abundance of year-round food. Over 30 mite species associated with honey bees (*Apis* spp.) are listed by Eickwort (1988), including those that are predatory, incidental, facultative, obligatory, or phoretic. The three major suborders of the mites on this list were the Astigmata (mostly detritus feeders), Prostigmata (mainly pollen feeders), and Mesostigmata (feeding on hive products such as bee bread as well as on bees) (Eickwort, 1988). Also on the list were three obligate parasitic mites; they are the Varroa mite (*Varroa* spp.), the tracheal mite (*Acarapis woodi*), and *Troplilaelaps* spp. (the latter is not yet reported in the Americas). The global spread of these parasites has had a profound impact on bees and beekeepers, due to massive losses and increased mite-associated pathologies (mostly bee viruses).

Bee mites, particularly Varroa mites, are now a major topic in scientific articles; over 1,700 journal articles on Varroa have been published since 1971, 44 in 2010 alone, compared to 102 articles on tracheal mites since 1934. *Troplilaelaps* mites appear in 28 articles and *Euvarroa* in seven.

Table 1. *(continued)*

	Mite Species				
Bee Host	*E. wongsirii*	*Troplilaelaps clareae*	*T. koengerum*	*T. mercedesae* n.sp.	*T. thaii* n.sp.
Apis florea					
A. andreniformis	X				
A. cerana					
A. koschenikovi					
A. nuluensis, Borneo					
A. nigrocincta, Sulawasi					
A. dorsata dorsata Asia, Indonesia, Palawan			X Sri Lanka	X Palawan, Sri Lanka	
A. d. breviligula		X Philippines (not Palawan)			
A. d. binghami Sulawesi				X**	
A. laborisoa Nepal			X	X Vietnam	X Vietnam
A. mellifera		X Philippines		X	
A. m. scutellata Africa					

X —Positive identification; ** — Currently unresolved; ## — Incidental visitor.

Tracheal Mite (Prostigmata: Tarsonemidae)

Female

Male

The first report of problems came when bees dying in the Isle of Wight between 1904 and 1919 were examined more closely. In 1921 the tracheal mite *Acarapis woodi* (Rennie) was identified in the breathing tubes or tracheae of bees where it feeds and reproduces (see Figure 1) (Eckert, 1961; Delfinado-Baker, 1982). Its detection led to the restriction of all live honey bee imports into the United States in 1922 (Phillips, 1923). Despite this precaution, the New World eventually became infested, perhaps via African bees that were transported to South America in the 1970s. By the early 1980s, this mite made its way to Mexico. The first report attributing mites to problems with bees came from beekeepers in Texas in 1984. Thereafter, Acarapis spread to all of the states, facilitated by commercial beekeepers transporting bees for pollination, and from the sale of mite-infected package bees from the southern states. In addition, infected swarms and drifting bees also contributed to the spread of the mite. Acarapic mites are probably distributed worldwide, wherever the Western honey bee (*A. mellifera*) has been imported. There are two external

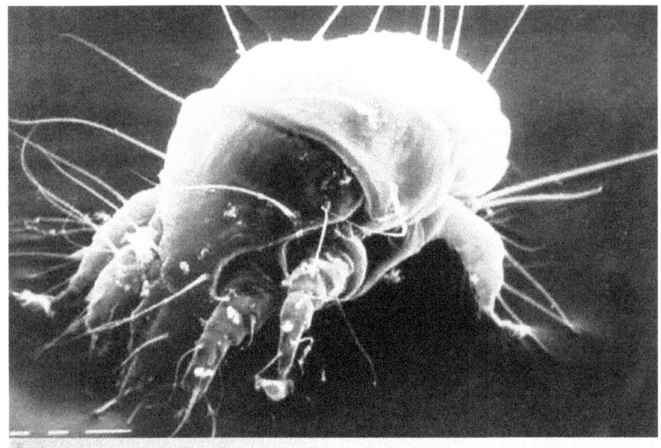

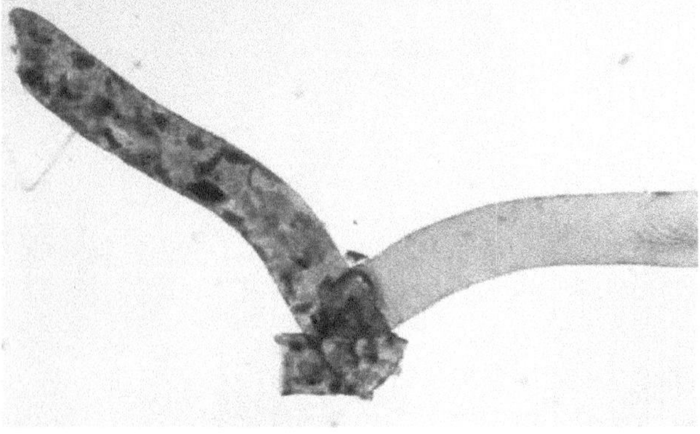

Figure 1. Top, scanning electron micrograph of tracheal mite. Bottom, oval shadows in tracheal tube (left) are mites, a clean tube is on the right. ×400. (*Photo A by W. Styer; photo B by D. Sammataro.*)

species in the genus *Acarapis* (*A. externus* and A. *dorsalis*), which are specific to the western honey bee (de Guzman et al., 2001). Currently, there are no reports of *Acarapis* species on other *Apis* bees.

Initially, *A. woodi* caused devastating losses throughout North America (see Figure 1, and 2). (For a complete life history, see Sammataro et al., 2000; Pettis, 2001; Wilson et al., 1997.) Tracheal mites interfere with the overwintering capability of colonies (bees are unable to form and maintain temperatures in the winter cluster) and are associated with paralyzed bees displaying disjointed wings (or K-wing) and crawling on the ground near hives. The actual cause for these early colony losses is still not clearly understood. Because of its small size, dissection of the honey bee is the only reliable way to determine infestation (Sammataro, 2006; and see video of bee dissection at: http://www.ars.usda.gov/pandp/docs.htm?docid=14370).

Control measures were vigorously studied and included vapors from menthol crystals, chemical acaricides, and oil or grease patties made from vegetable shortening and sugar (Sammataro et al., 1994; 2000). Breeding queens for resistance was another successful control method (de Guzman et al., 1998; Villa, 2006) and probably would have solved the problem. However, this endoparasite was soon overshadowed by the introduction of a second, more serious ectoparasitic mite, Varroa. However, *Acarapis* is still found but not readily seen because beekeepers are using multiple controls for Varroa.

Figure 2. Chart of the life cycle of tracheal mites. (*Illustration by Signe Nordin.*)

Age of Bee

1 to 3 days old — Female mite invades new bee 1 to 3 days old.

3 days — Mite feeds and lays about one egg per day.

8 days — Larvae hatch and feed on bee blood. Adult females hatch in 14 days, males in 12.

12 days — Daughter mites exit old bee, quest on bee hairs, and transfer to a new, young bee host; enter trachea to lay eggs.

Figure 3. Differences in appearance between Varroa, Euvarroa and Tropilaelaps. Bar is 1mm. (*Illustrations by G.W. Otis and J. Kralj.*)

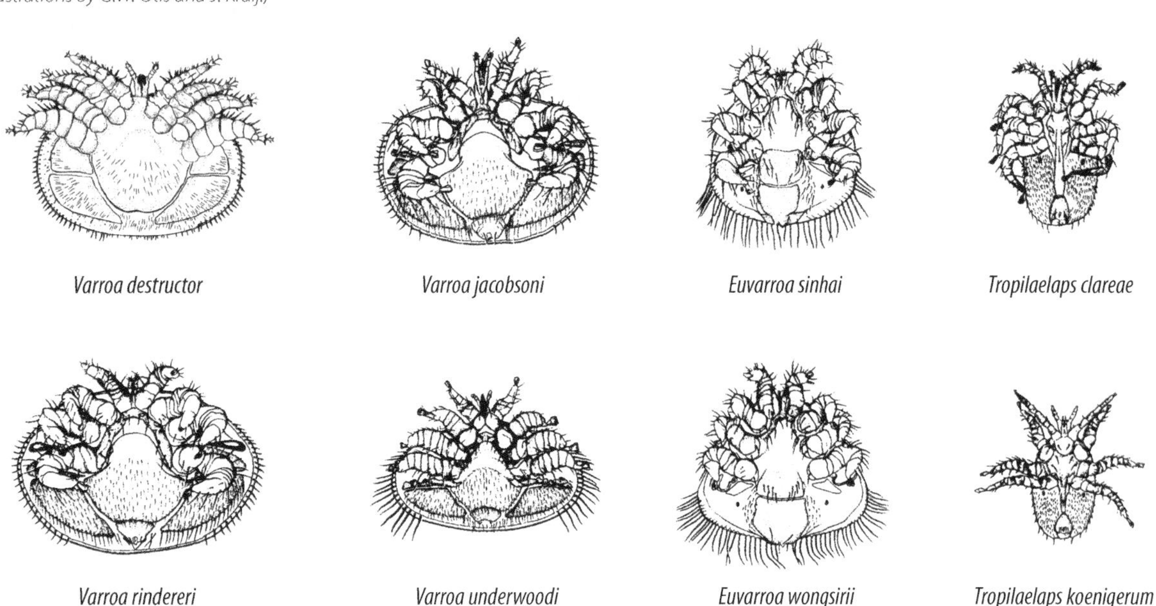

Euvarroa (Mesostigmata: Varroidae)

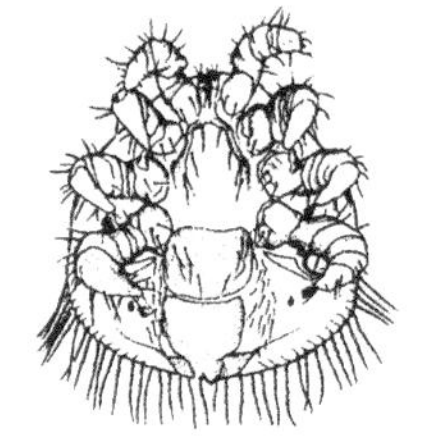

The ectoparasite *Euvarroa* was first identified in 1974 from *Apis florea*, the dwarf honey bee from India (Aggarwal and Kapil, 1988) and is reported to only parasitize drone brood. These mites are smaller than Varroa and currently there are two known species: *E. sinhai* on *A. florea* throughout its natural range, and *E. wongsirii* on *A. adreniformis* from Malaysia and Thailand (see Table 1 and Figure 3). *E. sinhai* has 34 to 46 long setae or hairs on the posterior edge of the pear-shaped body shield; *E. wongsirii* has more hairs and is more triangular in overall body shape (Otis and Kralj, 2001).

Their biology is not well known other than they seem to infest only drone larvae of *A. florea* and have little impact on *A. mellifera*. They are found on the comb and will feed on adult workers, though they seem to prefer adult drones. The seasonality of drone brood production is probably the significant limiting biological factor for *Euvarroa* reproduction. While seemingly an important pest, bees appear to abscond when mite populations become elevated. Grooming behavior to remove the mite has not been observed in *A. florea*. While *Euvarroa* has been collected from *A. mellifera* and *A. cerana* colonies, they appear unable to reproduce in them (D. Anderson, personal communication); for a complete life history, see Otis and Kralj (2001).

Varroa (Mesostigmata: Varroidae)

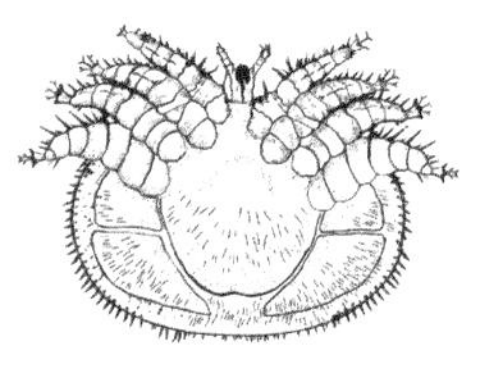

The Varroa mite was first identified in 1904 as *Varroa jacobsoni* (Oudemans) from specimens found on the Asian honey bee, *A. cerana* in Indonesia, where it reproduced only in drone brood and otherwise caused little damage to the bee

Figure 4. Chart of the life cycle of tracheal and Varroa mites. (*Illustration by Signe Nordin.*)

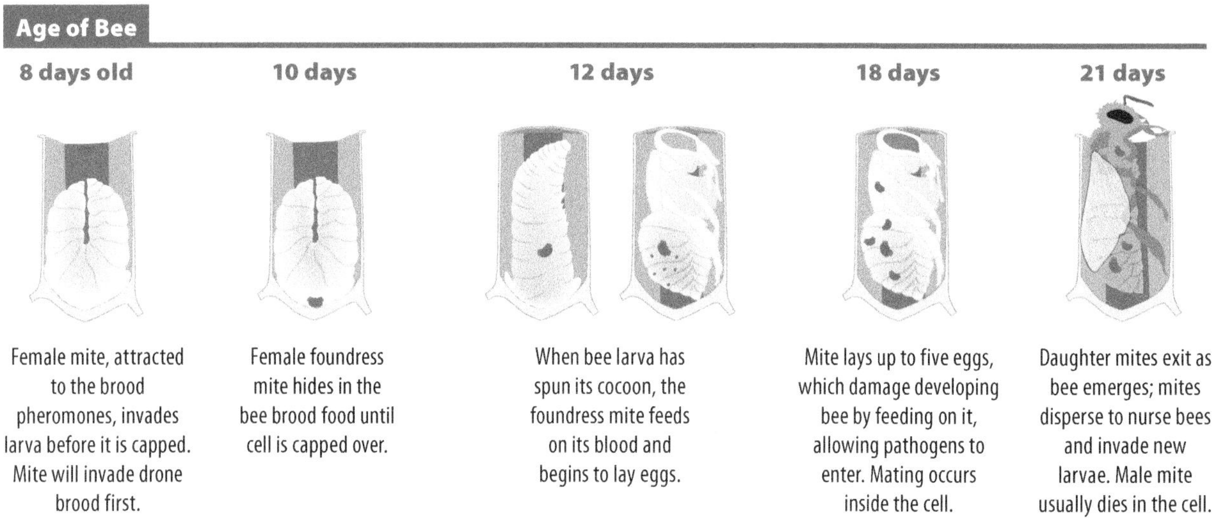

Age of Bee

8 days old	10 days	12 days	18 days	21 days
Female mite, attracted to the brood pheromones, invades larva before it is capped. Mite will invade drone brood first.	Female foundress mite hides in the bee brood food until cell is capped over.	When bee larva has spun its cocoon, the foundress mite feeds on its blood and begins to lay eggs.	Mite lays up to five eggs, which damage developing bee by feeding on it, allowing pathogens to enter. Mating occurs inside the cell.	Daughter mites exit as bee emerges; mites disperse to nurse bees and invade new larvae. Male mite usually dies in the cell.

colony. Since then, two other species have been described: *V. underwoodi* (Delfiando-Baker and Aggarwal, 1987) and *V. rindereri* (de Guzman and Delfinado-Baker, 1996; de Guzman et al., 1999; Warrit et al., 2006; Otis and Kralj, 2001); see Figure 3. Besides being morphologically distinct from *V. jacobsoni*, these two mite species are found on most Asian bee species (see Table 1). But once the European, or Western honey bee (*A. mellifera*) became established in Asia, *V. jacobsoni* was found to reproduce on both drone and worker brood, causing significant colony losses. In the 1980s, differences in *V. jacobsoni* were noted from diverse regions. Variations in the shape and size of adult females, in their reproductive biology, and in the parasite-host interface (Warrit et al., 2006; Rosenkranz et al., 2010) were recorded. Meanwhile, *A. mellifera* colonies in Europe, North America, and the Middle East were quickly succumbing to Varroa infestation, yet honey bees in the tropics of South America were not affected. Ultimately, molecular techniques revealed that Varroa mites from Asia were genetically distinct from those in the United States. In 2000, *V. jacobsoni* infesting the Western honey bee was found to be yet another species and was renamed *V. destructor* (Anderson and Trueman, 2000). After further study, 18 different haplotypes (mites with unique mitochondrial [mtDNA] sequences) have been recorded: nine in *V. jacobsoni*, six in *V. destructor*, and three that are still unresolved at this writing (Navajas et al., 2010).

The haplotypes are named after the country where the mite was first found. Two of the *V. destructor* haplotypes from the Asian honey bee, J1 and K1 (Japan 1 and Korea 1), have successfully adapted to *A. mellifera*. J1 shifted onto *A. mellifera* in Japan, Thailand, and Brazil, and finally into North America. K1 transferred to *A. mellifera* near Vladivostok, Russia, then to Europe and North America by 1987 (Navajas et al., 2010). More genetic variations of Varroa were also discovered off their native host *A. cerana*, but the K1 and J1 groups on the Western honey bee show little genetic diversity even though they are now widespread on *A. mellifera* in

Figure 5. Comparative sizes. **A.** *Acarapis* (arrow). **B.** Flea. **C.** A female *Varroa.*
(*Photos A–B by W. Styer; photo C by C. Pooley.*)

Asia. Currently there are at least two new haplotypes (J1-5 and K1-2) in Asia on *A. mellifera* (Navajas, et al., 2010).

The shift of these mites to the Western honey bee was facilitated by beekeepers introducing this susceptible species into Asia as far back as the 1950s. The accelerated dispersal of this relatively large, visible ectoparasite in North America was aided by colony movement for pollination or honey production (see also http://www.mylovedone.com/image/solstice/win10/SammataroandArlinghaus).

Physically, female Varroa are 1.1 mm long by 1.6 mm wide, weighing between 0.2 to 0.5 mg (live, fresh weight), depending on whether it is actively laying eggs or not (Sammataro and Finley, unpubl. data). Detailed information on the Varroa life cycle is found elsewhere (Sammataro et al., 2000; Martin, 2001; Rosenkranz et al., 2010); see Figure 3, 4, and 5. In general, female Varroa infest the 5th instar bee larvae just prior to capping (Day 7–8 of larval life). Drone larvae are preferred but workers are also parasitized. Once in the cell for about 70 hours, the female lays a male egg (Rosenkranz et al., 2010), followed by female eggs every 30 hours. Up to five eggs (average three) can be laid in worker larvae; but because of their longer development time, the mother mite can lay up to six eggs on drone larvae (which is why drone larvae are preferred).

Mites are able to find the correct-aged brood cells by means of chemical cues released by the bee larvae (Rickli et al., 1992; Rosenkranz et al., 2010). They will feed on larval hemolymph while going through their own developmental stages of proto- and deutonymphs, followed by the adult mites. This feeding weakens the bees, reducing hemolymph, transferring virus, and modifying their volatile profile (Salvy et al., 2001). Male mites are much smaller and once they mate with their sisters, die in their natal cell. The mature, mated females (daughters and the mother) will then emerge with the teneral bee and search for new larvae with which to begin the cycle anew. If no brood is present, as in winter months, mites can cling to adult bees, feeding intermittently on bee hemolymph until brood is once again available. Richards et al. (2011) found that the saliva from Varroa mites facilitates their ability to feed by suppressing the wound-healing capabilities of the bees. Further research on mite behavior and reproduction is sorely needed if new control tactics are to be developed and adaptive characteristics of Varroa on *A. mellifera* are to be understood.

Tropilaelaps (Mesostigmata: Laelapidae)

Tropilaelaps is the newest threat to global apiculture (Baker, 2010), but for now is confined to its Asian home range. Four species have now been described and, according to Anderson and Morgan (2007), the two species, *Tropilaelaps clareae* and *T. koenigerum*, are found on the giant honey bees of Asia, *Apis dorsata dorsata* and *A. laboriosa* (see Table 1 and Figure 3). *T. clareae*, an important pest of the introduced Western honey bee (*A. mellifera*) is found throughout Asia. However, *T. clareae* is now considered to be two species, consisting of haplotypes parasitizing the native *A. d. breviligula* and the introduced *A. mellifera* in the Philippines (except Palawan) and on Sulawesi Island in Indonesia, parasitizing the native *A. d. binghami* (Anderson and Morgan, 2007). The new *T. mercedesae* n. sp., (previously mistaken for *T. clareae*) and common in Sri Lanka, now includes haplotypes that parasitize the native *A. d. dorsata*, *A. laboriosa*, and the introduced *A. mellifera*. *T. koenigerum* is found on *A. d. dorsata* and *A. laboriosa*. Another new species, *T. thaii* n. sp., has been recently discovered on *A. laboriosa*; for complete details, see Anderson and Morgan (2007).

Tropilaelaps mites are around 1 mm in length, red-brown in color, and are elongated compared to the more oval Varroa (Figure 3). They hold their first pair of legs like antennae and quickly move on the comb surfaces. These mites are found on the combs or on adult bees and prefer drone brood (but also are found on worker brood) in which to lay eggs. T. *clareae* cannot feed on adult bees, nor can they survive more than a few days as phoretic passengers, which is why they thrive in tropical climates. In temperate climates there is a long broodless period when the bees are in their winter cluster. During this time, because Tropilaelaps cannot feed on adults, they may not survive the long winter. *T. koenigerum* is slightly smaller, and reported to be harmless to *A. mellifera*, as is *T. thaii* (Anderson, CSIRO personal communication).

The geographic range of this mite has expanded since it adapted to *A. mellifera* (Anderson and Morgan, 2007; see Table 1). Tropilaelaps are considered an emerging global threat to beekeeping, not only for their injury to bees but because they also carry the deadly Deformed Wing Virus (DWV) (Forsgren et al., 2009; Dainat et al., 2009). The United States Department of Agriculture is conducting a survey (2010–11) to search for this mite, as well as for *Apis cerana*, that may have come in with shipments of package bees from Australia. Currently the USDA has banned all imports of live bees from Australia into the United States. This threat makes it imperative to update the quarantine protocols for all countries that import and export live honey bees.

Mites and Virus

The biggest problem with bee mites is that they not only damage individual bees by the feeding activity and reduce their lifespan and learning capabilities (see Rosenkranz et al., 2010), but they also carry pathogenic RNA viruses. Virus particles have been found in Varroa's salivary glands (Cicero and Sammataro, 2010) as well as in other parts of their body (Zhang et al., 2007). To date, 18 different viruses have been identified from bees and many of them are vectored by mites, especially, but not limited to Varroa (Chen and Siede, 2007; Forsgren et al., 2009; Dainat et al., 2009; Boncristiani et al., 2009). These viruses include Kashmir (KBV),

Sacbrood (SBV), Israeli Acute Paralysis (IAPV), and DWV (Santillán-Galicia et al., 2010; Highfield et al., 2009; de Miranda et al., 2010; de Miranda and Genersch, 2010; de Miranda et al., this edition). Much of the blame for the incidence of Colony Collapse Disorder (CCD) in 2006 in which many bee colonies died in the United States was attributed to Varroa-vectored viruses (Berthoud et al., 2010; Ellis et al., 2010; Boecking and Genersch, 2008; vanEnglesdorp et al., 2009; Schäfer et al., 2010).

Monitoring Mites

When symptoms of mite infestation appear in bee colonies (mites seen on bees and larvae, bees with deformed wings crawling on the ground, bees pulling out diseased brood, colony collapse), there are several methods to determine what kind of ectoparasitic mites are present and to estimate their numbers. The optimal time to monitor mite levels is before they become a problem; twice per year (spring and fall) is often recommended. If mite populations are growing, monitoring colonies every two months may be necessary. Sampling 10% of colonies in an apiary is often mentioned, but more recent work by Lee et al. (2010a,b) showed that eight colonies per apiary gave enough information on Varroa mite densities.

Monitoring mites is useful because if mite populations are low, no treatments are needed and beekeepers can save money (on treatments) and time. Fewer treatments will also help diminish the buildup of resistance to acaricides, thus reducing the amount of product needed. Also, the reduced use of pesticides in the colony lessens contamination in the hive and its products. Collapsing colonies or those with high mite populations should be treated first and subsequently monitored, which will indicate whether control methods have been effective. The procedures listed below can help in estimating the number of mites in a particular colony and augment treatment times or options. They are broken into two broad categories, *active* (or invasive) and *passive* sampling.

Active Sampling for Mites

Ether Roll

Pick a colony and brush or shake 300 bees from one to three randomly selected frames containing brood, as mites usually are on the nurse bees that are on brood frames (Lee et al., 2010b). Bees should be brushed into a wide-mouth jar or shaken from frames into a larger container first, until 300 bees are collected. Queens should not be sampled. Knock the bees to the bottom of the jar with a sharp blow and measure a 2-inch (5.08 cm) line with a grease pencil or permanent marker to make this step more accurate. Note: A 1/3 cup (78.07 mL) of bees is between 238 to 300 bees; 1/4 cup (59.15 mL) is about 200 bees; a 100-mL beaker will hold 300 bees. Spray a 2-second burst of ether auto starter fluid and cover the jar with the lid. When the bees die, they will regurgitate the contents of the honey stomach onto the sides of the jar, making it sticky. Shake the bees vigorously for 1 to 2 minutes, then roll the jar horizontally to spread out the bees; the mites should stick to the sides of the jar. Count both mites and bees to determine the number of mites per 100 bees. CAUTION! Ether is HIGHLY flammable and should be used with vigilance near a lit smoker.

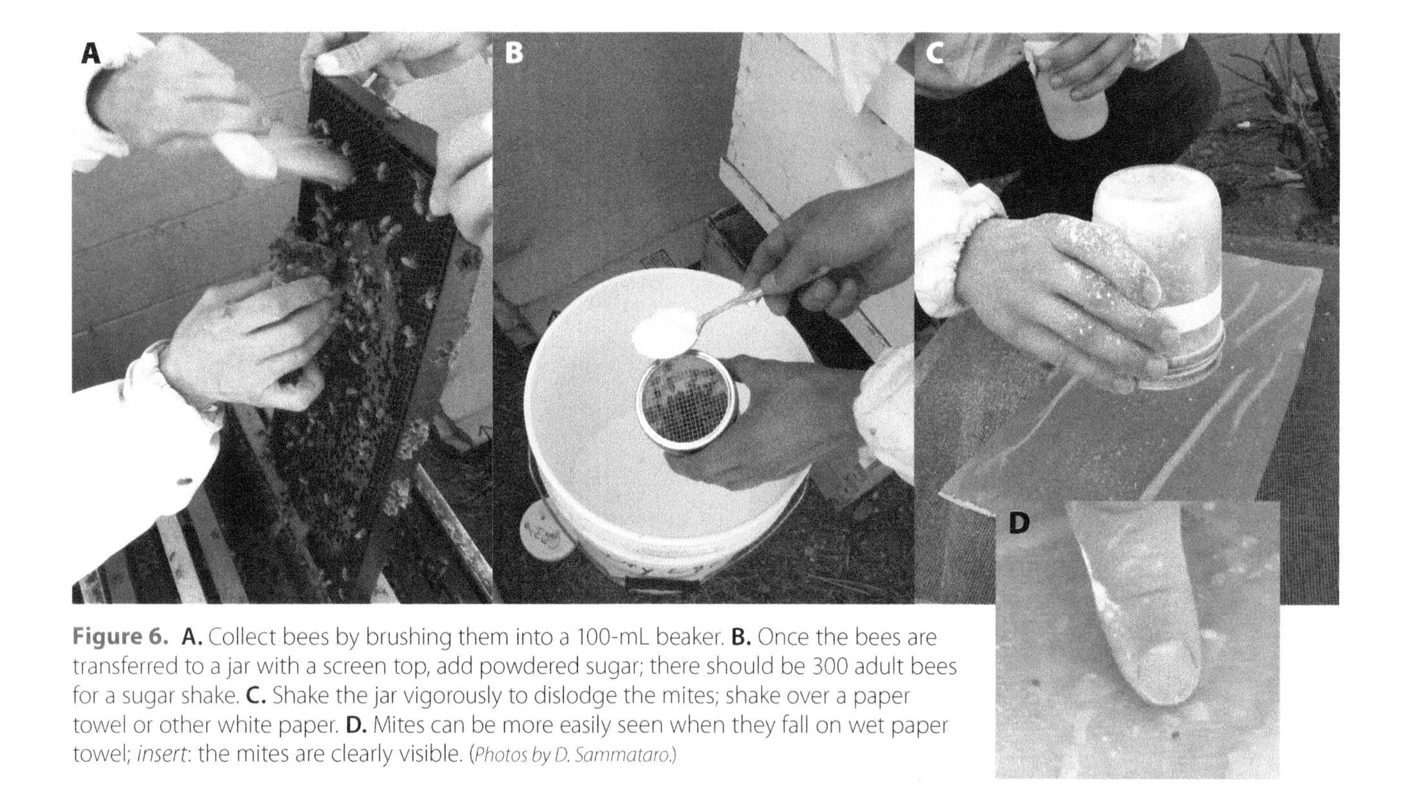

Figure 6. **A.** Collect bees by brushing them into a 100-mL beaker. **B.** Once the bees are transferred to a jar with a screen top, add powdered sugar; there should be 300 adult bees for a sugar shake. **C.** Shake the jar vigorously to dislodge the mites; shake over a paper towel or other white paper. **D.** Mites can be more easily seen when they fall on wet paper towel; *insert*: the mites are clearly visible. (*Photos by D. Sammataro.*)

The ether in this method can be replaced by 70% alcohol or soapy water (the latter is not a fire hazard); pour enough liquid into the jar to cover the bees (2 oz or 50 mL) and agitate the jar vigorously to dislodge mites. Strain out the liquid into another container, using a sieve or mesh to catch the bees; the mites will be in the liquid. Refill the jar with more liquid and shake again, repeating this at least 4 times to knock off all the mites. Filter the liquid and mites through a coffee or cloth filter to separate and count the mites and bees.

Sugar Shake

Instead of killing the bees, add 1 oz or 2 tablespoons (25 g) powdered or icing sugar (or flour) to the collecting jar with the 300 bees, and wait for about 2 minutes (Figure 6). Then put screening material (such as 8 × 8 mesh hardware cloth) on top of the jar and shake vigorously. The mites are best seen when shaken onto white paper to count; shake again and repeat at 2-minute intervals.

If an accurate mite per 100 bee count is desired, finish the sugar shake with an alcohol or soapy water wash to collect all the mites and count the bees in the samples (see Lee et al., 2010a,b). The number of mites in a colony can be estimated by doubling the number of mites/100 bees (Lee et al., 2010a,b; Fuchs, 1985); doubling the number of mites/100 bees accounts for the mites that are in the brood. Alternatively, after collecting and counting mites, use the chart (Table 2) to convert the number of counted mites to the percent of mites in a colony (colony infestation).

Colony and Brood Examination

Another method to monitor bee colonies for the presence of Varroa (or Tropilaelaps) is to examine both capped drone and worker brood. This can be done using a cappings scratcher (with fork-like tines) to pull up

# Mites per 300 Adult Bees	Colony Infestation	# Mites per 8 300-Adult-Bee Samples	Apiary Infestation
1	1%	8	1%
2	1%	16	1%
3	2%	24	2%
4	3%	32	3%
5	3%	40	3%
6	4%	48	4%
7	5%	56	5%
8	5%	64	5%
9	6%	72	6%
10	7%	80	7%
11	7%	88	7%
12	8%	96	8%
13	9%	104	9%
14	9%	112	9%
15	10%	120	10%
16	11%	128	11%
17	11%	136	11%
18	12%	144	12%

Table 2. Mite numbers per 300 adult-bee samples and the corresponding density in the colony; also the mites in eight 300-bee samples by apiary and the corresponding density in the apiary. After Lee et al. (2010b) and by permission of the author. This table is used instead of having to calculate mites per 100 bees.

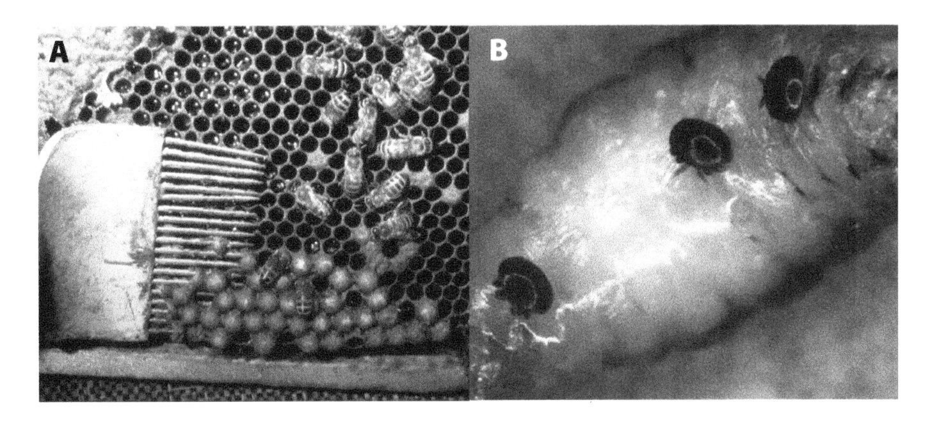

Figure 7. A. Pulling up drone brood (note the bullet-like capped cells). **B.** Varroa mites on a drone larva. (*Photos by D. Sammataro.*)

capped pupae. Examining drone brood is preferred because the mites favor drones (longer development time) and drone brood is larger and easier to pull up; the loss of some drones has little if any impact on hive health (Fuchs, 1990; Schulz, 1984). However, an equal number of drone and worker brood should be examined for a more realistic mite count and to determine if Tropilaelaps is present.

Sample between 100 and 200 (or as much as 2,000) capped purple-eyed pupae. Slide the fork horizontally along the comb surface to puncture as many capped cells as possible and then rotate it upward, along with the cappings, pulling up the brood from the cells (Figure 7). The mites will be clearly visible on the white pupae. Using a bright light source, examine the now-empty brood cells to determine if the white fecal material from Varroa is present on the cell walls; immature mites should be also visible inside the empty cells. Younger mite stages will be lighter in color; older mites are dark brown. Check several patches of brood within the colony to get a more complete picture of the mite distribution.

Another method to collect capped brood is to cut off the cappings (de-cap) with a serrated or heated knife (such as an uncapping knife used for honey) and wash out the brood with a stream of warm water. Brood can be rinsed in a coarse sieve (2 to 3 mm) and mites collected in a lower, fine sieve or screen (1 mm mesh) covered with filter paper to make them clearly visible. However, immature mites, using this method, may be overlooked or washed away.

Estimates of the infestation rate can be done by opening a predetermined number of brood cells to calculate a percentage of capped brood containing live mites. The accuracy of this method is, however, uncertain. Charriere et al. (2003) found that the mite numbers varied from one to six times even within a week of sampling, and Devlin (2001) reports that the method was best done during the fall season but not in spring and summer. Because of this, threshold levels using brood pulls have not been developed. The combination of the ether rolls and brood pulls may yield the most accurate estimate of mite density (Branco et al., 2006).

Treatment is usually called for if 10% to 12% of the colonies in an apiary (or a colony) are infested in the fall (see Table 2); but this can vary by region, bee population, and time of year. Keeping records of mite levels in colonies is a good way to determine if chemical treatments will be beneficial. Beekeepers in adjacent apiaries could cooperate in treatment schemes to keep mite levels low in a particular area.

Passive Mite Sampling

Figure 8. GL-IPM Varroa board used to count natural mite drop; see website references. (*Photo by D. Sammataro.*)

Sticky Board or Hive Debris

A precise diagnosis of mite infestation can be made using a sticky board that is made from a sheet of heavy cardboard, poster board, or other white stiff paper or plastic or laminated paper smeared with a sticky coating. The board should be covered with a wire mesh (Figure 8), such as screening material. The outside edges of the mesh can be folded under to raise it off the board and stapled or taped onto the board. This prevents the bees from removing the dislodged mites and keeps bees off the sticky material. The sticky board can be coated with Vaseline or another substance (or use a sheet of sticky shelf paper).

Boards can also be purchased that are already coated with a sticky material (good for hot climates) and that are visually easier to use when counting mites (Great Lakes IPM; Ostiguy and Sammataro, 2000; see Figure 8). The entire board needs to fit the full length of the bottom board of a hive. The board is left in the colony for up to 3 days, after which the debris can be examined for mites. It is important to use the full-size board to estimate mite populations; anything less will not give an accurate count. To determine that only mites are counted, a magnifier can be used to distinguish mites from other debris.

Boards are perfect as a noninvasive method to determine mite loads in a colony and, more importantly, the change of mite density over time (Sammataro et. al., 2002). However, it is difficult to predict the number of mites in a colony with this technique and the significance of the number of mites counted needs to be correlated with treatment requirements; how many mites constitutes a treatable number varies.

Table 3. Chemical controls for Varroa mites; compiled from Rosenkranz et al., 2010 and http://www.maf.govt.nz/biosecurity/pests-diseases/ animals/varroa/guidelines/control-of-varroa-guide.pdf.

Product Trade Name ®	Active Ingredient	Chemical Class
Apiguard	Thymol	Essential oil
Apilife VAR	Thymol, eucalyptol, menthol, camphor	Essential oils
Apistan **	Fluvalinate	Synthetic pyrethroid
Amitraz, Miticur, Api-warol (tablets)	Formamidine	Formetanate, methanimidamide
Apitol	Cymiazole	Iminophenyl thiazolidine derivative
Apivar **	Amitraz	Amadine
Bayvarol **	Flumethrin	Synthetic pyrethroid
Check-Mite+ **	Perizin, coumaphos	Organophosphate
Folbex	Bromopropylate	Chlorinated hydrocarbon
Sucrocide	Sucrose octanoate	Sugar esters
Hivestan	Fenpyroximate	Pyrazole (alkaloid)
Generic (e.g., Mite Away™)	Formic acid	Organic acid
Generic	Lactic acid	Organic acid
Generic	Oxalic acid	Organic acid

*** No longer effective in some areas.*

Delaplane et al. (2005) estimated a treatment threshold (on a 24-hour sticky board) of 71 to 224 mites in August in the southeastern states and 1 to 12 mites in the spring. In the northwestern states, these numbers were calculated as 12 mites in the spring and 23 in the fall; for ether rolls, the numbers were 3 mites in April, 14 in August, and 4 in October (Strange and Sheppard, 2001). Treatment thresholds in other regions need to be carefully researched.

Use of Chemotherapy to Control Mites

A list of some of the current chemical miticides (acaricides) used to control Varroa is outlined in Table 3; many of these may work for Tropilaelaps. Optional treatments to reduce Varroa levels are always being tested by beekeepers and researchers. Below are some methods that have been tried or are being developed; comments by the authors:

- Chlorfenvinphos (organophosphate), effective but residues may cause problems (Milani et al., 2009); Apiwarol tablets are burned, but are toxic to bee brood (Dzierzawski and Cybulski, 2010).
- Azadirachtin (Neem), needs more work (González-Gómez et al., 2006; Peng et al., 2000; Melathopoulos et al., 2000; Fassbinder et al., 2002).
- Plant-derived monoterpenoids, some report bee toxicity (Fassbinder et al., 2002; Ali et al., 2002); other essential oils are being tried (Gashout and Guzman-Novoa, 2009; Ruffinengo et al., 2006; Shaddel-Telli et al., 2008; Garcia-Fernandez et al., 2008; Sammataro et al., 2009) and other compounds (Emsen and Dodologlu, 2009).

- Food grade mineral oil, not effective (Elzen et al., 2004).
- Powdered sugar dust (not effective for control; Ellis et al., 2009a) but has potential (Fakhimzadeh, 2000) and is good for sampling, (Aliano and Ellis, 2005; Macedo et al., 2002).
- Screen wire bottom boards; not effective in keeping mite populations low (Ellis et al., 2001; Harbo and Harris, 2004) but does help some, especially when used with other methods (Delaplane et al., 2005).
- Smoke, burning different plant materials; reports of some success but may be harmful to bees (Çakmak et al., 2002; Çakmak et al., 2006; Eischen and Wilson, 1998; Elzen et al., 2001a,b; O'Meara, 2005; Romeh, 2009).
- Thermal treatments, such as heating frames in an "oven" (Rosenkranz et al., 2010; Tabor and Ambrose, 2001).
- Cell size modifications (Maggi et al., 2010a); small cell size does not work (Berry et al., 2010; Ellis et al., 2009b).
- Mite traps or killing with specialized frame, for example, Mite Zapper (Huang, 2001).
- Propolis (Daminai et al. 2010; Garedew et al., 2002).
- Antioxidants (Sammataro et al. 2010).

Problems with Chemotherapy

Resistance and Residue

Varroa mites developed resistance to the chemical acaricides within 10 to 15 years after they were first used. Since Varroa's introduction into the United States in 1987, the first report of resistance to fluvalinate surfaced by late 1994 (Milani 1994, 1995, 1999) followed swiftly in other areas (Elzen et al., 1998 and 1999a,b). By 1998, coumaphos (Elzen and Westervelt, 2002; Lodesani et al., 1995) and later amitraz (Ezlen et al., 1999c) were added to the list of resistant chemicals. By 2000, it was suggested to rotate different chemicals to keep ahead of the resistant mites (Elzen et al., 2001c). Reports of resistance vary by country but it appears to be a major problem, particularly in some newly infested areas. Increased resistance is driving commercial beekeepers to experiment with chemicals and other compounds in an effort to find a treatment that is economical, effective, and easy to use. Development of the pesticide "strip" has been described by some as a great disservice to the bee industry. The rationale is that "ease of use" of the early strips raised the expectations for an "instant solution" for control of the resistant mites. Preparation of untested acaricide cocktails by beekeepers to combat growing Varroa resistance has been an ongoing problem for years. Many of these recipes are harmful if ineffective and this kind of "experimentation" is discouraged by scientists.

Another common misconception is in the area of overdosing, such as doubling or tripling the amounts recommended on the label, which is both illegal and dangerous. Some products may kill the bees outright; the LD_{50} of the compounds should give a clue as to how toxic they are (LD_{50} indicates the individual dose required to kill 50% of a known population of honey bees). It does not indicate long-term effects, only short-term toxicity; the smaller the number, the more toxic the substance.

The LD_{50} (in μg/bee) of many pesticide compounds are now online at: http://www.ipmcenters.org/Ecotox/DataAccess.cfm.

Another area of concern is that some acaricides contaminate wax and other hive products. The first paper to discover that agrochemicals were found in honey was Ogata and Bevenue (1973). Now there are hundreds of journal articles on residues in bee colonies. The greatest residue risks are with acaricides, but other pesticides are also identified. Mullin et al. (2010) reported that 121 different pesticides and metabolites were found in hive products (see Chapter 14, this edition). Documenting what compounds and their breakdown products do to all bee life stages as well as the beneficial microbes living in bees and bee colonies is the subject of intense research (Johnson et al., 2010a,b) and how they may affect the overall health of bees (Le Conte et al., 2010).

Effect on Drones

Varroa has had a great impact on feral bees to the point that many "wild" honey bee colonies have now disappeared or are in low numbers. Varroosis has an additional negative effect on male fitness, decreasing flight duration and reducing the number of spermatozoa produced. Drone survival is also significantly reduced in Varroa-infested colonies (Rinderer et al., 1999) and the identification of Deformed Wing Virus (DWV) in the epithelial cells of the seminal vesicle of drones (could also be present in the semen) may have a negative effect on drone fertility (Fievet et al., 2006; Chapter 8, this edition) and could be passed on in the eggs the queens are laying. Reduced sperm counts and drone flight translate into less sperm available to queens, either because the drones are not able to take long mating flights or the spermatozoa are not as viable. Queens are therefore not well-mated and usually quickly superseded.

Other Control Measures: Biotechnical Controls and IPM

One result of Varroa's resistance to some chemicals is the search for different control tactics. Such measures can be used in an Integrated Pest Management (IPM) strategy that utilizes multiple tactics, as represented in Figure 9 (and see Chapter 6, this edition). One successful IPM strategy for Tropilaelaps is treating the mites during the phoretic, broodless periods because they cannot feed on adult bees and, more importantly, they cannot survive without access to brood (Calis et al., 1999; Wilkinson and Smith, 2002). Caging the queen while removing drone brood will also help control both Tropilaelaps and Varroa populations. Dusting with powders (such as icing or powdered sugar) can also suppress mite levels (Aliano and Ellis, 2005; Ellis et al., 2009b). Three other IPM techniques are outlined below. IPM strategies for Varroa were studied by Delaplane et al. (2005) and they found that screened bottom boards and mite-resistant queens show significant reduction, especially when used together. However, they also reported that these methods had a negative effect, such as reduced honey and pollen storage. In some instances, the more mite-resistant queens produced less brood. IPM methods are best for reducing chemical treatments and for keeping track of mite populations; new strategies need to be developed.

Figure 9. IPM tactics for mite control. (*Adapted by D. Sammataro with a template for IPM controls from Penn State University. Used with permission.*)

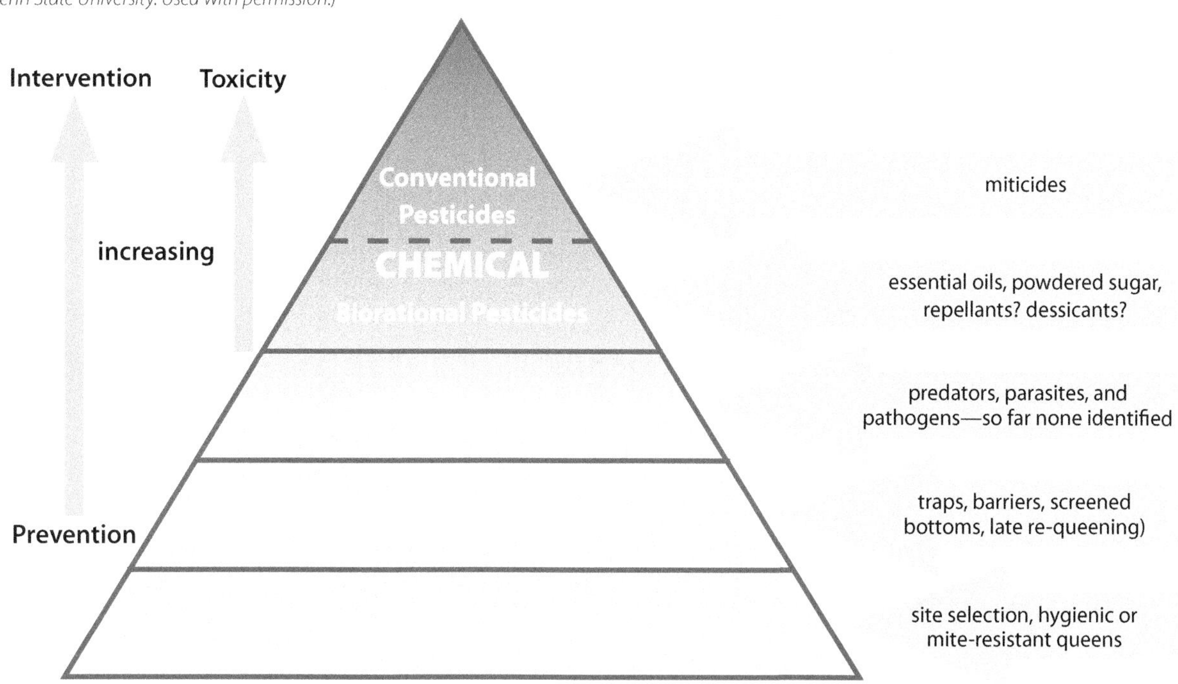

Open Bottom Boards or Open Mesh Floor

Mites cannot survive for very long without host contact. For this reason, mites that fall off bees or comb through wire mesh floors will generally perish. Overall, mite populations are reduced using open bottom boards and by limiting the buildup of debris on the hive floor, benefits can be realized, such as improving hive ventilation and discouraging other pests such as the wax moth from nesting in hive waste on bottom boards; wax moths successfully exploit bottom debris and can kill weakened colonies. Cold winters may preclude the use of open bottom boards, so another kind of bottom board with a cover for the screen should be used. In warmer climates, however, the open floor is an invitation to ant invasions unless other precautions are taken. Open floor bottoms can be an integral part of Varroa and Tropilaelaps monitoring and control. However, open bottom boards, with or without the sticky board, are not sufficient on their own to significantly reduce mite levels and work best when used in concert with other control methods (Ellis et al., 2001; Delaplane et al., 2005; Harbo and Harris, 2001, 2004).

Comb Trapping and Drone Brood Removal

Using sheets of drone foundation in the brood chamber or in a super frame encourages bees to draw out drone comb (Charriere et al., 2003; Rosenkranz et al., 2010). When the queen lays eggs in drone cells, the drone population can increase. Because drones are more attractive to the Varroa mite, once the brood is capped over, it can be removed and the contents disposed of (Wantuch and Tarpy, 2009), for instance by freezing the comb. However, there is a risk of increasing mite populations if the beekeeper fails to remove the sealed comb.

Breeding for Varroa Tolerance

Breeding bees to be resistant or tolerant to mites has been ongoing since mites were first discovered infesting the Western bees. The natural hosts of Varroa (and Tropilaelaps) are able to keep the mite–host relationship in balance by several mechanisms; for Varroa these are:

1. "Entombing" drone brood that is multiply infested, by not helping weakened drones emerge, thus containing the mites (Rath, 1992); this activity is found only in the *Apis cerana* bees.
2. Grooming behavior, where mites are removed by other bees (Peng et al., 1987); but this is difficult to establish in European colonies (Aumeier, 2001; Büchler et al., 2010; Stanimirovic et al., 2010).
3. Hygienic behavior, where the workers detect and remove infested brood (Spivak, 1996; Spivak and Reuter, 1998); interrupts the reproductive cycle of the mite. This trait has been refined into a line of bees called Varroa Sensitive Hygiene (Harbo and Harris, 2001; Harris et al., 2010). In addition, bees from eastern Russia that were imported into the United States. (Rinderer et al., 2001, 2010; Çakmak, 2010) show some tolerance to Varroa.
4. Nonreproductive or reduced fecundity of mites, which appears mostly in African or Africanized bees (Rosenkranz, 1999), is apparently too variable to be a factor in reducing mite levels in Western honey bees (Rosenkranz et al., 2010).
5. Mites damaged by bees is currently being re-evaluated as a heritable trait in bees and could be used in a selective breeding program (T. Webster, personal communication); for more information, see Corrêa-Marques et al. (2000).

Some reports on the downside of breeding mite resistance in honey bees document reduced honey production, elevated defensive behavior, increased brood disease incidence, small bee populations over winter, and increased swarming tendencies. In general, beekeepers want thriving bee populations with gentle behavior and high honey yields, so natural selection has a long, slow road ahead. For more information see Fries (chapter 3, this edition) and Rosenkranz et al. (2010).

New Treatments and Future of Varroa Research

Currently, acaripathogenic fungi are being evaluated for Varroa control, including the testing of *Beauveria bassiana*, *Metarhizium anisopliae*, *Verticillium lecanii*, and *Hirsutella thompsonii*. Two serious drawbacks to using fungi are that they take several days to infect and kill their target, and they may not be optimally adapted to the brood nest environment (Chandler et al., 2001; James et al., 2006; Kanga et al., 2002, 2003, 2006, 2010; Meikle et al., 2006, 2007, 2008a,b; Peng et al., 2002; Rodríguez et al., 2009a,b; Shaw et al., 2002; Steenberg et al., 2010).

Another new biological agent being tested is the bacterium *Serratia marcescens*, which was isolated from *A. cerana* bees; it secretes proteins that act as a biological degrader of chitin (Tu, 2010). While essential oils are efficacious, volatility issues make it problematic for primary mite control. Consequently, new delivery systems are being tested, including porous ceramics (Booppha et al., 2010) and encapsulating oils in starch (Glenn et al., 2010). RNA interference (or RNAi) and other molecular techniques may hold promise in controlling Varroa and even preventing bee diseases (Campbell et al., 2010; Liu et al., 2010; Maori et al., 2009; Paldi et al., 2010; Schlüns and Crozier, 2007). The idea is to silence particular genes or regulate gene expression in microbial diseases and mites that would be fatal to those organisms.

The attractiveness of brood and the release of brood volatiles are still under investigation (Piccolo et al., 2010). Varroa attractant and arrestment responses to brood volatiles are currently being investigated for use in bait traps (Carroll, M. J., USDA, personal communication). New controls are always a top priority for Varroa, but more research is needed to investigate the effects pesticides have on the Varroa life cycle, and the interactions of contaminants now found within the hive environment. Other areas of investigation to combat these mites, especially Varroa, and possibly Tropilaelaps, will be in overcoming resistant mites. Acaracide resistance is now a reality in the United States and most likely a global issue. We must understand the changes in this mite from the Varroa that was present 20 years ago, and to study its life cycle more thoroughly. New controls need to be developed that will not contaminate the colony or pose resistance problems. One important area is rearing Varroa off-host, on an artificial diet. In this way we will be able to understand the complete life cycle of the mite, their diet requirements, reproductive strategies, and weaknesses. At a recent COLOSS workshop (http://www.coloss.org/) on "Varroa, Viruses and Standardization of Methods" held in Magglingen, Switzerland (November 2010), the following list of Varroa research priorities was recorded:

- Biological control (pheromones, entomopathogens, endosymbionts) and IPM
- Trigger of Varroa reproduction (in original and new host, including the geographic and genetic variation)
- Development of Varroa *in vitro* rearing and reproduction
- Search for Varroa-tolerant bees and identify the tolerance mechanisms for breeding programs and the problem of narrowing genetic diversity
- Host parasite co-evolution; local adaptations, *V. jacobsoni* and *V. destructor* on *Apis cerana*, role and maintenance of genetic diversity
- New invasive mites (*Varroa* spp. and *Tropilaelaps* spp.)
- *V. destructor* genome
- Modeling of population dynamics
- Mite virulence thresholds (including virus transmission and replication)
- Investigation of Varroa invasion on virus presence in populations
- Virus transmission and virulence

There is still a lot to learn and research areas are changing fast. New molecular tools may help in our understanding of these mites and eventually may assist us in learning how to mitigate the bad things these mites do to bees. Even at this writing, new breakthrough research may have already solved the bee–mite problem.

Biological Control of Honey Bee Pests

CHAPTER 6

W. G. Meikle, D. Sammataro, and G. Mercadier

Abstract Biological control of bee pests is a small but growing field as beekeepers and bee researchers seek ways to reduce pesticide use. Of the arthropod pests of honey bees, those that have been targets of biological control on at least the laboratory level are the Wax Moths *Galleria mellonella* and *Achroia grisella*, the Varroa mite *Varroa destructor*, and the Small Hive Beetle *Aethina tumida*. Several organisms have been proposed as biological control agents against wax moth, including naturally-occurring parasitoids, and one, *Bacillus thuringiensis*, has been commercialized. Biological control of *V. destructor* has involved application of entomopathogenic fungi, and while some results have been encouraging, more work is clearly needed with respect to isolate selection, formulation, and application method. Fungal agents have likewise been used against *A. tumida* and elevated mortality has been observed, but no field tests have been reported thus far. The interaction of biological control agents, bees, and target pests needs further research.

Introduction

Biological control is considered one of the pillars of Integrated Pest Management (IPM), yet it is seldom considered with respect to the management of bee pests. Part of the reason for that likely stems from biological control lying outside "traditional" bee pest management strategies, and partly from feeling that the logic of controlling an undesirable arthropod (mite or insect) within the environment of a highly desirable arthropod (the honey bee) is suspect. In addition, honey bees are notoriously fastidious and live in a highly sophisticated environment (the hive). Many researchers have no doubt felt that natural enemies, be they parasitoids, predators, or pathogens, would have a very difficult time establishing themselves.

Biological control is defined as *pest control using natural enemies*, whether those natural enemies are arthropods or microbes, and includes predators, parasites and parasitoids, and pathogens (Perkins and Garcia, 1999). Biological control has been used in many agricultural systems with great success (Gutierrez et al., 1999). There are two major kinds of biological control:

1. Classical biological control, in which a new organism such as a parasite or a pathogen is released into an ecosystem where it did not previously occur in order to control a particular pest.
2. Augmentative biological control, in which the natural enemy to be used already exists in the ecosystem but at a density too low to sufficiently control the target pest (Perkins and Garcia, 1999).

Augmentative biological control includes *inoculative* control, in which small numbers or quantities of natural enemies are released that subsequently increase their density naturally, and *inundative* control, in which a very high density of natural enemies is applied to ensure a high kill rate.

The most likely targets for biological control are other arthropods, particularly the Greater Wax Moth (*Galleria mellonella* L.) and Lesser Wax Moth (*Achroia grisella* F.) (both Lepidoptera: Pyralidae); the Varroa mite (*Varroa destructor* Anderson and Trueman) (Mesostigmata: Varroidae); and the Small Hive Beetle (SHB) (*Aethina tumida* Murray) (Coleoptera: Nitidulidae). While the bee pests listed above are among the most important worldwide, they are not the only ones. Pests such as predatory wasps can cause considerable damage to bee hives; *Vespa velutina* (Hymenoptera: Vespidae) is an invasive species in France (Perrard et al., 2009). However, because these wasps do not live inside the hive environment, their control is implemented differently and has different constraints, and they will not be considered here. Mites such as *Tropilaeleps* spp. (Mesostigmata: Laelapidae) are not (yet) widespread enough in areas where the Western honey bee, *Apis mellifera*, occurs to have been the subject of many alternative control strategies. Similarly, biological control of Tracheal mites (*Acarapis woodii* [Rennie]) (Prostigmata: Tarsonemidae) is largely unexplored.

Wax Moth

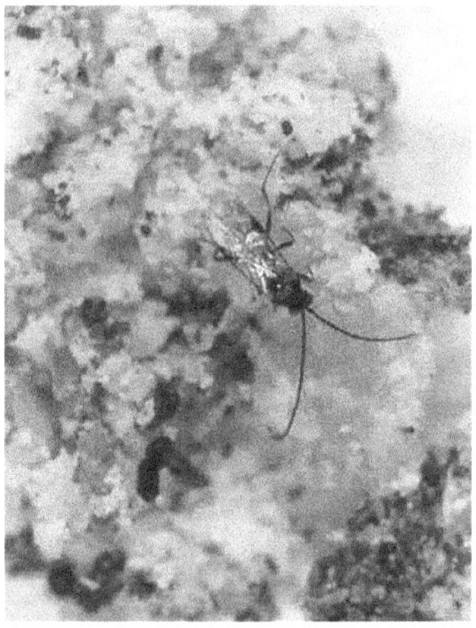

Figure 1. *Apanteles galleriae* Wilkinson, a wax moth parasitoid, lays its eggs in the larvae of wax moth. (*Photo by D. Sammataro.*)

Both *G. mellonella* and *A. grisella* have a number of natural enemies, including parasitoids and pathogens. Bollhalder (1999) proposed the use of an egg parasitoid, *Trichogramma* sp. (Hymenoptera: Trichogrammatidae). Larval parasitoids, however, have been examined more extensively. For example, *Apanteles galleriae* Wilkinson (Hymenoptera: Braconidae) is known to attack *G. mellonella* and several species of *Achroia* (Uçkan et al., 2004; Whitfield et al., 2001) (see Figure 1). Schöller and Prozell (2001) tested another braconid parasitoid, *Habrobracon hebitor*, against both *G. mellonella* and *A. grisella* as a potential commercial biocontrol agent to protect stored wax comb. Like *A. galleriae*, *H. hebitor* is a generalist parasitoid known to attack many species of moths. Blumberg and Ferkovich (1994) reported that *Microplitis croceipes* (Cresson), a braconid parasitoid whose usual host is the corn earworm *Helicoverpa zea* (Boddie) (Lepidoptera: Noctuidae), could reproduce successfully on *G. mellonella* as well. Harvey and Vet (1997) used *G. mellonella* as a host for *Venturia canescens* (Grav.) (Hymenoptera: Ichneumonidae).

If wax moths are suitable hosts to a number of parasitoid natural enemies in most if not all regions where they occur, then the question presents itself why the wax moth is not a problem. The answer may have to do with where the wax moth is found: inside silk and frass tunnels in honey bee comb (Figure 2). Honey bee worker activity likely keeps many parasitoids at bay and even those that get through must find and attack the larvae, which in healthy hives are at a low density. Uçkan et al. (2004) reported that *A. galleriae* was sensitive to high parasitoid densities and, in the absence of unattacked wax moth larvae, would hyperparasitize (i.e., attack parasitized) larvae, thus reducing the reproductive success of the parasitoids.

Bacterial products involving *Bacillus thuringiensis* (Bt) have been developed and tested against *G. mellonella* (Burges and Bailey, 1968), and come in two types: those involving the living bacteria (e.g., Burges and Bailey, 1968; Vandenberg and Shimanuki, 1990; Ellis and Hayes, 2009) and those involving just the endotoxin component of the bacteria (Burges and Bailey, 1968). Endotoxin products are not biological control in the usual sense because the active ingredient is not a living organism, but they are often described as "biopesticides." Tests using the live bacteria showed good protection of comb and low risk for bees (Burges and Bailey, 1968; Vandenberg and Shimanuki, 1990). Ellis and Hayes (2009) reported that comb made from foundation treated with Bt had 30% higher mortality of wax moth larvae than untreated comb a week after treatment. They qualified this by pointing out that the cost and labor associated with applying Bt were likely not offset by the benefits. Commercial products made with the bacillus included Certan®, which is available once more from Internet sites under the B401 Certan® label. This product was sprayed onto stored combs to protect them from wax moth. In commercial beekeeping operations, where thousands of honey supers were stored, it can be an important control measure.

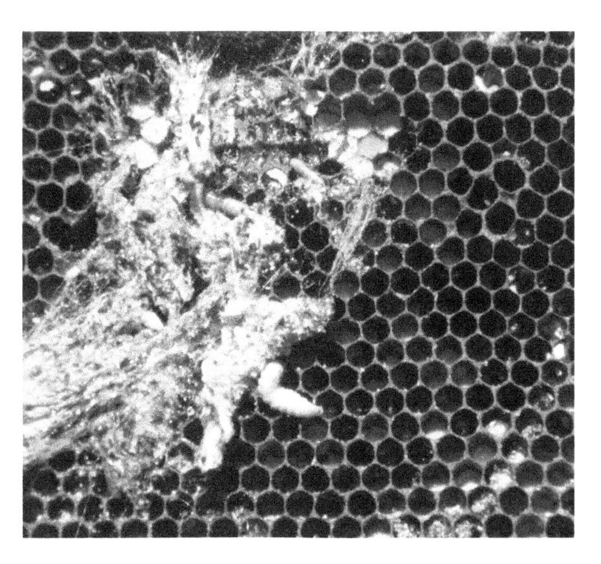

Figure 2. Larvae of the Greater wax moth tunnel through the wax comb consuming pollen and beeswax, leaving behind silken tunnels and fecal material. They can destroy a weakened colony in a short time. (*Photo by D. Sammataro.*)

Wax moths, particularly *G. mellonella*, are susceptible to many pathogenic fungi and nematodes (Chandler et al., 1997), probably because many of these pathogens are soil-dwelling and wax moths are seldom exposed to the soil environment. Indeed, their susceptibility is such that Zimmerman (1986) developed the "Galleria bait method," which involves simply exposing the larvae to soil to search for entomopathogenic fungi (EPF). EPF are recovered then from the infected larvae. This technique has been used by others for both fungi and nematodes (e.g., Chandler et al., 1997). Because *G. mellonella* is acceptable as a host to such a wide range of pathogens, it is often used as a factitious host in studies on pathogen ecology. Spence et al. (2011) used *G. mellonella* as a host for three species of entomopathogenic nematodes, *Heterorhabditis bacteriophora*, *Steinernema carpocapsae*, and *S. riobrave*; and Koppenhöffer and Fuzy (2008) used it as a host for four nematode species: *S. scarabaei*, *S. glaseri*, *H. zealandica*, and *H. bacteriophora*. To date, nematodes have not been reported as biocontrol agents against wax moths.

Galleria mellonella has been used extensively in studies on EPF. Shapiro-Ilan et al. (2003) used *G. mellonella* larvae to explore for naturally occurring insect pathogens in pecan orchards and, in addition to several nematode species, they recovered *Metarhizium anisopliae* (Metschnikoff) Sorokin (Hypocreales: Clavicipitaceae) and *Beauveria bassiana* (Balsamo) Viullemin (Hypocreales: Cordycipitaceae) from larval cadavers. Tseng et al. (2008) used *G. mellonella* to examine infection by the fungus *Nomuraea rileyi* (Hypocreales: Clavicipitaceae). However, little work has been done to date on developing a biological control program against wax moths using either nematodes or fungi. A nuclear polyhedrosis virus, GmMNPV, was obtained from *G. mellonella* (Shapiro, 2000), and a number of alphanodaviruses replicate well in *G. mellonella* larvae (Johnson et al., 2000), but, as with nematodes and fungi, these viruses have not been explored to date as biological control agents.

Wax moths have many natural arthropod and microbial enemies but few biological control options are currently available. This is probably due largely to the expense of developing and producing the products, and to the availability of low-cost approaches for most situations, such as removing infested comb from hives and freezing stored comb (Charrière and Imdorf, 1999). Wax moths are problematic for weaker hives (Nielsen and Bister, 1979) but are generally not counted among the worst problems in apiculture except for commercial beekeepers who typically have to store thousands of empty hive bodies. Annual losses can be upwards of $5 million (MAAREC, 2000). In most cases, chemical or thermal controls are used in their storage facilities.

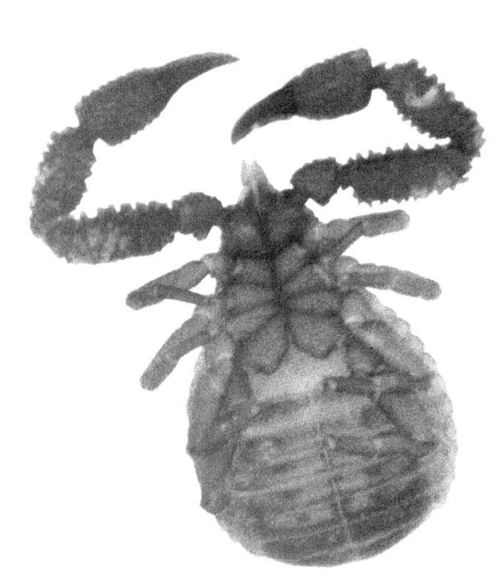

Figure 3. An unidentified pseudoscorpion, collected from an African bee colony in Kenya. Length of body is 6.35 mm. (*Photo by D. Sammataro.*)

Varroa Mite

Chandler et al. (2001) reviewed a wide range of taxa as candidate biological control agents against *V. destructor*. While parasitoids of *V. destructor* are unknown, the candidates did include a predator, pseudoscorpions (Pseudoscorpionida) (see Figure 3). Donovan and Paul (2005) suggested that the use of pseudoscorpions should be explored further because they are found naturally associated with feral honey bee colonies. Given that pseudoscorpions are active predators that would seek out mites, they probably are most affective against mites crawling on the comb. Pseudoscorpions would likely have little effect on phoretic mites or, owing to bee hygienic behavior, mites in brood cells, but they might impact mites on the comb. However, field evaluations have not been conducted as of this writing and the specificity is unknown.

Pathogens, especially bacteria and fungi, have been considered as potential biological control agents of *V. destructor*. Kleespies et al. (2000) reported a virus from a Varroa mite but no further work has been reported. Bacteria or bacterial products have also been evaluated and although the work did not always involve biological control in a strict sense, we include it here. Tu et al. (2010) isolated protein chitinases from a bacterium, *Serratia marcescens*, itself collected from the guts of the Asian honey bee workers (*Apis cerana*). They found that bees were not affected by the application of chitinases or the presence of chitinases in their food, and proposed the chitinases as a basis for genetically modified bee strains. Tsagou et al. (2004) isolated bacterial strains from *V. destructor* belonging to Bacillaceae (*Bacillus* sp.) and Micrococcaceae, in addition to three unidentified strains. The effect of these bacteria as whole cells, extracellular broth, and cellular extract, was tested on Varroa in laboratory bioassays. One *Bacillus* isolate in particular was found to decrease mite survivorship markedly. The bacteria have neither been tested for their effects on honey bees nor used in field trials.

Fungi

Most research on biological control of *V. destructor* has focused on the use of EPF, specifically facultative pathogens of the order Hypocreales. These fungi are preferred because they can be highly virulent against target pests, are easy to grow *in vitro*, and produce comparatively resistant spores called "conidia" that are easy to incorporate into a biopesticide (Goettel and Inglis, 1997). Davidson et al. (2003) examined growth rates of EPF in the genera *Beauveria* (see Figure 4A), *Hirsutella*, *Metarhizium* (see Figure 4B), *Paecilomyces* (now *Isaria*), *Tolypocladium*, and *Verticillium* in order to identify isolates which grew well at the higher temperatures encountered in bee hives. Shaw et al. (2002) reported the results of bioassays of isolates from these same genera—and many of the same isolates—against *V. destructor* and found several candidate isolates; no field tests were conducted. Published field trials have been largely confined to two species: *M. anisopliae* and *B. bassiana*. Both species have been used in other crop systems (Jaronski, 2010); neither are specific for *V. destructor* or even mites in general.

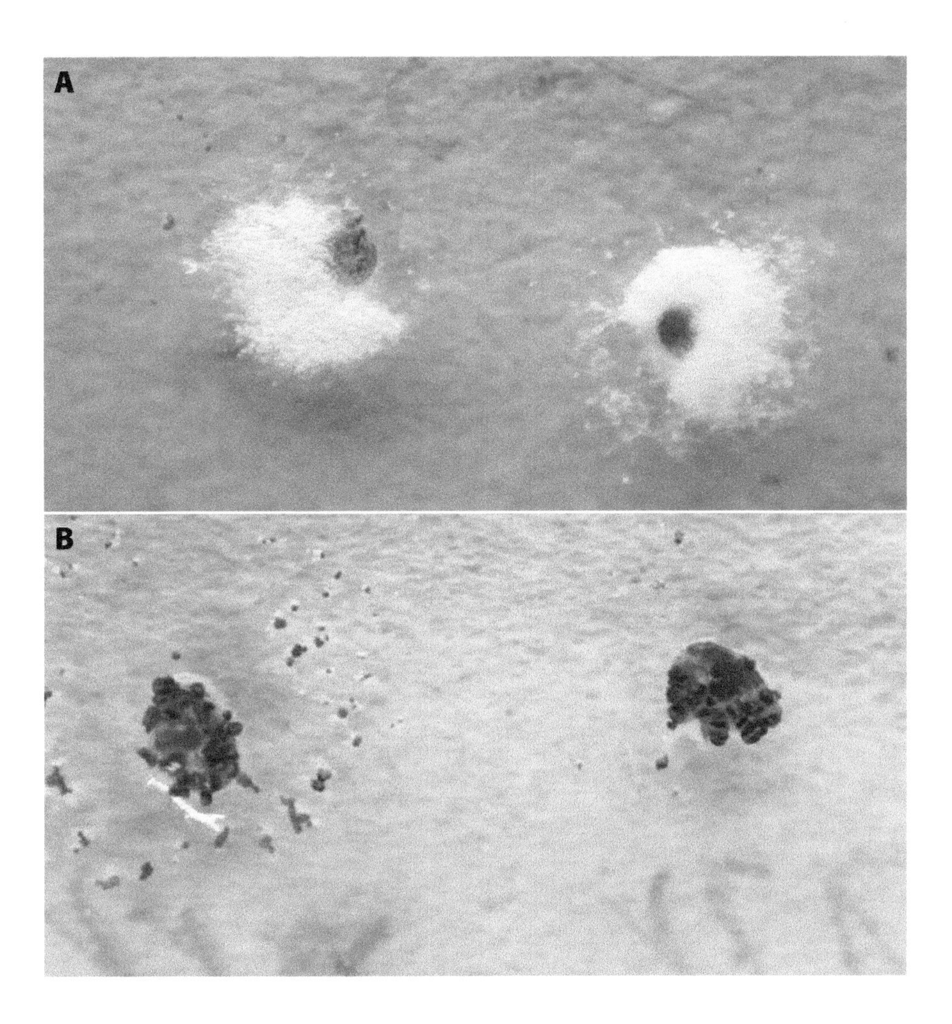

Figure 4. **A:** Varroa infected with *Beauveria* fungus. **B:** Varroa infected with *Metarhizium* fungus. (*Photos by G. Mercadier.*)

Isolates of *M. anisopliae* have been used to control many pests, including termites (Su et al., 2003) and locusts (Cherry et al., 1999). *Beauveria bassiana* is known to have a wide host range (Tanada and Kaya, 1993) and has been used against whiteflies (Islam et al., 2010) and mosquitoes (Farenhorst et al., 2009). Both species have been evaluated for use against acarines, including ixodid ticks (e.g., Stafford and Allan, 2011) and tetranychid mites (e.g., Bugeme et al., 2010). *Beauveria bassiana* is also, to date, the only EPF found naturally occurring on *V. destructor*, having been isolated from mites (see Figure 3) in Russia (Chernov, 1981 cited in Chandler et al., 2000), France (Meikle et al., 2006), Spain (García-Fernández et al., 2008), and Denmark (Steenberg et al., 2010).

Isolates of *M. anisopliae* have been evaluated against *V. destructor* by several research groups. Kanga et al. (2002, 2003, 2010) reported results of laboratory and field tests using *M. anisopliae* conidia introduced into the hives via several means, including strips, sprinkling between frames, and incorporated in protein patties. Kanga et al. (2003) reported good success in field trials, particularly when compared to *tau*-fluvalinate (Apistan©) but treated the hives with comparatively large amounts of material: 46.8 g of conidia powder (1010 conidia per gram) for the "low dose" treatment and twice that for the high dose. Given the cost of EPF conidia, it is questionable whether these dosage rates would be commercially feasible. Kanga et al. (2010) also reported good success using *M. anisopliae* conidia mixed with protein patties but did not include a "positive control" treatment to control for the effects of the protein patties themselves. Protein patties are a common hive treatment known to improve bee nutrition, increase worker bee production, and help colonies recover from diseases (Herbert, 1992). Thus, the results were inconclusive.

Other workers have reported equivocal results. James et al. (2006) tested *M. anisopliae* but without success. They tried a number of different application methods but observed no impact of treatment on mite densities and felt that the *M. anisopliae* conidia were possibly not coming into contact with enough *V. destructor* to reduce their density. Rodríguez et al. (2009a,b) first identified a desirable *M. ansisopliae* isolate based on temperature tolerances, then applied the conidia using several methods, including sprinkling the conidia between frames, attaching the conidia to strips hung between frames, and using an auto-application device at the entrance of the hive. While Rodríguez et al. (2009b) found that conidia sprinkled between frames gave the best results, bee mortality was highest in that treatment and, with respect to *V. destructor*, the biopesticide treatment needed further work to be considered a viable strategy.

Beauveria bassiana has not been tested as extensively as *M. anisopliae*. Meikle et al. (2007) mixed conidia of an isolate of *B. bassiana* collected from *V. destructor* in bee hives with carnauba wax powder and hydrated silica (8–12 × 10^9 conidia per g) and then applied that mixture between frames to control *V. destructor*. While high mortality rates of *V. destructor* and little or no impact on bee colony health were observed in treated hives, a single application did not sufficiently control the mite. Meikle et al. (2008a,b; 2009) reported further field experiments with different formulation ingredients (candelilla wax powder or wheat flour, or no ingredients at all), and in single or multiple applications. Some results were encouraging: Meikle et al. (2009) reported that, after three applications, mite fall onto sticky boards was much lower in treated hives than in untreated control hives. However, in that case, phoretic mite densities remained unacceptably high. A better understanding of the ecology of EPF in bee hives, particularly *B. bassiana*, was seen as crucial to designing more effective control strategies.

Figure 5. Larvae of the small hive beetle on the bottom board; when mature they will crawl outside to pupate in the soil. (*Photo by D. Sammataro.*)

In summary, development of an effective biological control program against *V. destructor* will likely take the form of an effective biopesticide formulation and application method. In addition, questions such as why *B. bassiana* has been found associated with *V. destructor* and no other, how *B. bassiana* gains entrance to bee hives, and what happens to EPF conidia in the hive are in need of further research.

Small Hive Beetle

Some workers have explored the possibility of biological control of the Small Hive Beetle (SHB) (see Chapter 13 of this edition for a review of SHB biology and ecology). While many of the same agents used against *V. destructor* have also been tested against SHB in laboratory sessions, SHB has life history traits that make it a very different kind of pest. This might have implications on its control. First, unlike *V. destructor*, SHB is not an obligate bee parasite; larvae feed on and contaminate pollen and honey stores in addition to bee brood. Second, SHB must spend part of its life, specifically pupation, outside of the hive (see Figure 5) and adults can be found both inside (see Figure 6) outside the hive. Taken together, these characteristics offer both a challenge in that control of the beetles

Figure 6. A Hood beetle trap positioned in a medium frame. The trap can be provisioned with a bait (such as apple cider vinegar) and a killing agent (mineral oil). Adult beetles (like those on the frame in the picture on the right) will enter the trap and be exposed to the killing agent. (*Photos by University of Florida. Used with permission.*)

may involve controlling them outside as well as inside the hive, and an opportunity in that SHB control outside the hive may have fewer restrictions than in-hive control. Successful control outside the hive would be expected to impact the ability of SHB populations to attack hives in large numbers.

As with *V. destructor*, biological control options thus far have been limited to a predator, pseudoscorpions (Donovan and Paul, 2005), and pathogens, in the form of nematodes (Cabanillas and Elzen, 2006; Ellis et al., 2010; Shapiro-Ilan et al., 2010) and EPF (Ellis et al., 2004; Muerlle et al., 2006; Richards et al., 2005). In the case of SHB, which spends different life stages in different environments, the choice of a biological control agent also rests on the target stage of the insect. Pseudoscorpions, which are expected to live inside the hive, would be either unlikely or unable to attack wandering larvae or pupae because they are buried in the soil. Likewise, adult beetles might not be suitable prey because they are active and have tough elytra. However, pseudoscorpions could attack eggs and young larvae in the hive. That would be highly desirable because the larvae would not have had a chance to cause much damage. To our knowledge, no field experiments have been conducted with biocontrol agents against SHB.

The use of nematodes would target SHB either in the wandering larvae stage or in pupation in the soil. Cabanillas and Elzen (2006), Ellis et al. (2010), and Shapiro-Ilan et al. (2010) all conducted experiments exploring the use of nematodes against SHB in the soil environment. Targeting the wandering larval and pupal stages of SHB removes difficulties such as bees being attacked by the nematodes, or bees removing the nematodes while cleaning. However, that method targets the beetles after they have done damage as larvae; its effectiveness will depend on factors such as the degree to which emerging beetles attack the hive from which they originated. A single small hive beetle can lay thousands of eggs (Ellis et al., 2002; Meikle and Patt, 2011), so even a few females would have the potential to cause serious problems for a weak or small hive that cannot maintain proper hygiene.

Finally, EPF have been explored as biological control agents of SHB. Ellis et al. (2004) and Richards et al. (2005) examined *Aspergillus niger* (van Tieghem) and *A. flavus* (Link: Grey) (both Eurotiaceae) as potential soil treatments against pupal SHB. Both teams of researchers found significant treatment effects where healthy larvae or pupae were exposed to the pathogens. However, as Ellis et al. (2004) noted, fungi of the genus *Aspergillus* are known to produce mycotoxins. While the potential for mycotoxin contamination from a human's point of view must be addressed, from the point of view of a honey bee the situation may be different. Niu et al. (2010) reported that propolis (bee-collected plant resins) helped bees break down mycotoxins such as those found in bee bread by enhancing the activity of enzymes involved in detoxification. An additional and possibly bigger problem with respect to *Aspergillus* fungi is that they can cause bee disease, including stonebrood (Muerle et al. 2006).

Muerrle et al. (2006) examined several commonly used EPF, including *M. anisopliae, B. bassiana,* and *Hirsutella illustris* (another hypocrealean EPF) for their effect on SHB in laboratory bioassays. Muerrle et al. (2006) collected an isolate of *M. anisopliae* from SHB in South Africa but found that the most virulent isolate in their tests was a *B. bassiana* isolate collected from termites.

Conclusion

Biological control of bee pests is still largely unexplored but interest is growing as beekeepers and bee researchers try to develop integrated methods of bee pest management and reduce the use of chemicals. Effective biological control methods have been developed for wax moth, but their use is apparently limited, probably owing to the lesser importance of wax moth compared to other honey bee threats. Results of Varroa biological control have been variable and limited to EPF; at the very least, further work is needed on identifying appropriate isolates and on developing appropriate formulations and application methods. In addition, field testing of SHB biological controls is needed to improve our understanding of SHB ecology.

Molecular Forensics for Honey Bee Colonies

CHAPTER

7

Jay D. Evans, Dawn Lopez, I. Barton Smith, Jeffery Pettis, and Yan Ping Chen

Abstract Recent declines in honey bee populations have inspired new efforts to diagnose honey bee diseases. One promising route is to use molecular traits to identify bee parasites and pathogens. This chapter describes gene-bassed efforts to find and quantify bee pathogens. These efforts allow for specific and sensitive detection of the microbes found in bees, aiding research efforts and giving new tools for improving bee breeding, management, and pest regulation. These methods can be used more generally to assess bee traits related to resistance, stress, and behavior. We provide protocols and cost estimates for a method now in place to screen U.S. bees for viruses and other pathogens.

Introduction

Genetic markers are available for each of the major honey bee pests and pathogens, and these markers offer new avenues for screening bees and colonies to predict ailments and causes of colony declines. In addition, the sequencing of the honey bee genome (Honey Bee Genome Sequencing Consortium, 2006) and associated efforts to define bee proteins have generated similar genetic tags for bee proteins involved with development, immunity, physiology, and behavior. These resources for bees and their disease agents can be exploited in order to improve bee breeding schemes, manage diseases or bee nutrition, and regulate the movement of viruses, bacteria, fungi, and other infectious agents. Interest in such "forensic" tools has surged in the past several years, triggered by enigmatic colony losses that have defied typical explanations. Currently, molecular diagnostics are a routine part of national surveys (Genersch et al., 2010) and research efforts (de Miranda and Fries, 2008; Johnson et al., 2009; vanEngelsdorp et al., 2009) aimed at understanding disease risk factors in the field (see Figure 1).

In this chapter, we will not carry out an extensive review of honey bee diseases or the markers used to describe them. The most up-to-date information on primers and protocols for individual disease threats will come from the primary literature, and readers are encouraged to seek out that literature. Two recent national disease surveys provide a snapshot of markers in use for the major honey bee pathogens. Genersch and colleagues (2010) describe extensive methods and target sequences for a national survey carried out in Germany for several years. Similarly, vanEngelsdorp and colleagues (2009) present pathogen sequences used in a national survey of U.S. honey bees, and Evans (2006) reviews markers for pathogens and honey bee health-related genes suitable for research on bee health. In the case of RNA viruses, there is still a need for consensus on genetic tags that will be useful for individual species and, in some cases, across close relatives, but we do not feel there is great value in "freezing" the genetic sequences useful for identifying specific targets in this chapter. Instead, we will present a start-to-finish set of protocols that we have in place for collecting, preserving, analyzing, and reporting on diagnostics aimed at bees and their microbes. We feel that this set of techniques can work as an economical and sensitive tool for identifying disease threats for bees.

We will focus on the four critical steps of molecular diagnostics:

- Sample collection
- Sample processing
- Analysis
- Reporting

Because budget and time constraints can limit any research or management effort, we will present both monetary costs and labor estimates for the techniques we currently use. Molecular diagnostics, like any technology, advance frequently and the individual scientist must decide when it makes sense to stay with a supported technology that just "works" or change to a new technology that will add information, deal with compromised samples, or save on costs or time. There are excellent recent advances in microfluidics and in techniques for diagnostics based on hybridization to fixed targets (e.g., microarrays) but they have yet to be applied broadly to honey bee samples. While it would be ideal for all labs to use the same diagnostics, many factors will determine when it is best for a specific lab to convert to a new technique. For now, assays using the polymerase chain reaction (PCR) have matured over two decades into the most sensitive and specific tests available. Finally, we do not mean to suggest that molecular techniques are the solution for all diagnostic needs. Diagnostics based on colony traits, parasite loads, microscopy, and lab cultures are still needed in evaluating bee health. In fact, nongenetic diagnostics are arguably still often the best way to provide immediate management or regulatory decisions (e.g., De Graaf et al., 2006).

Sample Collection and Transport

Sample collection is critical for both statistical issues and sample integrity. Statistically, an ideal diagnostic for a honey bee colony might strive to survey each of the thousands of colony members. Even if such a sample could be collected in a nonlethal way, it would require a great effort in terms of time and resources. Alternatively, colony diagnostics might be gathered via an aggregate resource such as honey, bottom board debris, or swabs of hive surfaces. Such samples have their role, e.g., in sampling for rare parasites (Ward et al., 2007) or for pooled honey samples (Nguyen et al., 2009) but we have not seen strong evidence that such environmental samples are sensitive to the microbes that are key for honey bee health. Instead, we and others (Cox-Foster et al., 2007; Genersch et al., 2010) favor diagnostics based on pooled samples of worker or larval bees. Statistically, the needed size of these samples can be predicted by variance across bees in disease presence. As one example, if a target-worthy pathogen is found in only 10% of adult bees, surveys for that pathogen will find it 96% of the time when 30 bees are sampled (i.e., the odds of receiving 30 consecutive negative results given a true frequency of 0.9 is 0.042). In the diagnostic below we use an aggregate of 50 bees, providing a 99.5% success rate in identifying targets found in 10% of sampled bees and a 92.5% success rate in identifying targets present in 5% of bees. Of course, these diagnostics are per sample and for any survey there will likely be a proportional number of false negatives.

The second major decision involves choosing the life stage that provides the most informative and consistent view of colony health. While there are few published studies to quantify this, older worker bees tend to carry a larger microbial contingent than do either larvae or newly emerged workers. Presumably, these bees also have normalized their microbial contingents by feeding and nest duties, and by acquiring environmental microbes. Collecting returning or departing foragers provides a consistent way of collecting older workers. When bees are not foraging, or simply to speed sample collection, arguably the next best method for selecting older bees is to scoop bees directly from the tops of frames from an upper box of bees or at the edge of an existing cluster. The main goal should be to collect bees from as far from brood frames as possible to minimize the collection of nurse bees (Figure 1). In all cases, a field sample of ca. 100 bees should suffice.

An overriding goal of our genetic tests is to measure both abundance and presence/absence of specific targets. Accordingly, we use quantitative-PCR (qPCR) methods as the diagnostic. We base most of our diagnostics on ribonucleic acid (RNA) rather than deoxyribonucleic acid (DNA). RNA gives a superior estimate of actual target activity (e.g., the process of making proteins by honey bees or by their parasites and pathogens) when compared to DNA. Further, while DNA is the heritable genetic component defining most organisms, many critical honey bee viruses never show a DNA stage and, consequently, would be invisible to DNA-based diagnostics. The cost of using RNA diagnostics comes in the form of needing slightly more stringent laboratory protocols and requiring an extra step during the actual diagnostic itself.

A major drawback of RNA diagnostics is the need for relatively pristine samples. While work continues to determine the limits of sample degradation on diagnostic accuracy (e.g., Benjamin Danait, Peter Neumann, and colleagues at the Swiss Bee Research Center and Chen et al., 2007), it is best to maintain sampled bees alive or in an ultra-cold state from the time of collection until RNA extraction. This presents obvious difficulties for field surveys and research projects. One solution is to ship field samples as live bees, a technique currently in use for a large-scale survey directed by the USDA Animal Plant Inspection Service (APHIS, http://www.aphis.usda.gov/plant_health/plant_pest_info/honey_bees/survey.shtml).

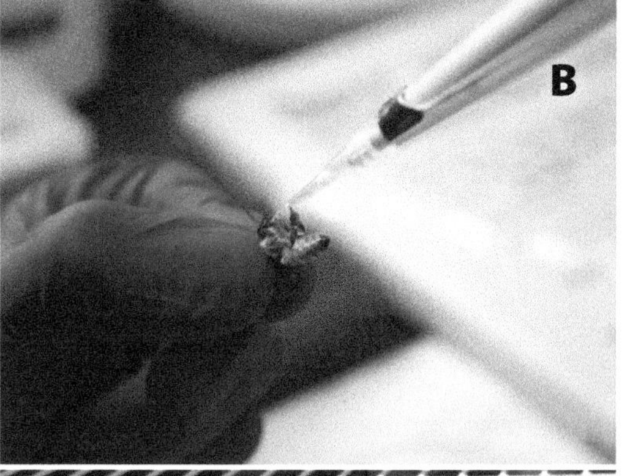

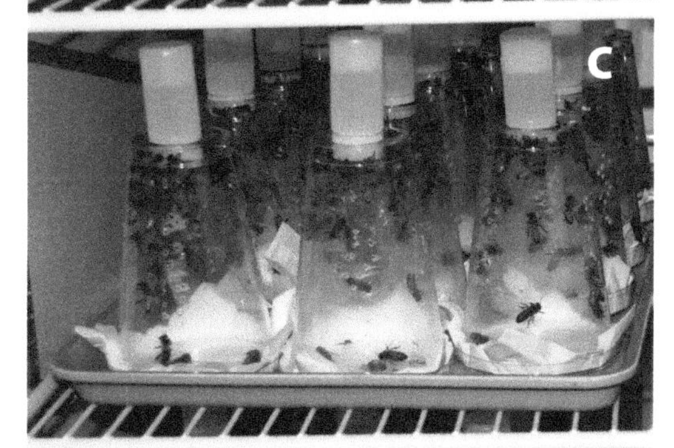

Figure 1. A. Sampling bees from honey frames to minimize collecting nurse (young) bees. **B.** Controlled inoculation of worker bee with pathogens. **C.** Maintaining experimental bees in an incubator prior to genetic diagnostics. (*Photo by J. Evans.*)

Alternatively, samples can be collected directly into vials (e.g., 50 ml plastic centrifuge tubes) and held *post mortem* in an ultra-cold freezer (–80°C) or on dry ice continuously until processing. Finally, the RNA preservative RNA*later*® (Ambion) can be used when samples will be at ambient temperatures for weeks (Teixeira et al., 2008). RNALater is a super-saturated salt solution and the drawback in using this preservative comes as much in the required subsequent dilution of samples for efficient RNA extraction as in the imperfect preservation of collected samples. New methods for capturing and stabilizing RNA from field samples are needed.

Molecular forensic techniques are also increasingly useful for experimental studies of bee infection, immunity, and gene expression.

Sample transport is rarely a problem here, as experiments can be carried out in close proximity to laboratories, or under controlled conditions within laboratories (e.g., Roetschi et al., 2008, and Evans et al., 2009, respectively).

Sample Processing and Target Quantification

The most widely used RNA extraction protocol involves the separation of nucleic acids from fats, proteins and body parts using an acid-phenol buffer that also contains reagents for preventing enzymatic degradation of RNA and for lysing target cells. One recipe built on this technique (Chomczynski and Sacchi, 1987) is widely available as TRIzol® (Invitrogen Corp.) or TRI Reagent® (Sigma-Aldrich), although other phenol-based extraction reagents are also widely available. There are related protocols from the primary literature that can be more economical and flexible for many lab applications. For example, we use a "hot phenol" extraction analogous to that described by Pinto et al. (2009) for isolating bacterial RNA. One advantage of self-made reagents is the capacity to change the extraction chemistry. We have found that delaying the addition of phenol until lysed samples have been divided into smaller aliquots does not affect extraction integrity and gives a >20-fold reduction in both reagent costs and in the production of hazardous phenol waste. A disadvantage of self-made reagents is increased exposure to potentially caustic and toxic chemicals, and as such all preparations must follow recommended safety warnings. This and all RNA extraction protocols are not foolproof in removing DNA, and extractions must be exposed to DNAse or another DNA degradation scheme prior to sensitive PCR assays.

The next step involves generating DNA copies of extracted RNA, and there are three distinct choices at this stage. In all cases, complementary DNA (cDNA) that matches RNA targets must be synthesized enzymatically with free nucleotides and reverse transcriptase (i.e., SuperScript®, Invitrogen). This synthesis can be primed by three methods:

1. A downstream "primer" (short DNA oligonucleotide) complementary to the diagnostic target itself
2. Primers complementary to polyadenylated ends found in many RNA transcripts
3. Random oligonucleotides long enough to anneal to RNA at the temperatures used in cDNA synthesis (generally oligos of 7, 8, or 9 nucleotides in length)

All three methods of cDNA priming have their merits; advantages of the second and third methods are that a single cDNA synthesis routine can generate material for dozens of assays. The advantage of using random primers is that this technique does not self-select for polyadenylated targets. Messenger RNAs from bees, and transcripts from most of the RNA viruses, are polyadenylated and a surprising number of additional microbial RNAs are either polyadenylated or carry sufficient poly-A stretches as to be identifiable using strategy 2. In fact, this strategy has proven successful for quantifying transcripts from ribosomal RNAs as well as coding-gene transcripts from bacteria and fungi, for which polyadenylation is idiosyncratic (Evans, 2006). Nevertheless, most RNA diagnostic assays currently rely on random priming at the cDNA synthesis stage, and this method is arguably less prone to biases in cDNA synthesis. Accurate PCR quantification of targets depends on using some form of "real-time" PCR, whereby the thermal cycler enacting PCR has an optical ability to screen products each thermal cycle for signs of target amplification. Many real-time PCR platforms are available, and the market for these changes sufficiently often that we will not attempt to review them. Reaction chemistries can be divided into those that are target specific (e.g., the signal from a specific PCR product results from binding to a DNA sequence within the specific amplified product) or general (signal is produced from all amplified products). The former strategy involves fluorescent reporters including varied "Taq-Man" (Applied Biosystems) reporters, "Black-Hole Quencher" probes (BioSearch Technologies), and other variants on fluorescent probes that report as they attach to a matching sequence internal to the amplified PCR product. This strategy is arguably most specific, as targets must match both their DNA oligonucleotide primers as well as an internal probe sequence. Fluorescent probes also allow for diagnostics that measure multiple targets simultaneously, and such "multiplex" reactions have proved useful for virus detection in bees (Chen et al., 2004; Grabensteiner et al., 2007).

The most widely used nonspecific reporter involves the intercalating dye SYBR Green, which binds most strongly to double-stranded DNA, including PCR products. SYBR Green reporting is widely used for studies of bees and elsewhere (Evans, 2004; Evans, 2006; Kukielka et al., 2008; Siede et al., 2008), despite a slightly higher risk from artifacts due to non-specific priming and generic fluorescence. These risks can be averted with proper primer development and post-PCR controls (notably the validation of amplified products by DNA sequencing and the routine use of "melt curve" dissociation assays to estimate product size and character). In our routine tests, SYBR Green fluorescence has been effective and accurate, and we use it most of the time. We have generally stopped using multiplex assays, after difficulties in maintaining efficient noncompetitive PCR conditions for all targets (see *Interpretation* below).

Other than the addition of SYBR Green or other reporters, qPCR is enacted much like standard PCR reactions, with a source of free nucleotides, target-specific primers, a thermostable polymerase enzyme, and a buffer selected to favor specific binding and efficient replication each cycle. There are many varied conditions; we provide one that we favor in the protocols at the end of this chapter, but readers can and should review the literature for advertised PCR mixes and conditions. The actual qPCR procedure is carried out on a "real-time" thermal cycler and, again, choices abound for specific machines, all of which will accept SYBR Green as a reporter and most of which will, through filter selection and light excitation where needed, accept the more common fluorescent reporter probes. As with PCR mixes, a review of current papers will provide insights into which machines are in use for which assays.

Interpretation of Real-Time PCR Output

Output from a real-time PCR run will reveal, for samples and controls, the minimum cycle number at which a fluorescent signal above a predefined threshold is generated (C_T). As mentioned before, C_T values generated by SYBR Green should be supported by *post-hoc* melt curve analyses. Prior to analysis, real-time PCR data should be transformed either by absolute quantification (standard curve method with strand abundances) or comparative quantification methods (e.g., the delta-delta C_T, $\Delta\Delta C_T$ method or the relative scale proposed by Pfaffl, 2001). Absolute quantification can be achieved by constructing a standard curve of known amounts of target DNA and then comparing unknowns to the standard curve and extrapolating a value. The standard procedure for determining relative abundances starts with normalization using one or more reference genes (eg., beta-actin or others found empirically to best reflect net messenger RNA abundance; Lourenço et al., 2008; Scharlaken et al., 2008). The relative measure ΔC_T is expressed as $C_T$reference − $C_T$target. Assuming the amplification efficiencies for target and reference genes are both high (a doubling each cycle), this estimated ΔC_T can be used as is, as a $\log_2$ statistic reflecting relative target abundance. The $\Delta\Delta C_T$ statistic is another popular means of presenting relative transcript abundances between samples via qPCR, and generally involves scaling ΔC_T estimates relative to the minimum value in a study set or sample (assuming each sample is tested against multiple targets). If the intent is to generate actual biological copy number estimates, the ΔC_T (or $\Delta\Delta C_T$) values must be transformed by using $2^{\Delta Ct}$ or $2^{\Delta\Delta Ct}$, respectively. These transformations also depend on ideal amplification of PCR products (a doubling each cycle) and this must be proven for each primer pair under each condition. If PCR efficiency is less than perfect, and differs between targets or controls, this must be accounted for under any normalization scheme and, arguably, the Pfaffl (2001) method remains the most robust way to deal with different amplification efficiencies. A normalized C_T or biological copy statistic can then be used for statistical analyses available for any scientific measure, under the same constraints regarding stochasticity, sample size, and skew faced by other tests.

Reporting Results

If there is one element with which we have struggled the most, besides choosing and validating specific targets and samples, it is over how best to present molecular forensic data to users. Most of our analyses are buffered by being in-house research projects; but when diagnostic summaries or individual reports are needed, we have had to decide which information is relevant and how best to present this. To be truly meaningful, diagnostics must be couched in terms of regional or national patterns for the presented targets. This is a truism for targeted bee genes, for example, "immune gene X was under-expressed relative to the norms of Californian bees," but is also true for target pathogens and pests. As one example, the viruses Black Queen Cell Virus (BQCV) and Deformed Wing Virus (DWV) are currently ubiquitous in the United States, as is the microsporidian parasite *Nosema ceranae*. Reports detailing presence/absence of these targets have little use, but their quantitative levels within bees might truly predict a health risk for sampled colonies. Consequently, reports on these targets should give the pathogen load (transcript copies/bee or relative abundances) alongside the survey mean for those values. Other pathogens are sporadic (most viruses, and *Nosema apis*) and it is often worthwhile to indicate their presence alone, although arguably it never hurts to say how their loads ranked relative to the other scarce records for their species.

One Lab's Protocol for Molecular Diagnostics

As with most protocols, ours is modular and some of the parts described below might work better than others for specific research or diagnostic needs. Below are protocol checklists and specific reagents for molecular diagnostics from RNA isolation to the generation of quantitative PCR data.

Acid Phenol RNA Extraction

Needed Reagents:

- Lysis buffer:
  0.8 M Guanidine thiocyanate,
  0.4 M Ammonium thiocyanate,
  0.1 M Sodium acetate,
  5% glycerol, 2% Triton-X100.
  For 620 mL: 94.53 g Guanidine thiocyanate, 30.45 g Ammonium thiocyanate, 33.4 mL 3M stock Sodium acetate, 50 mL glycerol, 20 mL Triton-X100. (read MSDS on thiocyanate safety).
  Mix in 300 mL water, bring up to 620 mL, then autoclave.
  Use within 30 days.
- Acid phenol (pH 4, Sigma-Aldrich)
- Choloroform
- Isopropyl alcohol
- 70% ethanol
- Sterile, DNAse/RNAse free water

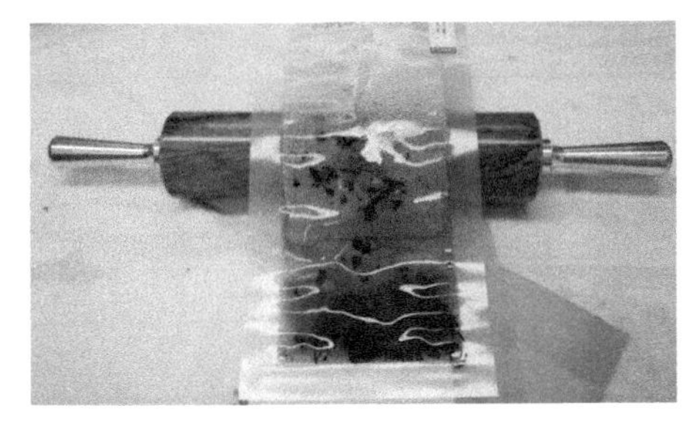

Figure 2. Collected bee samples in plastic sealable bags smashed by rolling pin. (*Photo by J. Evans.*)

Procedure

- Place 50 bees from vial into a zip-lock bag (1 quart Zip-Loc© or BioReba© sample bag).
- Add 25 mL (or 500 µL/bee) lysis buffer.
- Seal bag w/o air and smash with rolling pin until bees broken fully, for 30 sec (Figure 2).
- Massage liquid to mix fully, then draw off 620 µL, avoiding bubbles.
- Add to 380 µL acid phenol in a labeled microcentrifuge tube, vortex, incubate at 95°C for 10 min.
- Cool to room temperature (RT); can do on ice 5–10 min and add 200 µL chloroform.
- Shake tubes vigorously by hand for 15 sec and then incubate at RT for 2–3 min.
- Centrifuge the samples at 10,000 × *g* for 15 min at 4°C .
- Collect upper aqueous phase into fresh tube containing 500 µL isopropyl alcohol, vortex. Can stop at this point; sample is stable for days at 4°C.
- Incubate samples at 4°C for 30 min and centrifuge at 10,000 × *g* for 10 min at 4°C.
- Remove the supernatant. Wash the RNA pellet once with 1 mL 70% ethanol.
- Mix the sample by vortexing and centrifuge at 7,000 × *g* for 5 min at 2 to 8°C.
- Pour off liquid, centrifuge 30 sec, and pull off remaining EtOH with pipette.
- Briefly dry the RNA pellet (air-dry 5–10 min).
- Dissolve RNA in 50 µL RNase-free water by incubating for 10 min at 55 to 60°C.
- Store at −80°C.

Complementary DNA (cDNA) synthesis

Needed Reagents:

- Dnase I (2 U/µL solution)
- RNAseOut (Invitrogen)
- DNTP (100mM solution)
- DNAse I Buffer (10X)
- Random primer set (7-mer at 10 mM concentration)
- Supercript II enzyme (200 U/µL; Invitrogen)
- SuperScript Buffer (10X)
- DTT (0.1 M)

Procedure

- Prepare initial reaction mix containing, for each sample, 1 µL Dnase I solution, 0.4 µl RNAseOut, 0.4 µL dNTP mix, 1.1 µL DNAse buffer.
- Aliquot 2.9 µL reaction mix per well.
- Add 8 µL of 1 µg/µL total RNA per well.
- Incubate at 37°C for 1 hour, then 75°C for 10 min, cool to 4°C.
- Heat plate to 70°C for 10 min, place on ice.
- When cooled, centrifuge briefly.
- Incubate at 42°C for 2 min.
- Add 4 µL SuperScript mix to each sample (0.5 µL SuperScript II enzyme, 1.5 µL SuperScript buffer, 2 µL DTT).
- Incubate at 42°C, for 50 min, 70°C for 15 min, cool to 4°C.

Quantitative PCR (20 µL reaction)

Needed Reagents:

- Sterile H_2O
- Specific Forward and Reverse Oligo Primers
- Express SYBR Green ER qPCR SuperMix Universal (Buffer, Taq polymerase, DNTPs)*

Procedure

- Prepare Master Mix for each sample plus handling excess (calculate based on 100 samples for 96-well plate).
- Aliquot into wells of plate, 18 µl each on ice.
- Add 2 µL template per plate, on ice.
- Centrifuge plate to incorporate sample and remove bubbles.
- Program thermal cycler**, add plate, and run program.
- Collect post-run data for spreadsheet manipulation and statistics.

* It is of course common to mix qPCR reagents by hand rather than buying a prepared mix, and numerous published papers show those required reaction mixes (e.g., one "home-made" SYBR Green recipe that we have used for several years is described in Evans 2006.

** We use a Bio-Rad CFX-96 thermalcycler, programmed to start with a 50°C two-minute step (this safety step enables the degradation of any possible past PCR products as the SuperMix incorporates dUTP into all PCR products and also includes Uracil DNA Glycosylase which will degrade any pre-existing uracil residues). Next is a two-minute 95°C denaturing step, followed by 40 cycles of 95°C for 20 seconds, 60°C for 30 seconds, 72°C for 80 seconds. The fluorescence is measured at the 72°C step, as this provides another barrier against measuring short nonspecific primer products (Evans, 2006).

Rough Budget Estimate (in 2010 US$)

Quantitative PCR on Seven Bee Targets Plus One Control Gene

Sample Prep	Price/Unit	per Unit	Units/sample	Price/Sample
RNA extraction	$200/96	$2.08	1	2.08
Superscript II	$800/10000	$0.08	50.00	4.00
Random 7mers	0.0025/rxn	0.0025	1.00	0.0025
Disposable Plates	$220/200	1.1	0.01	0.01
Total Extraction/cDNA				$6.10 (per sample prep)

Quantitative PCR	Price/units	per Unit	Units/Reaction	Price/Reaction	Price/8 Targets
SuperMix Reagent	$1506/2000	0.75	1.00	0.753	6.02
Primer1	$5/2000	0.0025	1.00	0.0025	0.02
Primer2	$5/2000	0.0025	1.00	0.0025	0.02
PCR Plate/cover	$650/100	6.50	0.01	0.065	0.52
Total				0.83	$6.58

Labor: ca. 20 samples/person-day (trained undergraduate assistant)
$16.33/hr x 8 = $130.64 ($6.53 per sample labor)

Project 600 samples (total per sample cost is $19.23) x 8 targets				
	RNAprep	**PCR**	**Labor**	**TOTAL**
Total Cost	$3658.38	$3948	$3919.20	$11525.58

Honey Bee Viruses and Their Effect on Bee and Colony Health

CHAPTER

8

Joachim R. de Miranda, Laurent Gauthier, Magali Ribière, and Yan Ping Chen

Abstract Honey bee viruses are common in bee populations. Colonies may appear healthy even when several different viruses are present, causing little or no harm, and persist and spread harmlessly within and between bee populations. These viruses can cause severe, even fatal, diseases to bees as a result of being stimulated to replicate rapidly, infecting sensitive stages or organs. Misdiagnosis and management of honey bee diseases has been a problem due to the difficulty with virus identification. An overview of research and routine screening techniques involved with virus detection and disease diagnosis is presented in this chapter along with a summary of known honey bee viruses, management practices, and treatments, and direction of future research. Of particular interest is that frequently there is an association and interaction of viruses with other honey bee parasites and diseases, which is a significant aspect of honey bee viruses and their incidence of infection.

Acknowledgments JM was supported by grant 244956 from EU 7th Framework FP7-KBBE-2009-3 to the BeeDoc Research Network. YC was supported in part by the USDA-CAP grant (2010-85118-05718).

Introduction

All forms of life are infected by viruses, and there is consequently a large diversity of viruses in nature. This diversity includes viral particle shape and genome organization but can also be biological, in terms of the number of different hosts or tissues that can be infected. Virus particles are generally little more than genetic material (DNA or RNA) enclosed in a protective coat of proteins, with sometimes additional protective layers of lipid membranes, and they multiply only in the living cells of their host. Most are so small that they can be seen only by electron microscopy. The particles of many unrelated viruses, some responsible for different diseases, are indistinguishable by electron microscopy and can only be identified by indirect methods or occasionally by the symptoms they produce (Ball and Bailey, 1997).

In common with many mammalian viruses, most, if not all, of the honey bee viruses persist in the population as covert infection in live individuals (Hails et al., 2008). Consequently, many honey bee viruses commonly occur in bee populations from colonies that continue to appear healthy even when several different viruses are present. Such covert infections may be maintained in populations for many generations, causing little or no harm, yet in certain circumstances they may be stimulated or activated to replicate rapidly or to infect sensitive stages or organs and initiate overt (acute) and often fatal infections (Ribière et al., 2008).

Moreover, some bee viruses may be causally associated with other common bee parasites, such as the midgut microsporidian parasite *Nosema* spp. and the ectoparasite *Varroa destructor.* Honey bee viruses have also been detected in other honey bee parasites, predators, and pests such as the hive beetle (*Aethina tumida*; Eyer et al., 2009), ectoparasitic mites (*Tropilaelaps* spp.; Dainat et al., 2008; Forsgren et al., 2009), ants (Ribière et al., 2010), and wasps and hornets (*Vespa* spp.).

History

Approximately 18 distinct viruses are identified historically in the genus *Apis.* Some were isolated only once or were not considered sufficiently

Figure 1. Diagram of the picornavirus capsid structure and organization. Similar conformations also apply to the Dicistroviruses and Iflaviruses.
(*© 2008 ViralZone; Swiss Institute of Bioinformatics; used with permission.*)

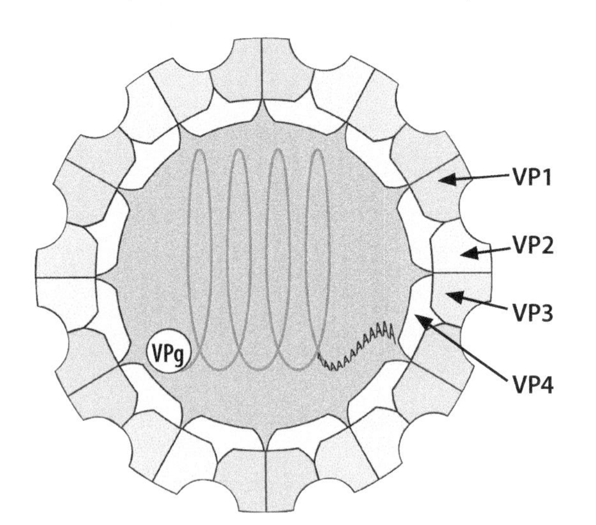

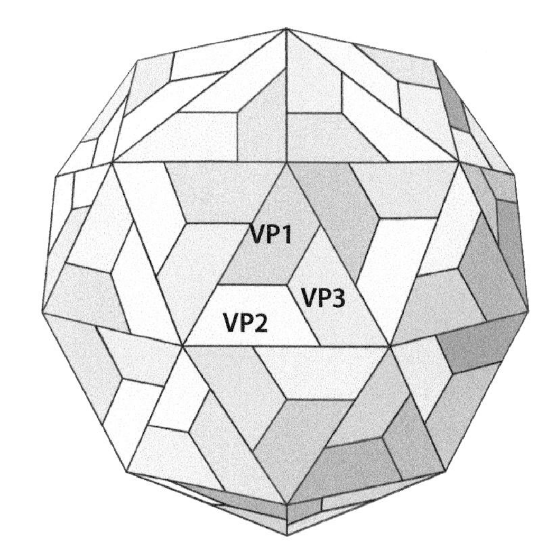

important for detailed study. With the exception of Filamentous Virus (*Am*FV) and Apis Iridescent Virus (AIV), which have DNA genomes, all are single-stranded RNA viruses with a globally isometric shape (Figure 1). Apart from minor differences in particle size, most are indistinguishable by particle morphology, with the exception of chronic bee paralysis virus (CBPV), whose particles are distinctly an-isometric. However, the viruses differ greatly in their genetic and protein composition and these properties form the basis of most of the diagnostic tests (Bailey and Ball, 1991; Ribière et al., 2008; de Miranda, 2008).

Many bee viruses are extremely common. Although able to cause severe, even fatal diseases at the individual or colony level, bee viruses usually persist and spread harmlessly at low titers within and between bee populations. There are more different types of bee viruses than there are other bee pathogens. This, together with the difficult laboratory virus identification, has historically probably been a major reason for the confusing diagnosis and management of honey bee viral diseases. Despite the welcome increase in detection accuracy afforded by the current molecular detection methods, the excessive sensitivity of these new methods presents a different problem: the mere *presence* of virus in an individual or population does not have inevitable pathological consequences. Virus detection therefore is still in a state of development and still needs to be adapted to thorough quantitative criteria for accurate evaluation of their natural history, epidemiology, and pathology (Ribière et al., 2008).

What follows is a review of the pathology, epidemiology, incidences, and distribution of the viruses, as well as the economic impact of honey bee viral infections.

What are Viruses?

Viruses are the ultimate parasites, entirely dependent on their hosts for reproduction. All of their physical, genetic, and biological properties are interpreted within the context of both their dependence on and exploitation of their hosts. Viruses are distinguished from bacteria primarily by having to rely on the host translation mechanism for their protein synthesis although with the more complex viruses this distinction is also becoming increasingly blurred. Viruses are distinguished from viroids and other sub-viral entities by the ability to replicate their own genome and even have their own molecular parasites (satellite or associate viruses, defective-interfering particles) that monopolize the viral replication apparatus at the expense of virus replication. They are therefore often associated with altering the symptoms of their "host" virus.

Particle Composition

Viruses come in all shapes and sizes, from the exotic mechanical landing structures of the phages to the mundane round particles of most RNA viruses. Some viruses have complex layers of membranes enveloping a core nucleoprotein (e.g., influenza), but most viruses consist of a nucleic acid genome encapsidated in a simple protein shell (e.g., the common cold virus). The most common nucleic acid is a single-strand RNA genome, either of "positive" polarity, if the RNA also functions directly as the messenger RNA for genome translation, or of "negative" or ambisense polarity, where the genomic RNA first needs to be replicated, generally by a viral polymerase also included in the particle, to generate the positive strand RNA for translation.

Less common are double-stranded RNA genomes, or single- and double-stranded DNA genomes. This genomic and particle diversity is also encountered to some extent among the honey bee viruses, most of which have a single-strand RNA genome packed in a highly structured fashion (Schneemann, 2006) into a small (~30 nm) icosahedral (round) particle composed of 12 pentameric facades of interlocking, viral structural proteins. The picorna-like Ifla- and Dicistroviruses, of which at least six distinct species infect honey bees, use four different structural proteins for this (Figure 1), with 60 units each of the three major proteins (VP1, VP2, and VP3) arranged in a pseudo T=3 symmetry while VP4 is located on the inside of the particle. Icosahedral RNA viruses with only a single structural protein (e.g., CWV, BVY/BVX, Macula-like virus) normally use 180 units of the single protein to produce a similar particle. The DNA viruses, Filamentous Virus (*Am*FV), and Apis Iridescent Virus (AIV) have much more complex genomes and particle composition, discussed in detail later.

Replication

Virus replication requires entry into a host cell, usually through some form of endocytosis following attachment of the virus particle to receptors on the cell surface. Some viruses progress further to the nucleus for replication, but most RNA viruses replicate in the cytoplasm. The particle disassembles releasing the genome, which then needs to be both translated (to produce the replication proteins) and replicated. Both these processes involve regulatory elements located in the nontranslated regions at the 5' and 3' termini of the genome, and usually also involve a number of host components, of which the ribosome is the most prominent. For most viruses, genome replication, translation, protein activation-modification, and particle assembly are highly coordinated processes that take place in a super-complex of viral and host proteins, suitably arranged in physical space and often anchored in a membrane. Often this leads to the appearance of organized lattices of virus particles visible by microscopy. Infectious virus particles are released from the cell either through exocytosis, leaving the cell structure intact, or more commonly through destruction of the cell releasing the contents, including the virus particles, into the environment.

Pathology

Pathogenesis is the molecular, physiological, or physical response of the host to the presence of a pathogen; viruses in this case. Sometimes these manifestations are highly characteristic for a particular virus (e.g., Deformed Wing Virus, Sacbrood Virus) in other cases the symptoms are less obvious, less consistent, or can have multiple causes (e.g. the paralysis viruses). Many honey bee viruses infect the brain tissues and are associated with behavioral changes, such as disorientation, learning difficulties, accelerated aging, reduced sensory ability, and foraging difficulties at subclinical level, as well as trembling, crawling, and flightless behavior at clinical level. Most viruses also cause a reduced life expectancy. If only few bees are thus affected, the rest of the colony can compensate and recover from the infection. However, if the disease becomes epidemic and the burden of dysfunctional bees with early mortality can no longer be compensated by the functional, healthy bees, the colony will dwindle, that is, enter a rapidly accelerating decrease in size and will ultimately die. Pathogenesis at the individual and colony level is related to the amount of virus. Most honey bee viruses are omnipresent at insignificant levels within the hive environment; in the bees, food stores, wax etc. without causing any damage. Generally, external factors (excessive confinement, stress, cold, humidity, starvation, poisoning, other pathogens/parasites, etc.) are required for converting a persistent, covert virus infection into an overt, lethal epidemic (Genersch, 2010).

Variability and Evolution

The single most defining property of viruses, especially RNA viruses, is their supreme ability to generate and maintain extraordinarily high levels of genetic variability. This variability lies at the core of their survival and adaptability. The variation is generated largely through error-prone replication, optimized to operate just below the threshold where deleterious mutations would accumulate too rapidly, and through extensive recombination between related variants. Viruses can also acquire large segments of genetic material from other viruses, the host, or any conceivable source, although these are generally evolutionary cataclysmic and rare events. Two processes act to regulate this perpetual creation and rearrangement of genetic diversity: fierce molecular competition between sequence variants for replication that ebbs and flows depending on the host condition (host species, infected tissue, molecular resistance mechanisms) and a molecular complementation process that allows temporarily unfavorable genotypes to persist within the virus population through complementation by functions provided by the locally dominant genotype, with the roles possibly reversed in another host or a future evolutionary constriction. The result is a dynamic, amorphous assemblage of variants connected by genetic and functional relatedness, perpetually changing yet globally constant, both competing and complementing. Occasionally a line of variants that can no longer be connected to the

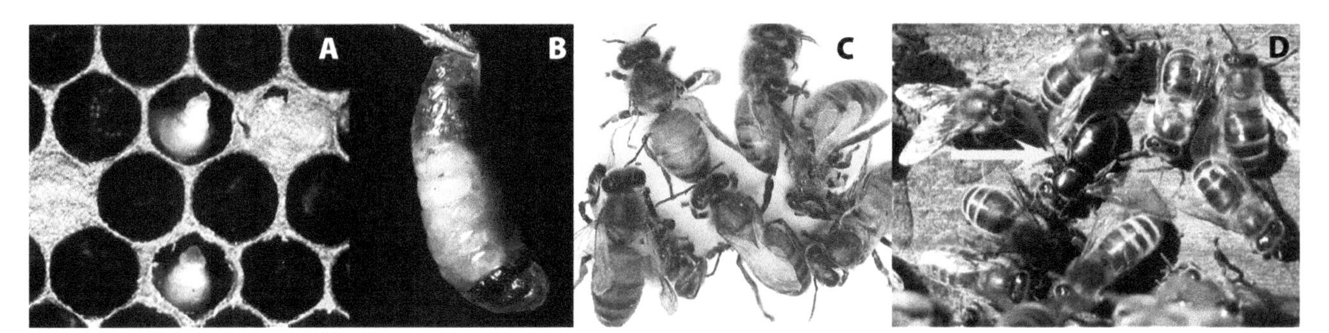

Figure 2. Typical symptoms of several honey bee virus diseases. **A–B.** Sacbrood Virus, 5×/20×. **C.** Deformed Wing Virus, 1.5×. **D.** Chronic bee paralysis virus—Type-2 syndrome (dark, shiny hairless bee; arrow), 1×. (*Photos: © S. Camazine (A–B), Y. P. Chen (C), M. Ribière (D); used with permission.*)

main group through complementation will branch off to generate a new swarm of variants, or quasi-species, around the principal sequence(s). Such quasi-species are like constantly shifting targets for host resistance mechanisms (and anti-viral drug engineers) whose response to the variability and adaptability of the viral genome is the size, complexity, and combinatorial capacity of the host genome: the molecular-evolutionary version of speed and mobility versus persistence and power.

Methods of Detection

Because of their small size, viruses are commonly detected using indirect methods, that is, with a labeled probe (an antibody or a nucleic acid sequence) targeting a specific component of the virus particle. The label is usually a chemical or fluorescent reaction that allows the probe bound to the virus particle to be visualized. Generally the target is either the viral genome (a nucleic acid sequence) or the protecting protein shell (capsid) of the virus. Nowadays, most of the methods employed for diagnostic purposes use either specific antibodies that recognize and bind with high specificity to protein motifs present on the surface of the viral particles (immunology-based methods) or nucleic acid probes that bind to a specific nucleic acid sequence in the viral genome (nucleic acid-based methods). The choice of method depends on the diagnostic aims of the project, as defined by parameters such as sensitivity, specificity, accuracy, speed, cost, and applicability, as well as the nature of the sample to be tested (for instance, some methods require fresh collected samples). We present below a brief overview of the techniques commonly used for detecting viruses or diagnosing viral diseases in honey bees, with some being more suitable for research purposes and others for routine screening surveys.

Symptoms

The cheapest, easiest, and quickest diagnosis of viral infections is through symptoms. This only works well for the few viral diseases where the symptoms are clear and unambiguous, for example, sacbrood virus (Figures 2A, 2B), black queen cell virus and deformed wing virus (Figure 2C). Paralysis is an easily recognizable symptom (Figures 2D, 3), but can have multiple causes, including several different viruses. The main problems with symptoms as a detection method is that symptoms only appear at very high virus titers, meaning that many persistent or asymptomatic infections that could have long-term implications for the colony go undetected, and that symptoms can be confounded by the simultaneous presence of multiple diseases, making the diagnosis inaccurate and unreliable.

Figure 3. Chronic bee paralysis virus—Type-1 syndrome (masses of dead bees in front of a hive during a paralysis epidemic). Circled area is an enlargement showing dead bees on the ground. (*Photos: © Magali Ribière and Elsevier; used with permission.*)

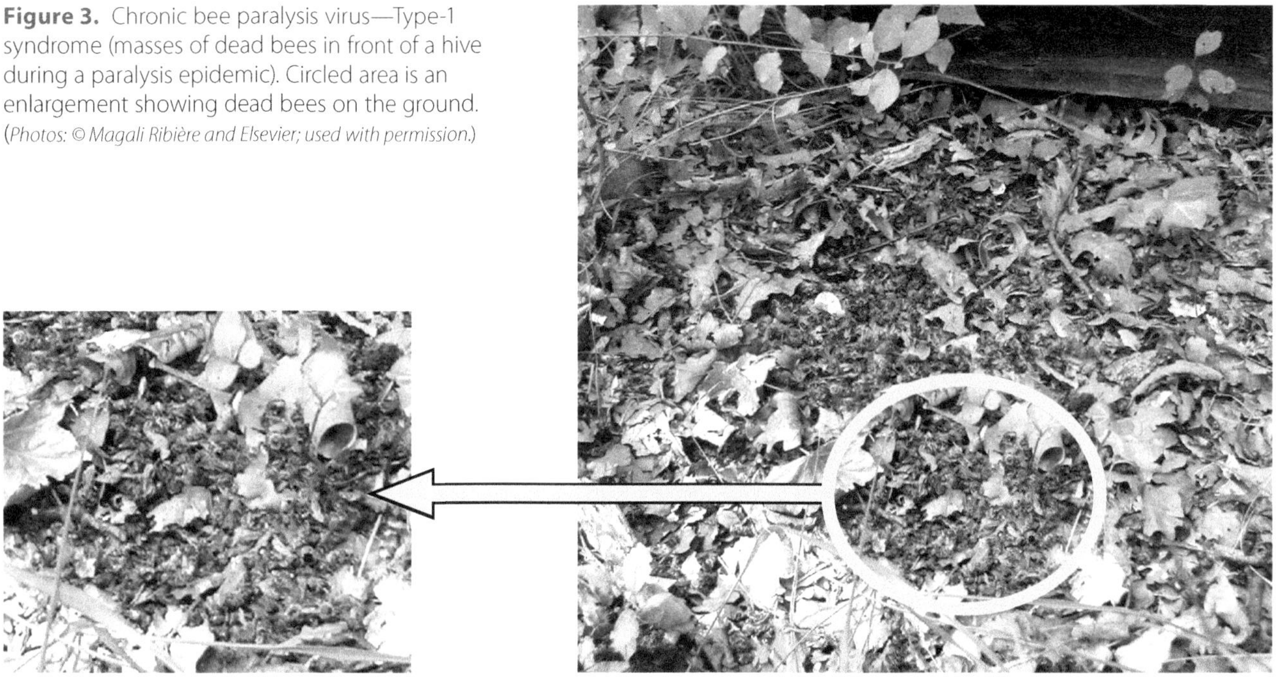

Microscopy

Most of the viruses found in *Apis mellifera* produce icosahedral (round) particles roughly 20 to 35 nanometer (nm) in diameter, with a few exceptions (e.g., filamentous virus, which is rod-shaped and 170 × 450 nm large). This means that an electron microscope must be used to observe directly the viral particles present in a sample. Usually this technique involves the adsorption of the viral particles in an extract on a carbon film and contrasting them with electron-dense material such as phosphotungstic acid or uranyl acetate so as to visualize them under an electron flow (Figure 4A). Alternatively it is possible to observe the viruses directly in cells by cutting tissues in thin slices. Although this method allows a relatively precise determination of the size and the shape of viruses, as well as their precise location (Figures 4B, 4C), it requires a high concentration of viral particles. Moreover, most bee viruses have a similar morphology (e.g., the Iflaviruses and the Dicistroviruses) such that individual virus species cannot be distinguished. This method also requires sophisticated and expensive equipment and cannot be applied for routine tests.

Immunology

Immunology-based methods require the production of antibodies specific for the viral proteins present in the sample. Such antibodies are generally produced by immunization of animals (most commonly rabbit, chicken, goat, rats, or mice) with a pure virus suspension or viral protein (called antigen, as they elicit an immune response). The antibodies are later recovered from the blood plasma and used for immune detection. Their main use is in routine enzyme immunoassays and lateral-flow devices. The antibodies are fixed onto a plastic surface where they bind the viral particles. The particles are then detected by similar antibodies linked to

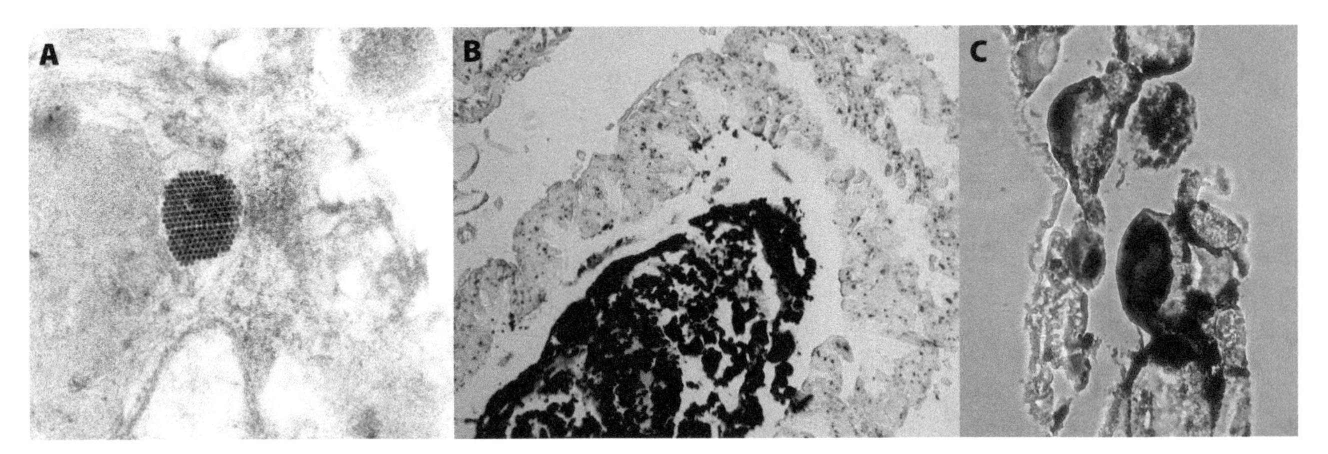

Figure 4. Microscopic images of DWV. **A.** Transmission electron micrograph of 30-nm DWV particles assembled in a crystalline array in queen ovary tissue. **B.** *In situ* hybridization image locating DWV RNA in drone midgut and rectal epithelial cells. **C.** *In situ* hybridization image locating DWV RNA in queen fat body cells, visualized by optical microscopy as a brownish-dark coloration contrasting with the pale coloration of tissues. (*Photos: © Laurent Gauthier, IBRA and BioMed Central Ltd.; used with permission.*)

an enzyme or other detectable marker molecule, which provides the basis for visualizing the detection (de Miranda, 2008). The viruses are then challenged with the specific antibodies that bind on the virus proteins. Immunological assays are generally cheap, once specific antibodies have been obtained, and can provide both qualitative and quantitative information. However, they are less sensitive, adaptable, and universal and are more cumbersome to develop and optimize than nucleic acid-based methods.

Nucleic Acid

Because all living organisms possess a unique genetic code made of a specific nucleic acid sequence, it is possible to identify precisely each species using nucleic acid-based detection methods. There are numerous different technologies available, but all rely on the capacity of nucleic acids to hybridize on their complementary sequences to form a stable duplex (de Miranda, 2008). Short nucleic acid sequences complementary to the genome of bee viruses can be synthesized chemically and used in different assays to detect and even quantify the viruses in a sample.

The most popular technologies are based on the polymerase chain reaction (PCR), which combines different temperature steps and enzymatic reactions in a cyclical, exponential amplification process to produce billions of artificial copies of a portion of the viral genome. The DNA fragments resulting from this amplification are separated by electrophoresis and visualized with a fluorescent DNA-binding dye. More recently, this amplification process has been adapted to allow very accurate quantification of the amount of target nucleic acid present in a sample (quantitative PCR, or qPCR). The advantage of PCR and other nucleic acid-based methods is that assay development is much easier, quicker, cheaper, and more uniform than serology-based methods. They are also more adaptable to different technologies, easier to interchange between laboratories, and have much greater potential for multiplexing reactions (multiple detections in a single reaction), leading to economies of scale and enhanced efficiency of detection. Nucleic acid-based methods are also much more sensitive and accurate than serological methods, although this can be a double-edged sword with respect to the high variability of RNA viruses.

Viral genetic sequences can also be visualized within bee tissues using labeled nucleic acid probes that bind on their complementary sequences. For such *in situ* hybridization, the bee organs are first embedded in paraffin or in a hydrophilic resin and shaved into microscopically thin slices. These slices are deposited on a glass slide and challenged with the nucleic acid probe, which binds to the viral sequences in the tissue. The complex is then detected using a chemical or a fluorescent reaction (Figures 4B, 4C). This is a very powerful technique for detailed pathological analysis, but is time-consuming, costly, and requires freshly collected material and is therefore not suitable for routine use.

Biotechnology is in a rapid growth phase and the next generation of biotechnological tools being developed will be able to analyze samples from a more global perspective. These methods include (meta)genomic, proteomic, transcriptome, and metabolite analyses. Their common feature is that they generate enormous integrated data sets for different biomolecules (DNA, RNA, proteins, metabolites, etc.). For nucleic acids, such analyses include DNA microarrays and increasingly high throughput sequencing (de Miranda, 2008), and their power lies in the simultaneous analysis of a high number of gene targets from a single biological extract, allowing an almost infinite number of comparisons between gene targets. Although these methods are still relatively expensive, and therefore restricted to research applications (e.g., Cox-Foster et al., 2007), they are very cost-effective in terms of cost per target, due to the large number of targets investigated simultaneously, and increase the analytical power of the sample exponentially. They will therefore probably replace the current diagnostic techniques in the future, once prices come down and there are enough different pathogens and host gene targets to be included in a detection-diagnosis. Such a dedicated, pathogen-based microarray, including also a selection of host response genes, has already been developed for research use on a semi-commercial basis.

Transmission Routes

The mode of transmission is an important step in the life cycle and the long-term persistence of viruses in honey bee host populations. Honey bees possess typical traits of social organisms, including cooperative brood care, overlap of generations, and reproductive division of labor. The large and dense populations with a high contact rate of the colony members create a highly suitable environment for disease transmission. Previous studies showed that honey bee viruses are transmitted through both horizontal and vertical transmission pathways (reviewed in Chen and Siede, 2007). Horizontal transmission refers to the spread of viruses between different individuals of the same generation in the same population. Horizontal transmission is further divided into direct (oral, contact, air, and venereal transmission) and indirect (vector-borne transmission) routes. Vertical transmission, on the other hand, is the transfer of the virus from mother queens to offspring via eggs during the reproduction. Vertical transmission can be further broken down into transovum transmission (on the surface of the egg), transovarial transmission (within the egg), and transspermal transmission (within the sperm, and therefore only transmitted upon fertilization).

Multiple transmission pathways are advantageous to honey bee viruses when considering adaptation options. When honey bee colonies are strong and healthy, the viruses remain in colonies as low titer persistent-covert infections, probably through a combination of vertical and direct horizontal transmission routes (Chen and Siede, 2007; Hails et al., 2008). However, the covert infection can become a symptomatic and overt infection. Overt or symptomatic infections are usually precipitated by environmental stresses, such as infestations of parasitic mites (e.g., *Varroa destructor*, *Tropilaelaps* spp., *Acarapis woodi*), co-infection with other pathogens or parasites (e.g., *Nosema* spp.), excessive confinement due to inclement weather, inadequate food stores, and toxicity of pesticides in the hive used to treat pests and diseases as well as in chemical-treated foraging crops. Overt infections are characterized by rapid production of high titers of virions, leading to recognizable disease symptoms, premature death of affected individuals, and possible collapse of the whole bee colony (Hails et al., 2008; de Miranda and Genersch, 2010). The adaptive flexibility of different transmission routes provides the viruses with multiple survival strategies under a wide range of epidemiological conditions. The evidence for the different transmission routes for the different bee viruses are summarized in Table 1, together with their prevalence and clear pathological effects in different life stages and seasonal incidences. This table is by no means definitive and is very likely to change as more evidence becomes available.

Table 1. Table outlining the current state of knowledge concerning the transmission routes, parasite associations, infection characteristics, and seasonal incidences of the honey bee viruses. For the transmission routes and parasite associations, the following marks are used: definitive positive proof (+); definitive negative proof (–); uncertain proof (~) or not known (?), with the major routes and associations also highlighted. For the bee life stages, a distinction is made between the infection of a life stage (Infect.) and whether or not pathological symptoms are found for that life stage (Pathol.). For the seasonal incidence, an indication is given as to the relative abundance or prevalence of a virus over the three major seasons: spring, summer and fall. (*© Joachim de Miranda and IBRA; used with permission.*)

	Transmission																	
	Horizontal				Vertical			Association				Lifestage Infect./Pathol.				Seasonal Incidence		
VIRUS	Oral-Fecal	Contact	Air	Varroa	Venereal	Transovarial	Transspermal	Varroa	Acaparis	Nosema	Malpighamoeba	Egg	Larva	Pupa	Adult	Spring	Summer	Fall
ABPV	+	–	?	+	+	+	?	+	?	?	?	+/–	+/–	+/~	+/+	+	+++	++
KBV	+	–	?	+	~	+	?	+	?	?	?	+/–	+/–	+/+	+/+	+	++	+++
IAPV	+	–	?	+	~	+	?	+	?	?	?	+/–	+/–	+/~	+/+	+	++	++
BQCV	+	–	?	~	?	+	?	+	?	+	?	+/–	+/–	+/+	+/–	+	+++	+
DWV	+	–	?	+	+	+	?	+	?	?	?	+/–	+/–	+/+	+/+	+	++	+++
VDV-1	+	–	?	+	+	+	?	+	?	?	?	+/–	+/–	+/+	+/+	+	++	+++
SBV	+	–	?	–	?	?	?	~	?	?	?	?/?	+/+	+/–	+/~	+++	++	+
SBPV	+	–	?	+	?	?	?	+	?	?	?	?/?	+/–	+/–	+/+	+	+	+
CBPV	+	+	?	–	?	?	?	~	~	?	?	~/–	+/–	+/–	+/+	++	++	+
CWV	?	~	~	–	?	?	?	~	?	?	?	–/–	~/–	~/–	+/+	+	+	+
BVX	+	?	?	?	?	?	?	?	?	–	?	–/–	–/–	–/–	+/+	+++	+	+
BVY	+	?	?	?	?	?	?	?	?	+	–	–/–	–/–	–/–	+/+	+++	+	+
ABV	?	?	?	?	?	?	?	?	?	?	?	?/?	?/?	~/?	+/?	?	?	?
BBPV	?	?	?	?	?	?	?	?	?	?	?	?/?	?/?	?/?	+/?	?	?	?
Macula	?	?	?	+	?	?	?	+	?	?	?	?/?	?/?	+/?	+/?	+	++	+++
AmFV	+	?	?	?	?	?	?	?	?	+	?	–/–	–/–	–/–	+/+	+++	+	+
AIV	?	?	~	?	?	?	?	?	?	?	?	–/–	–/–	–/–	+/+	+	++	+

Figure 5. Diagram outlining the major transmission routes for honey bee viruses, in relation to the honey bee developmental stages. **A.** Horizontal transmission among bees. **B.** Vector-mediated horizontal transmission. **C.** Vertical transmission. (*© Joachim de Miranda and IBRA; used with permission.*)

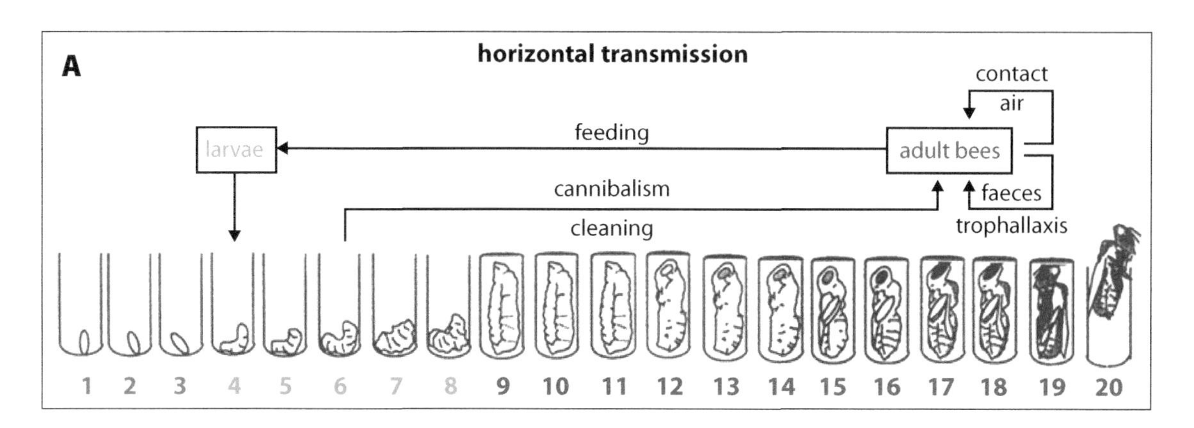

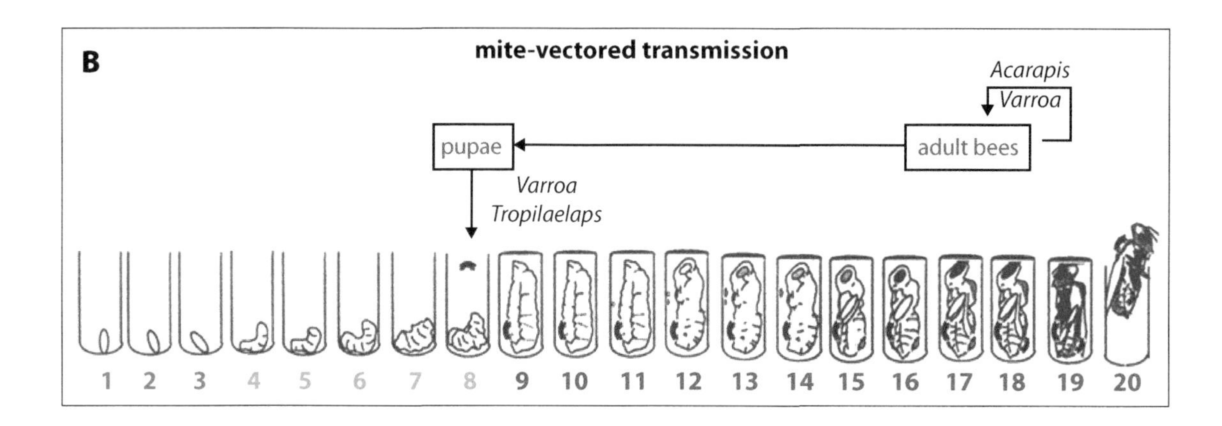

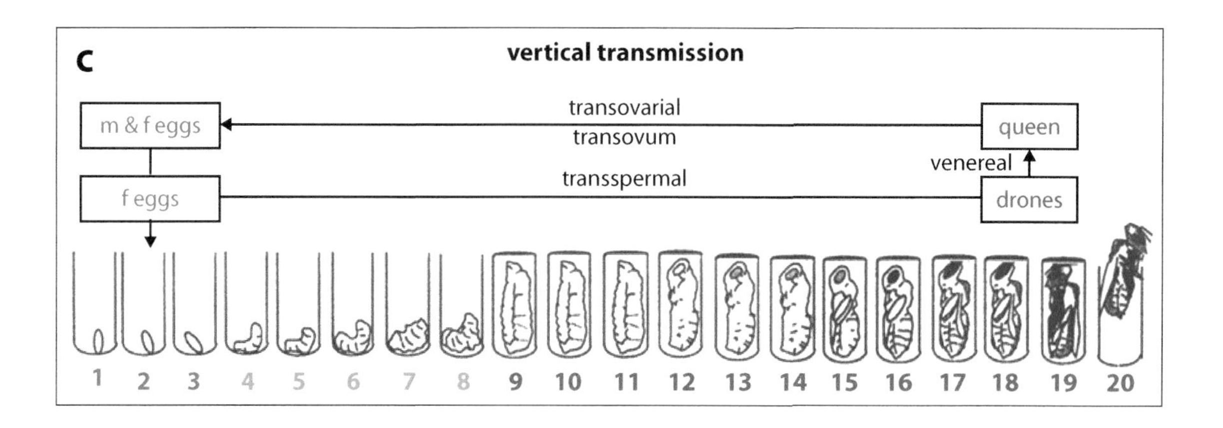

Oral Transmission

Oral transmission is a form of direct horizontal transmission that occurs after consuming virus-contaminated material. The principal forms of oral transmission in bees are the feeding of the larvae with royal jelly and pollen and through trophallaxis; and oral communication and nectar processing among adult bees (Figure 5A). In addition, honey bees also cannibalize eggs and young larvae, especially when deprived of protein sources, and use their mouth parts to remove diseased brood, clean cells, and remove feces from contaminated combs. The latter two sources often have high pathogen titers and therefore may be significant sources of oral transmission (Figure 5A). Several honey bee viruses have been detected in colony food stores, including honey, pollen, and royal jelly, providing evidence of virus oral transmission via virus-contaminated foods (Chen et al., 2006). Oral transmission of viruses has been further confirmed by detection of viruses in gut tissue and feces (Chen et al., 2006). The virus concentration in the gut was found to be significantly higher than in other bee tissues, suggesting that the gut is one of the principal sites of virus accumulation. While the trophallactic activities (nectar processing, pollen packing, communication, and queen attendance) are important for the distribution of food, they also provide chances for spreading pathogens among the adult bee population. However, the oral transmission routes in general are fairly inefficient, in that usually very large amounts of virus are required to establish an infection (Bailey and Ball, 1991).

Contact Transmission

Honey bee viruses can also be transmitted by contact (Figure 5A; Bailey and Ball, 1991; Bailey et al., 1983a). The virus probably leaves and invades the bodies of adult bees via the epidermal cytoplasm that is exposed when the hairs are broken. Transmission by contact is particularly evident in highly crowded colonies and during long spells of forced confinement when the bees normally should be active, which can be caused by a sudden decrease in nectar flows, periods of inclement weather, robbing, or when too many colonies are kept for the available nectar (Ball and Bailey, 1997). Artificial confinement of active bee colonies of course occurs in migratory beekeeping, during the long-distance transport of closed colonies, with additional increases in contact, and aggression through the jostling and bumping of the hives.

Airborne Transmission

Very little is known about the possibility of virus transmission via air and the tracheal system. This transmission route would therefore be limited to adult bees (Figure 5A). Most of the information available is speculative and concerns primarily cloudy wing virus (CWV), which on occasion was transmitted between groups of bees kept in cages but without direct contact between them (Bailey and Ball, 1991; Ribière et al., 2008).

Vector-Borne Transmission

The main reason that many honey bee viruses have become a serious problem for managed and wild honey bee populations is the parasitic mite *V. destructor* (Anderson and Trueman, 2000), which has been catastrophic for the beekeeping industry since it adapted from its original host, the Asian honey bee (*Apis cerana*), to the Western honey bee (*Apis mellifera*) during the first half of the 20th century, and spread around the world (Rosenkranz et al., 2010). The Varroa mite feeds on the haemolymph of adult bees during its phoretic phase and on developing pupae during its reproductive phase (Figure 5B). In addition to its direct negative impact on honey bee health (DeJong et al., 1982; Kovac and Crailsheim, 1988; Yang and Cox-Foster, 2005), the Varroa mite also acts as a vector transmitting the virus from infected bees to healthy bees by piercing the cuticle of the bee. The term "bee parasitic mite syndrome" has been used to describe the disease complex precipitated by the interaction between honey bee viruses with Varroa mites (Shimanuki et al., 1994). This vector-borne transmission route has become especially important to honey bee virus epidemiology because it is far more efficient than any of the other transmission routes, requiring several orders of magnitude fewer particles to establish infection, and also generates several orders of magnitude higher virus titers in affected individuals. Not surprisingly therefore, Varroa-mediated virus transmission has seriously destabilized the natural, relatively benign, evolutionary compromise between the viruses and their honey bee host. Viral disease outbreaks in colonies infested with *V. destructor* inevitably result in the collapse of the colony (Ball and Allen, 1988; Kulincevic et al., 1990; Allen and Ball, 1996). The transmission by Varroa mites has been proven experimentally for several honey bee viruses (Ball 1989; Bowen-Walker et al., 1999; Shen et al., 2005; Yue and Genersch 2005; Gisder et al., 2009; Santillán-Galicia et al., 2010; Santillan-Galicia et al., 2010), highlighting the importance of Varroa mites for virus disease phenomena in honey bee colonies. The timely and efficient control of the Varroa mite population will usually result in reduction of virus titers and keep honey bees healthy to survive winter (Sumpter and Martin, 2004; Martin et al., 2010; Locke et al., unpublished).

V. destructor is not the only parasitic mite of honey bees known to vector viruses. Mites of the *Tropilaelaps* species complex, also native to Southeast Asia, have a very similar life cycle to Varroa except that they feed and reproduce exclusively on pupae (Figure 5B), with no phoretic stage on adult bees. Tropilaelaps mites have been shown to transmit at least one honey bee virus, deformed wing virus, with similar efficiency and virulence as the Varroa mites in *Apis mellifera* colonies (Dainat et al., 2008; Dainat et al., 2008; Forsgren et al., 2009). However, because these mites require continuous brood-rearing for survival, they are of little risk to beekeeping in regions with long breaks in brood-rearing.

Finally, the tracheal mite *Acarapis woodi* has long been known to infest adult honey bees, where it feeds and reproduces in the trachea. It transfers only between adult bees, through close contact (Figures 5A, 5B) and is therefore associated with crowded conditions, especially when older infested individuals are together with younger susceptible ones, as occurs during a lack of nectar flow (Bailey and Ball, 1991) and during the winter months when bees are clustered together. Although *A. woodi* has long been linked to paralysis, a disease caused by viruses, and like Varroa and Tropilaelaps also feeds on bee hemolymph, there is no evidence that it is a vector for any of the bee viruses and the association with paralysis is essentially coincidental, as both conditions are strongly affected by crowding (Bailey and Ball, 1991; Ribière et al., 2008, 2010).

Vertical Transmission

There are several components to vertical transmission. Venereal transmission is actually a type of direct horizontal transmission in which pathogens are transmitted between two sexes during mating (Figure 5C). In honey bees this is a one-directional transmission because drones die immediately after mating. Mating represents a very efficient means for virus dispersal between colonies and over large geographic distances as virgin queens mate high in the open air in special drone congregation areas (DCAs) where drones and queens from colonies within a very large catchment area meet for mating. The queens furthermore mate with numerous drones over several days, possibly visiting different DCAs, before returning to the parental hive to start egg laying. Drone seminal vesicles and semen can contain very high titers of several honey bee viruses (Yue et al., 2006; Fievet et al., 2006). Semen is stored in queens in a spermatheca, a sperm repository organ for egg fertilization throughout the queen's lifetime. Virus-infected semen can infect both the queen's spermatheca and ovaries (Yue et al., 2007; de Miranda and Fries, 2008), resulting in further vertical transmission from queens to offspring via transovum and transovarial transmission. Transspermal transmission is a direct vertical transmission through the paternal line. It is distinct from venereal transmission in that the virus is carried within the sperm, rather than the seminal fluid, and only causes infection upon fertilization. Because drones develop from unfertilized eggs, transspermal transmission only applies to female (worker bee) offspring (Figure 5C). The frequent detection of multiple viruses in the ovaries of honey bee queens in conjunction with detection of viruses in corresponding surface-sterilized eggs (Chen et al., 2006; Gauthier et al., 2011) suggests that transovarial transmission could be a popular dispersal route for many honey bees.

Alternative Hosts

Very little is known about the host range of honey bee viruses outside of *Apis mellifera*, or the extent to which such alternative hosts represent an infection risk for honey bee colonies. Evidence is accumulating that most of the *Apis* species are probably also susceptible to most honey bee viruses. It is known that wasps and bumble bees can also be natural hosts for some of the honey bee viruses (Singh et al., 2010), which they almost certainly acquire through scavenging and robbing from honey bee colonies. However, it is highly unlikely that this represents a major infection risk for honey bees, because aside from guard bees there is no transmission-sensitive contact with wasps or bumble bees or their colonies. From the honey bee perspective, the transmission is therefore unidirectional, away from the colony, with wasps and bumble bees representing dead-end hosts. The same is also likely true for other pests of hive products (small hive beetle, wax moth, ants, earwigs), several of which have been shown to contain traces of bee viruses, as well as bacteria (Celle et al., 2008; Eyer et al., 2009; Ribière et al., 2010). By contrast, the reverse applies to certain tree aphids, who provide honeydew gathered by honey bees for forest honey, but who do not acquire food or other forms of transmission from honey bees. In this case, the viruses naturally present in these aphids could well be transmitted to honey bees although they may of course not be infectious in bees.

Dicistroviruses

Black queen cell virus (BQCV), acute bee paralysis virus (ABPV), Kashmir bee virus (KBV), and Israeli acute paralysis virus (IAPV) are members of the *Dicistroviridae*, an insect-specific virus family in the order *Picornavirales* (Le Gall et al., 2008). The *Dicistroviridae* are closely related to the *Iflaviridae*, the main distinction being a different arrangement of the viral proteins along the genome.

Genome Organization and Function

The viruses have a single-stranded RNA genome with a positive polarity, which means that the genomic RNA also serves as a messenger RNA for the translation of the viral proteins. The RNA genome is dicistronic, that is, it has two nonoverlapping open reading frames (ORFs) that are separated and flanked by untranslated regions (UTRs; Figure 6A). A small genome-linked virus protein (VPg) is covalently attached to the 5' end of the genome and the 3' end of the viral RNA genome is polyadenylated, both of which stabilize the RNA genome and protect it from degradation. The two ORFs are separated by an untranslated region known as the intergenic region (IGR). The 5'-proximal and 3'-proximal ORFs encode nonstructural and structural protein precursors, respectively. Translation of the RNA genome proceeds directly from two distinct Internal Ribosome Entry Sites (IRESs) located within the 5' UTR and the IGR (Figure 6A; Le Gall et al., 2008).

The Acute Bee Paralysis Virus Complex

Acute bee paralysis virus (ABPV), Kashmir bee virus (KBV), and Israeli acute paralysis virus (IAPV) are best regarded as a complex of closely related viruses (de Miranda et al., 2010a). These viruses have similar transmission characteristics and affect similar life stages, are widespread at very low titers, and can very quickly develop highly elevated titers with extremely virulent pathology, before subsiding equally rapidly (Bailey and Ball, 1991). These viruses are frequently implicated in honey bee colony losses, especially when the colonies are infested with the parasitic mite *Varroa destructor*, which is an active vector for these viruses (Table 1). This group of viruses is naturally highly variable, complicating both reliable diagnosis and classification (de Miranda et al., 2010a).

Acute Bee Paralysis Virus

ABPV was discovered as an unintended consequence of laboratory infectivity tests with chronic bee paralysis virus (CBPV; Bailey et al., 1963). Bees injected with purified CBPV survived for about 5 to 7 days with the trembling, flightless symptoms typical of CBPV, before dying. However, control bees injected with an extract of apparently healthy bees died much faster and also became flightless with symptoms of paralysis and contained large quantities of a distinct virus, which was named acute bee paralysis virus, in recognition of its quick lethality.

ABPV can be detected in both brood and adult stages of bee development. In the field it is a common virus in apparently healthy adult bees and colonies, and is especially common in Europe (de Miranda et al., 2010a). Prior to the arrival of Varroa, ABPV rarely caused disease or mortality of colonies (Bailey, 1965; Bailey et al., 1981) but its transmission by Varroa caused it to become a major factor in Varroa-associated colony

Figure 6 Genomic maps for the different virus families known to infect honey bees.

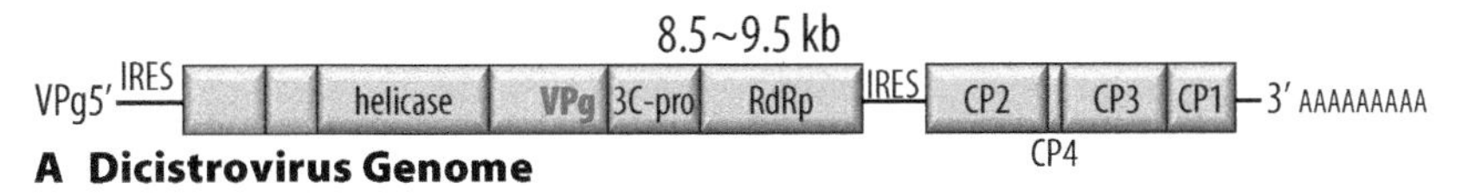

A Dicistrovirus Genome

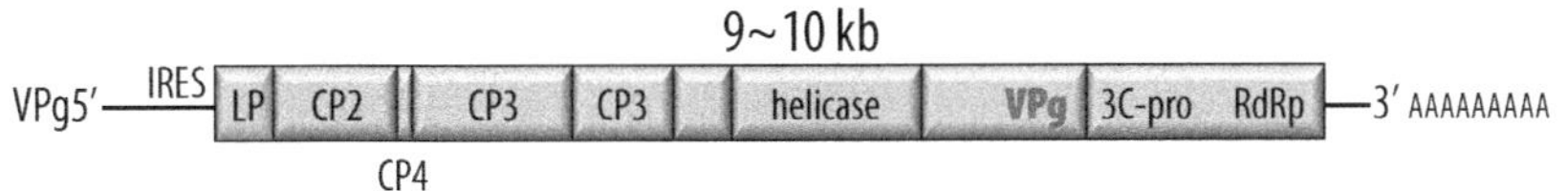

B Iflavirus Genome

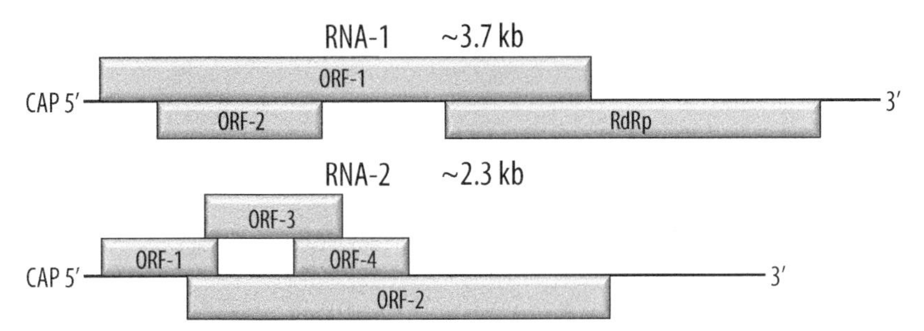

C Chronic Bee Paralysis Virus Genome

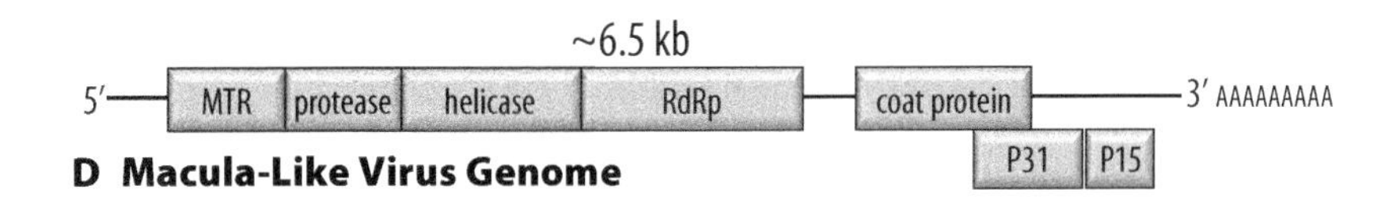

D Macula-Like Virus Genome

The RNA genomes are shown in the usual 5′-3′ orientation with the approximate size indicated in kilobases (kb). The major open reading frames (ORFs) encoding viral proteins are shown in grey blocks. The vertical separations mark the proteolysis sites for the polyproteins (**A, B, D**). Overlapping ORFs are shown above each other (**C, D**).
The following are also shown:
- Location where the 5′ terminus is modified by a VPg (**A, B**) or a CAP structure (**C**)
- Locations of the Internal Ribosome Entry Sites (IRES; **A, B**) and the poly-A tail (**A, B, D**)
- Known functional proteins
- Capsid proteins (CP1~CP4; coat protein),
- RNA-dependent RNA polymerase (RdRp)
- The Helicase
- The Viral protease (3C-pro; protease)
- Viral genome-linked protein (VPg; the probable position in the Iflavirus genome is indicated with red labels
- L-protein (LP)
- Methyl transferase region (MTR).

(© Joachim de Miranda and IBRA; used with permission.)

mortality (Ball, 1989; Sumpter and Martin, 2004; Genersch, 2010; Genersch et al., 2010). Like several other honey bee viruses, ABPV accumulates in the brain area and the glands that produce the royal jelly used to feed young larvae, and this is probably the principal means for maintaining persistent infections in the colony. Severely infected larvae either die before they are sealed in their brood cell or survive to emerge as inapparently infected (asymptomatic) adult bees (Bailey and Ball, 1991).

When transmitted by Varroa, ABPV quickly kills both adult bees and pupae. This premature mortality of pupae prevents the Varroa mite from completing its reproductive cycle (Figure 5B) but also causes the colony to dwindle toward death because the adult population is not replaced adequately (Sumpter and Martin, 2004).

One feature of the adaptation of Varroa to honey bees is that it is currently primarily associated with deformed wing virus (DWV), which is much less virulent than ABPV at individual pupae level, thus allowing the Varroa mite to complete its reproduction in pupae. At the colony level, this means that Varroa can reach much higher infestation levels with DWV transmission than ABPV transmission and, crucially, allow the colony to survive one, perhaps two winters before it ultimately succumbs to the DWV epidemic transmitted over 2 or 3 years, whereas a Varroa-transmitted ABPV epidemic would kill the colony within one season (Sumpter and Martin, 2004).

Kashmir Bee Virus

The host origin and geographical distribution of Kashmir Bee Virus (KBV) are obscure. KBV was first isolated from adult European honey bees, *Apis mellifera*, that were experimentally inoculated with an extract prepared from the diseased Asian honey bee (*A. cerana*) in Kashmir, hence the name. Since then, KBV has been found to have a worldwide distribution (Allen and Ball, 1996; Ellis and Munn, 2005). KBV may have been spread to honey bees in Australia and New Zealand via importation of bees from North American or other countries where KBV is endemic (Todd et al., 2007). Among all the viruses infecting honey bees, KBV is considered the most virulent under laboratory conditions. It multiplies quickly once introduced into the bee hemolymph and can cause mortality within 3 days. However, KBV does not cause infection when bees are fed food mixed with KBV particles. Like many other bee viruses, KBV commonly persists within brood and adult bees as a persistent covert infection. However, KBV can be activated to a lethal level by the feeding action of parasitic mites. KBV-induced mortality occurs in all stages of the bee life cycle without clearly defined disease symptoms. Because KBV is serologically and pathologically closely related to ABPV and IAPV, the diseases caused by these three viruses can easily be confused. Varroa mites are an effective vector of KBV (Chen et al., 2004; Shen et al., 2005), and may transfer the virus to other mites in the same brood cell using the bee pupa as an intermediary. Like ABPV, KBV can kill colonies in association with moderate mite infestation levels (Todd et al., 2007) and Varroa mite control is essential to prevent colony losses due to this lethal combination.

Israeli Acute Paralysis Virus

Israeil acute paralysis virus (APV) was isolated and characterized in Israel in 2004 (Maori et al., 2007), although it was almost certainly classified previously as a strain of KBV. The symptoms of experimental IAPV infection of adult bees are shivering wings, darkened and hairless thoraxes and abdomens, and progressive paralysis followed by death (Maori et al., 2007). These disease symptoms are similar to those of ABPV, hence the name. IAPV infects all developmental stages and castes of honey bees including eggs, larvae, pupae, adult workers, queens, and drones (Chen et al., unpublished data). While IAPV is both biologically and phylogenetically closely related to ABPV and KBV, sharing between 65% and 75% nucleotide identity, respectively, with each of these, there are also enough critical genetic differences among these viruses to regard IAPV as a distinct species within the complex.

IAPV came to national and international attention during the 2007 outbreak of Colony Collapse Disorder (CCD), when a metagenomic analysis of the pathological differences between CCD-affected colonies and nonaffected colonies showed IAPV to be strongly associated with the condition in the United States (Cox-Foster et al., 2007). Similar symptoms in other geographic regions had previously been associated with other members of the complex, either KBV or ABPV, again emphasizing the similarities of these three viruses (de Miranda et al., 2010a). However, while IAPV was better correlated with CCD than other pathogens examined, not all colonies with CCD were infected by IAPV, thus failing Koch's first postulate that "the pathogen should be present in every case of the diseased individuals." Furthermore, a causal relationship between

CCD and any of these viruses has not yet been shown. Currently other possible explanations are being sought, either based on one or two pathogens acting in combination (Higes et al., 2008; Bromenshenk et al., 2010) or based on a pathogen overload precipitating progressive and rapid decline (Johnson et al., 2009; vanEngelsdorp et al., 2009), although these studies are also largely based on associations and correlations rather than cause and effect. IAPV has since been found in different countries in Europe, Asia, and North America and in historical samples, and the entire CCD episode has thrown a spotlight on the decline of pollinators in general as well as the critical importance of pollination to ecosystem and crop health and productivity (vanEngelsdorp and Meixner, 2010), as well as the risks of honey bee virus infections in the international trade in bees and bee products.

Black Queen Cell Virus

Black queen cell virus (BQCV) was first isolated from dead queen larvae and prepupae sealed in queen cells whose walls had turned dark brown-black (Bailey and Woods, 1977). Diseased pupae are initially pale yellow and have a tough sac-like skin similar to those of sacbrood-infected larvae. The BQCV-infected pupae rapidly darken following death, eventually turning the walls of the queen cell dark brown to black, thus producing the characteristic symptom of BQCV infection. BQCV-symptomatic drone pupae have also been observed (Siede and Büchler, 2003). Like most honey bee viruses, BQCV has a worldwide distribution in *Apis mellifera* (Allen and Ball, 1996; Ellis and Munn, 2005) and persists within the colony as asymptomatic infections in worker bees and brood. BQCV can be propagated through injection into pupae, but not either larvae or adult bees. Oral BQCV infection of adults requires co-infection with *Nosema apis* and BQCV appears to be more dependent on Nosema for infection than either bee virus Y (BVY) or Filamentous Virus (*Am*FV), the other two viruses associated with Nosema (Ribière et al., 2008). Under experimental conditions, adult bees infected with both BQCV and Nosema have shorter life spans than those infected with Nosema disease alone. Nosema infects the adult bee's midgut tissues, thereby increasing the susceptibility of the alimentary tract to infection by BQCV. Not surprisingly therefore, the seasonal prevalence of BQCV closely follows that of Nosema, with a strong peak in spring which is also the time when queens are actively reared by queen breeders, who seem to suffer most from BQCV outbreaks. One theory is that the colonies used to rear queens in such operations are kept broodless, such that the developing queen larvae are fed by older workers bees, who are also more likely to be infected with Nosema, resulting in the oral co-infection required for BQCV infection (Allen and Ball, 1996; Ribière et al., 2008). BQCV can be detected in the ovaries of healthy queens and in about 25% of her larval offspring, suggesting a possible vertical transmission route (Chen et al., 2006), but not in the spermatheca, head, hemolymph, or body cavity. Although BQCV can occasionally be detected in *Varroa destructor*, there does not seem to be a very active vectorial relationship between BQCV and Varroa mites, at least at lower virus titers (Locke et al., unpublished) and its seasonal incidence and distribution is largely independent of mite infestation (Tentcheva et al., 2004; Ribière et al., 2008).

Iflaviruses

Iflaviruses are nonenveloped icosahedral RNA viruses, very similar in many respects to the Discistroviruses except for how the different genes are arranged on the genome (Figure 6B). Like the Dicistroviruses, the particles are about 30 nm, consisting of an icosahedral protein shell that protects the RNA genome inside. To date, this family counts three members among honey bee viruses: a species-complex that includes deformed wing virus (DWV), Kakugo virus (KV) and *Varroa destructor* virus 1 (VDV-1), sacbrood virus (SBV), and the recently sequenced slow bee paralysis virus (SBPV). Like the Dicistroviruses, the replication cycle of these viruses takes place in the cellular cytoplasm.

Genome Organization and Function

The genome of Iflaviruses consists of a positive strand RNA molecule that encodes a single large open reading frame flanked by untranslated regions (UTRs) that contain the regulatory elements involved in replication and translation (Figure 6B). The RNA is 3'-polyadenylated and its translation, driven by an Internal Ribosome Entry Site (IRES), produces a single polyprotein that is subsequently cleaved in functional units by proteolysis activity. Approximately half of the genome codes for proteins constituting the capsid of the virus and half for the proteins involved in genome replication (Le Gall et al., 2008).

The Deformed Wing Virus Complex

Prior to the sequencing of the DWV genome in 2006, two closely related viruses were described: Kakugo virus (KV) and *Varroa destructor* virus 1 (VDV-1). However, despite some interesting biological differences between DWV and KV (Fujiyuki et al., 2004, 2006, 2009), their nucleotide sequences are practically identical and from a taxonomic perspective only VDV-1 can be considered a unique virus. The analysis of samples isolated from different geographical origins suggests a recent global distribution of DWV (Berenyi et al., 2007) although locally highly distinct strain polymorphism can be observed when analyzing bee populations in greater detail (de Miranda and Genersch, 2010). This polymorphism may reflect the recent mixing of geographic DWV variants or may arise through selection of complementary biological features, such as strain-specific preferences for different transmission routes, tissue types (Fujiyuki et al., 2006; Fievet et al., 2006), or specializing in covert or overt infection strategies (de Miranda and Genersch, 2010).

Deformed Wing Virus

Deformed wing virus (DWV) was first isolated from symptomatic bees collected in Japan in 1979, although it is serologically related to a virus isolated from adult bees collected from dying colonies in Egypt in the 1970s (Bailey and Ball, 1991). DWV currently is highly prevalent with a worldwide distribution. This is almost certainly the result of its close association with *Varroa destructor*, as it was unknown as a pathological agent prior to the adaptation of Varroa to honey bees. Both the prevalence and the titer of DWV infections in colonies tend to follow the mite population development, with practically 100% of bees infected in highly mite-infested colonies. At this point, most of the workers also will have very high DWV titers, even if they do not have symptoms. However, treatment against *Varroa destructor* mites results in a gradual decrease of DWV titer in the colony, both in brood and adults through the natural turnover of the bee population (Martin et al., 2010; Locke et al., unpublished), illustrating that the pathological effects of DWV are entirely through its association with *Varroa*. DWV can also be transmitted, with similar pathological consequences, by *Tropilaelaps mercedesae*, whose life cycle closely resembles that of Varroa (Figure 5B; Dainat et al., 2008; Forsgren et al., 2009).

DWV infections can persist in the absence of obvious clinical signs in adult worker bees, drones, and queens, suggesting that this virus has little natural pathogenicity at the individual or colony level. DWV infects all the developmental stages of the bee, including eggs. However, for reasons that are not yet fully understood, DWV develops acute infections characterized by production of high viral titers that result in clinical symptoms. These acute infections are invariably linked to high *Varroa destructor* infestation rates in the colony. The clinical symptoms include early death of pupae and deformed wings, a shortened abdomen, and cuticle discoloration in adult bees (Figure 2C). Such deformed adult bees arise nearly always from mite-infested pupae, or very occasionally from noninfested pupae (Bowen-Walker et al., 1999; Tentcheva et al., 2006), although not all mite-infested pupae develop symptoms (Bowen-Walker et al., 1999; Yue and Genersch, 2005). Bees infected as adults can also generate very high DWV titers, but do not develop symptoms. DWV infection may also compromise adult bee learning ability (Iqbal and Müller, 2007) and affect aggression in bees. Kakugo virus (KV), a very close genetic variant of DWV, is associated with elevated aggressive behavior in guard bees (Fujiyuki et al., 2004, 2006), and elsewhere elevated aggression has also been associated with DWV (Terio et al., 2008). However, aggression in bees is a complex trait, subject to many genetic and environmental influences (Rortais et al., 2006), thus making it difficult to determine its precise relationship to DWV/KV infection (Fujiyuki et al., 2009). Since the original DWV isolate from Japan, which is genetically identical to KV (de Miranda and Genersch, 2010), was purified from deformed adult bees, both the biological and genetic differences between KV and DWV appear to be minimal.

Recent data suggest that DWV replication in mites prior to its transmission to pupae is a prerequisite for inducing such symptoms (Yue and Genersch, 2005; Gisder et al., 2009). This may be either a quantitative requirement (to produce enough initial virus inoculum for symptom development) or a qualitative effect, due to the possible selection of DWV variants that can replicate in the mite and have an elevated virulence in honey bees (Yue and Genersch, 2005). The first hypothesis, that of a purely quantitative effect, may be more likely because the DWV genetic composition is generally very similar between mites and their corresponding pupae (Fujiyuki et al., 2006; Forsgren et al., 2009). Furthermore, symptoms in worker bee pupae are almost invariably induced by simultaneous infestation with more than one mite and are closely related to virus titer (Bowen-Walker et al., 1999; Nordström, 2003; Chen et al., 2005; Tentcheva et al., 2006; Gauthier et al., 2007). Winter colony mortality is strongly associated with the amount of DWV present in the bees, independent of infestation by Varroa, indicating that Varroa should be removed from the colony well in advance of the production of overwintering bees, in order to bring DWV titers down sufficiently for winter survival (Highfield et al., 2009; Genersch, 2010; Genersch et al., 2010; Berthoud et al., 2010; Martin et al., 2010; Locke et al., unpublished).

Varroa destructor is a highly competent biological vector for horizontal transmission of DWV (Figure 5B; Bowen-Walker et al., 1999; Nordström, 2003; Yue and Genersch, 2005; Gisder et al., 2009). However, the mite acts also as an activator of DWV replication in bee tissues, presumably through the cuticle-piercing feeding action because this is similar to the highly efficient, laboratory injection-based inoculation in pupae and adult bees (Bailey and Ball, 1991; Santillán-Galicia et al., 2010; de Miranda and Genersch, 2010). There is possibly also further stimulation of DWV replication through Varroa-induced suppression of the host immune system (Yang and Cox-Foster, 2005). Horizontal oral transmission is inferred from the detection of DWV in pollen, feces, and the glandular secretions used to feed larvae (Yue and Genersch, 2005; Chen et al., 2005) and from frame-exchange experiments (Nordström, 2000) but is very difficult to confirm experimentally (Iqbal and Müller, 2007) due to the instability of the virus particle during purification (Bailey and Ball, 1991). DWV is also vertically transmitted from the queen to offspring. DWV is often highly abundant in drone seminal vesicles and in sperm (Fievet et al., 2006) and can be transmitted to queens by artificial insemination (Yue et al., 2007; de Miranda and Fries, 2008). DWV infects various tissues of the queen, and high viral titers are commonly detected in ovaries (Fievet et al., 2006; Gauthier et al., 2011).

Many studies are currently underway concerning the transmission and pathogenesis of DWV and its effect on honey bee molecular biology, physiology, and colony performance. However, it is clear that timely and efficient Varroa treatment is essential in order to limit winter colony mortalities due to high DWV titers (Bailey and Ball, 1991; Berthoud et al., 2010; Martin et al., 2010; Locke et al., unpublished).

Varroa Destructor Virus-1 (VDV-1) is a close genetic relative of DWV. It was first identified in *Varroa destructor* mites (Ongus et al., 2004) but has since also been found in worker bees and queens (Gauthier et al., 2011). Although its prevalence appears more restricted to Varroa compared to DWV (Ongus et al., 2004), it has been detected in long-term Varroa-free apiaries (Gauthier et al., unpublished), suggesting that VDV-1 was originally a bee virus. Moreover, both VDV-1 and DWV can replicate in bees and in mites (Ongus et al., 2004), both are capable of producing wing deformities in emerging adult bees (Zioni et al., unpublished) and naturally occurring, viable recombinants between DWV and VDV-1 have been found (Moore et al., 2010).

Sacbrood Virus

Sacbrood virus (SBV) was first described by White (1913) and was later proven to be the causative agent of a larval disease named sacbrood (Bailey et al., 1964). This virus has a global distribution in *Apis mellifera* colonies and has a distinct, but closely related variant in *Apis cerana* colonies in Asia. While SBV is a relatively unimportant and incidental brood disease for *A. mellifera*, limited mostly to early spring colony expansion, for *A. cerana* it is a major lethal disease that can readily kill colonies. SBV is a very common virus, although the prevalence in adult bees in Europe can vary greatly, depending on the survey (Tentcheva et al., 2004; Berenyi et al., 2006; Forgach et al., 2008; Nielsen et al., 2008; Kukielka and Sánchez-Vizcaino, 2009).

SBV causes a brood disease with a distinct mortality in larvae, but also infects adult bees asymptomatically. The infectious cycle starts with the feeding of young larvae with SBV-contaminated royal jelly produced by the infected hypopharyngeal glands of adult nurse bees. Two-day-old larvae are the most susceptible to the infection. The infected larvae carry on their development until the cell is capped, but they fail to pupate and remain stretched on their backs with their heads pointing toward the top of the cell (Figure 2A, 2B). The larvae fail to shed their skin during the final molt, causing the ecdysial fluid to accumulate beneath the unshed skin until the larva takes on the appearance of a little sac with a pale yellow coloration (Figure 2A). This fluid is full of SBV particles that start to lose their infectivity after a few days (Hitchcock, 1966). Dead larvae not removed from their cell by bees dry out, forming a dark brown gondola-shaped scale that, unlike the similar American foulbrood scales, will detach from the cell wall. Young adult bees are susceptible to oral SBV infection for just a few days after emergence. SBV-infected adults forage earlier in life, with a preference for nectar collecting, cease eating pollen, and have a reduced lifespan (Bailey and Fernando, 1972). Diseased brood is largely confined to spring and is usually a transient phenomenon that does not affect the health of the colony. A lack of food resources at certain times of the year (mostly spring) is suspected to induce the disease.

Slow Bee Paralysis Virus

Slow bee paralysis virus (SBPV) was discovered fortuitously in England in 1974 during propagation experiments with bee virus-X. The genomes of two coexisting SBPV strains have been sequenced recently (de Miranda et al., 2010b). In contrast to ABPV, which produces symptoms a few days after virus injection into adult bee abdomens, SBPV induces paralysis of just the anterior two pairs of legs, about 12 days after injection (Bailey and Woods, 1974), hence the name "slow bee paralysis virus." The virus accumulates in different bee tissues such as the head; hypopharyngeal, mandibular and salivary glands; fat body; crop; and forelegs, but less in the hindlegs, midgut, rectum, or thorax (Denholm, 1999). SBPV persists naturally as a persistent-covert infection but can be readily transmitted among adult bees and to pupae by *Varroa* (Santillán-Galicia et al., 2010), with lethal consequences at individual bee and colony levels (Carreck et al., 2010). The natural prevalence of this virus is very low in Europe and, to date, only in Britain has SBPV ever been associated with colony mortality (Carreck et al., 2010). A recent survey detected SBPV in only 4 out of 120 apiaries in the United Kingdom, and in a few dying, heavily mite-infested colonies in Switzerland. However, it was not detected in 36 apparently healthy apiaries in France, nor in highly mite-infested colonies in Sweden (de Miranda et al., 2010b).

Chronic Bee Paralysis Virus

Although the symptoms of "paralysis" were probably recognized more than 2,000 years ago by Aristotle as he described hairless black bees he called "thieves," the causative agent was not confirmed until 1963 when Bailey and colleagues isolated and characterized chronic bee paralysis virus CBPV (Bailey et al., 1963). Some 20 years earlier, Burnside (1945) in the United States had succeeded in reproducing the disease in caged bees following spraying, feeding, or injection with bacteria-free extracts of paralysed bees, and concluded that the responsible agent was a virus (Bailey and Ball, 1991; Ribière et al., 2008).

Genome Organization and Function

The morphology of the chronic bee paralysis virus (CBPV) particles and the multipartite organization of the RNA genome are exceptional, as most honey bee viruses are picorna-like viruses belonging to the *Iflavirus* and *Cripavirus* genera with symmetric particles and monopartite positive, single-stranded RNA genomes. CBPV is currently classified as an RNA virus but is not included in any family or genus. Recently, the genome of the virus was isolated from heads of experimentally infected bees. The analysis of the CBPV sequences obtained showed that RNA 1, with 3,674 bases, and RNA 2, with 2,305 bases encoded three and four putative overlapping open reading frames (ORFs), respectively (Figure 6C). Among the different putative ORFs predicted for this genome, ORF 3 on RNA 1 shows significant similarity with viral RNA-dependent RNA polymerases (RdRp), and especially with the conserved sequence domains of the RdRp of single-stranded RNA viruses. No significant similarities were found between the deduced amino acid sequences from the other ORFs and the known database sequences (Olivier et al., 2008a; Ribière et al., 2010).

Chronic Bee Paralysis Virus

Paralysis is the only common viral disease of adult bees whose symptoms include both visible behavior modifications and physiological modifications. Paralysis symptoms including trembling and clusters of flightless bees crawling at the hive entrance have long been described. Two distinct sets of symptoms (Type-1 and Type-2 syndromes) have traditionally been described in the available literature (Bailey and Ball, 1991).

The Type-1 syndrome (Figure 3) was described as seemingly the most common in Britain, including an abnormal trembling motion of the wings and bodies of affected bees (paralysis). These bees are unable to fly and often crawl on the ground and up plant stems, sometimes in masses of thousands of individuals. Frequently they huddle together on top of the cluster in the hive. They often have bloated abdomens and partially spread, dislocated wings. The bloated abdomen is caused by distension of the honey sac with fluid that accelerates the onset of so-called "dysentery." Sick individuals die within a few days following the onset of symptoms. Severely affected colonies suddenly collapse, particularly at the height of the summer, typically leaving the queen with a few workers on neglected combs. All these symptoms are identical to those attributed to the "Isle of Wight disease" in Britain at the beginning of the 20th century.

The Type-2 syndrome (Figure 2D) has been given a variety of names: "black robbers" and "little blacks" in Britain; *Schwarzsucht*, *mal noir*, *maladie noire*, or *mal nero* in continental Europe, and "hairless black syndrome" in the United States. At first, the affected bees can fly but they become almost hairless, appearing dark or almost black, which makes them seem smaller than healthy bees, with a relatively broader abdomen (Figure 2D). They are shiny, appearing greasy in bright light, and they suffer nibbling attacks by healthy bees of their colony, which makes them seem like robber bees. In a few days they become flightless, trembling, and soon die.

CBPV is the etiological agent of both syndromes, which can occur simultaneously in the same colony, producing a general syndrome of clusters of trembling, flightless, crawling bees with some individual black, hairless bees standing at the hive entrance, sometimes to be expelled by bees of their colony. One feature of CBPV is that often only some hives in an apiary are affected, even though they are all exposed to the same environmental conditions and forage in the same fields. Moreover, it is often the stronger colonies that are most affected, and masses of thousands of dead individuals can be observed in front of these hives (Figure 3; Ribière et al., 2010).

Trembling and crawling bees can also be caused by a number of other pathogens, by pesticide poisoning or chemical intoxication. Crawling bees are sometimes also attributed to the parasites *Nosema apis*, *Malpighamoeba mellificae*, or *Acarapis woodi*, or associated viruses such as BVX, BVY, or *Am*FV. Clinical diagnosis by symptoms alone is therefore often a source of confusion and error, leading to possible mismanagement of colony diseases. This is illustrated by the lingering misconceptions concerning the Isle of Wight disease. The tracheal mite *A. woodi* was quickly, but mistakenly, accepted as the cause of the widespread collapse of colonies with trembling and crawling bees, even though the evidence for this was largely anecdotal, entirely circumstantial, and frequently conflicting (Bailey and Ball, 1991, 1997; Ball and Bailey, 1997).

CBPV is the principal honey bee virus causing paralytic symptoms under natural conditions. For ABPV, KBV, IAPV, and SBPV, the induction of trembling symptoms usually requires experimental inoculation with large doses of virus. However, severe natural bee and colony mortality preceded by a rapidly progressing paralysis of bees has been associated with all members of the ABPV complex. Furthermore, the terminal phase of the Type-1 (trembling-crawling) syndrome, a sudden collapse of the affected colonies, leaving the queen with a few workers on neglected combs (Bailey and Ball, 1991), is reminiscent of Colony Collapse Disorder (vanEngelsdorp et al., 2009), which has also been associated with the members of the ABPV complex in different parts of the world (Cox-Foster et al., 2007; Forgach et al., 2008; de Miranda et al., 2010a) as well as with other pathogens and parasites (Higes et al., 2008; Bromenshenk et al., 2010) although no causal explanation for CCD has ever been presented. Given the diversity of possible causes, paralytic symptoms alone are not sufficient for an accurate diagnosis of the underlying cause, which requires the identification and quantification of the viruses and possible chemical intoxicants present (Ribière et al., 2010).

Diagnosis of CBPV-caused paralysis is based on relating the clinical symptoms at the colony level to laboratory quantification of CBPV titers. In CBPV-symptomatic colonies, CBPV titers were significantly higher in dead bees by the hive entrance, symptomatic bees, and guard bees ($\sim 10^{12}$ copies per bee) than in asymptomatic bees, drones, foragers, and nurse bees (10^{4}–10^{8} copies per bee; Blanchard et al., 2007; Ribière et al., 2010). Within an infected colony, CBPV is found in all age groups of adult bees. Although paralysis is a disease of adult bees, CBPV can be detected in all bee developmental stages, including eggs, larvae, pupae, and adults.

CBPV multiplies and accumulates in the bee central nervous system as well as in other tissues, especially in the neuronal cells of the higher-order integration centers of the brain and the regions involved in sensory information processing (Olivier et al., 2008b; Ribière et al., 2010). This intimate interaction of CBPV with the bee nervous system, also found for many of the other bee viruses, is the most likely explanation for the paralytic symptoms such as ataxia, trembling, crawling, and the inability to fly.

The covert persistence of CBPV in colonies is most likely through a combination of oral exchange, contact, and possibly transovarial transmission. Epidemic transmission is primarily through contact, when healthy bees are crowded with overtly infected individuals and when contaminated feces from sick bees can infect the inside of the hive. The risk of disease from CBPV is therefore increased by forced confinement during the active season, through sudden failures of nectar flows, inclement weather, or when too many colonies are kept for the available nectar (Bailey and Ball, 1991). The effects of these nonseasonal factors may explain the irregular incidence of the disease. This risk profile for CBPV epidemics also explains the higher prevalence of the disease in areas such as the Black Forest in Germany, where colony density is very high both in relation to the available nectar and in comparison to other areas of West Germany or Britain (Bailey and Ball, 1991; Ribière et al., 2008, 2010).

With its worldwide distribution, CBPV and its symptoms have been detected on all continents. Overt disease can occur at any time of the year, but the highest mortality rates usually occur in spring and summer. During the 2007, 2008, and 2009 apicultural seasons, numerous cases of severe adult bee mortalities were reported across France, many displaying paralytic symptoms that were significantly correlated to high CBPV titers and were not correlated to either the members of the ABPV complex or to *Nosema* spp.

CBPV may also persist outside the closed environment of honey bee colonies, as it was detected in two species of ants, *Camponotus vagus* and *Formica rufa*, that carried dead bees from hives with moderate to severe symptoms of paralysis (Celle et al., 2008). Although CBPV transmission between ants and bees still must be proven, the risk of ants acting as a major reservoir for honey bee virus infections is low. Ants may well acquire viruses from honey bees, through scavenging on dead bees and hive products, but are much less likely to transmit to bees, which do not normally feed on ants or their products. In fact, the ants may help bee colonies by removing infectious material from the hives, tasks that would otherwise be done by honey bees.

Chronic Bee Paralysis Virus Associate

One peculiarity of CBPV is that it appears to have an associated particle. Large amounts of a 17-nm isometric particle, serologically unrelated to CBPV but that multiplied only in its presence, was consistently observed during early CBPV studies (Bailey et al., 1980). This particle was initially named chronic bee paralysis virus associate (CBPVA; Overton et al., 1982) but has recently been renamed Chronic Bee Paralysis Satellite Virus (CBPSV) and is the first known satellite virus in insects. The CBPSV genome was reported to consist of three single-stranded RNA fragments of about 1,100 nucleotides (nt) each (Bailey and Ball, 1991). However, during the recent CBPV studies in France neither CBPSV particles nor RNA fragments were detected. The nature of the relationship of CBPSV to CBPV therefore remains unclear. The other uncertainty is whether CBPSV, which should at least encode for a capsid protein unrelated to that of CBPV, is a true satellite virus, that is, dependent on CBPV for replication only, or an abortive virus particle (Ribière et al., 2010).

Miscellaneous Viruses

This section encompasses a heterogenous group of viruses that still must be fully characterized at the molecular level and that generally are less damaging to bees and colonies than the viruses discussed above.

Cloudy Wing Virus

Cloudy wing virus (CWV) is a common virus that is often associated with a marked loss of transparency of the wings in adult bees, although these symptoms are not reliable for diagnosis. CWV is a small virus, both the virions (17-nm round particles, with a single 19kD capsid protein species) and the RNA genome, which is around 1,500 nucleotides. The virus is concentrated in the thorax and head, and crystalline arrays of virions can be found in the thoracic muscle fibers, although there are no obvious histopathological changes associated with such arrays (Ribière et al., 2008). Heavily infected individuals have a reduced life span but overall the virus is not particularly damaging, either to bees or colonies. The primary mode of transmission is still unclear. The infection of the thoracic muscles, close to the primary respiratory trachea, suggests airborne transmission but direct evidence for this is ambivalent. Detection of CWV in the immature stages suggests perhaps an oral or vertical infection route but again, direct evidence is unclear. The virus cannot be propagated through injection, which may explain why there is no direct evidence for Varroa-mediated transmission, despite the occasional association between CWV prevalence and *Varroa destructor* infestation (Carreck et al., 2010). About 15% of colonies in a number of countries are naturally infected but there is little evidence of seasonal variation in incidence, or a clear association with colony mortality (Bailey and Ball, 1991; Nordström et al., 1999; Carreck et al., 2010).

The Bee Virus X-Y Complex

Bee virus X (BVX) and Bee virus Y (BVY) are two closely related viruses of adult bees with similar physical, genetic, and pathological properties, and can probably best be regarded as members of a single species complex. Both form round virions of about 35-nm diameter with a single 50–52 kilodalton (kD) capsid protein species and have an RNA genome (Bailey and Ball, 1991; Ribière et al., 2008). The main point of interest of these viruses is that they are closely associated with two intestinal parasites, *Malpighamoeba mellificae* (BVX) and *Nosema apis* (BVY), whose infection is affected by the corresponding virus and vice versa. Their principal transmission route is fecal-oral, and it is thought that the damage to the adult gut by the parasite infections facilitates the viral infections (Bailey and Ball, 1991; Ribière et al., 2008).

Bee Virus X

Bee virus X (BVX) was first isolated from adult bees from Arkansas, United States. It is associated naturally with the protozoan parasite *Malpighamoeba mellificae*, with mixed infections reducing the adult life span more than single infections (Ribière et al., 2008). Both virus and parasite are independently infectious, so the association may be more coincidental, driven by a common transmission route (fecal-oral) rather than causally codependent. BVX is only infectious at 30°C, not 35°C, and infection proceeds slowly (Bailey and Ball, 1991). It is therefore typically a disease of winter, when bees survive for longer periods and at lower temperatures. Paradoxically, BVX infection may mitigate the spread of *M. mellificae*, as the accelerated death of winter bees with mixed infections curtails the shedding of *M. mellificae* cysts in the feces inside the colony. Fecal material is the primary source of infection, especially when the colony expands and cleans the fouled combs in the spring. Ultimately, however, the increased loss of winter bees to mixed infections is more damaging for colony survival than the reduced *M. mellificae* infection in the spring (Bailey and Ball, 1991). The prevalence of BVX may be on the decline, as most current winter losses are due to *Varroa destructor*, which has no natural or experimental relationship with BVX. The accelerated colony collapse may therefore reduce the accumulation of the feces required for BVX transmission and persistence in the spring (Bailey and Ball, 1991; Ribière et al., 2008).

Bee Virus Y

Bee virus Y (BVY) was first detected in Britain and, like BVX, is only infectious when fed to adult bees at 30°C and not at 35°C. However, although single BVY infections can be established in young bees, BVY is more dependent on co-infection with its associated parasite, *Nosema apis*, than BVX is on *Malpiybamoeba mellificae* infection. Consequently, BVY has a similar annual incidence cycle as *N. apis*, with a peak in (late) spring and may be a significant co-factor for Nosema virulence (Ribière et al., 2008).

Arkansas and Berkeley Bee Viruses

Arkansas bee virus (ABV) and Berkeley Bee Virus (BBPV) are two unrelated viruses isolated in the Uinted States that (in limited studies) are found together naturally and during propagation (Bailey and Ball, 1991), although this may be partly because the diagnostic antisera most likely contained antibodies for both viruses.

Arkansas Bee Virus

ABV has a 30-nm icosahedral particle consisting of a single 43-kD capsid protein and an RNA genome of ~6,000 nucleotides (Lommel et al., 1985), which makes it different from all other bee viruses. It was discovered by injecting pollen extracts into healthy bee pupae but was subsequently detected in dead bees from dwindling colonies in California, together with BBPV. It has only been detected in the United States.

Berkeley Bee Virus

BBPV has a ~9,000 nucleotide RNA genome packaged into a 30-nm icosahedral particle containing three distinct capsid proteins of 37, 35, and 32.5 kD (Lommel et al., 1985), all features that are highly characteristic of the insect picorna-like viruses (Ifla-Dicistroviruses). Nothing is known of its natural distribution or effects. It has only been found in association with ABV, and only in the United States (Bailey and Ball, 1991).

Macula-Like Virus

Nucleotide sequences of a macula-like virus were recovered incidentally during the molecular characterization of deformed wing virus in the United States (Lanzi et al., 2006). Re-analysis of historical survey samples from France showed this virus to be very common, with particularly high titers in Varroa samples and a clear seasonal distribution peak in the fall, in both adult bees and pupae. Its closest relative, *Bombyx mori* macula-like latent virus (*Bm*MLV), has an RNA genome of about 6,500 nucleotides (Figure 6D) and produces 30-nm icosahedral virions with a single capsid protein species of about 24 kD (Katsuma et al., 2005). Of the known honey bee viruses not yet characterized at the molecular level, only CWV and ArkBV are possible candidates for this virus. However, neither possesses the complete set of biological, virion, or genome attributes of this Macula-like Virus, which may therefore well be a completely novel virus of bees. Its principal association with Varroa, both in prevalence and titer, suggest that it may be primarily a virus of *Varroa destructor* and secondarily of honey bees. It should be noted that all the evidence so far for this virus is based only on nucleotide data; true virological studies still have to be completed.

Filamentous Virus

Apis mellifera filamentous virus (*Am*FV) was first described in 1961 in Switzerland and was initially thought to be a rickettsial disease (Bailey and Ball, 1991; Ribière et al., 2008). It derives its name from the long (3,150 × 40 nm) filamentous nucleoprotein that folds in three superimposed figure-8 loops into a 450 × 170 nm, rod-shaped, enveloped virion (Clark 1978; Sitaropoulou et al., 1989). The nucleoprotein contains a central core of dsDNA wrapped by two major nucleoproteins while the tri-laminate virion envelope contains lipids, two major proteins, and several minor proteins (Bailey et al., 1981). The virus multiplies in the fat body and ovarian tissues of adult bees, and a highly diagnostic feature is that the hemolymph of severely infected individuals becomes milky-white from the large quantities of enveloped virions. Only adult bees appear to be susceptible, and the virus has a sharp annual peak incidence in the spring (Bailey and Ball, 1991) that may coincide with the replacement of infected overwintering bees by newly produced worker bees (Clark 1978). Oral infection of *Am*FV may be dependent on co-infection with *Nosema apis* (Bailey et al., 1983b) but infection does not cause outward symptoms or affect the lifespan significantly (Ribière et al., 2008). Even so, dying bees and moribund colonies have been associated with *Am*FV and milky hemolymph on a number of occasions. The nucleoprotein/virion and biological properties of *Am*FV are strongly reminiscent of the baculo-like viruses that infect the ovarian tissues of many endoparasitic wasp species and are often active contributors to their parasitic habit, while the milky-white hemolymph is highly characteristic of Ascovirus infection in the Lepidopteran hosts of the wasps (Federici et al., 2009). These viruses typically have genome sizes of 120 to 150 kilo base pair (kbp). A partial sequence of the *Am*FV genome obtained recently (Hartmann et al., unpublished) suggests that *Am*FV is related to both the baculo- and ascoviruses.

Apis Iridescent Virus

Apis iridescent virus (*A*IV) is an Iridovirus, a group of viruses with large (~140 nm) icosahedral virions consisting of a core nucleoprotein containing a ~200-kbp dsDNA genome, an inner membrane and an outer protein shell. They are found in a wide range of insects and vertebrates. *A*IV is associated in India with clustering disease of *Apis cerana*, where small clusters of unusually inactive bees become detached from the colony, particularly during summer. This clustering ceases with increased foraging (Bailey and Ball, 1991). *AIV* can also infect *A. mellifera*, replicating in most tissues, but natural occurrences are only known from *A. cerana* in the Indian subcontinent (Bailey and Ball, 1991). *AIV* cannot replicate in wax moth larvae (*Galleria mellonella*), a common pest of bee colonies, although most other insect Iridoviruses readily infect wax moth larvae. A partial sequence of AIV has been obtained (Webby, 1998). An Iridovirus has also been identified in Varroa mites from *Apis mellifera* colonies in the United States (Camazine and Liu, 1998). Iridescent viruses are the latest pathogenic agents to have been added to the list of factors associated with Colony Collapse Disorder (Bromenshenk et al., 2010) although like all previous investigations of CCD the evidence is entirely correlational and therefore not proof of the cause of CCD.

Summary

In this section we briefly discuss a significant aspect of honey bee viruses, which is their frequent association and interaction with other parasites and diseases, what options are available to minimize the effects of virus infections through management practices and treatments, and what we can expect to develop from future research in improved understanding of the importance of viruses to honey bees as well as in practical developments.

Interactions with Other Pathogens and Parasites

Most pathogens invade the digestive system of bees and replicate in the midgut epithelial cells. However, the constant renewal of these cells provides a way to confine pathogen infection to the gut tissues, protecting the other organs. Physical barriers such as the peritrophic membrane in the midgut and the basal lamina to which the epithelial cells are attached act as filters and impair microorganism passage to the hemolymph (the bee blood) and subsequent invasion of other bee tissues. Consequently, parasites such as *Nosema* sp. or *Malpighamoeba mellifica* that complete their biological cycle in the gut can create lesions in the epithelium, providing a way for the virus to pass into the hemolymph. Such an association between *Nosema* sp. infections and viruses has been shown experimentally by Bailey for black queen cell virus, filamentous virus, and bee virus Y (Bailey et al., 1983b).

In contrast to parasites that multiply in the intestine, the ectoparasitic mite *Varroa destructor* feeds directly on the bee hemolymph after piercing the bee cuticle. This provides an unexpected opportunity for many viruses to have a direct access to the hemolymph and from there to the different bee tissues. Indeed, most viral infections are nearly impossible to initiate orally but usually just a few infectious particles are enough to infect a bee when injected directly into the hemolymph (Bailey and Ball, 1991). Several bee viruses can be transmitted by mites, especially those in the deformed wing virus complex (DWV, KV and VDV-1; Bowen-Walker et al., 1999; Yue and Genersch 2005; Gisder et al., 2009; de Miranda and Genersch, 2010), the acute bee paralysis virus complex (ABPV, IAPV and KBV; Ball 1989; de Miranda et al., 2010a), and slow bee paralysis virus (Santillán-Galicia et al., 2009; Santillan-Galicia et al., 2010). Other viruses that have been detected in mites are sacbrood virus, black queen cell virus (Chen et al., 2006), the macula-like virus (de Miranda et al., unpublished), and filamentous virus (Hartmann et al., unpublished). The differences of viral titers recorded in mite populations suggest that different types of interaction exist between honey bee viruses and Varroa.

For instance, certain viruses may be able to replicate in mite tissues prior to being transmitted to bees. To date, such a replication in mite tissues has only been shown for DWV, for both *Varroa* (Yue and Genersch 2005; Gisder et al., 2009) and Tropilaelaps (Dainat et al., 2008), and very high titers can be frequently detected in such mites (Tentcheva et al., 2006; Gauthier et al., 2007). However, the replication cycle of honey bee viruses when transmitted by Varroa is not clear. This cycle starts when the mite feeds on infected bee hemolymph. The virus possibly infects the mite epithelial gut, gaining entry to the mite hemolymph. From there the virus may multiply in other tissues or migrate directly to the mite salivary glands from which it is released into the bee hemolymph during its next feeding action. We have currently no clear evidence for any component of this transmission pathway.

Alternatively, viruses could also merely stick to the Varroa mouth parts for a passive transmission, or part of the mite's gut contents could be regurgitated into the bee hemolymph, effecting transmission. This way of transmission requires a great stability of viral particles in a different environment that have to remain infectious during their passage from bee to bee. Finally, these viruses could be present naturally in or on bees and be activated by the piercing action alone, as has been demonstrated by the induction of virus infections following *in vitro* injection of sterile solutions into healthy bees (Anderson and Gibbs, 1988; Bailey and Ball, 1991). Although the mechanism of viral reactivation and induction in honey bees is still not elucidated, it contributes significantly to the increase in viral titers in mite-infested colonies.

Management Strategies

The management of viral diseases falls into two categories: minimizing the risk of transmission and reducing virus titers within colonies. There is some overlap between these categories, as the probability of transmission is positively correlated with both virus titer and the proportion of infected individuals in a colony.

Transmission Risk Management

Without a doubt, modern apiculture with movable-frame hives and a global trade in bees and bee products is one of the most efficient vectors of bee diseases. Because these are entirely manufactured risks, through apicultural practices, they are also the ones that can be most easily modified to minimize transmission risk. The most practical operating unit for transmission risk management is an apiary, generally consisting of 5 to 20 colonies placed in a single location. The most effective way for managing transmission risk is to separate infected from noninfected material. This applies primarily to the bees themselves (package bees, queens, sperm, colonies, and apiaries) and their hives, especially the frames, as well as supplementary feed (pollen, royal jelly) that often harbors dormant forms of many bee diseases. The tools used for managing bee colonies can also be a transmission risk, mainly the hive tool and the extraction equipment, as these are usually shared between apiaries. Keeping things clean between apiaries minimizes this transmission risk.

1. **Apiary management.** Managing each apiary as a separate, self-contained unit for working, treatment, extraction, feeding, and storage of hive material minimizes the risk of disease transfer between apiaries due to management. This includes stocking the apiaries with bees/queens bought from different sources. Accurate record-keeping is essential for identifying problems and minimizing the risk of such problems spreading beyond apiary boundaries. Placing apiaries at least 1 kilometer apart provides a natural quarantine. This is outside most of the practical flight range, especially if the apiaries are located within abundant season-long foraging. The placement and stocking density of apiaries is therefore a top-level disease management tool. Partitioning the local foraging between multiple smaller apiaries is, from a disease-management perspective, preferable to fewer larger apiaries.

2. **Colony management.** The first step in identifying possible disease is through routine inspection of colonies. Thorough knowledge of the symptoms of the different diseases is essential to identifying potential problems. Colonies that develop slower than expected in spring or fail to accumulate honey in summer are suspect, as are frames with spotty or peppered brood, which is caused by the removal of diseased brood. Initial identification can often be made through symptoms and confirmation by sending a sample of brood or adult bees to a diagnostic laboratory. Field-ready ID kits are currently being developed for a range of bee diseases that will expedite identification. Diseased colonies should be removed from the apiary to prevent spread. Within-apiary disease spread can be minimized by placing colonies well apart and using unique entrance color schemes (to prevent drifting) and entrance reducers during nectar dearth to prevent robbing, which especially affects weak (diseased) colonies.

3. **Queen rearing.** A special transmission risk is venereal transmission from drones to virgin queens. Due to the central reproductive status of queens in a colony, venereal transmission is a highly efficient means for a virus to infect an entire colony and is therefore a common transmission route for bee viruses. To minimize this risk, either the semen for artificial insemination or the drone donor colonies for isolated mating stations should be checked for diseases.

4. **National-regional management.** The worldwide dispersal and trade in honey bees and bee products means that virtually every bee (virus) disease can be found in every country. However, there are often great national differences in disease prevalence and damage, even between neighboring countries, which are generally related to different national bee management practices. Regional preference for certain bee races with different susceptibilities also affects prevalence and damage. For these reasons, vigilance and certification across regional and national boundaries are still an essential disease-transmission risk management tool.

Virus Load Management

Viruses are the ultimate opportunistic pathogens. They persist in normal, healthy colonies through low-level, persistent infections and flourish into a more overt, epidemic mode during times of stress. Consequently, the first measure to reduce virus pressure within colonies is to keep them healthy and stress-free. This mostly means an abundance of season-long foraging (apiary location and stocking density), minimal disturbance (movement, inspection), adequate ventilation (colony placement), and room for expansion. In other words, Good Apicultural Practice. However, there are several additional practices that can be employed to help reduce the virus loads in colonies:

1. **Vectors, parasites and alternative hosts.** Several bee viruses are actively transmitted by *Varroa* and *Tropilaelaps* mites or are closely associated with other diseases, such as Nosemosis. Active management to reduce these other pests will therefore also reduce the virus pressure.

   Although a number of pests and predators of hive products (beetles, moths, ants, wasps, hornets, bumble bees) are either confirmed or potential alternative hosts for honey bee viruses, it is unlikely that they represent a major infection risk for honey bees, as transmission is unidirectional (predators feed on bee products but not the other way around) and away from the hive. Weak and diseased colonies are much more likely sources of external infection, as these are actively robbed and attacked by other bee colonies during dearth of nectar flow. By similar logic, viruses infecting the aphids visited by honey bees for their honeydew may present a new pathogen risk for honey bees.

2. **Frame rotation.** Virus traces can be detected in wax, wood, honey, and bee bread (stored pollen), with the latter in particular frequently containing large amounts of virus. The extent to which such detections represent infectious material is still very much in debate. However, also for the management of other bee diseases and the accumulation of pesticides in wax, it is expedient to replace frames every 5 years or so. When doing so, it is better to rotate out an entire box rather than individual frames, to avoid mixing new, clean frames with potentially contaminated frames.

3. **Treatment.** Conventional wisdom used to state that there was no true cure for virus diseases, like there could be for bacteria and parasites, but that is no longer true. The developments in molecular biology in particular have yielded an ever-growing range of antiviral products, particularly in the medical and veterinary fields where the cost of treatment can be justified. However, honey bee virology has belatedly caught up with the rest of the field and there are now clinical trials of antiviral drugs for honey bee (virus) diseases, based on RNAi technology (Maori et al., 2009; Hunter et al., 2010; Liu et al., 2010). How effective and practical such treatments will be remains to be seen.

4. **Breeding.** Honey bee breeding is becoming increasingly sophisticated, using highly developed pedigree analyses and molecular mapping techniques to identify and select for desirable traits, including disease resistance and tolerance. Genes associated with hygienic behavior, the suppression of Varroa reproduction, and for Varroa-sensitive hygiene have been identified and incorporated into bee breeding programs. Natural genetic resistance to Varroa and the viruses it transmits is also being investigated, both through natural selection studies and by the conservation and analysis of local bee species and races. Although viruses are not yet specifically included in such breeding considerations, plans to do so are currently in development.

Future Research

With the new molecular tools capable of analyzing not only a single microorganism, but also the larger community of microorganisms living in bees, there is no doubt that more honey bee viruses will be discovered. The diversity of viruses found in bees is not surprising however, especially in the context of the enormous virus diversity discovered by metagenomic analyses of entire ecosystems, such as the Sargasso Sea project (Venter et al., 2004) and the recent investigation of Colony Collapse Disorder (Cox-Foster et al., 2007). Such large-scale sequence analyses show that organisms and ecosystems contain a much larger diversity of viruses than previously realized, partly because such comprehensive analyses also recover nonpathogenic and latent viruses, whereas most viruses described to date are known largely because they cause pathologies in their hosts.

It is now widely accepted that viruses played a central role in evolution by providing a route for horizontal exchange of genetic information between unrelated organisms. The discovery of apparently nonpathogenic viruses has precipitated a reappraisal of the possible roles and functions of viruses in higher organisms. Some virus infections can even lead to beneficial relationships for the host, such as those displayed by viral endosymbionts that help the eggs and larvae of certain parasitic wasps avoid the host immune system. Most of the viruses infecting honey bees are also largely harmless to colonies and are best considered as opportunistic pathogens that only develop pathogenic habits when stressful environmental conditions are present.

One major challenge for honey bee virus research is to identify such environmental stresses, as has already been shown for the effect of *Varroa destructor* infestation on the pathogenic epidemiology of certain viruses. A second line of research is to determine the predictive contribution of virus infections and titers for colony losses, in combination with other factors such as colony size, parasites, and food stores. We already know that high viral loads of DWV or ABPV in worker bees in the fall are strongly predictive for winter losses. However, the molecular and physiological mechanisms involved in viral pathology in such bees are still completely unknown and this constitutes another clear direction for future research, one that would furthermore make excellent use of the recently elucidated genome sequences of the honey bee and of several of its pathogens and parasites and of the RNAi technology being developed commercially for antiviral treatment. Finally, the need for accurate and practical diagnostic tools for the field as well as for experimental use has long been recognized and these are currently being developed in partnership with industry.

PCR for the Analysis of *Nosema* in Honey Bees

CHAPTER

9

Brenna E. Traver and Richard D. Fell

Abstract This chapter is a technique paper outlining steps for the successful detection of Nosemia intracellular parasites using microscopic and molecular techniques. Species that infect honey bees are *Nosema apis* (only infects *Apis mellifera)* and *Nosema ceranae* (infects both *A. mellifera* and *A. cerana*). This is a widespread parasite that has been recognized and has caused problems in the beekeeping industry for over 100 years. Control is through use of the antibiotic fumagillin. This parasite kills bees by causing an infection to the midgut. The chapter offers a standardized method of detection because outward symptoms of nosemosis are not always present.

Introduction

Nosema is a genus of Microsporidia that infects insects. Microsporidia are eukaryotic obligate intracellular parasites (Weiss and Vossbrink, 1999) now classified as fungi (Edlind et al., 1996; Keeling et al., 1996; Germot et al., 1997; Hirt et al., 1997, 1999). They are spore-forming organisms characterized by a hollow polar filament within the spore (Fries, 1993). Because many microsporidia are human pathogens, research on their molecular biology, especially the nucleotide sequence for the ribosomal ribonucleic acid (rRNA) genes, has allowed the development of molecular diagnostic detection assays for species identification and also facilitated phylogenic analysis of this group (Weiss et al., 1999). Approximately half of the described genera of microsporidia have an insect host (Becnel et al., 1999).

Currently, there are two major species of concern that infect honey bees: *Nosema apis* and *Nosema ceranae. N. apis* only infects *Apis mellifera,* while *N. ceranae* infects both *A. mellifera* and *Apis cerana. N. apis* was first described in 1909 (Zander, 1909) and has caused problems in the beekeeping industry but has been controlled through the use of the antibiotic fumagillin (Katznelson et al., 1952; Gochnauer, 1953; Farrar, 1954; Moffett et al., 1969). *N. apis* was very widespread, infecting honey bees on all continents where beekeeping is practiced (Matheson, 1993). Both species cause nosema disease or nosemosis through the ingestion of spores that infect the midgut epithelial cells. Infections are initiated when bees ingest spores from infected food, during trophallaxis, through cleaning of the comb (Bailey, 1981) and/or through grooming (Bailey, 1972). The infection cycle of *N. apis* has been extensively studied (reviewed in (Fries, 1993). In *N. apis* infections, the midgut becomes white, distended, and fragile unlike the normal translucent brown color (Fries, 1993). Outward symptoms of *N. apis* can occur but are not always present. In serious cases, brown fecal streaking from dysentery can be observed on the comb and on the outside of the hive. The feces contain large numbers of spores that may facilitate disease transmission (Bailey, 1981). There may also be crawlers around the entrance or outside the hive, and heavy infections can lead to high bee mortality in the winter (Fries, 1993).

Nosema ceranae is a more recently described pathogen of honey bees and was first described in 1996 in *A. cerana* (Fries et al., 1996). It was shown to naturally infect *A. mellifera* in Spain in 2006 (Higes et al., 2010) and in Taiwan the following year (Huang et al., 2007). Since the initial discovery, *N. ceranae* has been found ubiquitously anywhere beekeeping is practiced (Chen et al., 2010; Fries, 2010; Higes et al., 2010). Unlike *N. apis*, there are no known outward symptoms (Fries, 1993; Fries et al., 1996; Higes et al., 2010).

Detection

Microscopy

Classical detection methods for *N. apis* rely on microscopy. Abdomens of bees are crushed and examined microscopically at 400× magnification using bright field or phase-contrast settings (Shimanuki et al., 2000). The extent of the infection level can be determined using a hemocytometer to estimate the number of spores per bee. It is suggested that older worker bees be used and a minimum of ten abdomens be examined per colony (OIE, 2008).

Once *N. ceranae* was described, distinguishing between *N. apis* and *N. ceranae* became problematic because spore differentiation is only possible using electron microscopy. Both species have oval spores that differ in size, but differences are small. *Nosema apis* spores range from 5 to 7 by 3 to 4 mm in size with a dark edge (Fries, 1993) while *N. ceranae* spores have an average size of 4.7 by 2.7 mm (Fries et al., 1996). Spores can further be identified to species by counting the number of polar filament coils (Burges et al., 1974) as *N. ceranae* has 20 to 23 polar filament coils, whereas *N. apis* always contains more than 30, ranging from 30 to 44 (Liu, 1984; Fries, 1989).

Molecular

Up until 1996, there was only one known *Nosema* species that infected *A. mellifera* and there was no need for a molecular assay to discriminate between species. Microsporidian taxonomy and identification has been based on morphological characteristics (Fries et al., 1996), and previous detection and species identification depended solely on electron microscopy. The difficulty in distinguishing species based on spores necessitated the development of molecular assays, and the gene coding for the small subunit (SSU) rRNA is now used to discriminate between *N. apis* and *N. ceranae*. The most widely used molecular detection techniques rely on polymerase chain reaction (PCR). Because members of *Nosema* are DNA-based organisms, extraction of genomic DNA is all that is required as a precursor to performing PCR. Molecular detection offers both specificity and sensitivity. There is an increased need for molecular detection techniques due to the ease of misidentifying a species based solely on spores and because false positives and/or false negatives are decreased when using molecular assays.

Polymerase Chain Reaction

Polymerase chain reaction (PCR) is a molecular technique first described in 1987 (Mullis et al., 1987). It utilizes a process similar to DNA replication and can be used to amplify nucleic acids, either DNA or RNA. PCR is the most widely used molecular technique and has numerous applications. Here we address how it can be used for species identification. PCR is an enzyme-mediated amplification procedure where a specific target area of the genome of interest is amplified. In a PCR reaction there are three basic steps: denaturing, annealing, and extension elongation (Figure 1). In the first step of the reaction, the double-stranded DNA in a sample extract is separated at high temperatures, usually above 90°C. Once the strands are separated, the temperature is decreased to allow primers, short DNA sequences complementary to a segment of each single strand of DNA, to bind or anneal to the DNA flanking the target area of interest. Primers complementary to each single strand of DNA serve as the initiation point for the synthesis of a new strand (Mullis et al., 1987). In the final step of PCR, extension elongation, the temperature is increased and a heat-stable DNA polymerase synthesizes a complementary strand of DNA by incorporating deoxynucleoside triphosphates (dNTPs) to the primers. The cycle is then repeated multiple times (20 to 45 times). In each cycle amplification is exponential, resulting in a doubling of the DNA present as long as sufficient reagents are available (Figure 1).

PCR techniques used in the identification of *Nosema* spp. have involved both conventional PCR and real-time quantitative PCR. Both methods utilize thermal cycling but differ in the manner in which the amplified DNA products (amplicons) are detected and measured. In conventional PCR, reactions are carried out in small tubes in a thermal cycler utilizing reaction volumes on the order of 20 to 100 μL. The thermal cycler heats and cools the reaction mixture as required for the amplification process. Upon completion of the PCR reaction, the products are typically separated on an agarose gel by electrophoresis. The amplicons are separated by size and visualized using ethidium bromide or similar stain. The size of the products can be determined by comparison to a DNA ladder, which contains DNA fragments of known size and is run on the gel at the same time. Known standards can also be run on the gels to identify specific amplicons of interest. Real-time quantitative PCR, on the other hand, allows for the measurement of DNA products after each round of amplification. Real-time PCR instruments combine a thermal cycler with a detection system that utilizes fluorimetry to measure changes in the level of fluorescence after each cycle. Either nonspecific fluorescent dyes or specific fluorescent reporters can be used to monitor the amplification process. Real-time PCR is well suited for diagnostic applications and allows for the quantification of DNA in the original sample. It is also more sensitive than conventional PCR.

Optimization of PCR protocols can be time consuming and frustrating. The required components for a PCR reaction are template (either DNA or RNA), primers, buffer, magnesium, dNTPs, and a heat-stable polymerase. Each component concentration must be carefully considered to maximize PCR results. Fortunately, preformulated mixes are available for use in both conventional and real-time PCR. Before starting, the concentration of the template (genomic DNA) needs to be assessed using a spectrophotometer. The amount of DNA required for a PCR reaction can vary (50 to 500 ng), but too much DNA can result in nonspecific amplification and too little DNA can result in low yields.

Primer design is very important and can influence the efficiency of a reaction. During primer design, the melting temperature (Tm) and the guanidine and cytosine (G-C) and adenine and thymine (A-T) base content need to be considered. An optimal pair of primers will have a similar Tm and will have 40% to 60% G-C composition. There should not be any internal homology within or between primers that could result in primer dimer, which occurs when the primers hybridize to each other (Chamberlain et al., 1994). Primers can vary in length but generally 20 to 25 nucleotides are used (Chamberlain et al., 1994). In each reaction, the primer

Figure 1. Diagram of the basic steps of PCR. **A.** DNA starts as a double-stranded molecule. Here a single DNA molecule is represented. The box indicates the target area of the genome to be amplified. **B.** The first step of a PCR reaction is the denaturation step. DNA is heated above 90°C and the strands separate into single-stranded molecules. **C.** Primer annealing is the second step. The temperature is decreased to allow the primers to anneal to the complementary sequence. **D and E.** The final step in PCR is the extension step. The heat-stable DNA polymerase (indicated by P) synthesizes a complementary new strand of DNA from the primer. **F.** After one complete cycle of PCR, there are two copies of DNA. **G.** Demonstration of how amplification during a PCR reaction is exponential. The amount of DNA will double each cycle until the reagents become limiting.

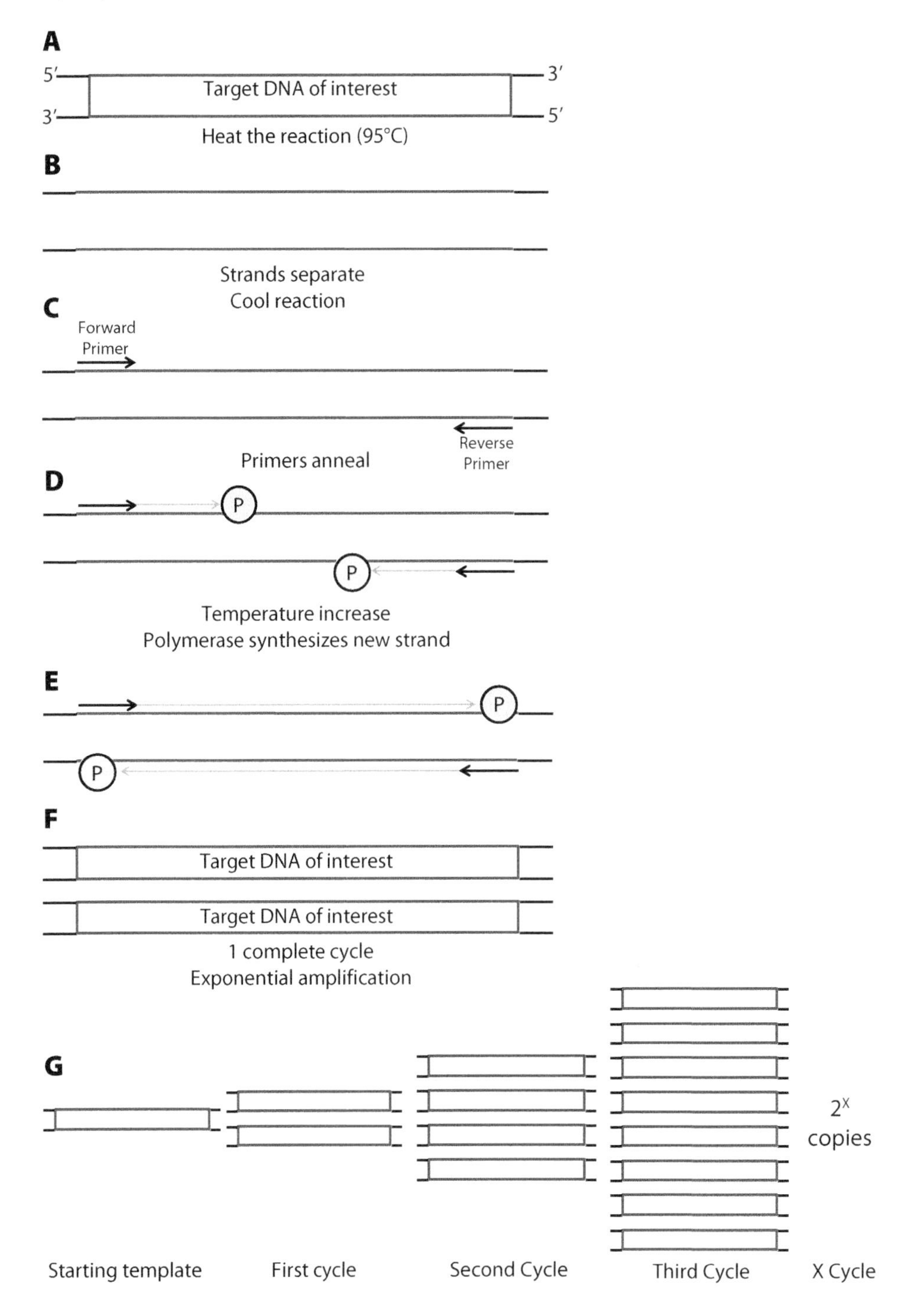

concentration should range from 0.1 to 0.5 mM with a rule of thumb being 5 pmol per primer per 25 mL PCR reaction (Viljoen et al., 2005). Higher primer concentrations can result in mispriming, which occurs when primers bind nonspecifically to the template. Primers can be nonspecific or specific for a species. For example, there are universal microsporidian primers that will amplify many different species because they are designed on a highly conserved and common genome sequence. There are also species-specific primers that will only amplify one species such as the DNA from *N. apis* or *N. ceranae*.

A buffer is required to provide an optimal environment for the polymerase; both activity and stability are influenced by the pH of the reaction. Buffers contain 10 to 50 mM Tris-HCl with a pH ranging from 8.3 to 8.8 and up to 50 mM potassium chloride, which helps with primer annealing (Viljoen et al., 2005). Buffers may also contain dimethyl sulfoxide to help lower the melting temperature of the primers (Chamberlain et al., 1994).

Magnesium is a divalent cation and a required cofactor for DNA polymerase. The concentration of Mg^{2+} can affect the specificity and yield of a reaction by influencing primer annealing, strand dissociation temperatures, template specificity, the formation of primer dimer, and the activity and fidelity of the polymerase (Viljoen et al., 2005). The concentration can range from 0.5 to 2.5 mM, with the most common concentration at 1.5 mM (Viljoen et al., 2005). High concentrations can result in a decrease in fidelity of the polymerase due to excess free magnesium leading to nonspecific amplification.

Deoxynucleoside triphosphates (dATP, dCTP, dGTP, and dTTP, all referred to as dNTPs) are the bases that are added to the reaction and required for the synthesis of a new complementary strand of DNA. The concentration of each dNTP is often supplied, or can be made, as a 10-mM solution (pH 7.0) with the final concentration ranging from 20 to 200 mM of each dNTP (Viljoen et al., 2005). High concentrations of dNTPs can result in a decrease in the specificity and fidelity of the polymerase (Viljoen et al., 2005) while lower concentrations can result in inconsistent amplification and formation of primer dimer most likely due to the excess free Mg^{2+}. Thus a balance between the Mg^{2+} and dNTP concentrations needs to be achieved (Chamberlain et al., 1994).

Finally, a heat-stable polymerase enzyme is required for the synthesis of a new, complementary strand of DNA. The polymerase must be heat-stable so that it will not become inactivated during the high temperature required to denature the two strands of DNA. Polymerases typically have three activities: 5'–3' polymerase activity, 3'-5' exonuclease proofreading activity, and 5'–3' exonuclease activity (Viljoen et al., 2005). Fidelity of a polymerase refers to the frequency of incorporating an incorrect nucleotide and how capable mismatched bases are able to be extended based on the proofreading capability (Viljoen et al., 2005). *Taq*, from *Thermus aquaticus* which is a bacteria from a hot spring, is the most common DNA polymerase used in PCR, but in its native form does not have a 3'–5' exonuclease proofreading capability (Viljoen et al., 2005). However, recombinant forms of *Taq* are commercially available with proofreading ability, in addition to other polymerases with high fidelity. The concentration of polymerase (reported as units) must also be considered. The amount used for a PCR reaction will depend upon the supplier and protocol provided for each polymerase. The polymerase is utilized as a limiting reagent to ensure a higher fidelity because high concentrations can result in nonspecific amplification.

The PCR program of the thermal cycler can also be optimized. An initial denaturation step for a few minutes, called a hold, is required to help ensure that all the DNA molecules are denatured. Cycling then occurs with a shorter denaturation step, an annealing step with a lower temperature, and an extension elongation step with a temperature optimal to the polymerase activity. Programs typically have from 25 to 40 cycles but will vary based on the target to be amplified. A final extension step may be used after the last cycle to ensure that any single strand of DNA is completely extended. Temperatures used will depend on the polymerase and primers. The annealing temperature is the most important parameter in a PCR reaction (Chamberlain et al., 1994). Typically an annealing temperature is 5 to 10°C below the lowest Tm of the primers, but for higher stringency a higher annealing temperature is preferred while using the fewest number of cycles. The length of the extension step depends on the size of the target to be amplified. A quick rule to start with is 1 minute per 1,000 bases of target DNA to be amplified.

A number of both conventional and real-time quantitative PCR reactions have been developed for the detection of *Nosema* and to differentiate between *N. apis* and *N. ceranae*. Each PCR reaction will be discussed below and end with a comparison of the different techniques. Specific reagents (i.e., magnesium, dNTPs, polymerase, buffer, reaction volume) used for each PCR assay can be found in the cited reference. All the primer sequences, and if applicable, the probe sequences, are listed in Table 1.

Conventional PCR Assays

The first PCR assay to distinguish between *N. apis* and *N. ceranae* was performed in 1996 (Fries et al., 1996) and the resulting amplicon was cloned, sequenced, and submitted to GenBank (Accession No. U26533). The 16S SSU rRNA was amplified from five different microsporidian species, four of which were from the genus *Nosema*, to create a phylogenetic tree demonstrating *N. ceranae* branches with *N. apis* and *Vairimorpha necatri* (formerly *N. necatrix*).

In 2007 the entire length of the 16S rRNA (the small subunit, internal transcribed spacer region, and the large subunit) was sequenced from a *N. ceranae* isolate (Accession No. DQ486027) obtained from Taiwan (Huang et al., 2007). By having the full sequence of the 16S rRNA (small and large subunits), additional PCR assays were developed to aid in species identification.

PCR-RFLP

Following successful amplification of a target DNA sequence, restriction fragment length polymorphism (RFLP) can be used to distinguish between different species present. RFLP, as the name suggests, is a technique that uses restriction enzymes to digest DNA, resulting in a species-specific banding pattern. The sequence of the target area of interest must be known and restriction enzyme sites found within the target region that will yield unique banding patterns for each species are selected. Following RFLP analysis, DNA banding patterns are visualized using gel electrophoresis and compared to a DNA ladder for size comparison. Sequencing of the PCR products is usually performed first but is not necessary once the assay is validated, making this method relatively inexpensive.

Type of PCR	Primer Name	Sequence (5′ to 3′)
Universal PCR	MICRO-F[1]	CACCAGGTTGATTCTGCCTGA
	MICRO-R	CCAACTGAAACCTTGTTACGACTT
PCR-RFLP	SSU-res-f1[2]	GCCTGACGTAGACGCTATTC
	SSU-res-r1	GTATTACCGCGGCTGCTGG
	Nos2990+[3]	TGGAGCAACGAGATTCCTAC
	Nos3426-	GCCTGCTACAAGCCAGTTAT
Species-specific	Nosema F[4]	GGCAGTTATGGGAAGTAACA
	Nosema R	GGTCGTCACATTTCATCTCT
	N. ceranae F	CGGATAAAAGAGTCCGTTACC
	N. ceranae R	TGAGCAGGGTTCTAGGGAT
	N. apis F	CCATTGCCGGATAAGAGAGT
	N. apis R	CACGCATTGCTGCATCATTGAC
	NOS-FOR[5]	TGCCGACGATGTGATATGAG
	NOS-REV	CACAGCATCCATTGAAAACG
Multiplex PCR	218MITOC-FOR[6]	CGGCGACGATGTGATATGAAAATATTAA
	218MITOC-REV	CCCGGTCATTCTCAAACAAAAAAACCG
	321APIS-FOR	GGGGGCATGTCTTTGACGTACTATGTA
	321APIS-REV	GGGGGGCGTTTAAAATGTGAAACAACTATG
Real-time PCR	N. apis-sense[7]	CCATTGCCGGATAAGAGAGT
	N. apis-antisense	CCACCAAAAACTCCCAAGAG
	N. ceranae-sense	CGGATAAAAGAGTCCGTTACC
	N. ceranae–antisense	TGAGCAGGGTTCTAGGGAT
	N. apis probe (VIC)	ATAGTGAGGCTCTATCACTCCGCTG
	N. ceranae probe (FAM)	CGTTACCCTTCGGGGAATCTTC
	For (N. apis)[8]	GCCCTCCATAATAAGAGTGTCCAC
	Rev (N. apis)	ATCTCTCATCCCAAGAGCATTGC
	N. apis probe (FAM)	ACTTACCATGCCAGCAGCCAGAAGA
	For (N. ceranae)	AAGAGTGAGACCTATCAGCTAGTTG
	Rev (N. ceranae)	CCGTCTCTCAGGCTCCTTCTC
	N. ceranae probe (JOE)	ACCGTTACCCGTCACAGCCTTGTT
	DQ486027 F[9]	GGTTGGGAGAAGCCGTTACC
	DQ486027 R	ACCTGATCCAACGCAAATGCTA
	N. ceranae probe (VIC)	CTTGCCAAACCCTCCC
	U97150 F	GGAACACCTTTTCTCCTACAAGCAA
	U97150 R	CCAAAAACTCCCAAGAGAAAAACAAAAC
	N. apis probe (FAM)	ACGCCAGCATACCTTT
Endogenous control	Apis-β-actin-sense[7, 10]	AGGAATGGAAGCTTGCGGTA
	Apis-β-actin-antisense	AATTTTCATGGTGGATGGTGC
	Apis-β-actin probe (FAM)	ATGCCAACACTGTCCTTTCTGGAGGTA

Table 1. Primer sequences for different PCR assays. Primers for different types of conventional PCR are listed first, followed by primer and probe sequences for real-time PCR.
References: [1](Fries et al., 1996); [2](Klee et al., 2007); [3](Tapaszti et al., 2009); [4](Chen et al., 2008); [5](Higes et al., 2006); [6](Matín-Hernández et al., 2007); [7](Chen et al., 2009a); [8](Bourgeois et al., 2010); [9](Traver and Fell, 2010, submitted); [10](Chen et al., 2005).

Two different PCR-RFLP assays have been used for the identification of *Nosema* spp. In 2007 (Klee et al., 2007), the first PCR-RFLP analysis for *Nosema* was performed. These researchers used universal *Nosema* primers to amplify three different species: *N. apis*, *N. ceranae*, and *N. bombi*. Primers were designed from an aligned 16S SSU rRNA gene consensus sequence from the three species using the following accession numbers from GenBank: for *N. apis* U26534 and U97150 (Gatehouse et al., 1998), for *N. ceranae* DQ078785 (Huang et al., 2007), and for *N. bombi* AY741110 (Tay et al., 2005). Primers SSU-res-f1 and SSU-res-r1 were used to amplify a 400 base pair (bp) amplicon (PCR program: 95°C, 4 minutes; 45 cycles of 95°C, 1 minute; 48°C, 1 minute; 72°C, 1 minute; final extension at 72°C, 4 minutes). Amplicons were resolved on a 1.4% agarose gel to ensure amplification before RFLP. Restriction enzymes *PacI*, *NdeI*, and *MspI* were all used in a triple digest. *MspI* digests amplicons from all three species at approximately base 239. *NdeI* only cleaves *N. apis* at base 151, while *PacI* cleaves *N. ceranae* only at base 119. *N. apis*-digested DNA would result in three fragments of 136 bp, 91 bp, and 175 bp; *N. bombi* with two fragments 226 bp and 177 bp; and *N. ceranae* with three fragments of 104 bp, 116 bp, and 177 bp. Digested products were resolved and visualized on a 3% agarose gel stained with ethidium bromide.

PCR-RFLP was also performed using the restriction enzyme *MslI* to differentiate between *N. ceranae* and *N. apis* (Tapaszti et al., 2009). Primers Nos2990+ and Nos3426− were used to amplify a 437 or 433 bp amplicon for either *N. ceranae* or *N. apis*, respectively (PCR program: 95°C, 10 minutes; 40 cycles of 94°C, 2 seconds; 49°C, 1 minute; 72°C, 1 minute; final extension at 72°C, 10 minutes). Amplicons were resolved on a 2% agarose gel to ensure successful amplification. If *N. ceranae* was present, *MslI* will result in two fragments—175 bp and 262 bp—and will not cleave *N. apis* because there was no restriction site present. PCR amplicons were sequenced to further confirm species identify and validate the use of this RFLP approach.

PCR with *Nosema*-Specific Primers

The previous PCR methods all used "universal" primers that are not species-specific but instead can amplify multiple microsporidian species. In 2006, a new assay was developed using primers specific to *N. apis* but also able to amplify *N. ceranae* (Higes et al., 2006). Primers NOS-FOR and NOS-REV were used to amplify a 340-bp region of *N. apis* (PCR program: 94°C, 2 minutes; 10 cycles of 94°C, 15 seconds; 62°C, 30 seconds; 72°C, 45 seconds; 20 cycles of 94°C, 15 seconds; 62°C, 30 seconds; 72°C, 50 seconds with a 5-second cycle elongation for each successive cycle; 72°C, 7 minutes). Amplified products were resolved on a 1.5% agarose gel, purified, and sequenced. Their sequences were compared to *N. ceranae* U26533 and *N. apis* U26534. In 11 of 12 sequences, they were identical to *N. ceranae*, while only one sample sequence was identical to *N. apis*. Their consensus sequence for *N. ceranae* was submitted to GenBank (Accession No. DQ286728). This same assay was used to detect *N. ceranae* in corbicular pollen (Higes et al., 2008).

An additional assay using generic primers for *N. apis* and *N. ceranae* (Nosema F and Nosema R) was designed based on a conserved region in both species in the 16S rRNA gene to allow for simultaneous amplification of both species (Chen et al., 2008). Species specific primers were also designed, *N. ceranae* F. and *N. ceranae* R. or *N. apis* F. and *N. apis* R. (from Accession Nos. U97150 and DQ486027 for *N. apis* and *N. ceranae*, respectively). Cycling conditions were based on the *Taq* polymerase manufacturer's protocol (95°C, 2 minutes; 45 cycles of 94°C, 15 seconds; 55°C, 30 seconds; 68°C, 30 seconds). DNA amplicons were sequenced directly instead of RFLP analysis. This approach was also used in two other studies (Chen et al., 2009a,b).

Multiplex PCR

Unlike the previous PCR assays, Matín-Hernández et al. (2007) developed a multiplex PCR assay that allows for the simultaneous detection and discrimination of *Nosema* spp. within the same reaction. In this assay, two different sets of primer pairs were used. Either *N. apis* sequences (Accession Nos. DQ235446, U76706. U97510, U26534, X73894, and X74112) or *N. ceranae* sequences (Accession Nos. DQ329034, U26533, DQ078785, and DQ286728) were aligned to form a consensus sequence for each species. Using the consensus sequence, primers specific for each species were designed. Either G or GC tails were added to the 5' end of the primers to obtain equal melting temperatures (indicated by underlined bases in Table 1). Primers 321APIS F and R and 218MITOC F and R, named based on the size of the amplicons generated in the reaction, were used to amplify either *N. apis* or *N. ceranae*, respectively (PCR program: 94°C, 2 minutes; 10 cycles of 94°C, 15 seconds; 61.8°C, 30 seconds; 72°C, 45 seconds; 20 cycles of 94°C, 15 seconds; 61.8°C, 30 seconds; 72°C, 50 seconds followed by an additional 5 seconds of elongation for each subsequent cycle; 72°C, 7 minutes). PCR amplicons were resolved on a 2% agarose gel and size discrimination of the bands was used to determine species.

Real-Time Quantitative PCR

Real-time quantitative PCR (qPCR) has become the method of choice for a number of different applications, including the quantitation of gene expression, DNA damage measurement, genotyping, and pathogen detection (http://www6.appliedbiosystems.com/support/tutorials/pdf/rtpcr_vs_tradpcr.pdf). As the name suggests, real-time qPCR detects and quantifies DNA amplification in "real time." As indicated earlier, these reactions are performed in a thermal cycler with a fluorescent detection system. There are several fluorescence-based technologies available for use with qPCR, and these have been recently reviewed by VanGilder et al. (2008). Here the focus is on specific reporter probes that require a hydrolysis step before fluorescence can be detected and allow for a high degree of specificity and sensitivity. There is also an assay that uses a nonspecific dye, SYBR green, for the detection of *N. apis* and *N. ceranae* (Cox-Foster et al., 2007). Also, because the probes are target specific, multiplexing becomes possible using different fluorescent reporter dyes with different emission spectra.

Hydrolysis probes (e.g., TaqMan or Molecular Beacons) are specific reporter probes that rely on the 5'–3' exonuclease activity of *Taq* DNA polymerase to cleave a hybridized probe during the extension step of PCR (Heid et al., 1996). These probes are sequence specific to the target DNA and are dual-labeled with a fluorescent reporter on the 5' end and a quencher molecule on the 3' end (Figure 2). When the probe is intact, the reporter and quencher are in close proximity, resulting in the suppression of fluorescence due to the fluorescence-resonance energy transfer from the reporter to the quencher (http://www3.appliedbiosystems.com/cms/groups/mcb_marketing/documents/generaldocuments/cms_041440.pdf). However, after the probe hybridizes to the target DNA, the DNA polymerase hydrolyzes the probe, separating the reporter and quencher molecules, allowing for an increase in fluorescence that is detected by the instrument (Figure 2; Heid et al., 1996).

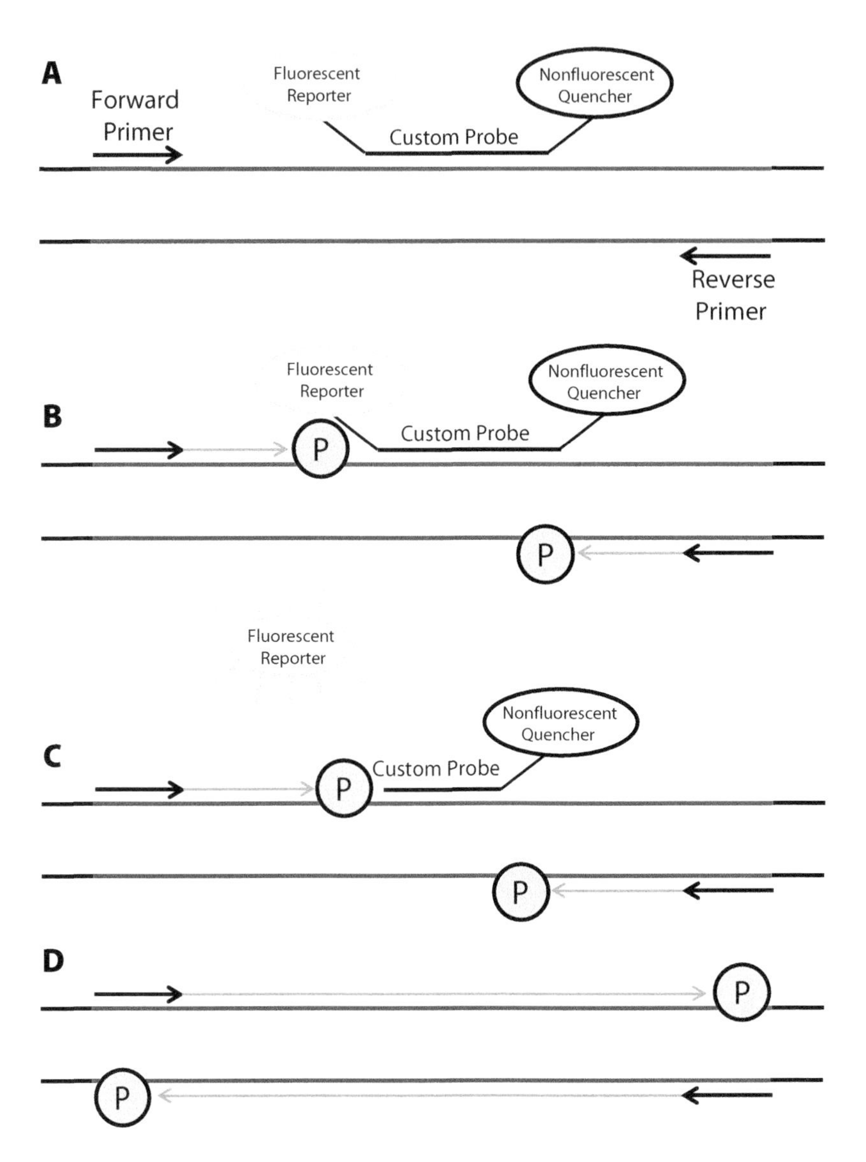

Figure 2. Cartoon depicting how real-time probes that require hydrolysis for fluorescence to be detected work. **A.** During the annealing step, the probe anneals in between the primers. **B.** As the DNA polymerase is synthesizing a new strand, it encounters the probe. The 5′–3′ exonuclease activity of the DNA polymerase cleaves the probe. **C.** Once the probe is cleaved, the fluorescent reporter and quencher are separated, resulting in an increase in fluorescence. The fluorescence detected is directly proportional to the amount of DNA present in the reaction. **D.** Synthesis of the complementary strand of DNA.

The amount of fluorescence is directly proportional to the amount of target DNA, generating an amplification curve that reflects the increase in fluorescence due to the synthesis of target DNA (VanGuilder et al., 2008). The amplification plot (Figure 3A) provides information that can be used for quantitative measurements of DNA (or RNA), a key component of which is the cycle threshold (C_T). The C_T value is the cycle in which fluorescence detected reaches a level above the background. C_T values are directly proportional to the amount of initial target DNA and can be used for comparative purposes (Figure 3).

One of the advantages of qPCR is that it can provide quantitative information, but in any real-time qPCR reaction, a choice must be made between using a standard curve or a relative method of quantitation before performing the analysis. In the standard curve method, designated copy numbers of the target DNA region, which has been cloned into a recombinant plasmid, are diluted to provide a range of copy numbers (i.e., 10^2 to 10^6 copies) to create a standard curve of copy number versus C_T (Creating Standard Curves with Genomic DNA or Plasmid DNA Templates for Use in Quantitative PCR, Applied Biosystems: http://www6.appliedbiosystems.com/support/tutorials/pdf/quant_pcr.pdf). Copy

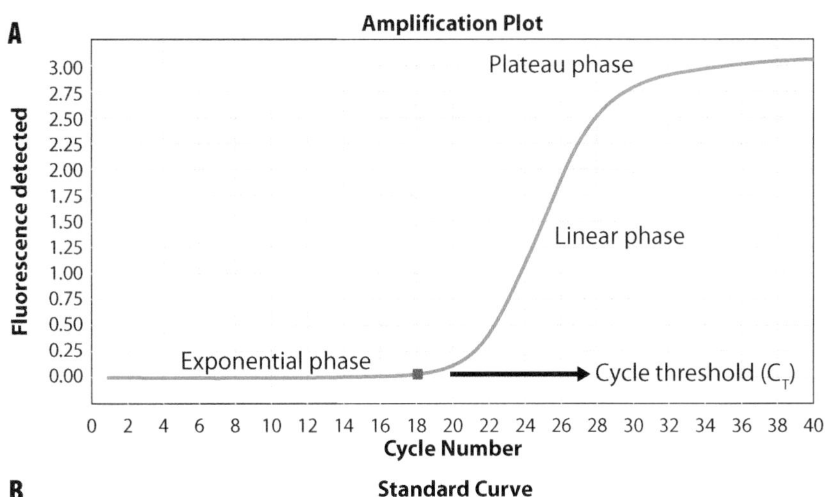

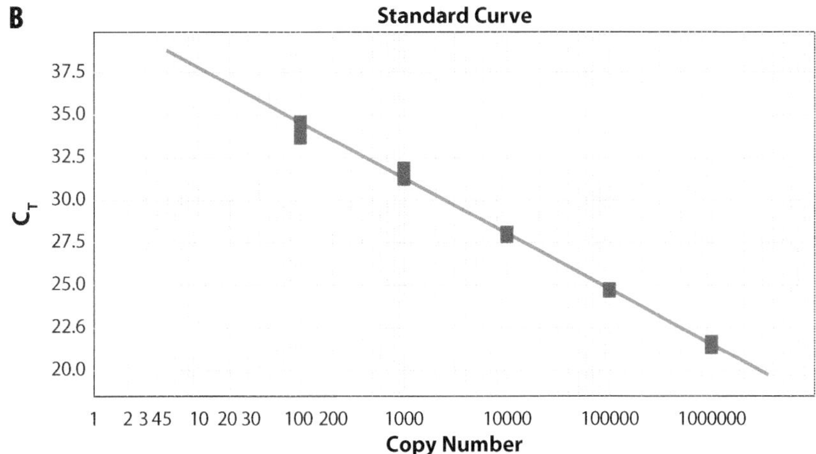

Figure 3. A. Amplification plot from real-time PCR. On the *y*-axis is the change in fluorescence detected by the instrument throughout the PCR reaction, and the *x*-axis shows the cycle number. The cycle threshold (C_T) is the cycle number in the PCR reaction where the fluorescent signal exceeds a pre-set threshold for detection above the background level. Here a curve is shown with a C_T value of approximately 18. **B.** Standard curve from real-time PCR. On the *y*-axis are the C_T values, and the copy numbers of target DNA are on the *x*-axis (log-scale). The amount of fluorescence detected is directly proportional to the amount of DNA present while the amount of DNA present is inversely proportional to the C_T value.

number and C_T are inversely proportional. Serial dilutions of the plasmid DNA are performed to obtain a range of copy numbers required for the standard curve. C_T values from the qPCR are then graphed against the log of the copy number to create the standard curve (Figure 3B). The amplification efficiency of the reaction can also be calculated from the following equation:

$$E = 10^{(-1/\text{slope})} - 1$$

(TaqMan Gene Expression Assays, Applied Biosystems Application Note: http://www3.appliedbiosystems.com/cms/groups/mcb_marketing/documents/generaldocuments/cms_040377.pdf).

Real-Time qPCR Assays

A duplex real-time qPCR assay to distinguish and quantify levels of *N. apis* and *N. ceranae* in *A. cerana* was developed using TaqMan probes (Chen et al., 2009a). *N. apis*-sense and *N. apis*-anti-sense were used to generate an 269 bp amplicon for *N. apis* while *N. ceranae*-sense and *N. ceranae*-antisense were used to amplify a 250 bp amplicon from *N. ceranae*. Two different fluorophores were used to distinguish between species: FAM (6'-carboxyfluorescein) with excitation and emission wavelengths of 495 nm and 515 nm and a VIC dye with excitation and emission wavelengths of 535 nm and 555 nm were used because both the absorption and emission wavelengths were separate enough for detection. The *N. apis* probe was labeled with VIC at the 5' end while the *N. ceranae* probe was labeled with FAM at the 5' end. Using two different probes allowed for the simultaneous detection of both species in a single reaction. Both 3' ends of the probes were labeled with TAMRA (6'-carboxy-tetramethylrhodamine) as a quencher dye having excitation and emission wavelengths between 535 nm and 605 nm. Beta-actin was used as an endogenous control (Chen et al., 2005). A total of 500 ng DNA per sample was used from individual *A. cerana* (PCR program: 95°C, 2 minutes; 45 cycles of 94°C, 15 seconds; 55°C, 30 seconds; 68°C, 30 seconds). Standard curve quantitation was used. *N. apis* and *N. ceranae* amplicons obtained as described above (Chen et al., 2008) were cloned into a recombinant plasmid to create standards of known copy number. This was the first assay that allowed for a one-step identification of species present as well as quantification of both *N. apis* and *N. ceranae* infections in honey bees.

In 2010, a real-time qPCR assay was designed to simultaneously detect both *N. apis* and *N. ceranae* in *A. mellifera* and to quantify spore levels for individual or pooled bee samples (Bourgeois et al., 2010). Sequences for *N. apis* and *N. ceranae* (Accession Nos. U97150 and DQ486027) were used to design primers and probes (from Molecular Beacons) specific for either *N. apis* or *N. ceranae*. Primers For and Rev for *N. apis* were used to amplify a 142-bp amplicon detected using a FAM-labeled probe. For *N. ceranae*, primers For and Rev were used to amplify a 104-bp amplicon detected using a JOE-labeled probe (similar to VIC). In each reaction, 300 ng total genomic DNA from whole abdomens of individual or pooled bee samples were used (PCR program: 95°C, 20 seconds; 40 cycles of 95°C, 1 second; 63°C, 20 seconds).

An additional duplex real-time qPCR assay has been developed to simultaneously detect *N. ceranae* and *N. apis* in *A. mellifera* (Traver and Fell, 2010, submitted). The sequences of the rRNA gene for *N. apis* and *N. ceranae* (Accession No. U97150 and DQ486027, respectively) were used to design primers DQ486027 F and DQ486027 R which amplify a 103-bp amplicon from *N. ceranae*. Primers U97150 F and U97150 R were used to amplify a 92-bp amplicon from *N. apis*. A VIC-labeled probe with a non-fluorescent quencher (NFQ) at the 3' end was used to detect *N. ceranae* while a FAM-labeled probe specific to *N. apis* was designed (TaqMan probes, Applied Biosystems). Each reaction had 50 ng total genomic DNA per reaction (PCR program: 50°C, 2 minutes; 95°C, 10 minutes; followed by 40 cycles of 95°C, 15 seconds; 60°C, 1 minute).

Conventional PCR versus Real-Time qPCR

Time, money, and ease of use must be considered for each molecular diagnostic approach. The number of samples to be analyzed as well as the type of information required for the research should also be evaluated. Conventional PCR followed by either RFLP analysis or direct sequencing of the PCR product is a reliable and inexpensive method to determine species, but is time consuming and not necessarily streamlined for high-throughput sample analysis. Real-time qPCR is more expensive and requires a thermal cycler capable of detecting and quantifying fluorescence levels.

However, once the initial cost of the machine is taken out of consideration, real-time qPCR is a much faster technique that not only allows for species identification, but can also provide quantitative data. Hydrolysis probes are more expensive yet species specific and save time as compared with other methods, such as PCR-RFLP (Table 1).

Nosema ceranae Detection by Microscopy and Antibody Tests

CHAPTER

10

Thomas C. Webster and Katherine Aronstein

Abstract *Nosema ceranae*, a honey bee pathogen now known worldwide, may be detected quickly by either light microscopy or by antibody tests. While these tests are less sensitive than polymerase chain reaction, they may be more practical for routine diagnosis. Phase contrast light microscopy allows one to distinguish between primary, environmental, and germinated spores. Fluorescent stains may demonstrate the maturity of spores and the integrity of the spore membranes. Polar filaments from the spores are also seen by proper microscopic techniques. Antibodies are also helpful in identifying Nosema spores and can be specific to the *Nosema ceranae* species. The antibodies attach to protein in the wall of the spore. This test is able to detect an infestation as few as 1,000 spores, a tiny fraction of the spores present in a highly infected bee. We hope to see this test commercialized so that it is available to beekeepers.

Introduction

The recent discovery of the Microsporidian *Nosema ceranae* in the European honey bee (Higes et al., 2006; Huang et al., 2007) came more than 10 years after its establishment in the United States (Chen et al., 2008). It is likely that *N. ceranae* had been undiagnosed in other countries for long periods also. Stored bee samples had been misdiagnosed due to the similarity in the appearance of spores of *N. ceranae* and those of the more familiar pathogen *Nosema apis*. *N. ceranae* spores may be smaller and narrower than *N. apis* spores (Fries et al., 1996), although this difference is not always reliable. When polymerase chain reaction (PCR) was employed to distinguish the two Nosema species, *N. ceranae* was found to be distributed worldwide (Klee et al., 2007).

N. ceranae is now present in a high proportion of American honey bee hives. It has been implicated in colony mortality, especially in conjunction with parasites and other pathogens. PCR is the method of choice to detect light infections and to distinguish between species (see Chapter 8, Traver and Fell). It is highly sensitive and accurate, and allows the detection of vegetative forms of Nosema where spores are not evident (Chen et al., 2009; Gisder et al., 2010).

However, the PCR method is expensive and laborious. It cannot distinguish between the spore form and the vegetative form of a particular Nosema species, nor between viable and dead spores. Inexpensive and rapid methods for detection and evaluation of Nosema infections would be valuable for routine diagnosis and scientific studies. Hence the development of techniques in light microscopy and antibody tests will be beneficial to all concerned with this disease.

Nosema Life Cycle

Like other Microsporidia, *N. apis* and *N. ceranae* undergo a complex life cycle. We are concerned here with *N. ceranae*, although much of our discussion pertains also to *N. apis*. When a spore is consumed by a bee,

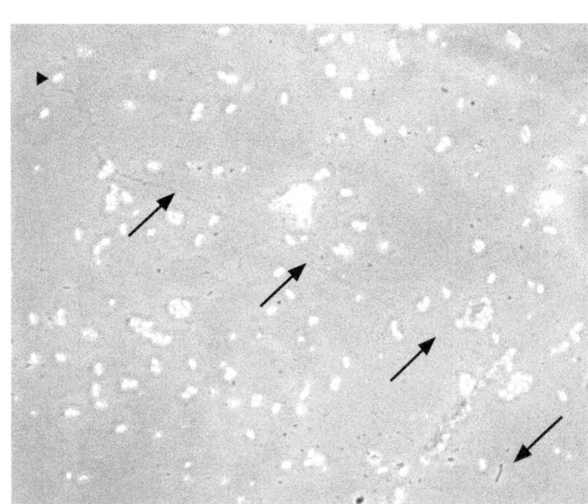

Figure 1. The Nosema spore (white oval, upper left) was stimulated to evert a very long polar filament. This is the fine line (arrows) that runs from the spore (►) to the lower right corner and out of view. It was about 50% longer than could be shown in this photo. (*Photo by T.C. Webster.*)

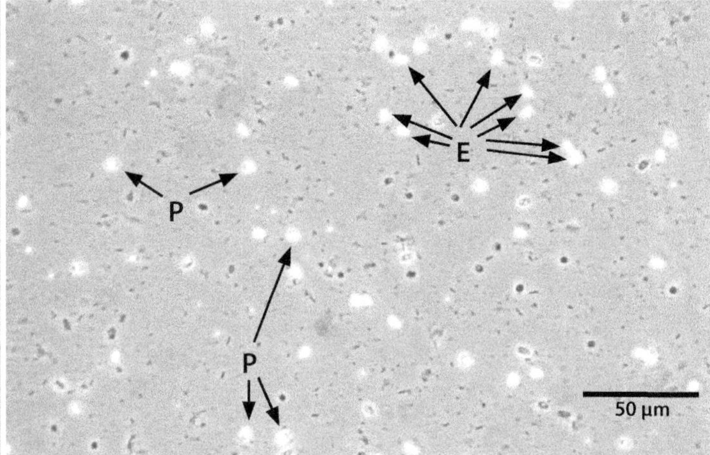

Figure 2. Primary spores (**P**) are often slightly tapered, with a dark area at the wide end. Environmental spores (**E**) are usually "race-track" or "almond-shaped" ovals, appearing solid white by phase contrast microscopy. Germinated spores appear dark by phase contrast optics. Note that spores will appear circular if seen end-on. (*Photo by T.C. Webster.*)

it passes through the crop to the midgut. A stimulus in the midgut causes the spore to germinate there. In the germination process, a very long polar filament everts from its coiled position inside the spore. This eversion is very rapid. Within several seconds, the filament stretches 200 microns or more, nearly 100 times the length of the spore (Figure 1). The filament also "telescopes" in the process, making it even longer than it was inside the spore. The tip of the filament may then penetrate a host cell, in the epithelium of the bee midgut. Then the sporoplasm, including the nucleus and cytoplasm of the spore, pass through the polar filament and into the bee's epithelial cell. Often a bulge in the filament is visible, where the sporoplasm has traveled part of the way from the spore. Inside the epithelial cell the Nosema parasite begins to form new spores. As an intracellular parasite, it depends entirely on the nutrients and metabolic machinery of the bee's cell.

Within several days, "primary" or "auto-infective" spores develop (Figure 2). These spores germinate within the bee epithelial cell, sending a filament to infect an adjacent epithelial cell. In the adjacent cell, the environmental spores develop over a period of 1 to 2 weeks. When the environmental spores mature, the epithelial cell ruptures and releases the spores into the midgut. Many of these spores pass into the rectum and then leave the bee in the feces. Some environmental spores may germinate shortly after they are released into the midgut, initiating a new round of infection. Fecal material left in the hive may be ingested by other bees as they clean the comb. In this way the disease cycle continues.

Methods

Assessment of Spores by Light Microscopy

Spores extracted from infected bees are typically examined at 400× using phase contrast microscopy. By these optics, spores often appear as distinct ovals, approximately 3 by 5 microns (Figures 1, 2).

Assessment of spore stage, maturity, and viability is often required for controlled studies of either Nosema species. Intact spores appear white and germinated spores are dark gray. It is possible to watch a spore darken as the filament everts. Primary spores are oval and often slightly wider and dark at one end. Environmental spores tend to be either "race-track" ovals (*N. apis*) or "almond-shaped" ovals (*N. ceranae*). However, both species show much variability in appearance.

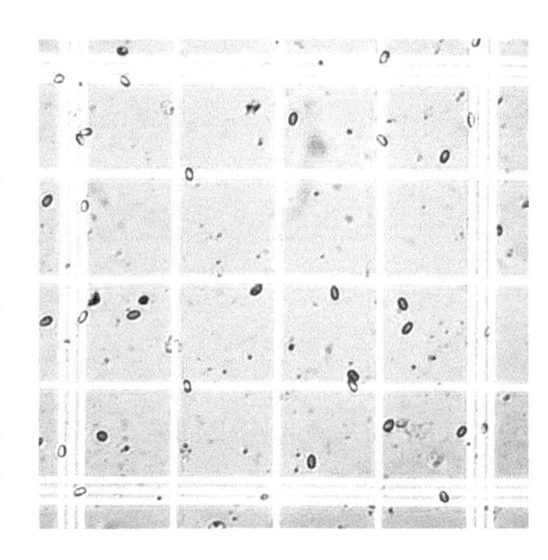

Figure 3. *N. ceranae* spores were counted using a hemocytometer (400x magnification). (*Photo by T.C. Webster.*)

Calculations for spores per bee require a hemacytometer. One mL of water per bee or midgut is usually appropriate for collecting spores from heavily infected bees. The sample should be mashed thoroughly, and then a single drop is placed on the hemacytometer (Figure 3). Most hemacytometers come with formulae to calculate the total number of spores per mL or per bee. A heavily infected bee will contain over 10 million spores.

The maturity of spores can be determined by the intensity of calcofluor white stain (also called optical brightener), viewed with a fluorescent microscope. Calcofluor fluoresces vividly as blue-white, with an excitation band of 395 to 415 nm. This stain binds to chitin, a major constituent of the spore wall. Vegetative forms of Nosema and incompletely developed spores do not have a heavy layer of chitin, and stain less intensely.

Spores with ruptured membranes cannot be viable, and are identified by their uptake of the stain trypan blue. A 0.4% solution applied to the spores will enter the spores over a 5-minute period. This stain fluoresces a rose-red under ultraviolet light. Of course, not all spores with intact membranes are viable.

Spore germination is an essential part of the life cycle of *Nosema*, and observations of germination can be useful. Environmental spores dried on a microscope slide can be stimulated to germinate by the application of buffer at pH 7.1 (Olsen et al., 1986). This can cause rapid extrusion of the polar filament. However, this buffer does not cause all environmental spores to germinate.

Observations of polar filaments are valuable for several reasons. The filament itself shows that the spore is in fact a microsporidian because other microbes do not have this structure. Some microbes such as yeasts are found in bees and resemble microsporidia superficially, but do not contain polar filaments. Also, the extrusion of the filament can indicate conditions that cause or inhibit germination. The filaments are so thin, about 0.1 micron, that they can be difficult to see in aqueous preparations. The filaments may rise slightly above or fall below the focal plane, so focusing to elevate the objective can be helpful. One easy technique is to let the slide dry. Filaments will then adhere to either the slide or the cover glass. The dried filaments are much more visible by phase contrast than filaments in liquid (Figure 4) and are held by the glass in one focal plane.

Collection and Purification of Spores

There are several reasons to separate Nosema spores from other substances in live or dead bees. These extraneous substances include other microbes,

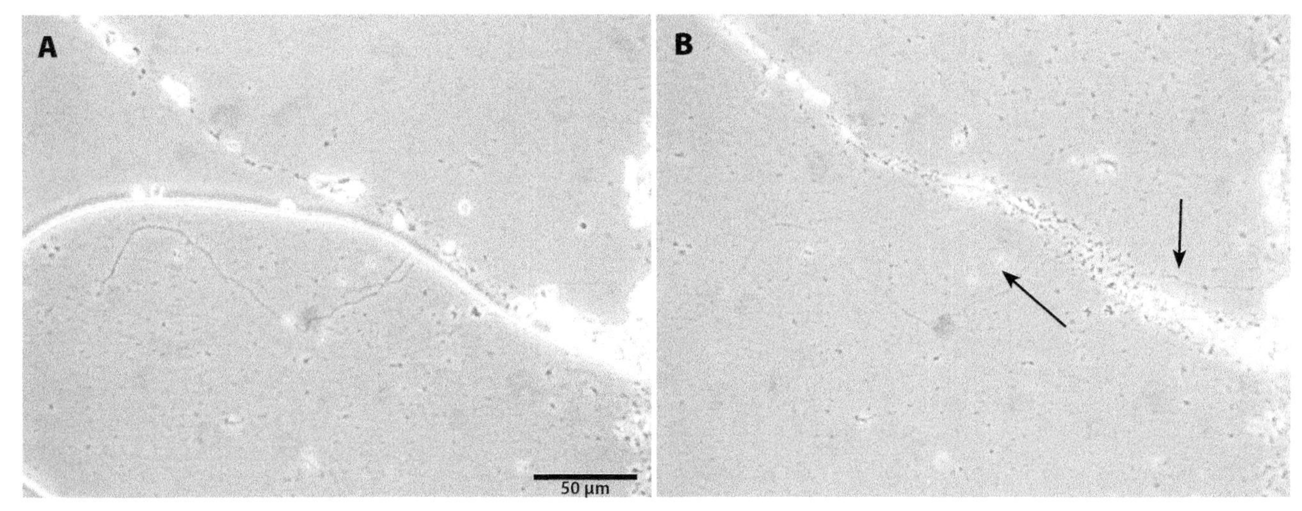

Figure 4 **A.** The area at the bottom of this photo has dried, allowing the polar filaments to become quite visible by phase contrast optics. These filaments are easy to see also because they are attached to the glass and hence supported in one focal plane. However, the filaments are much more difficult to see where they enter the water at the upper part of the slide due to the phase contrast. (Compare to Figure 1.) Filaments and other objects in water are often partly above or below the focal plane, and this also makes observation difficult. **B.** This photo is the same field of view as Figure 4A but after the slide dried. Now more of the polar filaments are visible (arrows). (*Photos by T.C. Webster.*)

pollen, and bee tissue fragments. First, the elimination of extraneous material can greatly simplify analyses. The extraction of DNA, for example, may be more effective. Also, studies involving the inoculation of bees with spores will be better controlled if debris and other pathogens are excluded. However, we should remember that a preparation of pure Nosema spores does not exist in the bee hive. The midgut and fecal debris are the natural environments to which the spores are probably well adapted.

Spores are most easily purified from live or recently killed bees. Digestive systems that have decomposed or been frozen are difficult to separate from the rest of the abdomen. Midguts from live bees can be removed from the bee intact, using forceps to pull simultaneously from the petiole and the last abdominal segment. The exposed midgut may then be cut free from the crop, ileum, and rectum. The midguts may be frozen collectively or individually until needed.

Spores collected from decomposed or frozen bees will include microbes, pollen, and pieces of the bee sclerites, which are difficult to eliminate from the sample. Filters may exclude some debris but will often retain many of the spores. Density gradient centrifugation is helpful, but does not eliminate all debris.

Collection of Spores from Hive Equipment

Spores in fecal debris are often a sign of serious infection in the colony. Feces may be deposited on the front of the hive, above the hive entrance, by bees crawling up from the entrance as they leave the hive. Fecal debris may be seen inside the hive also, on comb, top bars, and the inner cover. The color, location, and texture of fecal debris are distinct from other hive materials such as propolis. Feces inside the hive may be a sign of serious *N. ceranae* infection, even when the hive is populated by an apparently robust colony.

Feces may be scraped from the hive equipment, agitated vigorously in water, and examined by microscope as described above. Spores will be seen easily if the feces have come from heavily infected bees. However, not all fecal debris on equipment contains spores. Bees may suffer other maladies that cause them to defecate excessively.

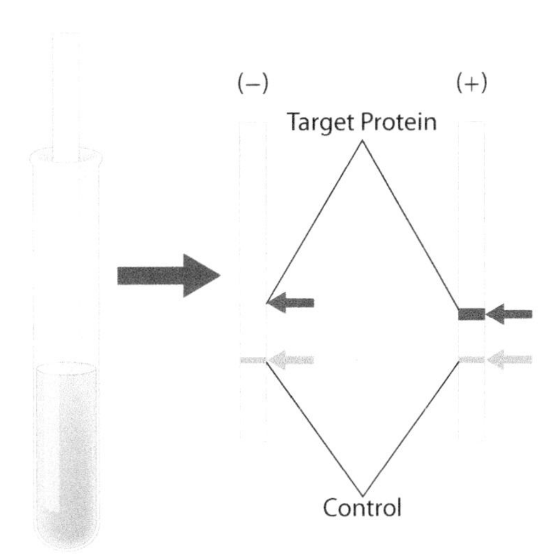

Figure 5. Schematic diagram of an antibody-based test. Dark staining on a (+) paper strip indicates the presence of *N. ceranae* spores in a tested sample (upper triangles). The lack of staining on a (−) paper strip indicates the absence of *N. ceranae* spores, but does not exclude the presence of other *Nosema* species. (*Illustration by Signe Nordin.*)

Antibody Tests for *Nosema ceranae* Spores

Antibody (Ab) tests for *N. ceranae* are a very different method for detecting this disease in bees. They depend on the use of antibodies that fit perfectly into a type of protein molecule unique to the *N. ceranae* spore wall. When the Ab molecules attach to the spores, they show their presence by developing a dark color on a paper strip (Figure 5). This research has been conducted at the Weslaco Honey Bee Research Unit, U.S. Department of Agriculture (USDA).

Generating Antibodies Specific to *Nosema ceranae*

To produce Ab, we have relied on the sequence of base pairs in the *N. ceranae* DNA molecule. The sequence is the precise description of the DNA molecule for this microbe. By interpreting this sequence, we are able to understand which part of the DNA instructs the microbe to construct the spore wall protein. This protein construction happens while environmental spores are developing inside the bee's infected midgut. Over a period of several days, the spore wall is assembled inside a midgut cell much as an egg shell is developed around an egg inside a bird. The *N. ceranae* DNA sequence was provided by Dr. Jay Evans of the USDA honey bee laboratory in Beltsville, MD. He played an important role in determining the entire *N. ceranae* DNA sequence (Chen et al., 2009).

With this information we created tiny circles of DNA called *plasmids*. The plasmids included the DNA instructions for the manufacture of *N. ceranae* spore wall protein. In order to produce Ab, we had these plasmids injected into rabbits. When injected into the rabbits, our plasmids produced a large quantity of *N. ceranae* spore wall protein fragments. That, in turn, triggered an immune reaction in rabbits. Like all mammals, rabbits react to injections of "foreign" materials (e.g., microbes or microbial proteins) by producing Ab as part of their overall immune response mechanism. Later, Ab were removed from the rabbit blood, purified, and tested using paper strips. When *N. ceranae* spore wall protein was applied and incubated with *N. ceranae* Ab, a very visible dark staining appeared. This reaction showed that antibodies recognized and attached to *N. ceranae* protein.

Figure 6. Small laboratory cages were used to study Nosema disease in bees and to produce a large quantity of Nosema spores. (*Photo by K. Aronstein.*)

Testing the System

To test this antibody system we wanted to know whether it is specific to *N. ceranae* and also to know how sensitive it is. We tested bees known to be infected with either *N. ceranae* or *N. apis*. Spores of a single Nosema species were produced under highly controlled laboratory conditions using small cages (Figure 6). Spore protein extracted from *N. ceranae*-infected bees produced a very strong positive reaction for samples containing 10^3 to 10^6 spores. This shows the extreme sensitivity of the test: a highly infected bee may contain over 10 million (10^7) spores, that is, 10,000 times the amount that can be detected by the test. When spores of *N. apis* were tested in comparison, no reaction was detected. This was expected, because the protein in the spore wall of this microbe differs from that in *N. ceranae*.

Production for Beekeepers

At this time the test has not been commercialized. The USDA Office of Technology Transfer is working toward developing a simple Ab-based test that would not require any laboratory equipment and can be used directly in the field. We hope that a suitable company will be able to manufacture Nosema detection kits that follow the above procedure so that they can be sold to beekeepers at a reasonable price.

Chalkbrood Re-Examined

CHAPTER

11

K. A. Aronstein and H. E. Cabanillas

Abstract The fungus *Ascosphaera apis* is the causative agent of chalkbrood disease, leading to heavy losses of honey bees and colony productivity. *A. apis* affects the brood of honey bees, turning them into a "mummy" that is regarded as a source of infection. Spores seem to always be present in bee colonies at a low level without causing disease symptoms and require predisposing conditions (cool, humid weather, or other stress factors) for the larvae to develop chalkbrood. This chapter consolidates the recent findings focusing on the pathogen's biology, disease symptoms, and management tactics. The most commonly used research methods, molecular techniques, and *in vivo* bioassay for culturing and diagnosis, supplemented with micrographs and illustrations, are provided. Because honey bee colonies infected with *A. apis* often have no visible signs of the disease, early detection is critical for the diagnosis and prevention of disease outbreaks. One important aspect for control of chalkbrood is the presence of a young, hygienic queen for maintaining a strong and healthy colony.

Introduction

Chalkbrood disease in honey bees is caused by the fungus *Ascosphaera apis* (Maassen ex Claussen) Olive and Spiltoir (Spiltoir, 1955). In bees, this disease was recognized as early as the 1900s (Maassen, 1913), but was not widely seen outside of Europe until the latter half of the 20th century. By the mid-1960s, chalkbrood was detected in the United States and it is now commonly seen in honey bee colonies around the world (Heath, 1985; Reynaldi et al., 2003). A detailed geographical distribution of this disease is described by Aronstein and Murray (2010). Chalkbrood can lead to heavy losses of honey bees in the diseased colonies and significant losses of colony productivity. The most serious outbreaks of the disease are recorded during cool and damp weather conditions, which commonly occur in most European countries, Canada, and the mid-western and northern United States (Heath, 1985). However, it is not uncommon to find chalkbrood-infected bee larvae in hot and dry environments, such as the southern states of the United States, Mexico, Australia, and sub-tropical Africa.

In this chapter we consolidate recent scientific findings focusing on the pathogen's biology, disease symptoms, and management tactics. We have also included the most commonly used research methods to assist beekeepers and bee scientists entering this area of research. Detailed micrographs and illustrations should help to familiarize readers with culture morphology, mode of reproduction, and host pathogenesis. For more advanced knowledge of this disease, see Aronstein and Murray (2010).

Disease Symptoms

Ascosphaera apis fungus affects exclusively the brood of honey bees. Both castes of bees (workers and queens) and drones are susceptible to this pathogen (Bailey, 1981; Gilliam et al., 1988). Spores seem to always be present in colonies at a low level without causing symptoms of the disease. However, these asymptomatic colonies can rapidly develop chalkbrood when larvae are exposed to certain predisposing conditions. In addition to cool and humid weather, other stress factors weakening bee colonies can increase the incidence of this disease.

Figure 1. A brood frame showing cadavers in their cells covered with white fungal growth (arrows).
(*Photo by K. Aronstein.*)

Infected larvae or pupae normally die in sealed cells. Cell cappings are then often perforated by the bees attempting to remove diseased brood from their cells. These small perforations in cappings are easily detectable during routine colony observations and indicate the presence of infection. More careful examination will help determine whether or not this infection is caused by a fungus. During more advanced stages of the disease, cadavers covered with a cottony white fungal growth can be found in their cells (Figure 1). This preliminary diagnosis based on clinical symptoms is sometimes misleading, and therefore must be confirmed using microscopy or DNA-based methods (see *Methods*).

As cadavers dry, they become hard and form the so-called chalkbrood mummies. Within time, the appearance of chalkbrood mummies can change from white to black (Figure 2), indicating production of fungal reproductive structures, or spore cysts. Although white mummies may contain few or no spore cysts, all white or black mummies should be treated as a source of infection. Provided that environmental conditions are conducive to reproduction, white mummies can form masses of spore cysts. This can be demonstrated by incubating them in the laboratory. Following incubation, white mummies develop visible dark brown or black patches, indicating completion of sexual reproduction (see *Fungal Reproduction,* below).

In the bee colony, mummies can be found in brood cells and on the bottom board. They can also be found in front of the colony, as bees will attempt to remove diseased brood from their cells. The extent of brood removal depends on the genetic make-up of the bees, and the presence of a young hygienic queen can be a key factor in maintaining a strong and healthy colony.

Spores can accumulate on all surfaces of the hive and in hive products; they can remain viable for a long time, at least 15 years (Gilliam, 1986; Gilliam et al., 1988; Gilliam and Taber, 1991; Anderson et al., 1997; Flores et al., 2005; Flores et al., 2005a). Spores are then passed onto larvae by nurse bees feeding them with the contaminated food. While adult bees are not susceptible to this pathogen, they can carry spores within and between colonies thus aiding disease transmission (Gilliam and Vandenberg, 1997). Dissemination of the disease between colonies,

Figure 2. Chalkbrood mummies are white (arrowhead), dark brown, or black in appearance (arrow). (*Photo by K. Aronstein.*)

apiaries, and even over long distances is a very serious problem and in most cases is related to the beekeeper's management practices (Gilliam and Vandenberg, 1997). Because any hive materials originating from infected colonies are contaminated with spores, they will serve as a long-lasting source of infection. Furthermore, any tools used in these colonies are also contaminated with spores and should be decontaminated before using them again. Improving management practices is absolutely vital in the prevention of this infectious disease and in controlling its spread when outbreaks occur.

Pathogenesis

A. apis is a highly specialized pathogen, adapted to the honey bee life cycle and particularly to the larval stage. Spores consumed by the larvae germinate in the lumen of the gut, probably activated by CO_2 (Heath and Gaze, 1987; Bailey and Ball, 1991). *In vitro* spore germination occurs within a wide range of temperature conditions (25 to 37°C/77 to 98.6°F) with the optimum range of 31 to 35°C (88 to 95°F). Our experiments showed that it takes 17 hours at 33°C on YGPSA medium to observe production of germ tubes (Figure 4D) greater in length than the spore diameter, which is generally considered an indication of germination. Recently, several enzymes identified in *A. apis* were implicated in assisting its penetration of the peritrophic membrane in the larval midgut (Theantana and Chantawannakul, 2008). Timing of this process coincides with the rapid reduction in food consumption by infected bee larvae, leading to starvation (Aronstein and Murray, 2010). After penetrating the gut wall, the fungal mycelium grows inside the body cavity, eventually breaking out through the posterior end of a dead larva (Figures 3A–3D). If the environmental conditions are conducive to fungal growth, *A. apis* mycelium extends from the posterior end to the anterior end, thus covering the entire cadaver. Later on, fungal growth becomes mottled with dark brown due to the production of spore cysts. Eventually, cadavers dry and form into chalkbrood mummies (Figure 2). It was estimated that each black mummy can produce about 10^8 to 10^9 ascospores (Hornitzky, 2001).

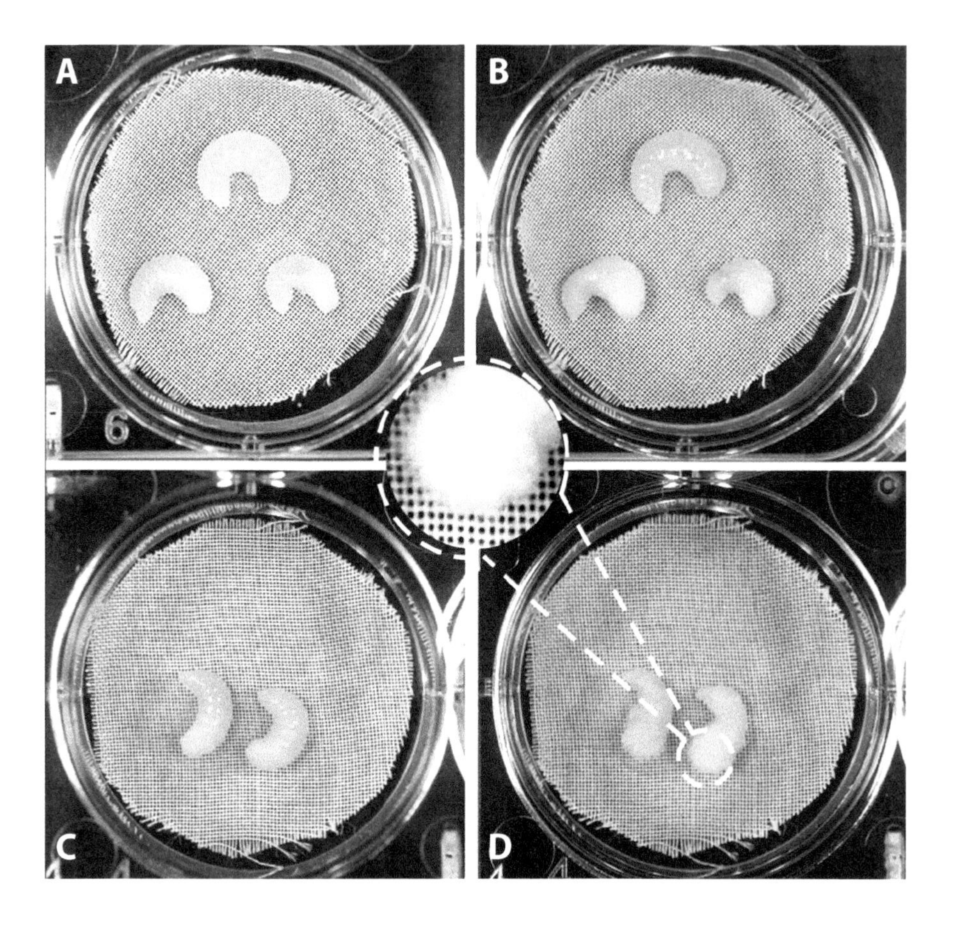

Figure 3. Clinical symptoms of the chalkbrood disease demonstrated using an *in vivo* bioassay. **A and C.** Three-day-old honey bee larvae were infected by feeding them a spore-containing diet; bee larvae 72 hours post infection. Fungal mycelium grows inside the body cavity and becomes visible as a white colored mass under the skin. **B and D.** By 78 hours post infection, fungal growth emerges outside the body cavity (see close-up view) and eventually covers the entire cadaver with white fluffy mycelium. (*Photos by K. Aronstein.*)

Fungal Biology and Reproduction

A. apis can be cultivated both on solid and in liquid mycological media. On solid media, *A. apis* grows as a dense white mycelium containing aerial, surface, and subsurface hyphae (Figure 4A–B) The hyphae are septate, 2.5 to 8.0 micrometers in diameter, and show pronounced dichotomous branching (Spiltoir, 1955; Skou, 1988). Septum gives some physical rigidity to hyphae and protects them from the loss of cytoplasm when they are physically damaged. Each septum has a pore that allows movement of small nuclei between vegetative cells (Spiltoir, 1955). The cytoplasm of the *A. apis* mycelium contains mitochondria and numerous ribosomes (Anderson and Gibson, 1998).

A. apis reproduces sexually (Spiltoir, 1955; Gilliam, 1978). Such reproduction occurs between morphologically identical and compatible haploid partners, distinguished only by their mating type (MAT) locus (see Figure 5A–B) (Poggeler, 2001). According to Heath (1982), cultures of the two *A. apis* mating type idiomorphs show no difference in pigmentation or fungal colony size. On the other hand, (Spiltoir, 1955) described the two *A. apis* idiomorphs as sexually dimorphic at the microscopic level, where hyphae of opposite mating types produce specialized structures. This initial confusion between male and female in filamentous ascomycetes was alleviated later on, during the genomic era. It was shown that heterothallism was independent of sexual differentiation and both opposite mating type strains can produce "female" organs (Coppin et al., 1997). Following mating, both MAT idiomorphs produced sexual reproductive structures (see Figures 6A–6C).

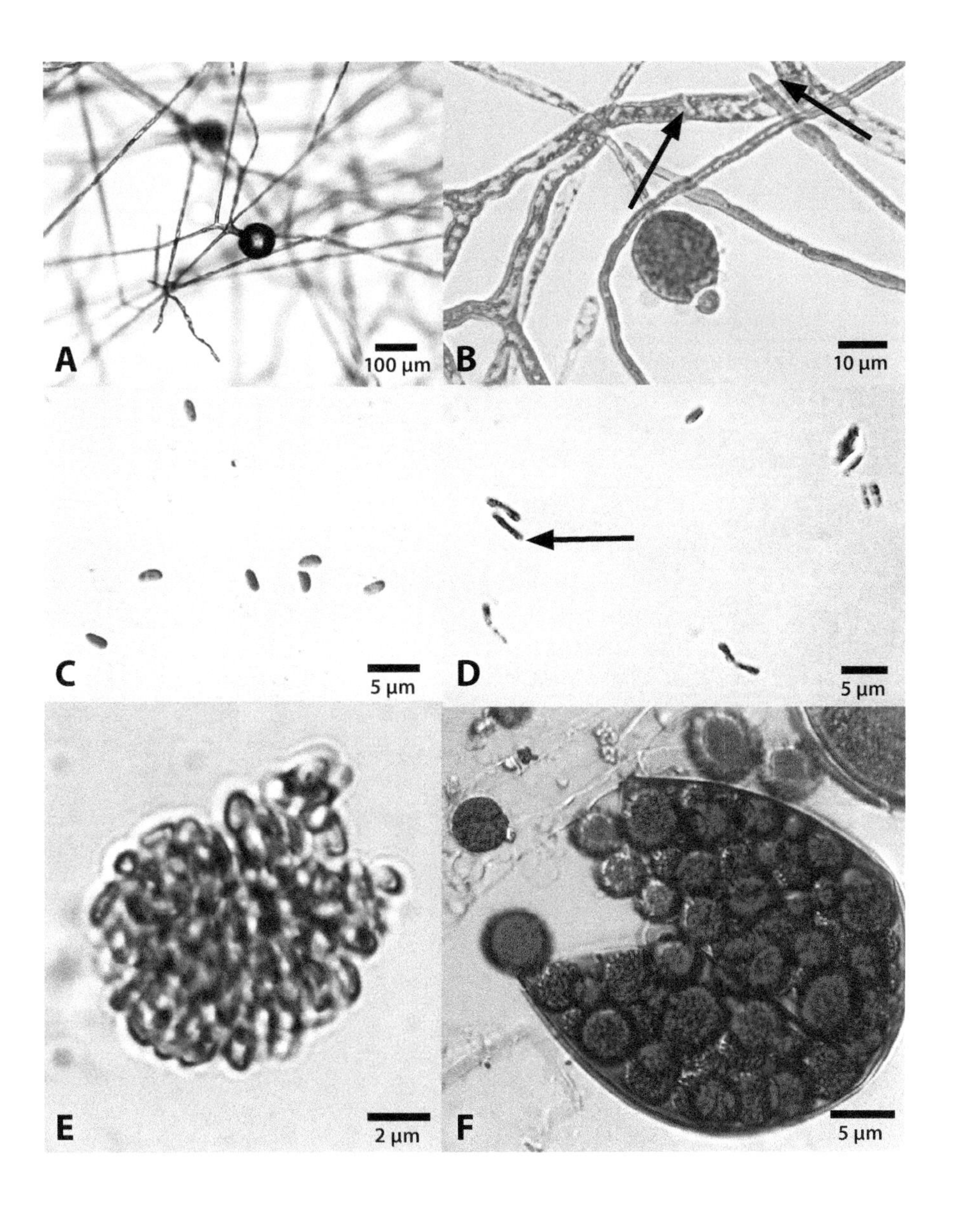

Figure 4. *A. apis* reproductive structures. **A.** Micrographs showing mixed idiomorphs *A. apis* culture. **B.** Fungal hyphae, dichotomous branching, and septum are indicated by the arrows. **C.** Mature ascospores. **D.** The germination process and germ-tube production in culture (arrow). **E.** Mature ascospores grouped into spore balls (asci). **F.** Spore balls tightly packed inside transparent spore cysts (ascomata).

A mature ascoma measures in the range of 47 to 140 μm in diameter. The micrographs presented in this chapter were produced with an Olympus BX-51 microscope fitted for differential interference contrast and a DP12 digital camera. (*Photos by K. Aronstein.*)

Mature ascospores are grouped into spore balls (asci) (Figure 4E) containing no visible outer membranes and are tightly packed inside transparent spore cysts (ascomata). A mature ascoma measures in the range of 47 to 140 micrometers in diameter (Christensen and Gilliam, 1983; Chorbinski and Rypula, 2003) and can be easily observed on a microscopic slide using 20× or 40× magnification. Ascospores have a thick spore wall and a spore membrane (Bissett, 1988) that protect spores from extreme environmental conditions. The size of the individual ascospore is in the range of 2.7 to 3.5 by 1.4 to 1.8 micrometers (Skou, 1972; Bailey and Ball, 1991; Anderson and Gibson, 1998) (Figure 4C). So far, asexual reproduction has not been documented in *A. apis*, though this is a common mode of propagation in the Ascomycota fungi.

Disease Management

Over the years, a number of effective and environmentally safe strategies have been developed and implemented to control diseases in bee colonies. These methods include improved management and sanitation practices,

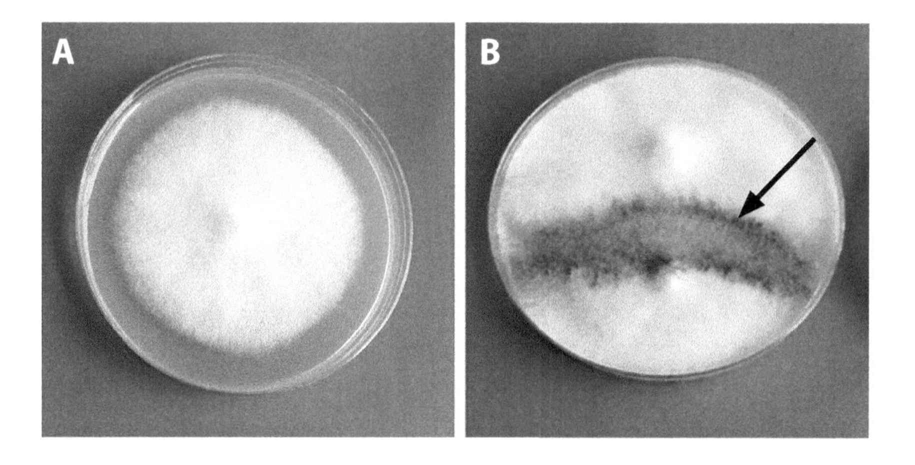

Figure 5. Mating of two *A. apis* idiomorphs in Petri dishes. **A.** *A. apis* cultured on solid medium. **B.** Mating type assay showing production of spore cysts (arrow); the MAT-1 (*bottom*) and the MAT-2 (*top*) idiomorphs were plated opposite to each other on a culture plate. (*Photos by K. Aronstein.*)

breeding disease-resistant bee lines, and use of alternative ecologically safe control methods (Arbia and Babbay, 2011).

Given the problems associated with the extended used of pesticides and antibiotics in honey bee colonies, disease management and sanitation must be the primary strategies in an attempt to improve bee health, to prevent dispersal of bee diseases, and to prevent infection from occurring in the first place. Such practices must include supplemental feeding to improve bee nutrition, using clean equipment, keeping colonies well ventilated, replacing old combs annually, and avoiding the transfer of combs between colonies. Because fungi grow best in humid conditions, providing good ventilation can slow down disease development, thus helping bees sanitize their colony more effectively. Old combs may harbor a wide range of bee pathogens, providing a continual source of infection. In addition, they often contain measurable levels of agricultural pesticides accumulated in beeswax and other hive products. This additional stress factor may decrease the bees' natural tolerance to diseases. Recent studies found that a cocktail of active compounds used in the same bee colonies could act synergistically, negatively affecting bee longevity, behavior, and possibly leading to increased colony losses (Bogdanov, 2006; Sheridan et al., 2008; Orantes-Bermejo et al., 2010; Arbia and Babbay, 2011).

Several different sterilization methods of old equipment have been tested over the years (e.g., fumigation, gamma irradiation from a Cobalt-60 source) in attempt to reduce the spore load in bee hives. However, none of them are widely accepted due to various limitations and/or accessibility of the radiation facilities.

Heat treatment of honey is an acceptable sterilization method. However, the temperature in water baths should be strictly controlled because overheating may result in the change of color, caramelization, and a decrease in the enzymatic activity of honey. Incubation regimens that showed good results have ranged from 8 hours at 65°C (149°F) to 2 hours at 70°C (158°F) (Anderson et al., 1997).

A broad range of chemotherapeutic compounds have been tested to inhibit *A. apis* growth in culture or to control chalkbrood disease in bee colonies (Hornitzky, 2001). However, none of the tested compounds have achieved the level of control required to fight this disease. Currently, there are no antifungal compounds registered for use against chalkbrood in bee colonies. Attempts to find natural products or microorganisms that control *A. apis* have also failed to produce usable applications (Gilliam et al., 1988; Aronstein and Hayes, 2004).

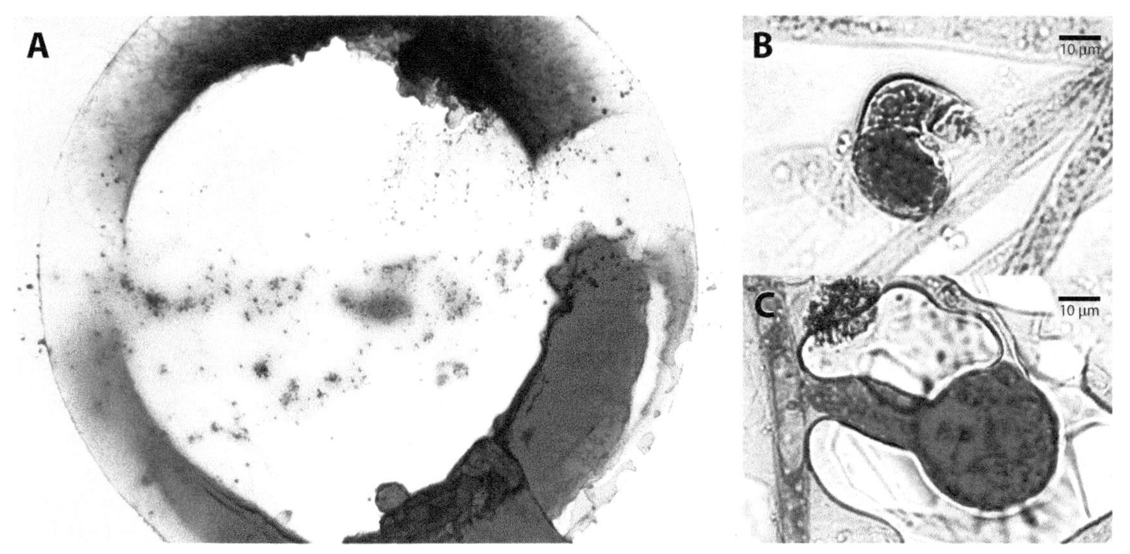

Figure 6. Micrographs showing sexual reproduction in *A. apis* using the coverslip method. **A.** Coverslip, mounted on a microscope slide. The two idiomorphs (MAT-1 and MAT-2) were differentially stained with CBLP (MAT-1; red staining) and LPAF (MAT-2; blue staining). **B.** Reproductive structures in MAT-2 (blue staining). **C.** Reproductive structures in MAT-1 (red staining) idiomorphs. (*Photos by K. Aronstein.*)

The anti-fungal activity of honey and bee bread (fermented pollen) has been tested and revealed that microorganisms, primarily *Rhizopus* sp., found in bee bread can inhibit *A. apis* growth in culture (Gilliam et al., 1988). More recent studies (Chapter 17 in this book) showed that bee bread harbors a large variety of beneficial microbes (some of which could have been introduced by the bees) that are significantly diminished in colonies exposed to fungicides. As a consequence, the reduced repertoire of beneficial microorganisms can have a detrimental effect on the nutritional status of bees and their ability to withstand infections.

Honey Bee Defenses against Chalkbrood

Replacement of a queen from good hygienic stock has become one of the most common practices for dealing with chalkbrood disease. The presence of a strong genetic component to chalkbrood resistance (Spivak and Reuter, 2001; Lapidge et al., 2002; Evans and Spivak, 2010) has become the basis for a number of successful breeding programs implemented to develop new, improved honey bee lines. Hygienic bees have the ability to detect and remove diseased brood very soon after infection (Spivak and Gilliam, 1993; Spivak and Reuter, 1998; Spivak and Downey, 1998; Spivak and Reuter, 2001). Therefore, colonies exhibiting significant hygienic behavior can reduce the numbers of fungal spores, an important factor in minimizing the likelihood of disease outbreaks. In most cases, hygienic bees performed at a level that did not require any additional treatments for control of chalkbrood (Evans and Spivak, 2010).

In addition to colony-level protection, individuals also possess mechanisms protecting them from diseases. Recent developments in honey bee genomics promoted numerous studies describing activation of a very complex network of innate immune reactions in bees infected with microbial pathogens (Aronstein and Saldivar, 2005; Evans et al., 2006; Aronstein and Murray, 2010; Aronstein et al., 2010). One of these studies investigated response mechanisms in honey bee larvae infected with chalkbrood fungus (Aronstein et al., 2010). All *in vivo* assays conducted in this study, using 1-day-old larvae resulted in 100% mortality within 48 hours. In comparison, 3-day-old larvae fed with fungal spores survived past 48 hours and showed signs of chalkbrood disease about 72 hours post infection. Even though this study demonstrated that fungal infection can trigger some level of immune activation, it was not enough to protect an individual bee larva from the disease. It is possible that a functionally mature immune system, in combination with a mild level of infection, may increase their chances for survival. Most importantly, this study showed that the activation of immune defenses comes with a cost. The up-regulation of immune and stress-related reactions negatively affects the level of maintenance of other physiological functions. In general, infection leads to nutrient wasting, and may trigger the nonreversible deterioration of the animal.

Insects have a wide range of innate immune reactions protecting them against diseases. Although immunological memory has been considered a prerogative of vertebrates, recent investigations have challenged this long-held idea. A revised understanding of insect immunity indicates that at least some form of immune memory may exist in insects that can protect them from repeated infections (Schmid-Hempel, 2005). Although the mechanisms of immune priming are still not fully understood, they are quite different from that underlining immune memory of vertebrates (Sadd and Schmid-Hempel, 2006). Nevertheless, priming insects by repeated exposure to microbial pathogens is a new and fascinating research direction (Kim et al., 2004; Sadd and Schmid-Hempel, 2006). If this controversial hypothesis is proven, it will open new horizons for treating insect diseases.

Methods

Fungal Culture

In culture, *A. apis* can grow on a wide variety of media in either aerobic or anaerobic conditions. Anaerobic conditions seem to facilitate germination of fungal spores (Bailey and Ball, 1991). A list of 19 different media supporting the growth of *A. apis* can be found in Heath (1982). Some of the most frequently used media are Potato-Dextrose agar (PDA), Yeast-Glucose-Starch agar (YGPSA), and Sabouraud Dextrose agar (SDA) (Bailey, 1981; Anderson and Gibson, 1998). For morphological examination of fungal cultures, Spiltoir (1955) recommended using PDA media supplemented with 0.4% yeast extract that supports a strong vegetative growth and abundant sexual reproductive structures. Malt agar (2%) is better for microscopic observations on the basis that it limits the growth of aerial hyphae. *A. apis* can also grow in a wide range of temperatures from 25 to 37°C with the optimum growth at 31 to 35°C (Heath, 1982; Anderson et al., 1997). Murray et al. (2005) reported that culturing *A. apis* on YGPSA (1% yeast extract, 0.2% glucose, 0.1 M KH_2PO_4, 1% soluble starch, 2% agar), a solid culture medium at 35°C under 6% CO_2 produced sufficient vegetative growth and abundant reproductive structures. The two *A. apis* isolates used in this study have been deposited in the USDA-ARSEF collection (Ithaca, New York; accessions 7405 and 7406) by Murray et al. (2005).

Storage conditions were tested using *A. apis* spores and hyphae (Jensen et al., 2009). Both freeze-dried and cryogenically stored *A. apis* spores preserved well. In contrast, *A. apis* hyphae preserved well only using cryopreservation in 10% glycerol, and remained highly viable (up to 98%) even after a year of storage at −80°C (−112°F).

Microscopic Observation of Fungal Structures: The Coverslip Method

This method can be used for the observation of fungal structures without disturbing them by transferring mycelia from a culture plate to a microscope slide (Nugent et al., 2006). Briefly, sterilized coverslips (18 or 22 millimeters) were placed on the bottom of a Petri dish and fixed using small amounts of sterile YGPSA medium. YGPSA medium is then gently poured over the top of the coverslip and allowed to set. Once set, a flamed cork borer is used to remove a plug from the coverslips. Inoculation is done by placing a 2.5-mm diameter inoculum disk from a culture plate of the MAT-1 type idiomorph at one edge of the coverslip.

Similarly, repeat this procedure but use the culture plate of the MAT-2 type idiomorph and place the disk at the opposite edge of the coverslip. Petri dishes, sealed with Parafilm, containing coverslips were then incubated at 25°C (77°F) and fungal structures were observed under a dissecting microscope with transmitted light at different time intervals. Within time, the fungus will gradually grow from the plug and extend onto the coverslip. The coverslip is then removed from the Petri dish and stained as described below. In general, the complete process of sexual reproduction in *A. apis* can be observed within 24 to 48 hours at 25°C using the coverslip method (growth distance ~1.0 centimeter). However, to observe mature ascomata using the Petri dish method (growth distance ~3 centimeters) takes ~5 days at 25°C on YGPSA.

Staining Method for Differentiation of Mating Type Idiomorphs

Similar to the coverslip method described above, growth and structures of the two different fungal strains (MAT-1 and MAT-2) can be examined microscopically without disturbing fungal features. Most taxonomically important structures can be detected at magnifications of 40× to 60×. This is achieved by staining mycelium of the two *A. apis* idiomorphs separately with two different dyes. We used Lactophenol-Acid Fuchsin (LPAF) to stain the MAT-1 idiomorph and Cotton Blue-Lactophenol (CBLP) to stain the MAT-2 idiomorph. Each stain is applied separately, by adding 20 microliters of each stain onto the side of a single idiomorph, then spreading it in a horizontal linear fashion (~10 millimeters). Once the stains are applied at both opposite sides of the coverslip, allow to set for 60 seconds. To mount the coverslip onto the microscope slide, the coverslip is carefully flipped upside down to place the stained tissues directly in contact with a microscope slide and then sealed with nail polish or other slide sealants (Figures 6A, B).

Disease Diagnostics

Diagnosis of the disease in a colony is generally made based on the presence of chalkbrood mummies (as described above). Mummies can be examined by microscopic observations of the smear preparations at 40× magnification. Staining slides with CBLP or LPAF can enhance contrast of the images. The presence of spore cysts in samples is diagnostic evidence of the disease (Figure 4E). To examine fungal growth, a black chalkbrood mummy is crushed in sterile water to produce inocula. A heat treatment of the inoculum is then performed prior to plating fungal spores to kill the non-spore-forming microbes that are routinely found in chalkbrood mummies (Johnson et al., 2005). Fungal growth is typically visible on plates in 2 to 3 days. After 4 to 5 days of incubation at 33°C (91.4°F), dark brown to black specks of the spore cysts appear on the fungal lawn.

Mating type assays in Petri dishes (9.0-cm diameter) are conducted to differentiate between the two opposite mating type strains. Briefly, fungal strains are inoculated onto culture plates at 3.0 cm apart. Reproductive structures will appear as characteristic black lines in the border region between the two opposite mating type strains (Figure 5B).

Very often, honey bee colonies infected with *A. apis* have no visible signs of the disease. Therefore, early detection could be critical for the diagnostics and prevention of disease outbreaks. Among a number of DNA-based methods (Anderson et al., 1998; Reynaldi et al., 2003; Chorbinski, 2004; Borum and Ulgen, 2008), we found two that were easy to use (Murray et al., 2005; James and Skinner, 2005).

Both of these methods utilize species-specific PCR amplification of fungal ribosomal DNA; the presence of a band from PCR amplification indicates the presence of DNA from that species.

We describe here an identification method developed by Murray et al. (2005). Briefly, fungal DNA is extracted using a simple but crude STE method (Aronstein and ffrench-Constant, 1995). This method allows processing a large number of samples in a very short period of time. However, the resulting DNA is not very stable and therefore should be used for PCR analysis within several days. Fungal DNA is amplified in 30-μm reactions using general *Ascosphaera* forward primer AscoF3 (5'–GCACTCCCACCCTTGTCTA–3') and species-specific reverse primer AapisR3 (5'–CCCACTAGAAGTAAATGATGGTTA–3') and the following cycling conditions: 35 cycles of 94°C for 1 minute, 64°C for 1 minute, and 72°C for 1 minute, followed by a final extension of 10 minutes at 72°C.

DNA-based application has also been developed to identify MAT-2 strains (Aronstein et al., 2007) of the fungus. A similar method for identification of MAT-1 strains is still under development (Aronstein, not published).

Larval Bioassay

The *in vivo* larval bioassay is a very useful tool and is now adapted in many research laboratories. In the past, the majority of investigators used "whole chalkbrood mummy" inocula (Gilliam et al., 1988; Starks et al., 2000; Tarpy, 2003) which in addition to *A. apis* contain various species of microbes (Johnson et al., 2005). Recent studies showed that cultured and purified *A. apis* spores can produce consistent results when used as inocula in larval bioassays (Aronstein and Murray, 2010), therefore eliminating all problems related to the use of crude spore preparations.

Another common problem in conducting *in vivo* bioassays is damage to larvae during collection from brood cells and during transfer from an old to a new plate. Stress to larvae during collection can be reduced by using the warm water removal technique (Aronstein et al., 2010). Using this technique, bee larvae are quickly washed out of their brood cells by a gentle stream of warm water. Also, to reduce injury to larvae, they can be manipulated using Chinese grafting tools (commonly used in queen rearing) instead of forceps. A high-efficiency bioassay incorporating these tactics that produced a reliable level of infection was described by Aronstein et al. (2010).

Critical Transition Temperature (CTT) of Chalkbrood Fungi and Its Significance for Disease Incidence

CHAPTER

12

Jay A. Yoder, Derrick J. Heydinger, Brian Z. Hedges,
Diana Sammataro, and Gloria DeGrandi-Hoffman

Abstract Fluctuating temperatures incite chalkbrood disease (*Ascosphaera apis*) in honey bee colonies. This chapter describes a novel technique for estimating the temperature where the incidence of chalkbrood is the highest, which may vary according to species. In the laboratory, growth rates for pure fungal isolates of *A. apis* were determined by measuring the spread of the mycelium in a Petri dish at different temperatures. Graphical analysis shows there is a critical transition temperature (CTT) where a rapid increase in growth rate occurs. CTT of *A. apis* matches the temperature where chalkbrood symptoms are seen. Thus, determination of CTT can be an effective tool for gauging the onset (and offset) of a fungal disease.

Acknowledgments This work was supported in part by the California State Beekeepers Associations and by the Almond Board of California.

Introduction

We report recent observations concerning the effects of temperature on the incidence of chalkbrood, caused by the fungus *Ascosphaera apis*, in honey bee colonies. In particular, this study shows that the temperature where the disease proliferates and pathogenic effects in the colony are seen can be estimated by examining the fungus in the laboratory and determining its critical transition temperature (CTT), the temperature where growth increases rapidly. This is done by measuring how rapidly the fungus grows at different temperatures—the CTT of the fungus matches the temperature where disease symptoms are seen in the colony. This technique of using the CTT can be used for any fungus in the bee colony that can be isolated from bee bread, honey, or the bees themselves.

Fungi have different strains of the same species, a variable that differs in geographic locations. Thus, the temperature where the disease proliferates may be different from apiary to apiary, especially if bee colonies are moved in and out of several locations, such as commercial beekeeping operations. If a fungal infection in the colony is suspected, a sample can be sent to a laboratory where the fungus can be isolated and the CTT determined. Remember, a CTT can be found for any fungus in the bee colony and at any point in time by analyzing bee-associated material (bee bread, honey, or bees). Because chalkbrood fungus (*A. apis*) is a regular inhabitant in the bee colony, symptoms of chalkbrood are not always seen (Gilliam et al., 1978), and the same applies to other fungi that reside in the bee colony. Thus, determination of the CTT of the bee colony fungus provides a tool that allows beekeepers to prevent chalkbrood, or other diseases, and can be applied to different fungus strains and species of fungi by determining which temperatures should be avoided.

Chalkbrood Temperature Requirements

Flores et al. (1996) suggested that 30°C (86°F) is a "critical temperature" for measuring chalkbrood in honey bees, representing a midpoint between the temperatures where the disease appears and disappears (25°C and 35°C; 77°F and 95°F, respectively). This 30°C (86°F) benchmark appears to be an average of the 32°C (89.6°F) (ranges 18°C–35.5°C; 64.4°F–95.9°F) of the brood area and colony environment, maintained by bee activity (Cooper, 1980), and the optimal 33°C (91.4°F) germination temperature for *Ascosphaera apis* as the etiological agent (Bamford and Heath, 1989). Triggering chalkbrood for highest incidence requires cooling (Flores et al., 1996) or warming (Gilliam et al., 1978) the brood, at extreme colony temperatures of 18°C (64.4°F) and 36°C (96.8°F). Therefore, a prevalence of mummies is often seen where temperatures fluctuate widely.

The goal of this study is to examine Flores' hypothesis and determine whether 30°C is a true CTT that applies to this fungus. We also analyzed the effect of pretreatment temperatures (cues known to incite chalkbrood) at 18°C and 36°C on the CTT. We also determined the CTT for *Ascosphaera aggregata*, an agent of chalkbrood in alfalfa leafcutting bees *Megachile rotundata*, extrapolating data from James (2005). We anticipate that CTT will be lower for *A. aggregata* than for *A. apis* because the highest incidence of infection for *A. aggregata* is at 20°C (68°F) rather than 25°C (77°F) for *A. apis*. The importance of this study is using the CTT of a fungus as a new tool for defining the onset (initial display of symptoms) and offset (low frequency of infection) of a fungal disease.

Methods

A. apis was recovered from Arizona mummies as described by Johnson et al. (2005) using Potato Dextrose agar containing 0.4% yeast extract in 100 × 15-mm Petri dishes (Fisher Scientific, St. Louis, MO) and subculturing hyphal tips to accomplish purity. Incubation conditions were 25 ± 0.5°C (77°F) transferring from 5% CO_2 to aerobic conditions after a day. Identification followed the techniques of Christensen and Gilliam (1983) using 1000× microscopy. Conditions used for experiments were similar, except the growth media was nonnutritional agar (Fisher) containing crushed autoclave-sterilized larval bees in 100 mg/100 × 15-mm Petri plates, to simulate natural infection, or using V-8 agar to permit comparison with *A. aggregata* (James, 2005). Temperatures were varied using programmable incubators (< ± 0.5°C; Fisher) and the relative humidity at 97% was maintained using glycerol-water mixtures (Johnson, 1940) measured with a hygrometer (SD ± 0.5% RH; Thomas Scientific, Philadelphia, Pennsylvania). Petri plates were inoculated with a 1-cm^3 block of the fungus inoculum centered on the plate.

Thirty measurements of mycelium spread over the agar surface were taken following the trisecting line method (Yoder et al., 2008) and fit to the equation: $K_r = (R_1 - R_0)/(t_1 - t_0)$ (where K_r is radial growth rate, R_0 and R_1 are colony radii at beginning of linear, t_0, and stationary, t_1, phases of growth). No growth was defined as the absence of fungal hyphae from the block of inoculum by 40× microscopy. CTT is defined as the temperature where the activation energy (E_a) changes as denoted by a difference in slope ($-E_a/R_{gas}$) on an Arrhenius plot (Thammavongs et al., 2000) in the equation: $k = Ae^{-Ea/(RT)}$ (where k is growth rate, A is steric factor, R is gas constant, and T is absolute temperature). Radial growth rate was determined as *ramp up* (same fungus culture transferred from

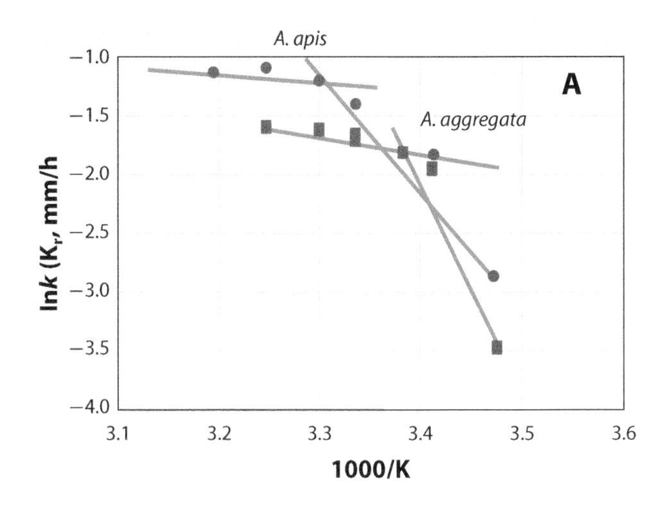

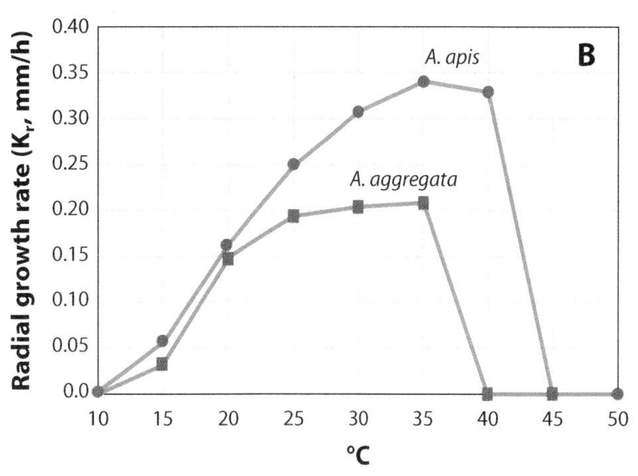

Figure 1. The importance of this figure is that it shows the graphical method (**A**) for determining critical transition temperature (CTT) and presence of a distinct CTT at 30°C (86°F) for *Ascosphaera apis* (circle) and 23°C (73.4°F) for *A. aggregata* (square) as indicated by a "kink" in the regression line; *k* is the radial growth rate (K_r) of the fungus in culture. The slopes of regression lines correspond to activation energies (E_a) and the "kink" indicates that activation energy (slope) changes, denoting a CTT at the intersection of the two lines. **B.** Linear plot of the same data set replotted from **A** to determine CTT. Similar results for *A. apis* were obtained using bee larva-supplemented media (shown), Potato Dextrose agar, and V-8© agar, permitting comparison to James (2005) using V-8© agar for *A. aggregata*. The temperature-growth data for *A. aggregata* are from James (2005).

low to high temperature), *ramp down* (same fungus culture transferred from high to low temperature), and *separate* (no transfer; one culture at each temperature). Each experiment was replicated three times (ten Petri plates/temperature/replicate) using *A. apis* from a different mummy. Data were compared by analysis of variance (ANOVA; SPSS 14.0; Chicago, Illinois) and a test for the equality of slopes of several regressions (Sokal and Rohlf, 1995).

Results

A. apis has a CTT at about 30°C where the radial growth rate increases abruptly, resulting in a change in activation energy (i.e., slope changes, represented by a "kink" in the regression line on the plot; Figure 1A) and matches the inflection on the curvature on a linear plot (Figure 1B). Radial growth rate increases proportionately in both low ($R \geq 0.89$; below CTT) and high ($R \geq 0.93$; above CTT) temperature ranges ($P < 0.001$). Lethal temperatures 10°C (50°F), 45°C (113°F), and 50°C (122°F) were excluded from analysis. Different long-term or "shock" exposure temperatures had no appreciable effect in suppressing or enhancing *A. apis* growth, as evidenced by the closeness of activation energies and where CTT occurs (Table 1), confirming the ability by *A. apis* to resist thermal stress (Hale and Menapace, 1980).

A. aggregata grows slower than *A. apis* (James, 2005; Figure 1B), reflected by the entire biphasic relationship being shifted down the ordinate and a lower steric factor *A* (*y*-intercept, Figure 1A). The CTT of *A. aggregata* is about 23°C (73.4°F) (Figure 1A) and has activation energies nearly two times higher than *A. apis* in both the low and high temperature ranges. This demonstrates that *A. aggregata* has a slower growth rate than *A. apis* (Figure 1B). So as the magnitude of E_a increases, *k* becomes smaller and rates decrease as the energy barrier increases (Thammavongs et al., 2000). *A. apis* appears to be more heat tolerant than *A. aggregata* (Figure 1B) based on higher growth response as the temperature rises and lethal threshold (i.e., temperature where the fungus dies): 40°C (104°F) for *A. aggregata* and 45°C for *A. apis,* and 10°C for both of these fungal species. *Ramp up*, *ramp down*, and *separate* experiments yielded nearly identical results. Because activation energy describes growth as a function of temperature over a broad range of temperatures, it is likely a good indicator of the fungus' response to naturally occurring temperature fluctuations.

Conclusions

We conclude that (1) bee colony temperatures do not suppress the growth of *A. apis* and *A. aggregata*; (2) the absence of changes in activation energy (Table 1) imply that cold or heat "shock" cues that promote chalkbrood do not trigger *A. apis* to proliferate; (3) CTT is an unreliable ecological indicator of the spread of *A. apis* (i.e., low CTT of *A. aggregata* does not restrict it to more northern climates or *A. apis* to warmer ones, because *A. apis* is also distributed in northern environments and has a higher CTT); and (4) CTT compares to the temperature where there is greatest incidence of chalkbrood infection (20°C [68°F] for *A. aggregata*; 25°C [77°F] for *A. apis*) that corresponds simultaneously with the peak sporulation activity of the chalkbrood fungus (25°C [77°F] for *A. aggregata*; 30°C to 35°C [86°F to 95°F] for *A. apis*).

These conclusions indicate that CTT may help in identifying pathogenic effects of different fungal species (or strains) and in predicting colony temperatures where disease symptoms occur. Cooling and warming cues that stimulate chalkbrood appear to be bee related, suggesting that chalkbrood is related more to a *decrease* in bee resistance, and not an *increase* in growth and proliferation of the fungus. Preventative measures for chalkbrood should focus on maintaining bees resistant to chalkbrood and in sustaining proper bee health and nutrition.

Table 1. This table shows that the CTT matches the temperature for measuring *A. apis* (about 30°C), as reported by Flores et al. (1996), and A. *aggregata* (about 23°C). Thus, temperature for chalkbrood can be determined by calculating CTT to know what temperatures should be avoided. Lower activation energies for *A. apis* indicate that it grows faster than *A. aggregata*. Also, chalkbrood is related more to a decrease in bee resistance rather than an increase in fungal growth, because a previous chilling and warming "shock" known to incite chalkbrood in the colony, when applied to *A. apis* in culture, does not act as a trigger to promote fungus growth (i.e., no changes in activation energies).

Fungus Pretreatment (Exposure at °C for # Days)	CTT (°C)	E_a < CTT (kJ/mol)	E_a > CTT (kJ/mol)	Chalkbrood Incidence (% Infected Larvae)*
A. apis				
15°C/6 days	31.4 ± 1.3[a]	81.33 ± 7.2[a]	2.48 ± 0.39[a]	
20°C/6 days	33.7 ± 2.3[a]	76.27 ± 3.1[a]	2.33 ± 0.53[a]	
*25°C/6 days	31.6 ± 0.8[a]	83.61 ± 4.8[a]	1.94 ± 0.29[a]	77.62 ± 20.88[a]
*30°C/6 days	29.1 ± 1.4[a]	85.92 ± 6.8[a]	2.12 ± 0.19[a]	15.31 ± 9.98[b]
*35°C/6 days	34.0 ± 2.1[a]	79.04 ± 5.1[a]	2.23 ± 0.43[a]	2.22 ± 3.01[c]
40°C/6 days	31.2 ± 2.6[a]	84.64 ± 7.3[a]	2.27 ± 0.24[a]	
Chilling stress				
*18°C/1 day, 25°C/6 days	31.1 ± 3.2[a]	83.71 ± 9.6[a]	1.89 ± 0.31[a]	94.99 ± 5.21[d]
*18°C/1 day, 30°C/6 days	30.5 ± 1.6[a]	86.29 ± 4.9[a]	2.37 ± 0.27[a]	94.99 ± 5.21[d]
*18°C/1 day, 35°C/6 days	29.8 ± 0.9[a]	82.36 ± 6.2[a]	2.17 ± 0.19[a]	29.00 ± 37.70[f]
Warming stress				
36°C/1 day, 25°C/6 days	32.7 ± 1.1[a]	77.08 ± 4.1[a]	2.41 ± 0.25[a]	
36°C/1 day, 30°C/6 days	31.3 ± 2.7[a]	85.11 ± 3.7[a]	2.39 ± 0.33[a]	
36°C/1 day, 35°C/6 days	30.2 ± 2.5[a]	84.72 ± 7.6[a]	2.08 ± 0.23[a]	
† *A. aggregata*				
32°C/7 days	22.8 ± 1.4[b]	140.67 ± 8.4[b]	5.73 ± 0.34	

* From Flores et al. (1996); † Obtained from James (2005) in Figure 1A. Mean values (± SE) followed by same letter are not different (P > 0.05).

Small Hive Beetle (*Aethina tumida*) Contributions to Colony Losses

CHAPTER

13

James D. Ellis

Abstract Small hive beetles (*Aethina tumida*) can kill honey bee colonies and impact colony health and productivity significantly. They typically inhabit colonies of African races or subspecies of Western honey bees (*Apis mellifera*), probably identifying the host colony by a suite of olfactory cues. Small hive beetles depend on bee colonies as their primary host. Molecular techniques have helped us understand the introduction, presence, and spread of the beetles in North America. The majority of losses occur in the southern United States, particularly the Southeast . The beetles appear to be a more regional than global pest. The largest threat to the bee colony posed by this beetle are the beetle larvae that feed on honey, pollen, and bee brood. The focus of this chapter is on how small hive beetles impact honey bee colonies and contribute to colony losses in the United States.
I emphasize the biology, researchable topics that remain poorly understood, particularly in the area of control (chemical, cultural, biological, genetic), and the need for further work. Control of small hive beetles seems to be best accomplished by attacking all beetle life stages simultaneously and maintaining bee colonies in a populous and healthy state.

Introduction

Small hive beetles (*Aethina tumida*) are members of the coleopteran family Nitidulidae and are native to sub-Saharan Africa (Lundie, 1940; Schmolke, 1974; Neumann and Ellis, 2008). In their native range, small hive beetles typically inhabit colonies of African races, or subspecies, of Western honey bees (*Apis mellifera*). In 1996, small hive beetles were found in colonies of European races of honey bees in the United States (Figure 1). Since that time, they have been found in Australia (Neumann and Ellis, 2008), where they have established populations, and in a few other countries where they do not seem to have established populations [such as Portugal (Neumann and Ellis, 2008) and Canada (Lounsberry et al., 2010).

Techniques in molecular genetics have provided a way to understand beetle presence in and dispersion through areas in which they are introduced. In North America, small hive beetles (SHB) have at least two distinct mitochondrial DNA haplotypes, thus initiating some discussion of the number of separate introductions into and their subsequent spread within the United States (Evans et al., 2003; Lounsberry et al., 2010). Furthermore, genetic markers have been improved, allowing a similar discussion on the global scale (Evans et al., 2007). Lounsberry et al. (2010) examined genetic variation in adult beetles collected from the United States, Australia, Canada, and Africa. They demonstrated that beetles in Canada were from at least two separate introductions, one from the United States and the other from Australia. SHB in North America and Australia appear to have different African origins (Lounsberry et al., 2010).

A number of good reviews on the biology, behavior, ecology, and control of SHB exist (Hood, 2000, 2004; Neumann and Elzen, 2004; Neumann and Ellis, 2008; Ellis 2005a,b,c; Ellis and Hepburn, 2006). In this chapter, I focus primarily on how SHB impact honey bee colonies and contribute to colony losses in the United States, though it is important to appreciate the biology of SHB to understand how they impact honey bee colonies specifically. I include within each section of this chapter a brief discussion of researchable topics that remain poorly understood for SHB.

Figure 1. The global distribution of small hive beetles (modified from Ellis and Munn, 2005; Neumann and Ellis, 2008). Shaded countries are where the beetle occurs, though their exact distribution within a given country may be unknown. For example, small hive beetles are in Australia so the entire country is shaded. However, the beetles are not distributed across the entire country.

This inclusion hopefully will demonstrate the need for further work on this significant honey bee pest.

Small Hive Beetle Life Cycle

SHB pupate in the soil around bee colonies (Lundie, 1940; Schmolke, 1974). Upon emerging from the ground, adult beetles (Figure 2) search for colonies, probably identifying the host colony by a suite of olfactory cues. Though SHB will feed and reproduce on fruit (Ellis et al., 2002b; Buchholz et al., 2008; Arbogast et al., 2009)—a behavior similar to that of other nitidulid beetles—they likely are dependent on bee colonies as their primary host. Some have suggested that SHB may reproduce on fruit *in vivo* (Arbogast et al., 2009; Arbogast et al., 2010) but this certainly is a topic that needs further investigation. If SHB are able to reproduce on fruit and do so readily, their populations could be sustained in areas where honey bees are not present.

Investigators have shown that SHB fly before or just after dusk (Elzen et al., 1999) and that odors from adult bees and various hive products (honey, pollen) are attractive to flying SHB (Elzen et al., 1999; Suazo et al., 2003; Torto et al., 2005; Torto et al., 2007a). Some authors have suggested that SHB may find host colonies by detecting honey bee alarm pheromones produced either by the bees themselves or by a yeast (*Kodamaea ohmeri*) carried by the beetles that produces components of the bee's alarm pheromone when deposited on pollen reserves in the hive (Torto et al., 2007b).

Figure 2. The life cycle of the small hive beetle. **A.** adult beetle (dorsal and ventral view), **B.** beetle eggs, **C.** beetle larvae (dorsal view), **D**. beetle pupae (ventral view).
(*Photos courtesy of the University of Florida.*)

Upon locating a colony, adult beetles are met with some resistance by the bees as they try to enter the nest. Once inside, the beetles hide in cracks and crevices around the periphery of the nest where they escape bee aggression (Ellis, 2005c). It remains unknown if beetles actively seek these sites as part of their normal behavior in colonies (i.e., they "want" to be in these sites) or if they find the sites passively as they try to escape bee aggression (i.e., they are "forced" into these sites by the bees). Regardless, honey bees station guards around the cracks/crevices where SHB hide. The sites are called "confinement sites" because the beetles are confined by the bees at these locations (Ellis et al., 2004b). By confining SHB, the guard bees limit beetle access to the brood combs where there is an ample supply of honey, pollen, and brood on which adult beetles reproduce. SHB do not starve while confined because they are able to solicit food trophallactically from their bee captors (Ellis et al., 2002a). The beetles accomplish this by using their antennae to rub the bees' mandibles, thus inducing the bees to regurgitate; the beetles then feed on the regurgitated substance.

Mating behavior of SHB (including whether female beetles mate once or multiple times) is not understood well, but beetles do not appear to be sexually mature until about one week post-emergence from the soil (Lundie, 1940; Schmolke, 1974). If able to escape confinement and reproduce, female beetles oviposit directly onto food sources such as pollen (Lundie, 1940) or on developing bee brood (Ellis et al., 2003a). Alternatively, female beetles may deposit irregular masses of eggs (Figure 2B) in crevices or cavities away from the bees; the ovipositors are long and flexible, and are perfectly designed to lay eggs in tiny and concealed places. A female beetle may lay 1,000 eggs in her lifetime. The majority of these eggs hatch within 3 days (Lundie, 1940); however, some eggs remain viable and hatch after 5 days. Humidity appears to be a crucial factor influencing hatch rates because small hive beetle eggs are prone to desiccation if exposed to circulating air and a relative humidity below 50% (Jeff Pettis, personal communication, unpublished data). Though beetle reproduction is not clearly visible in all hives, low-level, cryptic reproduction occurs in most colonies in which beetles are present (Spiewok and Neumann, 2006a).

Newly hatched beetle larvae (Figure 2C) immediately begin feeding on whatever food source is available including honey, pollen, and bee brood (Lundie, 1940; Schmolke, 1974), yet they have demonstrated a preference for bee brood (Elzen et al., 1999). Maturation time for larvae is generally 10 to 14 days although some may feed longer than a month (Lundie, 1940; Ellis et al., 2002b). Once the larvae finish feeding, a "wandering" phase is initiated where they leave the food source and migrate out of the colony to find suitable soil in which to pupate (Lundie, 1940). It is believed that most larvae do this at night.

Larvae in the wandering stage may wander some distance from the hive to find suitable soil in which to pupate, though most pupate within 90 cm (~3 feet) of the hive (Pettis and Shimanuki, 2000). Nearly 80% of the larvae burrow down into the soil less than 10 cm (~4 inches) from the soil surface but not generally more than 20 cm (~8 inches; Pettis and Shimanuki, 2000). They pupate (Figure 2D) best in moist soil (Ellis et al., 2004c). Once larvae cease burrowing, they construct a smooth-walled, earthen cell in which they pupate. The period of time spent in the ground pupating can vary greatly, depending on factors such as soil temperature (de Guzman and Frake, 2007); however, the majority of adults emerge from the soil in 3 to 4 weeks (Lundie, 1940). Upon adult emergence, the entire life cycle begins again. The turnover rate from egg to adult can be as little as 4 to 6 weeks; consequently, there may be as many as six generations in a 12-month period under moderate United States climatic conditions.

Much about SHB biology remains unknown. For example, no one is sure how far they can fly when searching for a host colony or if soil pH affects pupation success. Furthermore, we know very little about how confinement behavior is initiated within a colony, be it from the beetles themselves or from a sub-cohort of bees who "imprison" small, invading arthropods. Finally, the trophallactic relationship between SHB and guard honey bees remains unclear. Though we know SHB are fed honey (Ellis et al., 2002a), we do not know if they also pass food or other substances to their bee captors.

Colony Damage Typically Attributed to Small Hive Beetles

SHB can kill honey bee colonies. Though they are not responsible for universal, widespread colony losses as are other stressors (such as Colony Collapse Disorder, disease, or poor nutrition), they can plague beekeeping operations in areas where they are present. In their native range of sub-Saharan Africa, SHB regularly occur in honey bee colonies but typically do not damage strong, healthy hives (Lundie, 1940; Schmolke, 1974), even though low-level, cryptic beetle reproduction occurs (Spiewok and Neumann, 2006a). That said, African subspecies of honey bees are not immune completely to SHB as they can kill colonies, cause them to abscond, and may hurt other areas of colony productivity (Ellis et al., 2003b). Overall damage attributable to SHB in colonies south of the Sahara likely is minimal compared to the damage beetles can cause in European bee colonies that are in the beetle's introduced range. Why this is the case remains unknown.

It is possible that African races of honey bees are more aggressive toward SHB than are European races of bees (Elzen et al., 2001). That said, European bees also confine SHB (Ellis, 2005c) and bite/attack the beetles whenever possible. Regardless, more research should be conducted to determine how African races of honey bees tolerate small hive beetle invasion when European bees often do not.

In the United States and Australia, European bee colonies often succumb to SHB damage. Typically the process follows a characteristic pattern:

1. Adult beetle invasion into colonies
2. Population buildup of adult beetles
3. Reproduction of adult beetles
4. Significant damage to brood, pollen, and honey stores by feeding beetle larvae (Figure 3)
5. Exodus of larvae from the colony
6. Beetle pupation in the soil
7. Beetle emergence as adults and subsequent reinfestation of colonies (Ellis and Ellis, 2008).

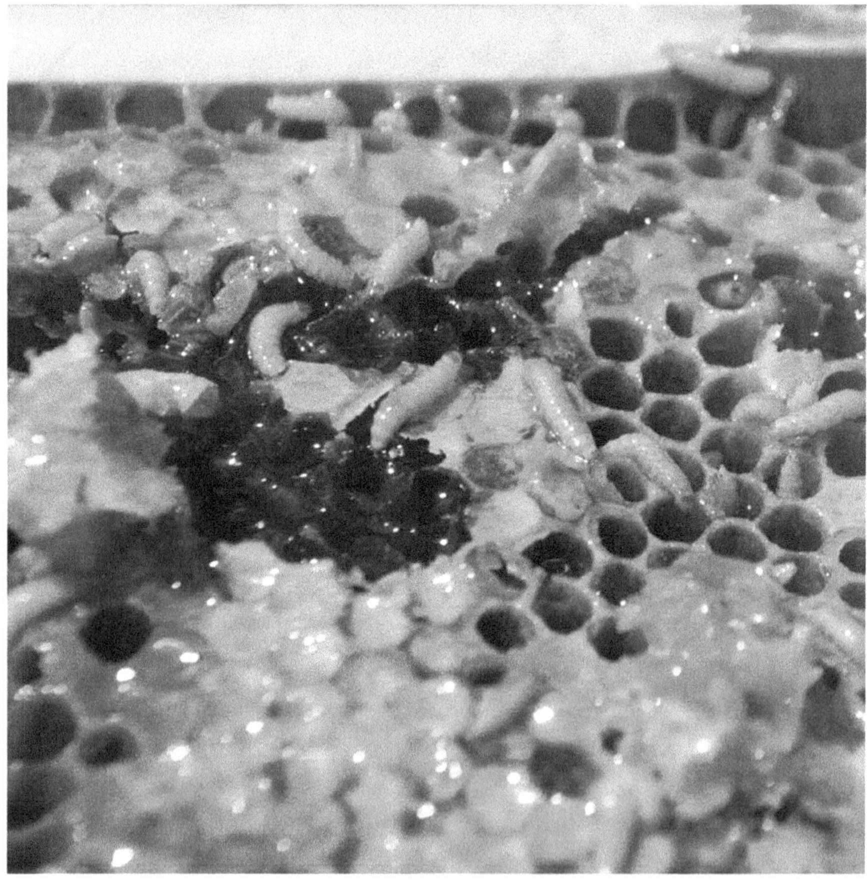

Figure 3. Two views of honey bee comb damaged by feeding small hive beetle larvae. Notice how the comb is "slimed" due to the fermentation of honey. (*Photo courtesy of University of Florida.*)

Though this pattern is somewhat predictable, the amount of time it takes to complete this cycle, which ultimately leads to colony loss, is unpredictable, as are the triggers that turn ordinary beetle inhabitance of colonies into damaging beetle presence. For example, SHB tend to occupy many/most colonies in an area where they are present. Yet, only a certain percentage of infested colonies ultimately succumb to beetle pressures. Furthermore, beetle damage in colonies does not appear completely related to beetle density within colonies (Delaplane et al., 2010). As such, step two in the typical pattern of colony loss (population buildup of beetles) does not appear critical if other conditions/stressors are present in the colony.

Likely, many colony losses attributable to SHB were initiated by other stressors. For example, queenless colonies, nucleus or other small colonies, and colonies stressed with parasites/pathogens all appear hypersensitive to SHB and associated depredation. To that end, SHB increasingly are considered a secondary pest of bee colonies, much like the greater (*Galleria mellonella* L.) and lesser (*Achroia grisella* Fabricius) wax moths. On the other hand, it is possible that not all stressors predispose colonies to SHB damage. Varroa mite (*Varroa destructor* Anderson and Trueman) infestations, for example, may not be linked directly to small hive beetle damage as investigators failed to correlate colony varroa mite populations with colony susceptibility to SHB (Delaplane et al., 2010). Regardless, colonies appear to be vulnerable to SHB more when they are stressed in other ways first. Though not always easy to investigate, synergisms between SHB and other colony stressors should be investigated further.

SHB may not be problematic only to honey bee colonies. In controlled studies, SHB have been shown to infest and cause significant damage to managed bumble bee colonies. Ambrose et al. (2000) and Stanghellini et al. (2000) found that commercial *Bombus impatiens* colonies artificially infested with SHB suffered significantly from beetle damage, with the beetles being able to complete a full life cycle within the colonies. Spiewok and Neumann (2006b) found that commercial *B. impatiens* colonies maintained close (~100 or ~500 m) to beetle-infested honey bee colonies became infested with SHB, which reproduced successfully. Hoffman et al. (2008) placed four commercial *B. impatiens* colonies and four honey bee colonies in a closed greenhouse and released 1,000 SHB. All colonies of both bee types were invaded by the beetles, which oviposited in the colonies and showed no apparent preference to honey bee over bumble bee colonies. Interestingly, the stingless bee *Trigona carbonaria* has been shown to host SHB but can avoid damage attributable to the beetles by mummifying them alive. The bees do this by coating the beetles with a mixture of mud, wax, and resin (Greco et al., 2010).

Other Effects of Small Hive Beetles on Colonies

SHB do not only cause colony collapse or death but they can affect honey bee colonies in other ways. First, there is a fermentation of hive products (particularly honey) associated with the feeding beetle larvae (Lundie, 1940; Schomolke, 1974). This likely occurs due to specific yeasts associated with the SHB. Affected honey can "bubble" out of the combs, resulting in a frothy mess inside the hive. Honey damaged by SHB is rendered foul and unfit for human or bee consumption.

Second, honey bee colonies heavily infested with adult SHB may abscond or completely abandon the nest (Ellis et al., 2003b). Admittedly, the number of adult beetles to elicit this behavior in European honey bee colonies likely must be high (>1,000 beetle adults/colony). However, colonies of African bees may abscond in response to lower beetle populations because African races of bees are prone to absconding anyway (Hepburn and Radloff, 1998). Although adult beetles are known to promote absconding behavior, it remains unclear if the presence of beetle larvae (particularly high populations of beetle larvae) can do the same.

As predicted (Ellis and Hepburn, 2006), SHB have been shown to transmit pathogens within (and possibly between) honey bee colonies mechanically (Eyer et al., 2009ab; Schäfer et al., 2010). This is the third way beetles can affect colonies without killing them directly. The bodies of SHB are covered with hair and other anatomical features that permit them to acquire bee pathogens. Beetle movement within diseased colonies and subsequent relocation to healthy colonies is a possible method of mechanical transmission of pathogens between colonies; although this has not been shown with certainly (Koch's Postulates must be satisfied first). SHB are known to harbor Sacbrood Virus (Eyer et al., 2009a), Deformed Wing Virus (Eyer et al., 2009b), and *Paenabacillus larvae*, the American foulbrood bacterium (Schäfer et al., 2010). Furthermore, it is possible, though untested, that SHB can acquire pathogens from bees trophallactically when confined (Ellis et al., 2002a). Regardless, beetle contributions to the movement of bee pathogens within and between honey bee colonies (mechanical transmission) must be investigated further as should their role as possible vectors of bee pathogens (i.e., the pathogen reproduces within the small hive beetle, making the beetle a true vector).

There is a fourth cost to honey bee colonies when hosting SHB: the cost of mounting an immune response to the beetle. The two primary immune responses of bee colonies to SHB include confinement behavior (Ellis, 2005c) and adult bee hygienic behavior toward beetle eggs/young larvae (Ellis et al., 2003a, 2003d, 2004a; Neumann and Härtel, 2004; Ellis and Delaplane, 2008; de Guzman et al., 2008). Confinement behavior involves honey bees being diverted from jobs they would do ordinarily to guarding confined SHB. Ellis et al. (2003c) showed that European and African races of honey bees are ~19 to 21 days old, respectively, when they begin guarding confinement sites. At this age, worker bees should begin foraging for colony resources. As such, increased populations of SHB within colonies may lead to a reduction in foraging activity in affected colonies (Ellis et al., 2003b).

Additionally, SHB can oviposit directly into brood cells (Ellis et al., 2003a), a behavior that elicits hygienic responses from adult worker bees. The bees are able to detect eggs (Ellis et al., 2003d, 2004a; Ellis and Delaplane, 2008; de Guzman et al., 2008) and young larvae (Lundie, 1940; Schomolke, 1974; Neumann and Härtel, 2004) present in the brood and remove the affected brood and beetle eggs/larvae. Although this trait may benefit colonies overall, expending energy to remove young bees that would otherwise contribute to the workforce seems costly to the colony.

A final effect of SHB is that their presence may make colonies susceptible to invasion by other beetles. "Small" hive beetles—so named to distinguish them from a variety of scarab beetles invading bee colonies (i.e., "large" hive beetles, Swart et al. 2001)—are not the only nitidulid beetles known to invade honey bee colonies. *Cychramus luteus* (Neumann and Ritter, 2004), *Glischrochilus fasciatus, Lobiopa insularis,* and *Epuraea corticina* (Ellis et al., 2008) all have been found in bee colonies, though they are presumed innocuous to honey bees. Regarding *G. fasciatus*, *L. insularis*, and *E. corticina*, all three have been found in honey bee colonies also hosting SHB (Ellis et al., 2008). It is possible, though untested, that the presence of SHB in bee colonies has opened an ecological niche for other invading nitidulid beetles (beetles never before documented in honey bee colonies prior to the occurrence of SHB) via their fermentation of hive products.

Small Hive Beetle Contributions to Colony Losses

It has been difficult to assess the overall impact of SHB on bee colonies in the United States. In surveys of colony losses, fewer than 1% (vanEngelsdorp et al., 2008) and 1.9% of beekeeper respondents (vanEngelsdorp et al., 2010) reported losing colonies to SHB during the 2007–2008 and 2008–2009 beekeeping seasons in the United States. The same surveys showed that the losses attributable to SHB were insignificant compared to losses due to colony starvation, queen loss, weather, mites, Colony Collapse Disorder, Nosema, management, and other colony issues.

Most losses attributable to SHB occur in the southern United States, particularly the Southeast where beetle populations are highest and the climate is conducive to beetle reproduction. Their presence in Australia is relatively localized as well (Neumann and Ellis, 2008). Consequently, SHB seem to play a minor role in overall colony losses in areas where they are introduced, although under certain circumstances their presence can impact colony health and productivity significantly.

Small Hive Beetle Control

Since the introduction of SHB into the United States, little progress toward developing chemical control methods has been made. Two pesticides are available, permethrin and coumaphos. Permethrin as an active ingredient is used as a ground drench and kills beetle larvae and pupae in the soil around infected colonies. As an in-hive control, the organophosphate coumaphos is embedded into a plastic strip that is placed on the bottom board of a colony. The strip is placed under a piece of cardboard where the beetles can hide, thus exposing them to the active ingredient. Neither product is very efficacious, though some beetle control is possible with both (Ellis, 2005a,b).

Because of the lack of quality chemical control options, more attention has focused on integrated pest management (IPM) using cultural, biological, and genetic controls. Cultural/mechanical controls result from a change in practice with the intention of limiting, but not eradicating, a pest. Practices such as removing honey, bits of comb and cappings from around the honey house will minimize foodstuffs to which SHB may be attracted (Hood, 2000, 2004; Ellis, 2005a,b). It is also important to extract supers of honey quickly to reduce the damage that adults beetles and larvae do to standing, unprotected honeycombs. Reducing the relative humidity to 50% in honey houses and other places where honey is stored inhibits SHB eggs from hatching (Jeff Pettis, personal communication). In the apiary, eliminate, requeen, or strengthen weak colonies to reduce colony stress and to make the colony better able to deal with the beetles. Avoid other conditions that might lead to colony stress such as brood diseases, mite problems, wax moth activity, failing queens, excessive swarming, and over-supering.

Multiple, moderately effective in-hive trapping devices have been developed for SHB control. These traps often are filled with an attractant (usually apple cider vinegar) and/or a killing agent (mineral oil, vegetable oil, etc.) and typically capture adult beetles. Some traps, such as the Hood Beetle Trap (Hood and Miller, 2003; Hood, 2006; Nolan and Hood, 2008), are placed within a frame while others, such as the West Beetle Trap (West, 2004), go under colonies, thus providing a place for beetles to hide.

Other control measures are being developed. For example, the yeast *Kodamaea ohmeri,* when mixed with pollen, produces volatiles that attract other SHB to weakened colonies (Torto et al., 2007b). Researchers are taking advantage of this relationship and are developing traps using the yeast mixed with pollen as bait for adult SHB (Noland and Hood, 2008). Second, multiple species of soil-dwelling nematodes have demonstrated activity against pupating SHB (Cabanillas and Elzen, 2006; Ellis et al., 2010; Shapiro-Ilan et al., 2010). Two species in particular that have shown promise as controls include *Heterorhabditis indica* and *Steinernema riobrave* (Ellis et al., 2010). Additionally, some fungal species have shown activity against pupating SHB, though the efficacy of fungi as biological controls remains questionable (Ellis et al., 2004d; Richards et al., 2005; Muerrle et al., 2006). Finally, researchers have shown that some honey bee colonies are able to detect and remove brood that has been oviposited on by SHB (Ellis et al., 2003d, 2004a; Ellis and Delaplane, 2008; de Guzman et al., 2008). This behavior, called hygienic behavior, possibly can be selected for in-breeding programs and may help reduce SHB problems.

SHB control probably is best accomplished by attacking all beetle life stages simultaneously. For example, one can use in-hive traps for adult beetles, nematodes for wandering larvae and pupal beetles, and hygienic breeds of bees for beetle eggs and young larvae. That said, the best defense to date against SHB is maintaining colonies in a populous and healthy state. Without question, the control of SHB is a research topic that still needs considerable further attention.

Conclusion

In conclusion, SHB likely do not contribute significantly to global losses of honey bee colonies, though they can be an important regional menace. That may change over time as the beetle disperses into areas where it is not indigenous. More research needs to be conducted to determine more subtle effects of their presence in bee colonies and to develop control measures for this pest.

Pesticides and Honey Bee Toxicity in the United States

CHAPTER 14

Reed M. Johnson, Marion D. Ellis, Christopher A. Mullin, and Maryann Frazier (edited by Yves Le Conte)

Abstract Until 1985 discussions of pesticides and honey bee toxicity in the United States were focused on pesticides applied to crops and the unintentional exposure of foraging bees to them. The recent introduction of arthropod pests of honey bees, *Acarapis woodi* (1984), *Varroa destructor* (1987), and *Aethina tumida* (1997) to the United States have resulted in the intentional introduction of pesticides into beehives to suppress these pests. Both the unintentional and the intentional exposure of honey bees to pesticides have resulted in residues in hive products, especially beeswax. This review examines pesticides applied to crops, pesticides used in apiculture, and pesticide residues in hive products. We discuss the role pesticides and their residues in hive products may play in Colony Collapse Disorder and other colony problems. Although no single pesticide has been shown to cause CCD, the additive and synergistic effects of multiple pesticide exposures may contribute to declining honey bee health.

Acknowledgments This chapter is reprinted with permission from *Apidologie* 41 (2010) 312–331.

Pesticides Applied to Crops

Despite the dependence on honey bees for the pollination of crops in the United States, colony numbers have declined by 45% over the past 60 years (NAS, 2007). Most honey bee losses from 1966 to 1979 were attributable to organochlorine, carbamate, organophosphorus, and pyrethroid pesticide exposure (Atkins, 1992). Efforts to restrict pesticide application during bloom provided some relief; however, the residual activity of some pesticides was never effectively addressed (Johansen and Mayer, 1990). Previous reviews and extension publications are available concerning the protection of honey bees from these four classes of pesticides (Johansen, 1977; Crane and Walker, 1983; Adey etal., 1986; Johansen and Mayer, 1990; Ellis et al., 1998).

Colony losses were especially severe from 1981 to 2005 with a drop from 4.2 million to 2.4 million (NAS, 2007) although some of the decrease is attributable to changes in how colony numbers were estimated. The introduction of parasitic honey bee mites, *Acarapis woodi* (1984) and *Varroa destructor* (1987), contributed to dramatic bee losses. At the same time, the control of crop pests in United States agriculture was rapidly changing. Genetically engineered (GE) crops were developed and extensively deployed, and two new classes of systemic pesticides, neonicotinoids and phenylpyrazoles, replaced many of the older pesticides described above.

The rapid development and deployment of these two new insect control techniques distinguish United States agriculture from agriculture in other regions of the world. In Europe a more cautious approach to the adoption of new agricultural practicices has been taken. Because the registration and regulation of GE crops and neonicotinoid and phenylpyrazole pesticides are major shifts in insect control in United States agriculture, they are emphasized in this section of our review.

The recent sequencing of the honey bee genome provides a possible explanation for the sensitivity of bees to pesticides. Relative to other insect genomes, the honey bee genome is markedly deficient in the number of genes encoding detoxification enzymes, including cytochrome P450 monooxygenases (P450s), glutathione-S-transferases, and carboxylesterases (Claudianos et al., 2006).

Genetically Engineered Plant Varieties

Genetically engineered (GE) plant varieties that have herbicide tolerance or insecticidal properties were first introduced into the United States in 1996. Soybeans and cotton are genetically engineered with herbicide-tolerant traits and have been the most widely and rapidly adopted GE crops in the United States, followed by insect-resistant cotton and corn. In 2007 these GE crops were planted on more than 113 million hectares worldwide (USDA-Biotech Crop Data, 2009). The United States leads the world in acres planted with GE crops with most of the plantings on large farms (Lemaux, 2008). Insect resistance is conferred by incorporating genes coding for insecticidal proteins produced by *Bacillus thuringensis* (Bt), a widespread soil bacterium (ISB, 2007). While Bt is also delivered through traditional spray application, plants benefit from continuous production of Bt toxins through genetic engineering. Bt δ-endotoxins are activated in the insect gut where they bind to receptor sites on the midgut epithelium to form pores. These pores allow gut contents to leak out of the lumen and cause osmotic stress to midgut cells, leading to the eventual destruction of the midgut and the death of the insect (Soberon et al., 2009). To date, Bt genes have been incorporated into corn (*Zea mays*), cotton (*Gossypium hirsutum*), potato (*Solanum tuberosum*), and tomato (*Lycopersicon esculentum*), and GE seeds of these crops are available to producers (ISB, 2007). Precommercial field tests of 30 different plant species with Bt genes were conducted in 2008, including apples, cranberries, grapes, peanuts, poplar, rice, soybeans, sunflowers, and walnuts (ISB, 2007).

Numerous studies have been conducted to determine the impact of GE crops on honey bees. Canadian scientists found no evidence that Bt sweet corn affected honey bee mortality (Bailey et al., 2005). Studies conducted in France found that feeding Cry1ab protein in syrup did not affect honey bee colonies (Ramirez-Romero et al., 2005). Likewise, exposing honey bee colonies to food containing Cry3b at concentrations 1,000 times that found in pollen resulted in no effect on larval or pupal weights (Arpaia, 1996). Feeding honey bees pollen from Cry1ab maize did not affect larval survival, gut flora, or hypopharyngeal gland development (Babendreier et al., 2005–2007). A 2008 meta-analysis of 25 independent studies concluded that the Bt proteins used in GE crops to control lepidopteran and coleopteran pests do not negatively impact the survival of larval or adult honey bees (Duan et al., 2008).

There is no evidence that the switch to Bt crops has injured honey bee colonies in the United States. To the contrary, it has benefited beekeeping by reducing the frequency of pesticide applications on crops protected by Bt, especially corn and cotton. On the other hand, the switch to GE crops with herbicide resistance has eliminated many blooming plants from field borders and irrigation ditches as well as from the crop fields themselves. The reduction in floral diversity and abundance that has occurred due to the application of Roundup® Herbicide (glyphosate) to GE crops with herbicide resistance is difficult to quantify. However, there is a growing body of evidence that poor nutrition is a primary factor in honey bee losses. Eischen and Graham (2008) clearly demonstrated that well-nourished honey bees are less susceptible to *Nosema ceranae* than poorly nourished bees. Because honey bees are polylectic, the adoption of agricultural practices that provide greater pollen diversity has been suggested, including the cultivation of small areas of other crops near monocultures or permitting weedy areas to grow along the edges of fields (Schmidt et al., 1995). A detailed review of management of uncropped farmland to benefit pollinators was reported by Decourtye et al. (2010).

Neonicotinoid and Phenylpyrazole Pesticides

Another major shift in United States agriculture has been the development and extensive deployment of neonicotinoid and phenylpyrazole pesticides. These pesticides are extensively used in the United States on field, vegetable, turf, and ornamental crops, some of which are commercially pollinated by bees (Quarles, 2008). They can be applied as seed treatments, soil treatments, and directly to plant foliage. Neonicotinoids are acetylcholine mimics and act as nicotinic acetychloline receptor agonists. Neonicotinoids cause persistent activation of cholinergic receptors which leads to hyperexcitation and death (Jeschke and Nauen, 2008). One neonicotinoid, imidacloprid, was applied to 788,254 acres in California in 2005 (CDPR, 2006), making it the sixth most commonly used insecticide in a state that grows many bee-pollinated crops. The phenylpyrazoles, including fipronil, bind to γ-amino butyric acid (GABA)-gated chloride ion channels and block their activation by endogenous GABA, leading to hyperexcitation and death (Gunasekara et al., 2007).

Neonicotinoid and phenylpyrazole insecticides differ from classic insecticides in that they become systemic (Trapp and Pussemier, 1991) in the plant, and can be detected in pollen and nectar throughout the blooming period (Cutler and Scott-Dupree, 2007). As a consequence, bees can experience chronic exposure to them over long periods of time. While some studies have shown no negative effects from seed-treated crops (Nguyen et al., 2009), acute mortality was the only response measured. Desneux and his colleagues (2007) examined methods that could be used to more accurately assess the risk of neonicotinoid and phenylpyrazole insecticides, including a test on honey bee larvae reared *in vitro* to test for larval effects (Aupinel et al., 2005), a proboscis extension response (PER) assay to access associative learning disruption (Decourtye and Pham-Delègue, 2002), various behavioral effects (Thompson, 2003), and chronic exposure toxicity beyond a single acute dose exposure (Suchail et al., 2001; Decourtye et al., 2005; Ailouane et al., 2009). Pesticide exposure may interact with pathogens to harm honey bee health. Honey bees that were both treated with imidacopríd and fed *Nosema* spp. spores suffered reduced longevity and reduced glucose oxidase activity (Alaux et al., 2010).

Registration Procedures and Risk Assessment

In the United States risk assessment related to agrochemical use and registration follows specific guidelines mandated by the Federal Insecticide Fungicide and Rodenticide Act (EPA, 2009a). Despite the importance of honey bees, the effect of pesticide exposure on colony health has not been systematically monitored, and the Environmental Protection Agency (EPA) does not require data on sublethal effects for pesticide registration (NAS, 2007).

For many years, the classical laboratory method for registering pesticides was to determine the median lethal dose (LD_{50}) of the pest insect. In a second step, the effects of pesticides on beneficial arthropods were examined by running LD_{50} tests on the beneficial species to identify products with the lowest non-target activity (Croft, 1990; Robertson et al., 2007). The honey bee has often served as a representative for all pollinators in the registration process, though the toxicity of pesticides to non-*Apis* species may be different (Taséi, 2003; Devillers et al., 2003). In the United States this protocol remains the primary basis for risk assessment in pesticide registration. However, this approach to risk assessment only takes into account the survival of adult honey bees exposed to pesticides over a relatively short time frame (OEPP/EPPO, 1992). In Europe, where the standard procedures do not provide clear conclusions on the harmlessness of a pesticide, additional studies are recommended; however, no specific protocols are outlined (OEPP/EPPO, 1992). Acute toxicity tests on adult honey bees may be particularly ill-suited for the testing of systemic pesticides because of the different route of exposure bees are likely to experience in field applications. Chronic feeding tests using whole colonies may provide a better way to quantify the effects of systemics (Colin et al., 2004).

Registration review is replacing the EPA's pesticide re-registration and tolerance reassessment programs. Unlike earlier review programs, registration review operates continuously, encompassing all registered pesticides. The registration review docket for imidacloprid opened in December 2008. To better ensure a level playing field for the neonicotinoid class as a whole and to best take advantage of new research as it becomes available, the EPA has moved the docket openings for the remaining neonicotinoids on the registration review schedule (acetamiprid, clothianidin, dinotefuran, thiacloprid, and thiamethoxam) to fiscal year 2012 (EPA, 2009b). The EPA's registration review document states that "some uncertainties have been identified since their initial registration regarding the potential environmental fate and effects of neonicotinoid pesticides, particularly as they relate to pollinators" (EPA, 2009b). Studies conducted in Europe in the late 1990s have suggested that neonicotinoid residues can accumulate in pollen and nectar of treated plants and represent a potential risk to pollinators (Laurent and Rathahao, 2003). Adverse effects on pollinators have also been reported in Europe that have further heightened concerns regarding the potential direct and/or indirect role that neonicotinioid pesticides may have in pollinator declines (Suchail et al., 2000).

Recently published data from studies conducted in Europe support concerns regarding the persistence of neonicotinoids. While the translocation of neonicotinoids into pollen and nectar of treated plants has been demonstrated, the potential effect that levels of neonicotinoids found in pollen and nectar can have on bees remains less clear. Girolami et al. (2009) report high levels of neonicotinoids from coated seeds in leaf guttation water and high mortality in bees that consume it. While the frequency of guttation drop collection by bees under field conditions is not documented, the authors describe the prolonged availability of high concentrations of neonicotinoids in guttation water as "a threatening scenario that does not comply with an ecologically acceptable situation." The pending EPA review will consider the potential effects of the neonicotinoids on honey bees and other pollinating insects, evaluating acute risk at the time of application and the longer-term exposure to translocated neonicotinoids (EPA, 2009b; Mullin et al., 2010).

Pesticides Used in Apiculture

The Varroa mite, *Varroa destructor*, is one of the most serious pests of honey bees in the United States and worldwide. It injures both adult bees and brood, and beekeepers are frequently compelled to use varroacides to avoid colony death (Boecking and Genersch, 2008). Varroacides must be minimally harmful to the bees while maintaining toxicity to mites, which is a challenge given the sensitivity of honey bees to many pesticides (Atkins, 1992). The varroacides used in the United States can be broadly divided into three categories: synthetic organic, natural product, and organic acid pesticides.

Synthetic Organic Pesticides

The pyrethroid *tau*-fluvalinate, a subset of isomers of fluvalinate, was the first synthetic varroacide registered for use in honey bee colonies in the United States. It was first registered as a Section 18 (emergency use label, state by state approval) in 1987 (Ellis et al., 1988). The Section 18 label allowed plywood strips to be soaked in an agricultural spray formulation of *tau*-fluvalinate, (Mavrik®), and treatment was made by suspending the strips between brood frames. In 1990, plastic strips impregnated with *tau*-fluvalinate (Apistan®) replaced homemade plywood strips (PAN, 2009) with a Section 3 label (full registration for use in all states). According to the label, a single strip contains 0.7 g *tau*-fluvalinate, as much as 10% of which may diffuse from the plastic strip formulation into hive matrices over the course of an 8-week treatment (Bogdanov et al., 1998; Vita Europe Ltd., 2009). While the agricultural spray formulation of *tau*-fluvalinate (Mavrik®) is no longer legal to use in the United States, its low cost and history of legal use in beehives make it vulnerable to misuse and may contribute to *tau*-fluvalinate residues detected in beeswax (Bogdanov, 2006; Wallner, 1999; Berry, 2009; Mullin et al., 2010).

As a pyrethroid, *tau*-fluvalinate kills mites by blocking the voltage-gated sodium and calcium channels (Davies et al., 2007). While most pyrethroids are highly toxic to honey bees, *tau*-fluvalinate is tolerated in high concentrations due in large part to rapid detoxification by cytochrome P450 monooxygenases (P450s) (Johnson et al., 2006). However, *tau*-fluvalinate is not harmless to bees and does affect the health of reproductive castes. Queens exposed to high doses of *tau*-fluvalinate were smaller than untreated queens (Haarmann et al., 2002). Drones exposed to *tau*-fluvalinate during development were less likely to survive to sexual maturity relative to unexposed drones, and they also had reduced weight and produced fewer sperm (Rinderer et al., 1999). However, the practical consequence of *tau*-fluvalinate exposure on drones may be limited, as drones exposed to *tau*-fluvalinate produced as many offspring as unexposed drones (Sylvester et al., 1999).

Tau-fluvalinate was initially very effective in controlling Varroa mites, but many Varroa populations now exhibit resistance (Lodesani et al., 1995). Resistance to *tau*-fluvalinate in Varroa is due, at least in part, to a mutation in the voltage-gated sodium channel, which confers reduced binding affinity for *tau*-fluvalinate (Wang et al., 2002). Despite diminished effectiveness, *tau*-fluvalinate continues to be used for Varroa control in the United States (Elzen et al., 1999; Macedo et al., 2002).

As the efficacy of *tau*-fluvalinate against Varroa was beginning to wane, the organophosphate pesticide coumaphos was granted Section 18 approval in the United States in 1999 as a varroacide (Federal Register, 2000) and as a treatment for the small hive beetle, *Aethina tumida* Murray. Coumaphos is administered as Checkmite+® strips, each containing approximately 1.4 g coumaphos, which are hung between brood frames for 6 weeks. Coumaphos, or its bioactivated oxon metabolite, kills through the inactivation of acetylcholinesterase, thereby interfering with nerve signaling and function. While coumaphos initially proved effective at killing *tau*-fluvalinate-resistant Varroa populations (Elzen et al., 2000), coumaphos-resistant mite populations were found as early as 2001 (Elzen and Westervelt, 2002). The mechanism of resistance to coumaphos in Varroa is unknown, though esterase-mediated detoxification may be involved (Sammataro et al., 2005). Resistance likely follows the mechanisms of coumaphos resistance observed in the southern cattle tick, *Boophilus microplus*, which include acetylcholinesterase insensitivity and enhanced metabolic detoxification (Li et al., 2005). Honey bees tolerate therapeutic doses of coumaphos, at least in part as a consequence of detoxicative P450 activity (Johnson et al., 2009). However, honey bees can suffer negative effects from coumaphos exposure. Queens exposed to coumaphos were smaller, suffered higher mortality, and were more likely to be rejected when introduced to a colony (Haarmann et al., 2002; Collins et al., 2004; Pettis et al., 2004). Drone sperm viability was lower in stored sperm collected from drones treated with coumaphos (Burley et al., 2008).

Amitraz, a formamidine pesticide, was once registered (1992–Section 18 label) in the United States under the trade name Miticur® with the active ingredient incorporated in a plastic strip that was suspended between brood frames (PAN, 2009). However, the product was withdrawn from the market in 1994 when some beekeepers reported colony losses following treatment (PAN, 2009). While no conclusive evidence was presented that the product had harmed bees, the registrant decided to withdraw the product from the market (PAN, 2009). Amitraz is available in the United States as a veterinary miticide (Taktic®), but the label does not allow for use in honey bee colonies. However, the frequency with which amitraz metabolites are found in beeswax suggests that it continues to be used (Mullin et al., 2010; Berry, 2009). Amitraz strips (Apivar®) were granted an emergency registration for Varroa control by the Canadian PMRA for 2009 (PMRA, 2009) but they are not available to beekeepers in the United States.

Amitraz is an octopaminergic agonist in arthropods (Evans and Gee, 1980) and as such has the potential to influence honey bee behavior. High levels of octopamine in the honey bee brain are associated with increased foraging behavior, and young bees fed octopamine are more likely to begin foraging than untreated bees (Schulz and Robinson, 2001). Forager honey bees treated with octopamine increased the reported resource value when communicating via the dance language (Barron et al., 2007). Amitraz has also shown acute toxicity, with larvae showing increased apoptotic cell death in the midgut when exposed to an amitraz solution (Gregorc and Bowen, 2000).

Despite the status of amitraz as an unregistered varroacide, Varroa mite populations in the United States exhibit resistance to amitraz, possibly through elevated esterase-mediated detoxification (Sammataro et al., 2005). The mechanism of Varroa resistance may be similar to the detoxicative resistance to amitraz that has been observed in some populations of Southern cattle ticks (Li et al., 2004).

Fenpyroximate is a pyrazole acaricide that was introduced for use in the United States in 2007 as Hivastan® under Section 18 registration (Wellmark, 2009). Hivistan® is formulated as a patty containing 675 mg of fenpyroximate. Fenpyroximate presumably kills mites through the inhibition of electron transport in the mitochondria at complex I, thereby interfering with energy metabolism (Motoba et al., 1992). While resistance to fenpyroximate in Varroa has not yet been observed, the eventual emergence of resistance is likely as it has been observed in other mites, including the 2-spotted spider mite (*Tetranychus urticae*) that achieved resistance through elevated detoxicative P450 and esterase activity (Kim et al., 2004). The mechanism of tolerance to fenpyroximate in honey bees is unknown, but it is likely through the same detoxicative mechanisms, P450-mediated hydroxylation followed by transesterification, that occurs in vertebrates and other insects (Motoba et al., 2000). One potential consequence of chronic exposure to fenpyroximate, as an inhibitor of complex I mitochondrial activity, is the increased generation of reactive oxygen species (Sherer et al., 2007). Increased adult mortality has been observed with fenpyroximate use during the first days after application (CDPR, 2008).

Natural Product Pesticides

Natural product-based varroacides have come into widespread use as synthetic pesticides have dwindled in effectiveness. Thymol and menthol, monoterpenoid constituents of plant-derived essential oils, are used for control of Varroa and tracheal mites, respectively. Thymol is the chief constituent in the fumigants Apilife Var® (tablets) and Apiguard® (gel), both of which are registered under Section 3. Essential oil-based varroacides were exempted from extensive testing for EPA registration because they are common food additives and "generally recognized as safe" for human consumption (Quarles, 1996). However, monoterpenoids such as thymol and menthol may not necessarily be safe for honey bees as these compounds play a role in plants as broad-spectrum pesticides (Isman, 2006). Indeed, thymol and menthol were found to be among the most toxic of all terpenoids tested when applied to honey bees as a fumigant (Ellis and Baxendale, 1997). These monoterpenoids likely kill Varroa by binding to octopamine (Enan, 2001) or GABA receptors (Priestley et al., 2003). Despite being naturally derived, these compounds may harm honey bees: thymol treatment can induce brood removal (Marchetti and Barbattini, 1984; Floris et al., 2004) and may result in increased queen mortality (Whittington et al., 2000).

Organic Acid Pesticides

Two organic acids, formic acid and oxalic acid, are attractive options as varroacides because both are naturally present in honey (Bogdanov, 2006; Rademacher and Harz, 2006). Formic acid is registered with Section 3 approval in the United States under the trade name MiteAway II™ (NOD, 2009). MiteAway II is a fumigant varroacide that is packaged in a slow-release pad. Formic acid likely kills Varroa by inhibiting electron transport in the mitochondria through binding of cytochrome c oxidase, thereby inhibiting energy metabolism (Keyhani and Keyhani, 1980) and may produce a neuroexcitatory effect on arthropod neurons (Song and Scharf, 2008). Formic acid can harm honey bees by reducing worker longevity (Underwood and Currie, 2003) and harming brood survival (Fries, 1991).

Oxalic acid is registered for use as a varroacide in Canada and Europe, but not in the United States. In Canada it is trickled over honey bees in a sugar syrup solution (Canadian Honey Council, 2005) or sublimated using a vaporizer (Varrox, 2007). Research has shown it to be highly effective against Varroa in cool climates when brood is not present (Aliano and Ellis, 2008). The mode of action of oxalic acid against Varroa is unknown, but direct contact between Varroa and oxalic acid is required (Aliano and Ellis, 2008). Oxalic acid treatments administered in water are ineffective (Charrière and Imdorf, 2002), but administration in sugar water improves efficacy by adhering the active ingredient to the bees (Aliano and Ellis, 2008). In mammals, oxalic acid interferes with mitochondrial electron transport, probably through interaction with complex II or IV, leading to increased production of reactive oxygen species and to kidney toxicity (Cao et al., 2004; Meimaridou et al., 2005). Repeated treatment of colonies with oxalic acid can result in higher queen mortality and a reduction in the amount of sealed brood (Higes et al., 1999). The midguts of honey bees fed oxalic acid in sugar water exhibited an elevated level of cell death (Gregorc and Smodisskerl, 2007), though in field conditions bees will generally avoid consuming syrup with oxalic acid (Aliano and Ellis, 2008). Oxalic acid is readily available and inexpensive in the United States for use as a wood bleach, but it is not labeled for use in controlling Varroa. Its easy availability from many sources has limited the willingness of suppliers to undergo the expensive and time-consuming registration process.

Interactions

With the large number of varroacides available to beekeepers in the United States, there is potential for interactions between compounds, a problem made worse by the fact that many synthetic varroacides are lipophilic and may remain in the wax component of hives for years following treatment (Bogdanov et al., 1997; Wallner, 1999; Mullin et al., 2010). The overlapping modes of action and mechanisms of tolerance in honey bees are also cause for concern. Interactions have been observed between *tau*-fluvalinate and coumaphos at the level of P450 detoxification (Johnson et al., 2009), and it seems likely that all varroacides depending on P450-mediated detoxification will display similar interactions. Fenpyroximate and the organic acids all interact with components of the mitochondrial electron transport chain (Keyhani and Keyhani, 1980; Motoba et al., 1992), where interactions could also be possible. Synergistic interactions between formamadines and pyrethroids occur in other insects (Plapp, 1979) and may occur in honey bees between amitraz and *tau*-fluvalinate.

Interactions between in-hive varroacides and out-of-hive insecticides and fungicides are also of concern, particularly interactions between the P450-detoxified varroacides and the P450-inhibiting ergosterol biosynthesis inhibiting fungicides (Pilling et al., 1995).

Pesticide Residues in Hive Products

Need for Sensitive Pesticide Analysis

Pesticide contamination of hive products is expected when honey bee colonies perish due to pesticide exposure. Colony mortality is often accompanied by part-per-million (ppm) residues in wax, beebread, honey,

Figure 1. Fungicides are applied to some crops, in this case almonds, during bloom to control plant diseases. The compounds are collected by bees when they are foraging. (*Photo by D. Sammataro.*)

and dead bee samples. However, part-per-billion (ppb) and occasionally ppm residues levels can be detected in hive matrices when honey bees forage in any conventional agricultural or urban setting (Figure 1). Since honey or pollen contaminated at ppb levels with newer classes of insecticides such as neonicotinoids (e.g., imidacloprid) or phenylpyrazoles (e.g., fipronil) are known to impair honey bee health (Decourtye et al., 2004; Halm et al., 2006; Desneux et al., 2007), it is important to use sensitive analytical technologies. Many pesticide contaminants, such as lipophilic pyrethroids and organophosphates, can be monitored in the hive using gas chromatography-mass spectrometry (GCMS). The more recently developed liquid chromatography-tandem mass spectrometry (LC/MS-MS) analytical capability is essential for newer systemic pesticides, particularly the neonicotinoids (Bonmatin et al., 2005; Chauzat et al., 2006). Older systemic residues, like the toxic sulfur-oxidation metabolites of aldicarb, many modern fungicides and herbicides, and the polar degradates of newer fungicides and herbicides cannot be analyzed at ppb limits of detection without LC/MS-MS (Alder et al., 2006).

During the last decade, some European beekeepers have reported heavy losses of honey bee colonies located near crops treated with the neonicotinoid imidacloprid (Rortais et al., 2005). Although Bonmatin et al. (2005) and Chauzat et al. (2006) found low ppb levels of the imidacloprid in a high percentage of pollen samples collected from maize, sunflower and canola, when pesticide residues from all matrices were pooled together, analysis did not show a significant relationship between the presence of pesticide residues and the abundance of brood and adults, and no statistical relationship was found between colony mortality and pesticide residues (Chauzat et al., 2009). Another way to associate pesticides and honey bee mortality is to examine dead bees, but obtaining samples can be difficult if bees die away from the hive as it is necessary to use recently dead or dying bees (Johansen and Mayer, 1990).

Figure 2. Pollen pellets from forager bees can be collected and sampled from pollen traps attached to bee hives. (*Photos by D. Sammataro.*)

Major Incidences of Pesticide Residues in the Beehive

This review will focus on pesticide residues studies done during the past 20 years. Smith and Wilcox (1990) report residues found in beehives in the United States prior to the period covered by this review. Over 150 different pesticides have been found in colony samples (Mullin et al., 2010). In recent years, the highest residues of pesticides in colonies are from varroacides that accumulate in the wax (Mullin et al., 2010). Varroacides found in beeswax, pollen, and bee bread include amitraz, bromopropylate, coumaphos, flumethrin and *tau*-fluvalinate. Residue levels of these varroacides generally increase from honey to pollen to beeswax (Lodesani et al., 1992; Wallner, 1995; Bogdanov et al., 1998; Bogdanov, 2004; Tremolada et al., 2004; Martel et al., 2007; Frazier et al., 2008).

Varroacide residues in honey have been found to reach as high as 2.4 ppm acrinathrin, based on its 3-phenoxybenzaldehyde degradate (Bernal et al., 2000); 0.6 ppm amitraz (Mullin et al., 2010); 2 ppm coumaphos (Martel et al., 2007); and 0.75 ppm fluvalinate (Fernandez et al., 2002). Bee bread was also found to be contaminated with up to 1.1, 0.01, 5.8, and 2.7 ppm, respectively of amitraz, bromopropylate, coumaphos and fluvalinate (vanEngelsdorp et al., 2009b; Mullin et al., 2010). Levels in brood and adult bees can be higher than in the food, with 14 ppm amitraz, 5.9 ppm fluvalinate (vanEngelsdorp et al., 2009b; Mullin et al., 2010), 2.8 ppm coumaphos (Ghini et al., 2004), and 2.2 ppm bromopropylate (Lodesani et al., 1992) being reported. Nevertheless, wax remains the ultimate sink for these varroacides, reaching 46, 94, and 204 ppm, respectively, of amitraz, coumaphos, and fluvalinate (vanEngelsdorp et al., 2009b; Mullin et al., 2010), 135 ppm of bromopropylate (Bogdanov et al., 1998), and 7.6 and 0.6 ppm, respectively, of the miticides chlorfenvinophos and acrinathrin (Jimenez et al., 2005).

Pesticide residues of agrochemicals acquired by foragers are equivalent or higher in pollen (stored and trapped at the hive entrance; see Figure 2), adult bees, and occasionally honey, than in wax. Major pollen detections include the insecticides aldicarb (1.3 ppm), azinphos methyl (0.6 ppm), chlorpyrifos (0.8 ppm), and imidacloprid (0.9 ppm); fungicides boscalid (1 ppm), captan (10 ppm), and myclobutanil (1 ppm), and herbicide pendimethalin (1.7 ppm; see Table 1, pp 166–168 herein). The carbamates carbofuran and carbaryl, and the organophosphate parathion methyl have been frequently found at up to 1.4 (Bailey et al., 2005), 94 (cited in Chauzat et al., 2006), and 26 ppm (Rhodes et al., 1979), respectively. High levels of pyrethroids, including cyhalothrin and cypermethrin at 1.7 and 1.9 ppm, respectively, have been reported in mustard pollen along with up to 2.2 and 2.1 ppm, respectively, of the cyclodiene endosulfan and the new lipid-synthesis inhibitor insecticide spiromesifen (Choudhary and Sharma, 2008b).

Fungicides often account for most of the pesticide content of pollen. Unprecedented levels (99 ppm) of the widely used fungicide chlorothalonil were found in honey bee-collected pollen (Table 1, Mullin et al., 2010). Chorothalonil, a contact and slightly volatile fungicide, was found to be a marker for entombing behavior in honey bee colonies associated with poor health (vanEngelsdorp et al., 2009a). Entombing may be a defensive behavior of honey bees faced with large amounts of potentially toxic food stores. Kubik et al. (1999) noted high residues of the fungicides vinclozolin and iprodione up to 32 and 5.5 ppm, respectively, in bee bread.

High residues in the honey bees themselves (Table 1) are often associated with direct kill from the respective pesticide application, such as with 19.6 ppm of permethrin (LD_{50}-1.1 ppm) and 3.1 ppm of fipronil (LD_{50}-0.05 ppm) (Mullin et al., 2010). Anderson and Wojtas (1986) linked dead honey bees to high residues of the insecticides carbaryl (5.8 ppm), chlordane (0.7 ppm), diazinon (0.35 ppm), endosulfan (4.4 ppm), malathion (4.2 ppm), methomyl (3.4 ppm), methyl parathion (3.6 ppm), and fungicide captan (1.7 ppm). Walorczyk and Gnusowski (2009) found exceptional amounts of the organophosphates dimethoate (4.9 ppm), fenitrothion (1 ppm), and omethoate (1.2 ppm), and up to 1.2 ppm of the systemic fungicide tebuconazole in honey bees from other poisoning incidences (Table 1). Similarly, elevated residues of the organophosphates bromophos methyl (1.7 ppm) and fenitrothion (10.3 ppm) were associated with high honey bee mortality (Ghini et al., 2004). In contrast to systemic fungicides, systemic neonicotinoid residues are generally absent from bee samples, although present in pollen and wax.

Notable honey residues in Europe include 0.65, 0.66, and 4.3 ppm, respectively, of carbofuran, DDT, and lindane (Blasco et al., 2003), and 0.6 ppm of methoxychlor (Fernandez-Muino et al., 1995). A recent broad sampling of United States honey following reports of CCD (USDA-PDP, 2008) showed only a few detections of low ppb amounts of external pesticides like dicloran and dicofol but also revealed more frequent low levels of coumaphos and fluvalinate up to 12 ppb. A standard treatment of synthetic piperonyl butoxide-synergized pyrethrum to kill managed and feral honey bees in a hive (Taylor et al., 2007) can leave high residues in both honey (up to 3, 0.6 ppm, respectively) and wax (470,237 ppm).

Very high amounts of the fungicide chlorothalonil (54 ppm) and substantial levels of chlorpyrifos (0.9 ppm), aldicarb (0.7 ppm), deltamethrin (0.6 ppm), and iprodione (0.6 ppm) were found in comb wax (vanEngelsdorp et al., 2009b; Mullin et al., 2010). Elevated levels of the acetylcholinesterase-inhibitors azinphos methyl (0.8 ppm), fenitrothion (0.5 ppm, Chauzat and Faucon, 2007), carbaryl (0.8 ppm), parathion methyl (3.1 ppm, Russell et al., 1998), and malathion (6 ppm, Thrasyvoulou and Pappas, 1988) have been reported. Bogdanov et al., (2004) detected up to 60 ppm of *p*-dichlorobenzene and Jimenez et al. (2005) up to 0.6 ppm of the miticide tetradifon in beeswax.

High Diversity of Pesticides in Beehive Samples

The recent phenomenon of CCD triggered a close look at the role of pesticides as a possible contributing factor to honey bee decline in general and CCD specifically. Mullin and Frasier used LC/MS-MS and GC/MS and a modified QuEChERS method to analyze for pesticide residues in honey bees and hive matrices in the United States and Canada to examine colonies exhibiting CCD symptoms (vanEngelsdorp et al., 2009b; Frazier et al., 2008). One hundred twenty-one different pesticides and metabolites were found within 887 wax, pollen, bee, and associated hive samples (average of 6.2 detections per sample) from migratory and stationary beekeepers. These included 16 parent pyrethroids, 13 organophosphates, 4 carbamates, 4 neonicotinoids, 4 insect growth regulators, 3 chlorinated cyclodienes, 3 organochlorines, 1 formamidine, 8 miscellaneous miticides/insecticides, 2 synergists, 30 fungicides, and 17 herbicides. Over 40 of the pesticides detected are systemic (Table 1). Only one of the wax samples, 3 pollen samples and 12 bee samples had no detectable pesticides.

Overall, pyrethroids and organophosphates dominated total wax and bee residues followed by fungicides, systemics, carbamates, and herbicides, whereas fungicides prevailed in pollen followed by organophosphates, systemics, pyrethroids, carbamates, and herbicides. By comparing these residue levels across the matrices, in-hive varroacides were more concentrated in wax than in pollen, whereas externally derived pesticides were higher or equivalent in pollen compared to wax. This is consistent with chronic use and long-term accumulation of these lipophilic varroacides in the wax as a source to contaminate pollen.

All foundation (beeswax pressed into sheets and used as templates for comb construction) sampled from North America is uniformly contaminated with *tau*-fluvalinate, coumaphos, and lower amounts of other pesticides and metabolites (Mullin et al., 2010). The broad contamination of European foundation, especially with varroacides, has been reviewed previously (Wallner, 1999; Bogdanov, 2004; Lodesani et al., 2008). The uniform presence of these acaricides in foundation is particularly disturbing because replacement of frames is the main avenue currently used to purge a colony of accumulated pesticide contaminants. Fluvalinate residues in beeswax were the best correlative with the French honey bee winter kill of 1999–2000 (Faucon et al., 2002), although disease factors were emphasized in the report.

High levels of the pyrethroid fluvalinate and the organophosphate coumaphos are co-occurring with lower but significant levels of 119 other insecticides, fungicides, and herbicides in hive matrices. Fluvalinate and coumaphos, but not amitraz, are highly persistent in the hive, with an estimated half-life in beeswax of 5 years (Bogdanov, 2004). Chronic exposure to high levels of these persistent neurotoxicants elicits both acute and sublethal reductions in honey bee fitness (Stoner et al., 1985; Lodesani and Costa, 2005). The direct association of any one of these varroacides with CCD remains unclear, although higher coumaphos levels may benefit the colony presumably via mite control (vanEngelsdorp et al., 2009b).

Externally derived, highly toxic pyrethroids were the most frequent and dominant class of insecticides samples (Mullin et al., 2010). Contact pyrethroids, and systemic neonicotinoids, and fungicides are often combined as pest control inputs, and many of the latter may synergize the already high toxicity of neonicotinoids and pyrethroids to honey bees (Pilling and Jepson, 1993; Iwasa et al., 2003). Pyrethroids frequently are associated with honey bee kills (Mineau et al., 2008), as has been the case with neonicotinoids (Halm et al., 2006), although the latter with less documentation of acute residues in bees. The effects of toxic chronic exposure to pyrethroids, organophosphates, neonicotinoids, fungicides, and other pesticides can range from lethal and/or sublethal effects in the larvae and workers to reproductive effects on the queen (Thompson, 2003; Desneux et al., 2007). These chemicals may act alone or in concert in ways currently unknown to create a toxic environment for the honey bee.

Conclusions

The widespread planting of transgenic crops appears to be a net benefit for honey bees in the United States, as the pesticidal toxins produced by these plants do not appear to harm honey bees. Additionally, such crops do not require as many applications of traditional pesticides, most of which are known to be toxic to bees.

The systemic nature of the neonicotinoids and phenylpyrazoles present a trade-off from the standpoint of honey bee health. New methods of application help to minimize direct exposure of bees to these compounds during application. The downside is that honey bees may instead be exposed to these pesticides, or their metabolites, in pollen, nectar, and plant exudates over extended periods of time. Further research is needed to assess the true dangers posed by extended low-dose exposure to these systemic pesticides.

Beekeepers searching for the primary source of pesticides contaminating bee hives need only to look in a mirror. Unfortunately, the regulatory system governing the veterinary use of pesticides in bee hives in the United States may be perversely contributing to the problem. Two of the handful of pesticides registered for legal use in the United States, coumaphos (CheckMite+®) and *tau*-fluvalinate (Apistan®), both of which seriously contaminate wax, have become largely useless against the primary pest of honey bees, the Varroa mite. Another effective varroacide used in Europe and Canada, oxalic acid, is not registered in the United States because it is low in cost, readily available, and potential registrants are deterred by the cost of the registration process. There are three registered in-hive pesticides that provide effective Varroa control in the United States., fenpyroximate (Hivistan®), formic acid (Miteaway II™), and thymol (ApiGuard® and Api-Life Var®). Other effective pesticides, including amitraz and oxalic acid, are used by some beekeepers in the absence of any regulatory approval. A change in the regulatory system needs to occur to make effective and safe veterinary pesticides available to beekeepers and to spur research into the effects of candidate compounds on honey bee health. Likewise, beekeepers need to realize that honey bee pests and parasites are community as well as individual problems, and that pesticide labels are crafted to protect the sustainability of pesticides. The use of unregistered products is a serious threat to the beekeeping community and should not occur.

Honey bees are being exposed to high levels of in-hive varroacides and agrochemical pesticides. Chronic exposures to neurotoxic insecticides and their combinations with other pesticides, in particular fungicides, are known to elicit reductions in honey bee fitness, but direct association with CCD and declining honey bee health remains to be resolved.

Acknowledgments We gratefully recognize the major input of Dennis vanEngelsdorp, Pennsylvania Department of Agriculture, and Jeffery S. Pettis, USDA-ARS, for providing samples for the CCD analyses reviewed here; Roger Simonds, USDA-AMS National Science Laboratory, Gastonia, North Carolina, for conducting the pesticide analyses; and of Jim Frazier, Sara Ashcraft (PSU), Lizette Peters, and Alex Heiden (UNL) for technical support. We thank the National Honey Board, the Florida State Beekeepers, the Tampa Bay Beekeepers, Penn State College of Agriculture Sciences, Preservation *Apis mellifera* (P*Am*), The Foundation for the Preservation of Honey Bees, the USDA Critical Issues Program, the USDA CAPS Program, and the EPA for the financial support that made some of the work included in this review possible.

Table 1. Maximum pesticide incidence in apiary samples of wax, pollen, bee, and honey.

Total Pesticide[a]	Class[b]	Maximum Detection in ppb (ref.)[c]			
		Wax	Pollen	Bee	Honey
Acephate	S OP	n.d.	163	n.d.	52
Acetamiprid	S NEO	n.d.	134	n.d.	n.d.
Acrinathrin	PYR	590 (1)	—	—	2400 (2)
Aldicarb	S CARB	693	1342	n.d.	n.d.
Aldrin	CYC	5 (3)	—	—	150 (4)
Allethrin	PYR	139	11	24	n.d.
Amicarbazone	HERB	n.d.	98	n.d.	n.d.
Amitraz	FORM	46,060	1117	13,780	555
Atrazine	S HERB	31	49	15	81 (5)
Azinphos ethyl	OP	—	—	94 (6)	—
Azinphos methyl	OP	817 (7)	643	91 (6)	n.d.
Azoxystrobin	S FUNG	278	107	n.d.	4 (27)
Bendiocarb	S CARB	22	n.d.	n.d.	n.d.
Bifenthrin	PYR	56	13	12	3
Bitertanol	S FUNG	—	—	—	0.1 (8)
Boscalid	S FUNG	388	962	33 (9)	n.d.
Bromophos ethyl	OP	—	—	—	12 (10)
Bromophos methyl	OP	—	—	1733 (6)	—
Bromopropylate	MITI	135,000 (11)	11	2245 (12)	245 (12)
Captan	PS FUNG	400 (13)	10000	1740 (13)	19 (14)
Carbaryl	PS CARB	820 (13)	94,000 (15)	5800 (13)	42
Carbendazim	S FUNG	133	149	14	27 (27)
Carbofuran	S CARB	22	1400 (16)	669 (6)	645 (17)
Carfentrazone ethyl	PS HERB	17	3	n.d.	n.d.
Chlordane	CYC	60 (13)	—	690 (13)	—
Chlorfenapyr	PS MITI	12	1	3	n.d.
Chlorfenvinphos	OP	7620 (1)	11	n.d.	0.2 (18)
Chlorothalonil	FUNG	53,700	98,900	878	10 (19)
Chlorpyrifos	OP	890	830	57 (6)	15 (5)
Chlorpyrifos methyl	OP	—	—	36 (6)	0.2 (18)
Clothianidin	S NEO	n.d.	2.6 (20)	n.d.	0.9 (20)
Coumaphos	OP 94	131	5828	2777 (6)	2020 (21)
Cyfluthrin	PYR	45	34	14	9 (19)
Cyhalothrin	PYR	17	1672 (22)	2	0.8 (23)
Cymiazole	MITI	—	—	—	17 (24)
Cypermethrin	PYR	131	1900 (15)	26	92 (5)
Cyproconazole	S FUNG	—	8 (15)	—	—
Cyprodinil	S FUNG	106	344	19	n.d.
DDT-p,p	OC	>40	45	7658 (17)	
Deltamethrin	PYR	613	91	39	7 (23)
Dialifos	OP	—	—	—	92 (4)
Diazinon	OP	4	29	350 (13)	35 (24)
***p*-Dichlorobenzene**	OC	60,000 (25)	n.d.	n.d.	112 (25)
Dichlofluanid	FUNG	—	—	—	11 (26)
Dichlorvos	OP	—	—	—	8 (19)
Dicloran	FUNG	—	—	—	2 (27)
Dicofol	OC	21	143	4	90 (27)
Dieldrin	CYC	35	n.d.	12	13 (4)
Difenoconazole	S FUNG	n.d.	411 (14)	n.d.	0.9 (14)
Diflubenzuron	IGR	n.d.	128	n.d.	n.d.
Dimethomorph	S FUNG	133	166	56	n.d.
Dimethoate	S OP	—	—	4864 (9)	9 (23)

[a] Acrinathrin is based mostly on 3-phenoxybenzaldehyde degradate, Aldicarb based on sulfoxide and sulfone metabolites; Amitraz based on total DMA and DMPF metabolites; Bromopropylate based on 4,4-dibromobenzophenone; Captan includes THPI; Carbaryl includes 1-naphthol; Carbendazim is also a degradate of benomyl; Carbofuran based on parent plus 3-hydroxy metabolite; Coumaphos includes oxon, chlorferone, and potasan; DDT includes DDD and DDE; Endosulfan includes isomers and sulfate; Heptachlor includes heptachlor epoxide; Imidacloprid includes 5-hydroxy and olefin metabolites; Thiabendazole is a degradate of thiophanate methyl.

[b] Class: CAR = carbamate, CYC = cyclodiene, FORM = formamidine, FUNG = fungicide, HERB = herbicide, IGR = insect growth regulator, INS = misc. insecticide, MITI = miticide, NEO = neonicotinoid, OC = organochlorine, OP = organophosphate, PS = partial systemic, PYR = pyrethroid, S = systemic, SYN = synergist.

[c] Numbers in parentheses denote references, shown at end of table.

n.d. = not detected

Table 1. *(continued)*

Total Pesticide[a]	Class[b]	Maximum Detection in ppb *(ref.)*[c]			
		Wax	Pollen	Bee	Honey
Diphenamid	S FUNG	n.d.	1	n.d.	n.d.
Diphenylamine	FUNG	n.d.	32	n.d.	n.d.
Endosulfan	CYC	800 (13)	2224 (22)	4400 (13)	24 (5)
Endrin	CYC	—	—	—7 (4)	
Esfenvalerate	PYR	56	60	9	0.7 (23)
Ethion	OP	131	n.d.	n.d.	n.d.
Ethofumesate	S HERB	560	n.d.	n.d.	n.d.
Etoxazole	MITI	n.d.	n.d.	n.d.	1
Famoxadone	FUNG	n.d.	141	n.d.	n.d.
Fenamidone	FUNG	138	74	n.d.	n.d.
Fenbuconazole	S FUNG	183	264	n.d.	n.d.
Fenhexamid	FUNG	9	129	n.d.	n.d.
Fenitrothion	OP	511 (7)	—	10,330 (6)	—
Fenoxaprop-ethyl	S HERB	n.d.	n.d.	15	n.d.
Fenoxycarb	IGR	—	—	157 (6)	—
Fenpropathrin	PYR	200	170	37	n.d.
Fenthion	S P	—	—	38 (6)	—
Fipronil	INS	36	29	3060	n.d.
Fluoxastrobin	S FUNG	45	n.d.	n.d.	n.d.
Fluridone	S HERB	7	24	7	n.d.
Flusilazole	S FUNG	—	71 (15)	—	0.03 (8)
Flutolanil	S FUNG	105	n.d.	n.d.	n.d.
Flumethrin	PYR	50 (28)	—	—	1 (28)
Fluvalinate	PYR	204,000	2670	5860	750 (24)
Fonofos	OP	—	—	—	15 (10)
Heptachlor	CYC	31	2	n.d.	57 (4)
Heptenophos	S OP	—	—	162 (6)	230 (17)
Hexachlorobenzene	FUNG	1	1	n.d.	270 (17)
Hexaconazole	S FUNG	—	12 (15)	—	—
Imidacloprid	S NEO	14	912	n.d.	2 (29)
Indoxacarb	INS	n.d.	330	n.d.	n.d.
Iprodione	FUNG	636	5511 (30)	n.d.	266 (30)
Lindane	OC	290 (1)	7 (29)	11 (29)	4310 (17)

Total Pesticide[a]	Class[b]	Maximum Detection in ppb *(ref.)*[c]			
		Wax	Pollen	Bee	Honey
Malathion	OP	6000 (31)	61	4200 (13)	243 (5)
Metalaxyl	S FUNG	1	n.d.	n.d.	n.d.
Methamidophos	S OP	—	—	38 (6)	—
Methidathion	OP	79	33	32	68 (17)
Methiocarb	CARB	—	—	346 (6)	27 (17)
Methomyl	S CARB	140 (13)	—	3400 (13)	—
Methoxychlor	OC	—	—	—	593 (4)
Methoxyfenozide	IGR	495	128	21	3 (27)
Metolachlor	PS HERB	n.d.	103	n.d.	n.d.
Metribuzin	S HERB	8	44	n.d.	n.d.
Myclobutanil	S FUNG	n.d.	981	n.d.	n.d.
Norflurazon	S HERB	38	108	n.d.	n.d.
Omethoate	S OP	—	—	1156 (9)	—
Oxamyl	S CARB	n.d.	43	n.d.	n.d.
Oxyfluorfen	HERB	34	5	5	n.d.
Parathion ethyl	OP	99 (7)	19 (15)	5 (6)	—
Parathion methyl	OP	3085 (32)	26 000 (33)	3600 (13)	50 (33)
Penconazole	S FUNG	—	126 (15)	8 (29)	—
Pendimethalin	HERB	84	1730	28	n.d.
Permethrin	PYR	372	92	19,600	11 (27)
Phenothrin	PYR	n.d.	84	n.d.	n.d.
Phenthoate	OP	—	—	1 (6)	—
Phorate	S OP	—	—	—	0.9 (18)
Phosalone	OP	n.d.	31	66 (9)	n.d.
Phosmet	OP	209	418	96 (6)	n.d.
Phosphamidon	S OP	—	—	50 (6)	—
Phoxim	OP	—	—	355 (6)	—
Piperonyl butoxide	SYN	470,000 (34)	n.d.	3000 (34)	10 (27)
Pirimiphos ethyl	OP	—	—	30 (6)	—
Pirimiphos methyl	OP	57	n.d.	62	19 (10)
Prallethrin	PYR	7	8	9	n.d.
Prochloraz	FUNG	—	—	412 (9)	—
Procymidone	S FUNG	28 (7)	—	—	—

Table 1. *(continued)*

Total Pesticide[a]	Class[b]	Maximum Detection in ppb *(ref.)*[c] Wax	Pollen	Bee	Honey
Profenofos	OP	—	—	17 (6)	—
Pronamide	S HERB	23	378	2	n.d.
Propanil	HERB	n.d.	358	n.d.	n.d.
Propiconazole	S FUNG	227	361	n.d.	n.d.
Pyraclostrobin	FUNG	438	265	9	17
Pyrazophos	S OP	—	—	53 (6)	6 (10)
Pyrethrins	PYR	237,000 (34)	62	600 (34)	n.d.
Pyridaben	MITI	5	27	n.d.	n.d.
Pyrimethanil	FUNG	28	83	n.d.	4
Pyriproxyfen	IGR	8	n.d.	n.d.	n.d.
Quinalphos	OP	—	—	70 (6)	10 (23)
Quintozene = PCNB	FUNG	3	n.d.	n.d.	n.d.
Sethoxydim	S HERB	n.d.	173	n.d.	8
Simazine	S HERB	n.d.	54	n.d.	17 (5)
Spinosad	INS	—	320 (16)	—	—
Spirodiclofen	MITI	29	2	n.d.	n.d.
Spiromesifen	S INS	n.d.	2101 (22)	n.d.	n.d.
Tebuconazole	S FUNG	n.d.	34	1146 (9)	5 (5)
Tebufenozide	IGR	28	58	23	n.d.
Tebuthiuron	S HERB	22	48	n.d.	n.d.
Tefluthrin	PYR	3	n.d.	n.d.	n.d.
Temephos	OP	—	—	689 (6)	7 (10)
Tetradifon	MITI	580	n.d.	n.d.	19 (19)
Tetraconazole	S FUNG	—	—	17 (29)	—
Tetramethrin	PYR	n.d.	6	23	n.d.
Thiabendazole	S FUNG	76	6	n.d.	n.d.
Thiacloprid	S NEO	8	115	n.d.	33
Thiamethoxam	S NEO	n.d.	53	n.d.	n.d.
Triadimefon	S FUNG	2	n.d.	n.d.	n.d.
Triallate	HERB	—	—	—	4 (26)
Triazophos	OP	—	—	9 (6)	—
Tribufos = DEF	SYN	59	4	n.d.	n.d.
Trifloxystrobin	PS FUNG	22	264	n.d.	0.3 (8)
Trifluralin	HERB	36	14	n.d.	9 (19)
Vamidothion	S OP	—	—	24 (6)	—
Vinclozolin	FUNG	27	31,909 (30)	657 (9)	173 (30)

Data from Frazier et al. (2008), vanEngelsdorp et al. (2009b), or Mullin et al. (2010), unless otherwise referenced.

1 Jimenez et al. (2005)
2 Bernal et al. (2000)
3 Estep et al. (1977)
4 Fernandez-Muino et al. (1995)
5 Rissato et al. (2007)
6 Ghinietal. (2004)
7 Chauzat and Faucon (2007)
8 Nguyen et al. (2009)
9 Walorczyk et al. (2009)
10 Blasco et al. (2008)
11 Bogdanov et al. (1998)
12 Lodesani et al. (1992)
13 Anderson and Wojtas (1986)
14 Kubik et al. (2000)
15 Chauzat et al. (2006)
16 Bailey et al. (2005)
17 Blasco et al. (2003)
18 Balayiannis and Balayiannis (2008)
19 Rissato et al. (2004)
20 Cutler and Scott-Dupree (2007)
21 Martel et al. (2007)
22 Choudhary and Sharma (2008b)
23 Choudhary and Sharma (2008a)
24 Fernandez et al. (2002)
25 Bogdanov et al. (2004)
26 Albero et al. (2004)
27 USDA-AMS (2009)
28 Bogdanov (2006)
29 Chauzat et al. (2009)
30 Kubik et al. (1999)
31 Thrasyvoulou and Pappas (1988)
32 Russell et al. (1998)
33 Rhodes et al. (1979)
34 Taylor et al. (2007)

Cellular Response in Honey Bees to Non-Pathogenic Effects of Pesticides

CHAPTER

15

Aleš Gregorc, Elaine C. M. Silva-Zacarin, and Roberta C. F. Nocelli

Abstract Different broad-spectrum pesticides applied in agricultural settings or used in controlling bee pests could have drastic effects on bee colonies. Exposure to pesticides impacts foragers, shortens worker longevity, decreases queen survival, and affects colony vitality. Sublethal effects can lead to physiological modifications and changes in bee behavior and cellular physiology consistent with chemically induced stress responses. We present a summary of cellular damage to bees by pesticide exposure showing that pesticide use subjects pollinators to severe stress that can cause economic damage evidenced by a decrease in bee density. Research focuses on damage that can be repaired, cells that remain viable after intermediate level of damage, and cells that undergo apoptosis or necrosis after a high level of damage primarily to brain and gut. Cellular biomarkers have been developed to evaluate chronic exposure of bees to pesticides to understand the effects of synergistic action of xenobiotics in the environment and to separate the effects of pathogens and pesticides. These studies can bring substantial benefits to agro-ecosystems.

Determination of Environmental Stressors on Honey Bees

Honey bees are the most important insect pollinator worldwide, especially in the areas of large agricultural monocultures. The honey bee, *Apis mellifera* L., provides pollination services for diverse crop plants, and these services are at risk due to exposure of bees to parasites, pathogens, and environmental chemicals, including pesticides and other anthropogenic chemicals (vanEngelsdorp et al., 2009). Honey bees forage and collect nectar and pollen from flowers to sustain the colony and support healthy brood development; thus pesticides in the environment could potentially be transmitted to the hive through pollen and nectar contamination. There have been several reports in recent years that the use of pesticides could result in a drastic reduction of beneficial insects, including honey bees. They are affected severely by different broad-spectrum pesticides applied in the field. Physiological responses of bees to chemicals, including insecticides, fungicides, and herbicides, have been studied. Some of them are often used by beekeepers to control *Varroa* mites and/or small hive beetles (*Aethina tumida*) in colonies; others are commonly used in agricultural settings, and thus have been found in honey bee colonies (Ellis and Munn, 2005).

There are several ways in which *Apis mellifera* L. foragers are exposed to pesticides: field treatment during peak of honey bee flight activity; treatment of nonflowering crops when nearby cover crops, weeds, and wildflowers are in bloom (pesticide drift); and treatment of colonies by beekeepers for pest and disease control. Exposure to pesticides can impact foraging bees (Vandame et al., 1995), shorten worker longevity, decrease queen weights and their survival (Pettis et al., 2004), and affect colony vitality (Beliën et al., 2009). Sublethal effects caused by various pesticides can lead to physiological modifications (Papaefthimiou et al., 2002), changes in individual honey bee behavior (Weick and Thorn, 2002; Aliouane et al., 2009), and alterations in cellular physiology consistent with chemically induced stress (Gregorc and Bowen, 1999; Gregorc and

Bowen, 2000; Gregorc et al., 2004; Silva-Zacarin et al., 2006; Gregorc and Smodiš Škerl, 2007; Smodiš Škerl and Gregorc, 2010). Because honey bees can be exposed to multiple chemical agents at once, synergistic or antagonistic interactions among these pesticides could also play a role in bee and colony health (Johnson et al., 2009).

Bees are very sensitive to most of the pest control chemicals, and the reasons and mechanisms for this sensitivity are mostly unclear. Resistance mechanisms in honey bee colonies are complex and depend on several factors that could influence slow resistance development. The most important mechanism is the structure of the colony, where only workers are directly exposed to pesticides and the reproductive queen is not under direct selection pressure; the colony reproductive behavior ensures slower population growth in comparison to a majority of target pests. Beekeeping practice with frequent requeening is another element in reducing the development of potential resistance where resistant stocks are possibly discontinued. All these factors contribute to preventing bees from developing resistance on a colony level. Resistance on the individual bee level is attributed to several factors: behavioral, physiological, morphological, and biochemical are the most widely studied. The latter one is a very important and well-studied detoxification mechanism on the cellular level that encompasses the metabolism of xenobiotics with elevated activities of a series of enzymes, including cytochrome P450-linked microsomal oxidases, known also as mixed-function oxygenase (MFO) system and other less-studied glutathione transferases, carboxylesterases, and epoxide hydrolases.

The first studies about insecticide toxicity to bees date back to the 1940s and were initiated in the United States and Europe. In South America, the first research studies about insecticide toxicity to bees have been conducted in Brazil since 1970 (Malaspina, 1979). Due to the global economic importance of *A. mellifera,* the studies of the insecticide toxicity to bees tend to prefer this species as a model, while studies with native bees from each country are still scarce.

The concern with the growing demand for agricultural chemical substances on crops mobilized many organizations around the world that tried to create standardized methodologies for studying the effects of these chemicals on bees. Currently, risk assessments of pesticides are performed according to guidelines published by the International Protocol for Testing of Chemicals issued by the Organization for Economic Cooperation and Development (OECD, 1998). It comprises guidelines for testing acute toxicity by oral administration and by contact based, mainly, on the guidelines of the European and Mediterranean Organization for Plant Protection (EPPO, 1993) and on the proposals prepared by the International Commission for Plant-Bee Relations (ICPBR, 1993). The tests are performed, initially to determine the LD_{50} with adult workers that are exposed to different doses of the test substance added to food (oral toxicity) or applied directly to the thorax (contact toxicity). The experiments are conducted in greenhouses, with controlled temperature and humidity, according to the conditions presented by the colonies of each species.

Pesticides, beyond the effect of acute toxicity leading to death in low concentrations, cause sublethal effects that result in behavior alteration and cognitive disorders that trigger serious harm in maintaining the colony. According to Medrzycki et al. (2003), in some circumstances, the effect of pesticides on bees may not be noticed immediately, requiring assessments of sublethal doses to be able to observe its influence on

Figure 1. Diagram showing how honey bees are exposed to contaminants.

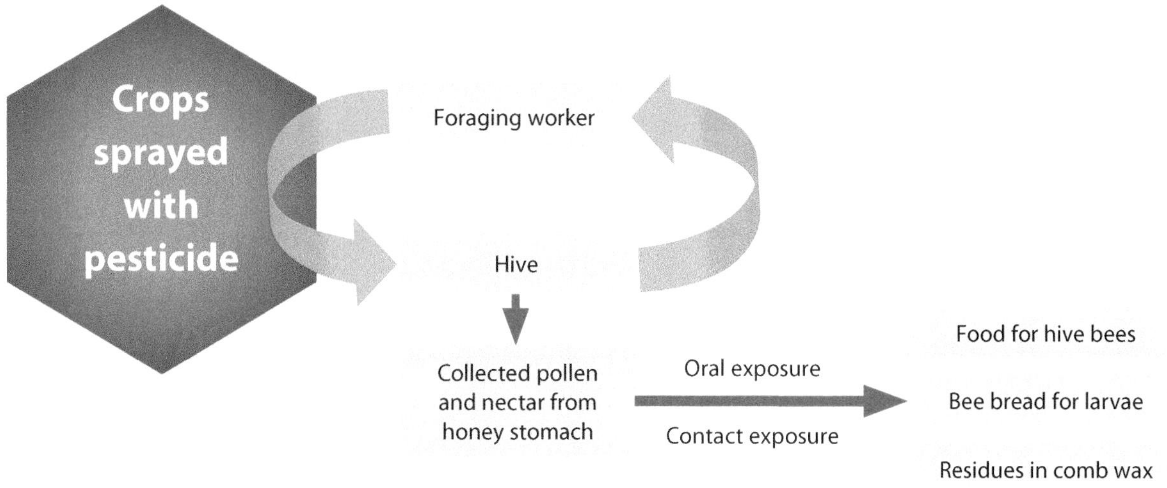

Crops sprayed with pesticide
Foraging worker
Hive
Collected pollen and nectar from honey stomach
Oral exposure
Contact exposure
Food for hive bees
Bee bread for larvae
Residues in comb wax

survival and behavior. Reduction of motion and mobility, reduced capacity for communication and learning difficulties of foraging bees returning to the colony have been observed in bees treated with sublethal doses of insecticides (Bortolli et al., 2003; Decourtye et al., 2005).

When the standard procedure for evaluating risks to the bees (EPPO, 1993) is insufficient to determine the effect of a particular agrochemical product, there is an official recommendation for the use of additional studies that should provide data for the final verdict on the level of toxicity. Among these additional studies, behavioral changes can be evaluated through assays of the proboscis extension reflex (PER) and displacement. These show how the compounds can affect the activities of foraging and, consequently, the pollination process (Lambin et al., 2001; Decourtye et al., 2005). The PER method aims at reproducing interaction between bee and plant, being that this reflection leads the bee to gather the nectar and to memorize the floral odor presented (Decourtye et al., 2005). Lambin et al. (2001) studying the effects of imidacloprid, a neonecotinoid and neurotoxic insecticide, found deficiencies in learning and memory and, in addition, bees were moving with difficulty.

Using imidacloprid, Decourtye et al. (2004) verified that after 30 minutes of treatment with sublethal oral doses, bees showed a deficiency in olfactory learning. Deficiencies in of memory, learning, olfactory ability and spatial orientation, and mobility difficulties may, depending on the number of affected individuals, have a major impact on the functioning of the colony as a whole because it is based on learning ability and the orientation of foragers that, by their own external activity, are the individuals most exposed to contamination.

Foragers are not the only ones exposed to pesticides. Nurse bees, newly emerged bees (and larvae) that feed on pollen and nectar stored in the combs are also being exposed to them (Figure 1). It is likely that the concentration of this exposure may vary because the amount of chemical compounds, the persistence of the residuals, and the frequency of the pesticide's application in the field will not be informative enough to state precisely this concentration. The chemical compounds may have accumulated for a long time within the colony where they are subject to degradation/concentration processes (Chauzat et al., 2006).

Figure 2. Diagram showing effects induced by environmental stressors on honey bees.

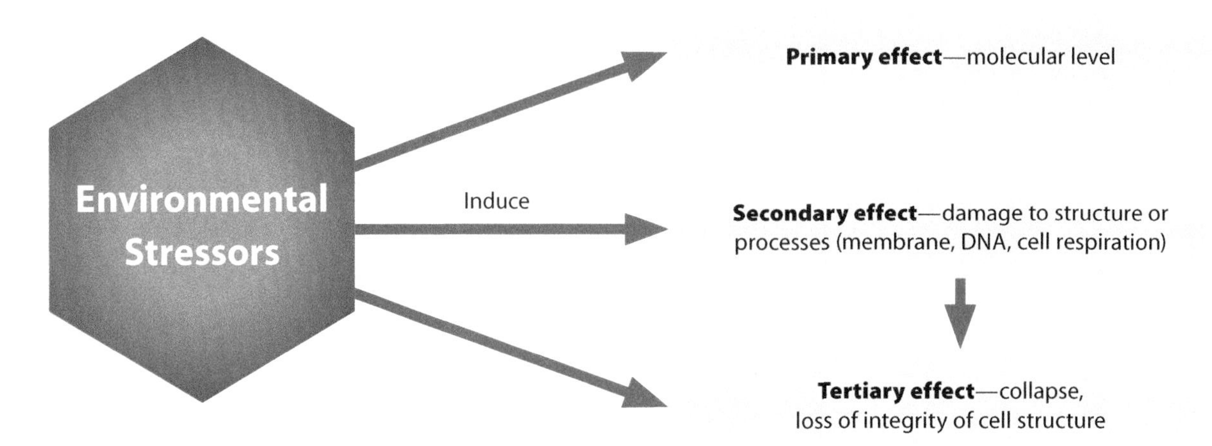

To cause such reactions, the pesticides or their metabolites must penetrate the bee's body and affect certain cellular mechanisms that may reflect morphological, biochemical, and/or physiological alterations, both mediated by changes in the presence, or absence, and in the amount of certain proteins.

There is increasing agreement among beekeepers that the sublethal effects of pesticides havea a significant impact on bee colonies (Pajot, 2001). Kevan (1999) considered that these effects, particularly resulting from chronic exposure to pesticides, are poorly understood and often overlooked. In the literature review carried out by Pham-Delégue et al. (2002), the lack of data on the sublethal effects of pesticides on bees is evident. The few data on these effects are described, mainly, for behavioral changes (Thompson, 2003). Morphological and immunohistochemical evidence needed for the detection of cellular biomarkers in the bees' organs is still rare. The studies done by Gregorc and Bowen (1998, 1999) were an important start in this area of research. Carvalho et al. (2009) studied enzymes as possible biomarkers for the assessment of environmental contamination with pesticides. Together, these efforts will add important information for understanding the sublethal effects of pesticides on bees. Stressors from the environment may be harmful on cellular and tissue levels and affect the whole organism. Cellular damage can vary, depending on the type of stress and its severity and duration and thus result in three levels: (a) damage that can be repaired and cellular homeostasis that is restorable, (b) cells that remain viable after intermediate level of damage, and (c) after a high level of damage, cells undergo apoptosis or necrosis (see Figure 2).

Organs and Tissues of Bees as Targets of Pesticides

The vast majority of nonsystemic pesticides affect the nervous system, the brain being its main target. However, pesticides can induce varying degrees of cytotoxicity in the organs involved in absorption, metabolism, and excretion and are able to exert additional effects beyond those provided by its mode of action.

Among the target organs for toxicological evaluation in bees, the ventricle and the Malpighian tubules are the most exposed (Malaspina and Silva-Zacarin, 2006; Silva-Zacarin et al., 2010). These organs are involved in absorption and excretion of xenobiotic compounds, respectively, and the evaluation of their morphology can uncover histological changes and/or ultrastructural ones induced by environmental contaminants (Cruz et al., 2010; Silva-Zacarin et al., 2010). The ventricle or midgut, where most of the digestion and absorption of nutrients and chemical substances occur (Chapman, 1998), is an organ of extreme importance in the study of the pathology and toxicology of insects. Additionally, its morphology is well described in the literature (Snodgrass, 1956; Cruz-Landim and Melo, 1981; Jimenez and Gilliam, 1990; Cavalcante and Cruz-Landim, 1999), which facilitates the interpretation of the results.

Despite their importance in histopathological studies, there are relatively few available studies on the action of chemical compounds or pathogens on the midgut morphology of bees. Liu (1984) examined ultrastructurally the intestines of bees infected by *Nosema apis*, where he found signs of cellular lysis. Gregorc and Bowen (1998) showed that during the infestation by *Paenibacillus larvae*, an increase in the rate of cell death in the ventricle of larvae of *Apis mellifera* occurs. This level of cellular death in the ventricle also increased in the larvae treated with oxalic acid or formic acid, both used to treat colonies infested with Varroa mites (Gregorc et al., 2004). In workers treated with boric acid, morphological analysis of the ventricle suggested that this compound caused impaired production of enzymes in digestive cells (Jesus et al., 2005). Cellular death was also observed in the ventricle of the larvae treated with the same compound (Cruz et al., 2010). As for morphological studies conducted with other insects, Cochran (1994) evaluated the toxicity of boric acid in the German cockroach (*Blatella germanica*) and Sumida et al. (2010) evaluated the same chemical compound in ants.

Morphological changes are not always observed in the ventricle if it is first detoxified (Yu et al., 1984). In honey bees, for example, there is an important group of enzymes involved in the detoxification metabolism which gives bees a mechanism of resistance or tolerance to insecticides of the pyrethroid group (Johnson et al., 2006). This detoxification metabolism even provides a biochemical adaptation that protects the foragers in contaminated ecosystems (Smirle and Winston, 1987). Chemical compounds that were absorbed in the midgut are transported to the hemolymph (Figure 3). The substances in the hemolymph that were not metabolized are eliminated by its excretory system, the Malpighian tubules, which also performs reabsorption of useful components to the organism, ensuring ideal ionic conditions for the proper performance of cells (Berridge and Oschmann, 1972). The Malpighian tubule is also the osmoregulator of the insects, analogous to nephrons of vertebrates, and also has the role of removing the chemical compounds of endogenous and exogenous hemolymph (Pannabecker, 1995).

The Malpighian tubules of insects possess active transport mechanisms for the removal of toxic compounds and metabolic products from the hemolymph (Bradley, 1985). The mechanism of excretion by the cells of the tubules was described by Snodgrass (1956), and the histology of the Malpighian tubules of bees was described by Cruz-Landim (1998). The excreta produced by the tubules flow into the intestine and accumulate in the rectum, where absorption can occur by the rectal papillae (Wigglesworth, 1974). Due to their excretory function, the Malpighian tubules represent one of the key organs in the detoxification of insects (Sorour, 2001).

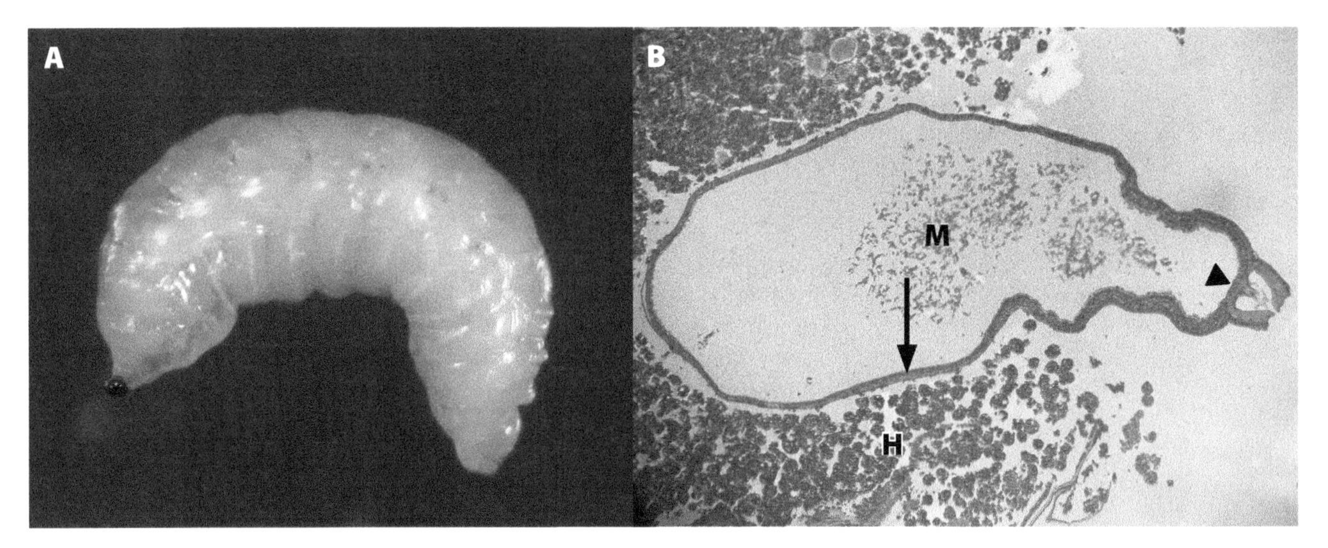

Figure 3. **A:** Three-day-old larva. **B:** Histological cross section showing larval midgut lumen (M). The arrow indicates midgut epithelium, the arrowhead indicates proventriculus, entrance of the larval diet into the midgut, and H indicates larval hemocoel. Histological section photographed at 100× magnification. (*Photos by A. Gregorc.*)

During their transit through the insect's hemolymph, the chemical compounds absorbed by the ventricle and not excreted by the Malpighian tubules may be indirectly absorbed by other organs and tissues of the insect. One example is larval salivary glands (silk glands). Once these glands remove precursors from the hemolymph to synthesize its secretion, they may be exposed to low doses of pesticides and therefore they may also be useful in assessing the toxic effects of chemical compounds in bee larvae. Because the silk glands of *Apis mellifera* have a defined secretory cycle, well characterized morphologically and histochemically (Silva-Zacarin 2002; Silva-Zacarin et al., 2003), changes in the cells induced by chemicals absorbed by the secretory cells via the hemolymph may be detected. Beyond that, as the process of cell death in the silk glands is also well characterized morphologically and histochemically (Silva-Zacarin, 2007; Silva-Zacarin et al., 2007; Silva-Zacarin et al., 2008), any acceleration of this process or histophysiological changes induced by such chemicals may be diagnosed in the glandular secretory cells (Silva-Zacarin et al., 2006).

The fat body cells of bees are also of interest because they play a role, in part, in the organism's detoxification process, allowing it to bioaccumulate the chemical compounds. The trophocytes of the fat body are the main headquarters of the intermediary metabolism of insects and thus they are considered the most significant part of the hemolymph proteins that are synthesized by these cells; they are also the precursors of vitellogenin. Fat body localized in the haemocoel remains immersed in the hemolymph; thus it can respond quickly to changes of metabolites regulating the composition of the surrounding environment of the insect, assisting its immediate metabolic requirements (Cruz-Landim, 2009). The enocytes, cells associated with body fat, may also be involved in detoxification processes of the adult insect although there is no ultrastructural proof of such participation (Cruz-Landim, 2009).

Many insecticides mainly target the nervous system and, after passing through several organs as described above, are still capable of causing damage in bees. The nervous system of insects is organized in a central and peripheral nervous system (Cruz-Landim, 2009), basically consisting of two cell types: *neurons* that are the fundamental unit, with specialized morphology and function in the perception and conduction of stimuli, and *neuroglia* or glial cells that occupy the spaces between neurons and serve, among other things, to support and isolate neurons (Kretzschmar and Pflugfelder, 2002). Beside the brain located dorsally on the head, the central nervous system consists of the ventral nerve chain and the estomogastric system (Chapman, 1998). Despite the modifications taking place during metamorphosis, adult *A. mellifera* preserve their general pattern of nervous system organization from their larval phase. The brain, the ventral nervous chain, and the stomagastric system are thus all part of the central nervous system (Cruz-Landim, 2009).

The brain consists of regions containing neurons called *somata*, and of regions that contain cellular extensions called *neuropiles* (Chapman, 1998) that are spread throughout the cerebral structure. In addition to the neurons, glial cells are also found. These cells are important for development, support, and other physiological functions within nervous tissue (Kretzschmar and Pflugfelder, 2002). The brain of adult bees is divided into three regions resulting from the fusion of three ganglia. These regions are the protocerebrum, the deutocerebrum, and the tritocerebrum (Chapman, 1998).

The region of the protocerebrum has lateral projections that correspond to the optical lobes. They have a complex structure and contain nerve elements that form the compound eyes and are responsible for the capture of visual stimuli. Also included are the mushroom bodies that consist of three main divisions: First, a region in the form of cup, stem or stalk, and lobes. The second region or the deutocerebrum presents the antennal lobes, which are spherical structures that receive olfactory stimuli via sensory neurons present in the antennae and structures of the mouth (Ribi et al., 2008). The third region or tritocerebrum is small, composed of a pair of lobes located ventral to the antennal lobes and parallel to the esophagus. This portion of nervous system is located outside the brain structure and provides motor and sensory elements that bind the system estomogastric (Cruz-Landim, 2009). Both mushroom bodies as well as the antennal lobes are key structures because they are important in processing information related to olfaction, learning, and memory and, consequently, they are structures important for understanding the action of pesticides on the nervous system of bees.

Cellular Response Induced by Pesticides in Organs and Tissues

Detoxification Systems

The mechanism of enzymatic metabolic detoxification is the main route for the insect's resistance to xenobiotics as suggested, for example, on workers of Africanized honey bees exposed to organophosphate insecticides (Attencia et al., 2005) and workers of European bees exposed to pyrethroid insecticides (Johnson et al., 2006). It happens, mainly, by the protective action of glutathione S-transferase, a detoxification enzyme that presents greater expression in forager bees, precisely those individuals that perform tasks outside the colony and therefore have more contact with chemicals in the environment.

Pesticides in sublethal doses, while not killing the bees, penetrate the body and enter their cells, triggering a series of cellular responses, including detoxification responses. Many detoxification enzymes were described in bees by Yu et al. (1984). They differ in their action according to the colony examined, worker age, population, and the quantity and quality of food received by the larvae (Smirle and Winston, 1987, 1988; Smirle, 1990) Among these enzymes, glutathione S-transferase and glutathione peroxidase are expressed in forager bees and help protect bees from various environmental pollutants when they start to perform activities outside the colony (Smirle and Winston, 1988; Diao et al., 2006).

The glutathione transferases belong to a family of enzymes with diverse functions, including detoxification of cells contaminated with pesticides. They act on the molecules of pesticides, catalyzing reactions that lead to the formation of less toxic compounds that are easier to excrete. Many studies have shown variations in the ratio of the expression of detoxification enzymes with resistance to insects (Rodpradit et al., 2005), including bees (Yu et al., 1984; Johnson et al., 2006).

Due to the importance of these processes in the maintenance of cell viability and of the resistance to pesticides, it is extremely important to understand the functions of each. This may give us new tools in ecotoxicological monitoring. With the completion of the genome of *Apis mellifera*, the study of genes related to detoxification processes may help us understand the phenomena of tolerance and resistance to insecticides (The Honeybee Genome Sequencing Consortium, 2006).

Oxidative Stress Response on the Cellular Level

Oxidative stressors may generate reactive oxygen species (ROS), that is, chemically reactive molecules containing oxygen. They are a byproduct of the normal metabolism of oxygen with cell-signaling roles, and occur during environmental stress. The activity is upregulated and can cause damage to cell structures (known as oxidative stress). Environmental stressors can induce elevated amounts of ROS, which could accumulate to toxic levels. Oxidative stress results in cytotoxicity and damage to cellular structures and can lead to cell apoptosis or necrosis (Fuchs et al., 1997; Richter and Schweizer, 1997).

One way to map the activities of different brain areas is through the technique of immunohistochemistry to detect cytochrome oxidase (CO), commonly used in vertebrates as a marker of neuronal activity, to assess pathological changes or chronic effects after pharmacological treatment (Armengaud et al., 2001). Cytochrome oxidase, also known as complex IV respiratory chain mitochondrial, and is a terminal enzyme of the electron transport chain in the process of mitochondrial respiration and generates energy that is linked to neuronal activity. This enzyme catalyzes the last reaction of oxidative metabolism, that is, the primary means of energy production in the brain. The distribution of CO activity can be visualized in sectioned tissues and are an indicator of the metabolic capacity of the tissue and varies depending on location in the body (Hevener and Wong, 1991).

Decourtye et al. (2004) observed an increase in the activity of CO in mushroom bodies in worker honey bees after treatment with the insecticide imidacloprid. Besides this, other researchers (Déglise et al., 2003; Armengaud et al., 2001) used this technique to evaluate the effects of short-term cholinergic ligands in the metabolism of different brain structures, focusing their investigations on the regions of the antennal lobes and mushroom bodies because these structures are related to the processes of learning and memory. Studies performed by Roat (2010) using immunohistochemistry showed that the CO-induced neural activity at certain ages in the cells of mushroom bodies in response to the insecticide fipronil. Similar results were obtained by Decourtye et al. (2005), also working with imidacloprid. Mushroom bodies have three different types of Kenyon cells: the inner and outer compact cells and non-compact cells, which have different morphology and different protein profiles (Farris et al., 1999; Strausfeld et al., 2000). These structures are responsible for processing information from a variety of receptors and have similar functions in the cortex of vertebrates. Thus, these cellular effects are directly related to behavioral changes, loss of spatial orientation, learning ability, and reduced memory in individuals treated with pesticides.

Mixed-Function Oxygenase System Alteration and Xenobiotic Degradation

The mixed-function oxygenase (MFO) system represents a diverse class of enzymes. The so-called cytochrome P450 enzymes are the main enzymes in the group. The MFO system is directed toward deactivation and excretion of endogenous or exogenous compounds, pesticides, and other pollutants from the environment and can be an indicator of the detoxification capacity in different organisms (Yu et al., 1984; Kezic et al., 1989). Cells possess an array of enzymes capable of biotransforming a range of chemicals. The enzymatic detoxication of xenobiotics proceeds in a tightly integrated manner (Sheehan et al., 2001).

Phases of detoxification involve the conversion of a lipophilic, nonpolar xenobiotic into a more water-soluble and therefore less toxic metabolite that can then be eliminated more easily from the cell. This phase is catalyzed mainly by the cytochrome P450 system. The family of microsomal proteins is responsible for a range of reactions, of which oxidation appears to be the most important (Guengerich, 1990). In the second phase, enzymes catalyze the conjugation of activated xenobiotics to an endogenous water-soluble substrate, such as reduced glutathione (GSH), UDP-glucuronic acid, or glycine. Quantitatively, conjugation to GSH, which is catalyzed by the GSTs, is the major reaction. GSTs are dimeric, mainly cytosolic enzymes that have extensive ligand-binding properties in addition to their catalytic role in detoxification (Barycki and Colman, 1997). They have also been implicated in a variety of resistance phenomena involving insecticides (Ranson et al., 1997) or herbicides (Edwards et al., 2000).

The harmful effects of environmental pollutants require early assessment in order to monitor sublethal effects. Measurement of MFO activity can be performed by monitoring the activity of aryl hydrocarbon hydroxylase (AHH) known as benzo(a)pyrene-3-hydroxylase, also an enzyme of the microsomal MFO group. The enzyme can be induced by a number of exogenous air, water, and food pollutants such as pesticides (Busbee et al., 1978). The relationship between insecticide resistance, detoxifying enzyme capacity, and seasonal fluctuation in these enzyme levels in bees was also studied, and thus the relationship between colony polysubstrate mono-oxidase activity and intercolony variation in susceptibility to pesticides was established (Smirle and Winston, 1988). It has been recognized that microsomal cytochrome P450-dependent MFO is involved in the metabolism or biotransformation of xenobiotics in bees.

Specifically, interactions were observed between *tau*-fluvalinate and coumaphos at the level of P450 detoxification (Johnson et al., 2009). MFO activity in bees is induced to its maximum as a response to the presence of xenobiotics and, over a period of days, the activity is slowly decreased (Kezic et al., 1983; Smirle and Winston, 1987).

The microsomal mono-oxygenase and cytochrome P450 system, as an important system in the detoxification of xenobiotics such as drugs, pesticides, allelochemicals, and other lipophilic substances, has been studied extensively in a variety of insect tissues (Hodgson, 1985), including the honey bee midgut (Gilbert and Wilkinson, 1974, 1975). The microsomal mono-oxygenase studies are well established, and xenobiotic catabolism may be only a small fraction of the subcellular activity localized in the insect microsomal tissue.

Whole bees and bee tissues have been used to study oxidase activity in the microsomal fraction as a susceptible organism to common pesticides; it was found that the midgut had the highest MFO activity (Gilbert and Wilkinson, 1974). The presence of peroxisomal marker enzymes in microperoxisomes of the honey bee midgut columnar cells and their involvement in both intermediary metabolism and a holocrine secretory process were established (Jimenez and Gilliam, 1988).

Pyridaben, an inseticide, is rapidly metabolized in mammals to a large number of metabolites, and the sulfoxide of pyridaben readily reacts with glutathione and GST to give multiple adducts (Schuler and Casida, 2001). As such, the low cellular activity of GST that was measured with the *Spodoptera exigua* cell line may be an effector of the high *in vitro* toxicity identified as mitochondrial electron transport inhibitors (METI). In contrast, the GST activity in the whole third-instar larvae is high and the activity of this important enzyme is thus about nine times less in the cell line compared with the larvae (Francis et al., 2002). The variation in GST activity was not the only biochemical difference between the two systems tested in *S. exigua* cell cultures or third-instar larvae, as seen with the great variation in patterns of esterases, isozymes, and G3P dehydrogenases (Nims et al., 1998). In addition, because all the above-mentioned enzymes are in some way related to specific invertebrate processes, a nonspecific enzyme, namely malate dehydrogenase was also tested and showed the same expression in all samples of third-instar larvae and of cell lines Se1, Se4, and Se5.

It seems that differences are true indicators of biochemical variation between larvae and insect cell lines (Decombel et al., 2005). Specific tests performed in honey bees or on their organs and tissues are thus important in order to evaluate the threshold effect. While the results obtained by the cell bioassay using cell cultures are not an effective screening technique, there are other types of assays that are more specific. Results obtained using the insect model can give more meaningful results that are better correlated with *in vivo* activity. We may also conclude that cell culture bioassay is not able to spot insecticides that trigger a neuronal signal. When applied pesticides exhibit *in vivo* toxicity but do not exhibit any activity in the cell bioassay, a predictive conclusion could be that its action is neuronal (Decombel et al., 2005). Beekeepers are facing the potential presence of several pesticides in beehives. The study of the mode of action in bee organs and tissues is thus crucial in order to recognize sublethal and subclinical changes at the individual and at the colony level, and may lead to individual or colony mortality.

Fos-Gene and Its Expression in the Bee Brain

The optical lobes located in the protocerebrum are complex structures that form the composite eyes where different types of cells dealing with the capture/perception of visual stimuli are organized. The deutocerebrum houses the antennal lobes. These are spherical structures that receive olfactory stimuli captured by the sensitive neurons located in the antennae and mouth structures (Hansson and Anton, 2000).

These two structures play important roles in the daily functioning of bees because they are responsible for the interaction between insects and their environment. This is especially true in worker bees older than 21 days, the age at which they start to venture away from the colony. The transition between the intra-colonial life and life outside the colony comes with a series of new experiences for the bees, which receive a variety of new stimuli during field activities. These stimuli may halt the expression of genes not required to perform the intra-colonial activities (Robinson, 1992). Because it has been extensively studied both in vertebrates (Herrera and Robertson, 1996; Hoffman and Lyo, 2002) and in invertebrates (Bidmon et al., 1991; Fonta et al., 1995; Cymborowski, 1996, Cymborowski and King, 1996; Renucci et al., 2000; Rousseau and Goldstein, 2001; Giesen et al., 2003), the *fos*-gene can be selectively expressed in accordance with the age polyethism and consequently with a wide range of stimuli received by the workers while foraging.

With the appropriate stimulus, the *fos*-gene is immediately expressed and together with another gene product (*jun* gene) forms the AP-1 complex. This complex is a transcriptional factor leading cells to change their morphology in response to the stimuli. This expression continues as long as the neurons are stimulated. When the stimuli stop, the translation of both genes stops too. This gene was isolated by Curran et al. (1982) from the virus-causing murine osteogenic sarcoma (*v-fos* gene). The *c-fos* gene found in vertebrates was isolated using the *v-fos* gene as a probe (Curran et al., 1983). Both *fos*-genes codify phosphoproteins with nuclear localization (Curran et al., 1983) and reveals itself to be sufficiently preserved. In invertebrates, transcription factors analogous to the product of *fos*-gene expression in mammals were identified in embryos of *Drosophila* by Perkins et al. (1988). The structure of the gene homologous to the mammal *c-fos* gene was identified by Rousseau and Goldstein (2001), who designated it as *d-fos*.

Today it is known that in invertebrates there are many different isoforms of the *fos*-gene, but in *Drosophila* there is only one form of this gene present, known as *kayak* (kay) (Kockel et al., 2001; Souid and Yanicostas, 2003). Many studies have demonstrated the involvement of AP-1 complex in different cellular processes at all developmental stages of invertebrates.

In *A. mellifera*, studies performed by Fonta et al. (1995) demonstrated the existence of a similar *fos*-gene being expressed during the development and maturation of the nervous system in worker bees. Recently, Nocelli et al. (2010) showed that the *fos*-gene is differently expressed in the bee brain of newly emerged nurse and worker bees, which can be explained by different tasks that bees play in different life stages and by different environments that bees live in: inside and outside the colony. In other research, the same authors observed different patterns of *fos* expression in mushroom body cells. Compact inner and outer cells show higher expression levels when compared to the noncompact cells through the age polyethism, reflecting the differential functions of these cells. Because of its characteristics and plasticity, *fos*-gene could be an excellent marker for cell activity under chronic exposure to pesticides.

Gene Expression in Stress Situations and Heat Shock Proteins (HSPs)

The accumulation of genetic changes in an organism represents evolutionary adaptation to enhance the survival potential of the species in the environment. Genetic changes that are heritable are expressed on the transcriptional level or in the function of the encoded protein. The homeostasis of the organism can be conserved through the complexity of the responses to tissue or system level in multicellular organisms in contrast to unicellular organisms. Stress responses in an organism can be expressed as tissue-specific transcriptional responses (Gracey et al., 2001). Biotic and abiotic stressors, including ecotoxicants can be studied by the analysis of the repressed genes in tissues; they reveal that genes encode components of the protein translation machinery, together with genes coding for the abundant tissue-specific structural proteins. Environmentally induced stress or the effects of disease states can thus be studied in honey bees in the range of tissues. This will contribute to the knowledge of the role of transcriptomes in honey bee organisms under stress conditions and will give new insight into environmental adaptation.

Stress proteins can be constitutively present in tissues or they can be induced during stress conditions. High expression of cellular stress proteins may contribute to the tolerance of bees to xenobiotics by functioning as molecular chaperones (Feder et al., 1995) and they are involved in intracellular protein maturation (Georgopoulos and Welch, 1993). The stress protein response in cells is often very fast when the cell is exposed to stress and the function of the proteins is directed toward cell survival and restoration of homeostasis. Then, these proteins participate in a number of diverse biological processes and play a central role in the protection and maintenance of many cellular functions (Lindquist and Craig, 1988; Morimoto et al., 1994; Becker and Craig, 1994; Yokoyama et al., 2000). These stress proteins named by Sanders (1993) are best known as HSPs and are evolutionarily conserved from bacteria to humans (Ashburner, 1982). According to their molecular weight, they are classified into HSP100, HSP90, HSP70, HSP60, and small HSPs 15-30 families (Garrido et al., 2001).

Among the stress protein families, the HSP70 family is the most important. Eukaryotic cells contain eight to ten different proteins in this family. The diverse functions of HSP70 are illustrated by their activity in the nucleus, cell organelles, and cytosol. They stimulate protein transport into the endoplasmic reticulum (ER), mitochondria and the nucleus, and mediate lysosomal degradation of cytosolic proteins (Chiang et al., 1989). They also participate in protein folding and together with specific soluble or membrane-bound partner proteins are involved in protein traffic, cell growth and development, translocation, and gene regulation (Rassow et al., 1995).

They are upregulated in response to induction by many stressors (Hendrick and Hartl, 1993). HSP70 is important in the maintenance of cellular functions under stress situations (Beckmann et al., 1992). It has been reported that HSP70 induction by certain environmental chemicals is generally correlated with certain initial cytotoxic events and is a secondary consequence of damage that affects cellular integrity (Neuhaus-Steinmetz and Rensing, 1997; Ait-Aisa, 2000).

HSP90 is reported to be highly specialized in binding protein and has important intracellular chaperone properties. It is an abundant type of eukaryotic stress protein whose function has remained largely enigmatic (Jakob and Buchner, 1994). Nuclear HSP90 was found in the *Xenopus* (frog) oocytes that bind tightly to purified histones (Hendrick and Hartl, 1993). Thus, the stress proteins act as a "buffer system," which generally protects the cell from the induction of cell death process.

Nadeau and Landry (2007) reported that HSPs induce specific signaling cascades that promote and regulate cellular homeostasis. The authors argue over the direct connection of these signaling pathways to HSPs with the mechanisms of survival and programmed cell death. HSPs could participate in an alternative route for autophagic degradation of alterative proteins called chaperone-mediated autophagy (Cuervo and Dice, 1996), where proteins with a sequence-specific signal are transported to the lumen lysosome, a process mediated by receptors associated with HSPs. Additionally, HSPs can act in multiple points of intracellular signaling pathways involved in programmed cell death, like apoptosis (Garrido et al., 2001).

Because family members of HSPs are often expressed following environmental stress and play a central role in the protection and maintenance of many cellular functions (Becker and Craig, 1994), it has been proposed that HSP proteins are used as cellular biomarkers to monitor the impact of environmental factors in various organisms, including several invertebrates (Köhler et al., 1992; Eckwet et al., 1997; Kar Chowdhuri et al., 1999; Lewis et al., 1999; Bierkens, 2000; Mukhopadhyay et al., 2002; Nazir et al., 2003a,b; Mukhopadhyay et al., 2006).

HSP Localization in Tissues

Monoclonal antibodies have been used to determine the presence of HSPs in different tissues (Chiang et al., 1989). Anti-HSP antibodies were used as markers of the effects caused by toxic metals on terrestrial isopods and terrestrial and marine mollusks (Köhler et al., 1992). HSP70 and HSP90 proteins are stabilizing polypeptide chains active during *de novo* folding and under stress conditions. They also participate in specific processes such as the initiation of DNA replication (Langer et al., 1992).

Immunocytochemical studies of the localization of heat-shock proteins in the larval tissues of honey bees after acaricide applications have helped to better understand the possible adverse effects acaricides may have on bees (Gregorc and Bowen, 1999, 2000). The sensitivity of the HSP system to induction by a wide variety of chemical or physical stressors makes it attractive as a biomarker to evaluate the biological effects of exposure to any given toxic agent. In bees, the use of cell markers in ecotoxicological studies is more recent than in other invertebrates. In their pioneer work Gregorc and Bowen (1998 and 1999) evaluated the expression of HSPs in the organs of bee larvae infested with *Paenibacillus larvae*. Then Silva (2002) and Silva-Zacarin et al. (2006) evaluated the expression of HSPs in the larval salivary glands of *A. mellifera* treated with antibiotics and acaricides used in commercial apiaries (Figure 4). Gregorc et al. (2004) and Gregorc and Smodiš Skerl (2007) evaluated the rate of cell death in the ventricle of larvae treated with acaricides used to treat hives infested with *Varroa*.

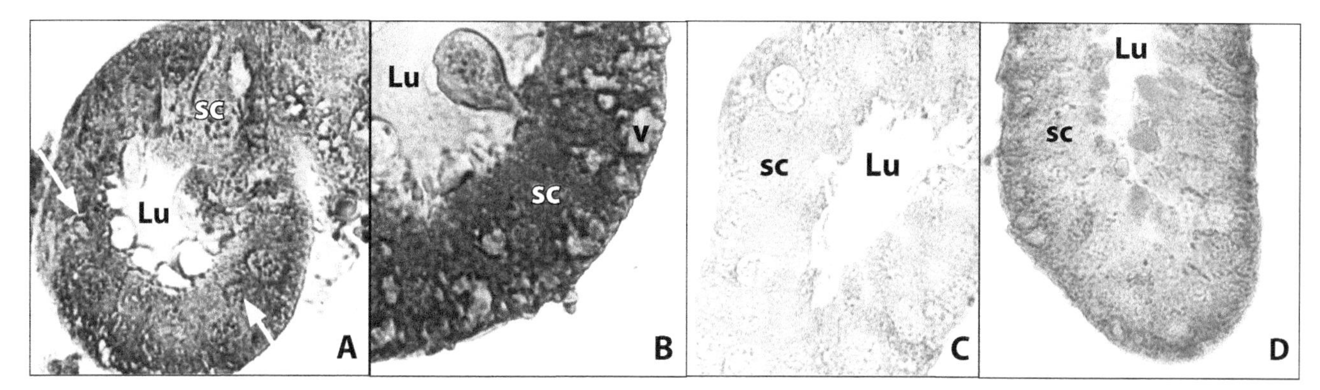

Figure 4. Secretory portion of formalin-fixed, paraffin-embedded salivary glands of bee larvae at the beginning of their 5th instar; collected 48 hours after the oxalic acid (OA) treatment. **A.** Control group; mag. 200×. **B.** Treated group; magnification 400×. Detection of HSP70 by monoclonal antibody. Employed secondary antibody is conjugated with peroxidase and the diaminobenzidine (DAB) is used as a substrate. DAB, which gives a brown staining reaction product, is localized in the nuclei and throughout the cytoplasm. DAB-reaction product is absent from the vacuolated regions (v) of the cells. Mag. 300×. **C.** Control group; mag. 400× and **D.** Treated group; mag. 300x. Detection of HSP90 by monoclonal antibody. Intensive DAB-reaction product (arrow) is localized throughout the cytoplasm of cells, which have also suffered morphological alterations. Lu = gland lumen; sc = secretory cell. (*Photos by E. C. M. Silva-Zacarin.*)

Increased expression of HSPs seen in the cells of bees exposed to stress such as exposure to pesticides (Silva, 2002; Silva-Zacarin et al., 2006) or infestation by pathogens (Gregorc and Bowen, 1998, 1999) suggests that this phenomenon could be a defense mechanism to prevent large-scale cell death in targeted organs. Immunohistological techniques can be used to detect HSPs associated with cell death labeling, and the tests could help diagnose bee infections and evaluate the effects of acaricide applications on bees (Gregorc and Bowen, 1999). An increased localization of HSP70 in the basal cell cytoplasm appears within 6 hours after the application of rotenone or oxalic acid solution (Silva-Zacarin et al., 2006). The cytoplasmic localization of stress-induced HSP70 coincides with the mitochondria-rich cellular region (Silva, 2002). Wong et al. (1998) suggested that HSP70 may protect cells from energy deprivation and/or ATP depletion associated with cell death. In rotenone-treated larvae we observed an intensive localization of HSP70 in the nuclei and basal cytoplasm of salivary gland cells, due to its suppressive effect of mitochondrial damage and nuclear fragmentation (Buzzard et al., 1998) and its maintenance of the features of secretory cells. This suggests that HSP70 has an anti-apoptotic effect.

The concentration of applied oxalic acid (OA) as an acaricide used in the experiments, simulating acaricide application to the bee hive, promoted a higher degree of stress than the cells could support. In this way, the overexpression of HSP70 and HSP90 could not have been sufficient to inhibit massive cell death. It was also found that some larvae tested are either more sensitive to OA application than others, or that the quantity of OA consumed by the larvae varied during the longer exposure to OA present on the comb (Silva-Zacarin et al., 2006).

It was shown that in biological and toxicological studies of honey bees, the determination of HSP localization and *in situ* labeling of DNA strand breaks to detect cell death can be useful immunohistochemical methods for demonstrating and understanding any possible adverse, sublethal effects on bees. Such a combination of the immunohistochemical assays may help detect the cellular responses of larval tissue suffering from different chemical stressors from within their environment. These methods are, thus, potentially powerful tools for diagnosis in biological models and can be explored in research to evaluate the threshold effects of substances used in bee colonies or their source in the environment (Silva-Zacarin et al., 2006).

Apoptotic and Necrotic Cell Deletion as a Response to External Influences

During metamorphosis, extensive degradation of several larval tissues takes place. Ecdysteroid is responsible for the induction of metamorphosis and causes degradation of larval tissues and juvenile hormone for qualitative changes during metamorphosis (Rachinsky et al., 1990). Lysosomal and free acid phosphatase activity as an indicator of cell death has been determined in honey bees by Jimenez and Gilliam (1990). Cell death can occur either by accident, referred to as necrosis, or by design, which is described as programmed cell death or apoptosis (Bowen et al., 1996). It has been shown that apoptosis can be induced by either genetic or nongenetic means (Ellis et al., 1991). Necrotic cell death appears to be induced under extreme conditions such as ischaemia, hypoxia, exposure to toxins, and hyperthermia (Bowen et al., 1996). Disturbances in cell respiration, energy generation, and other metabolic processes may lead to DNA damage that is indicative of active apoptosis, also called programmed cell death. When the apoptotic process cannot be finished properly, stress response in tissues may lead to an uncontrolled deletion process referred to as necrosis. Necrotic pathways including uncontrolled degradation of DNA are normally a result of the high severity of environmental stress. When stress has a local effect in some tissues, it may also influence the function of the whole organism, culminating in effects such as fecundity, survival, or life span, which is determined by the aging procress (Korsloot et al., 2004). Cell death has been revealed in the regressive hypopharyngeal glands of worker bees (Moraes and Bowen, 2000), in their midgut after a *Paenibacillus larvae* infection, and after acaricide (amitraz) application (Gregorc and Bowen, 1998, 2000), and in the isolated atria of the bee's heart after exposure to the herbicide 2,4-dichlorophenoxyacetic acid (Papaefthimiou et al., 2002). The DNA breakdown that precedes the nuclear collapse of apoptotic nuclei can be tested using terminal deoxynucleotidyl transferase-mediated dUTP nick end labeling, which is generally termed a TUNEL assay (See Figures 5 and 6; Levy et al., 1998; Sgonc and Gruber, 1998).

It was found that honey bee larvae treated with oxalic (OA) or formic acid undergo subclinical changes that are detected using immunohistochemical methods. Apoptosis and necrosis coexist (Matylevitch et al., 1998) and an apoptotic process might become necrotic (Bell et al., 2001). In our previous experiments, up to 5% apoptotic cell death in the larval midgut is indicative of a normal level of tissue turnover (Gregorc and Bowen, 1997, 1999).

An OA application to bee larvae affects the columnar cells of their midgut and it appears to show apoptotic and necrotically induced DNA changes (Gregorc and Bowen, 2000), indicating that apoptosis and accidental cell death were occurring simultaneously after the acaricide application (Gregorc et al., 2004). In OA-treated larvae, the increased cell death is accompanied by morphological characteristics of cytoplasm vacuolization, nuclear envelope expansion, and intercellular detachments.

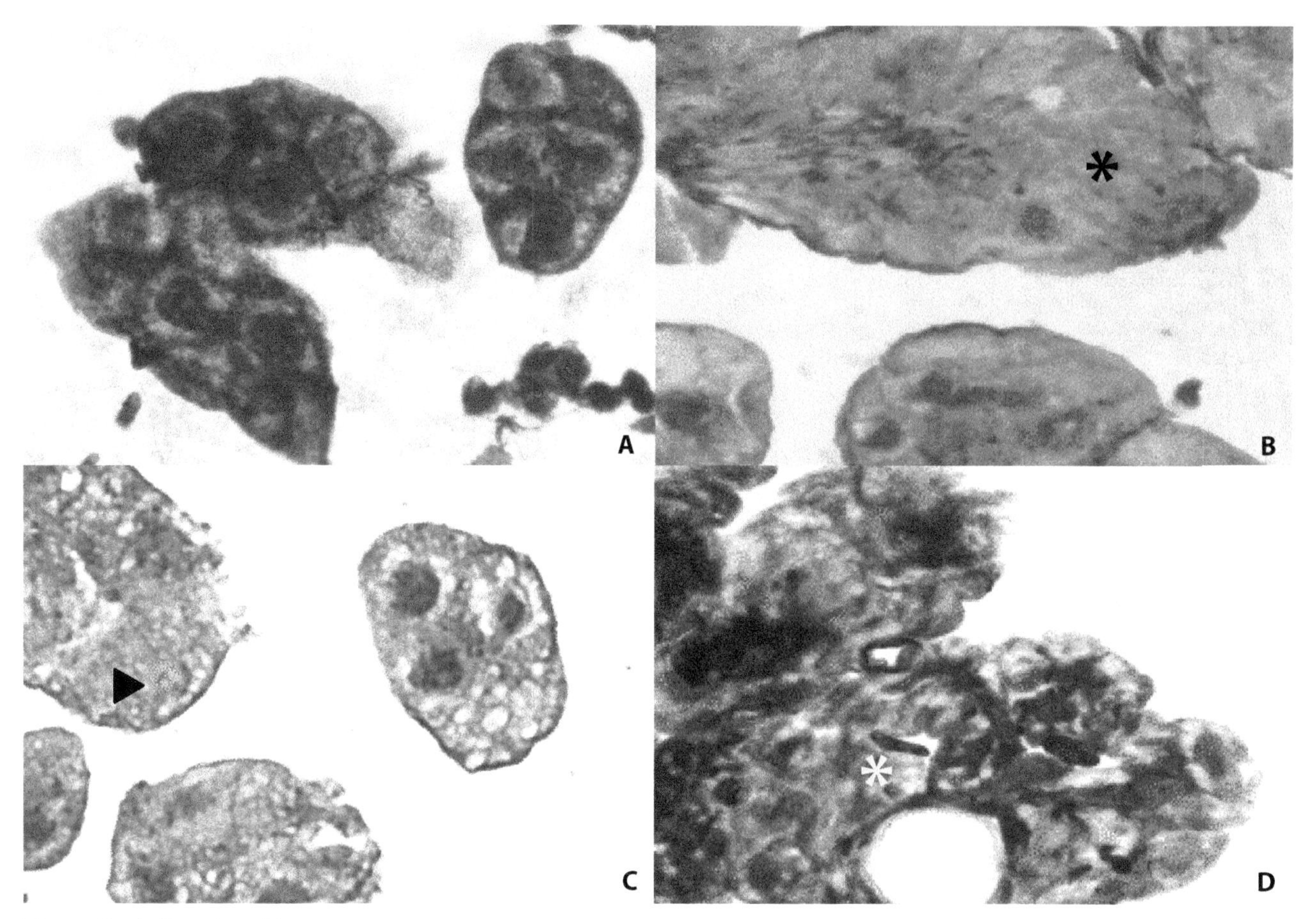

Figure 5. **A.** Two-day-old worker bee after 24 hours of imidacloprid treatment. Stained with HSP70 monoclonal antibody. Red azo-dye reaction product dispersed throughout the cytoplasm, cell nuclei (arrow), and ducts in glandular cells. Mag. 400×. **B.** Six-day-old worker bee after 24 hours of imidacloprid treatment. Stained with HSP90 monoclonal antibody. Activity found in cytoplasm (*). **C.** Hypopharyngeal glands (HPG) of 9-day-old bee after 48 hours of coumaphos treatment with sporadic HSP70 positive nuclei (arrow). Note negatively stained nuclei (▶). Mag. 400×. **D.** Twenty-day-old bee after 48 hours of coumaphos treatment. Stained with HSP70 monoclonal antibody. Fast red reaction product is diffused throughout the cell cytoplasm (*) and nuclei remain HSP 70 negative (arrow). Mag. 200×. (*Photos by M. I. Smodiš-Škerl.*)

The increased cell death in the OA-treated larvae is due to necrosis. The study of cell death in larvae after oxalic or formic acid treatments has helped us understand the possible adverse effects they may have. In addition, these studies have resulted in establishing the different features of cell death, necrosis, and apoptosis that could be detected using kits with different specificity (Sperandio et al., 2000). Quantification of cell death could be used in monitoring the effects of organic acids on larval tissue when applied in a bee colony. These methods are thus useful for evaluating the detrimental thresholds of larval tissue.

The acaricide coumaphos, often applied to bee colonies to reduce Varroa mite numbers, triggered an increased level of necrosis in workers up to 6 days old and also showed some level of apoptosis. Imidacloprid also induced extended necrosis in hypopharyngeal glands (HPG). Necrotic and apoptotic cell death thus increased with the prolonged time of imidacloprid treatment (Smodiš Škerl and Gregorc, 2010). *In situ* biological and toxicological studies of worker organs treated with different pesticides and determination of HSP70 and HSP90 reveal that localization can be a useful immunohistochemical method. It is possible now to demonstrate and better understand any adverse (sublethal) effects on tissue *in vivo*. A combination of the immunohistochemical assays may help to detect the cellular responses of honey bee tissues to widely used pesticides. These methods are also potentially powerful tools for the detection of subclinical changes and the evaluation of threshold effects of other chemical substances.

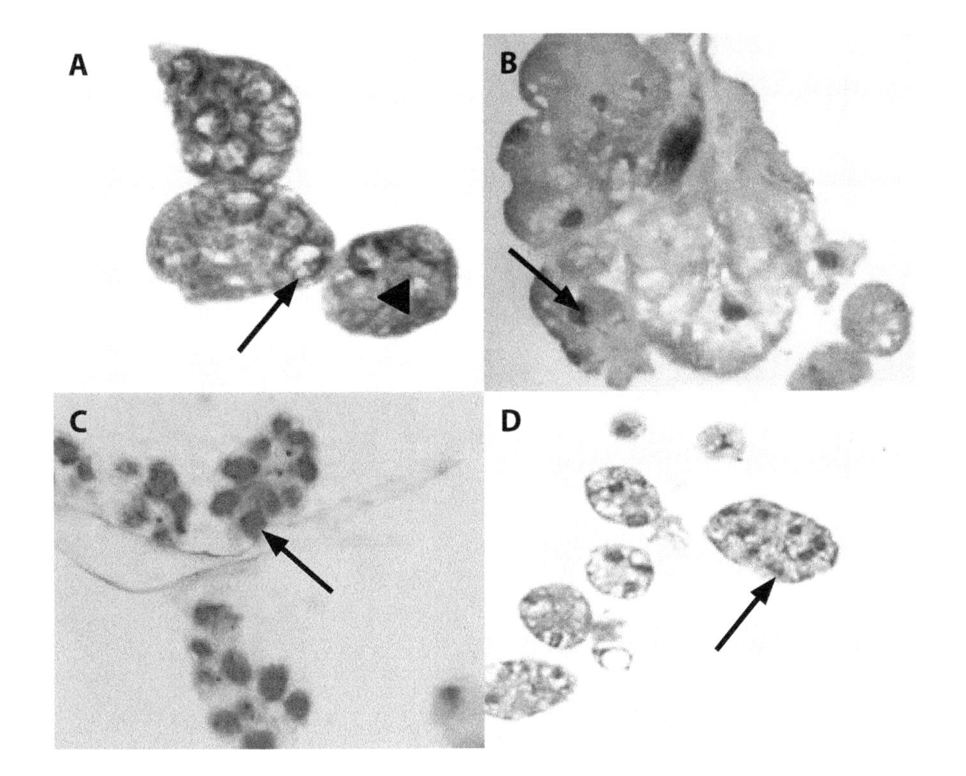

Figure 6. **A:** Four-day-old bee after 48 hours of imidacloprid treatment. Cell death was detected by the TUNEL technique using terminal deoxynucleotidyl transferase (TdT)-mediated dUTP for DNA labeling, and anti-fluorescein alkaline phosphatase conjugated antibody (ISCDDK); fast red was used for visualization, and counterstaining with haematoxylin. The figure shows dense red azo-dye staining localized to the nuclei (arrow) of the glandular cells. Note some negatively stained cell nuclei (triangle). **B:** Six-day-old bee after 48 hours of imidacloprid treatment. Staining of formalin-fixed, paraffin-embedded HPG and the terminal deoxynucleotidyl transferase (TdT)-mediated dUTP nick end labeling (TUNEL) technique using "ApopTag" kit. Peroxidase-conjugated anti-digoxigenin secondary antibody and DAB as a substrate was used to obtain specific brown reaction product. DAB reaction product is localized in sporadic cell nuclei (arrow). **C:** Seven-day-old bee after 72 hours of imidacloprid treatment, with all dense red azo-dye staining localized to the nuclei (arrow). **D:** Eleven-day-old bee after 48 hours of coumaphos treatment. Cell death was detected by the TUNEL technique (ISCDDK). Dense red azo-dye staining localized to the nuclei (arrow). (*Photos by A. Gregorc.*)

Biomarkers as Useful Tool for Pesticide Exposure Evaluation

The employment of morphological analysis of the bee's organs associated with the detection of cell markers (Malaspina and Silva-Zacarin, 2006) could be added to the risk assessment when standard protocols are not sufficient to provide the effect of chronic exposure of bees to low doses of the chemical. Additionally, the study of proteins of cellular stress associated with the analysis of cell death contributes to understanding the mechanisms involved in the toxicity of pesticides to bees, in sublethal doses, which until now was unknown.

The analysis of morphological changes in target organs of bees associated with the evaluation of biomarkers of cellular stress and cell death are of great importance for the ecotoxicological study of bees (Malaspina and Silva-Zacarin, 2006), especially when conventional protocols are not sufficient to answer questions about the effect of sublethal doses over the long term.

The development of cellular biomarkers to evaluate chronic exposure of honey bees to sublethal doses of pesticides is necessary in order to understand the effects of the synergistic action of xenobiotics in the environment. In other aspects, the evaluation of the effect of stress caused by the action of pathogens in bees must be integrated into the study of cellular biomarkers with the intention to discriminate the effects of pathogens and pesticides.

Köhler and Triesbskorn (1998) developed a protocol based on ultrastructural responses of invertebrate midgut epithelium, induced by exposure of chemical compounds, in order to identify the compromising degree of this target organ in soil invertebrates (see Table 1).

Table 1. Protocol developed by Köhler and Triesbskorn (1998) based on ultrastructural responses of invertebrate midgut epithelium induced by exposure to chemical compounds.

Cell Structure	Compensatory Response	Non-Compensatory Response
Microvilli	Shortening, irregular morphology and numeric reduction	Absence or destruction by lyses
Mitochondria	Alteration of the typical electron density of the matrix (increase or decrease), moderate swelling, and increasing in the amount of cristae	Membrane disruption, disorganization of cristae, myelin figures in the matrix, exacerbated swelling combined with decreasing the amount of cristae
Rough endoplasmic reticulum	Vesiculate cisternae or with dilatation, degranulation, alterations in the amount of cisternae (increase or decrease)	Membrane disruption and myelin figures formation
Nucleus	Morphological alteration in nucleus and/or nucleolus, electron density changes (chromatin compaction), dilatation of perinuclear space	Blebbing of condensed nucleus or swelled nucleus with karyolysis

Some of these compensatory responses indicate a defense mechanism of the midgut epithelium, such as the alteration in quantity and height of microvilli that decreases the absorption of the toxic compounds and increases the quantity of secretion vesicles containing glycoprotein, described in diplopod midgut by Fontanetti et al. (2001). Compensatory responses described in Table 1 show the first signs of cell death in order to eliminate damaged cells but these signs are compatible with the reversible stage of cell death, referred to as programmed cell death (PCD). Features of cell death are observed as noncompensatory responses that could induce PCD and/or necrosis.

In addition to these ultrastructural responses described in Table 1, the cytoplasm of bees is a good indicator of noncompensatory (non-reversible) response to pesticides, so that vacuolation is observed in several organs exposed to different concentrations of chemical compounds (Cruz et al., 2010). During the compensatory response, vacuolation could represent macroautophagy of damaged organelles or chaperone-mediated autophagy of damaged proteins (similar to that described in mammals) in order to promote cell survival in a stress condition; this last hypothesis has not yet been studied in insects. The intensification of vacuolation could indicate a noncompensatory response that probably is related to PCD features, if other parameters that indicate cell death such as DNA fragmentation will be present. Figure 7 exemplifies the compilation of possible biomarkers to identify pesticide exposure in bees.

Considerations and Perspectives

It is of utmost importance to identify the range of pesticides in the environment and understand the impact they have on the diversity of pollinators and, consequently, on the pollination process.

The indiscriminate and irrational use of pesticides is subjecting all pollinators to situations of severe stress, which can cause economic

Figure 7. Biomarker indicatives of cellular responses to pesticide exposure that can be detected by immunohistochemistry.

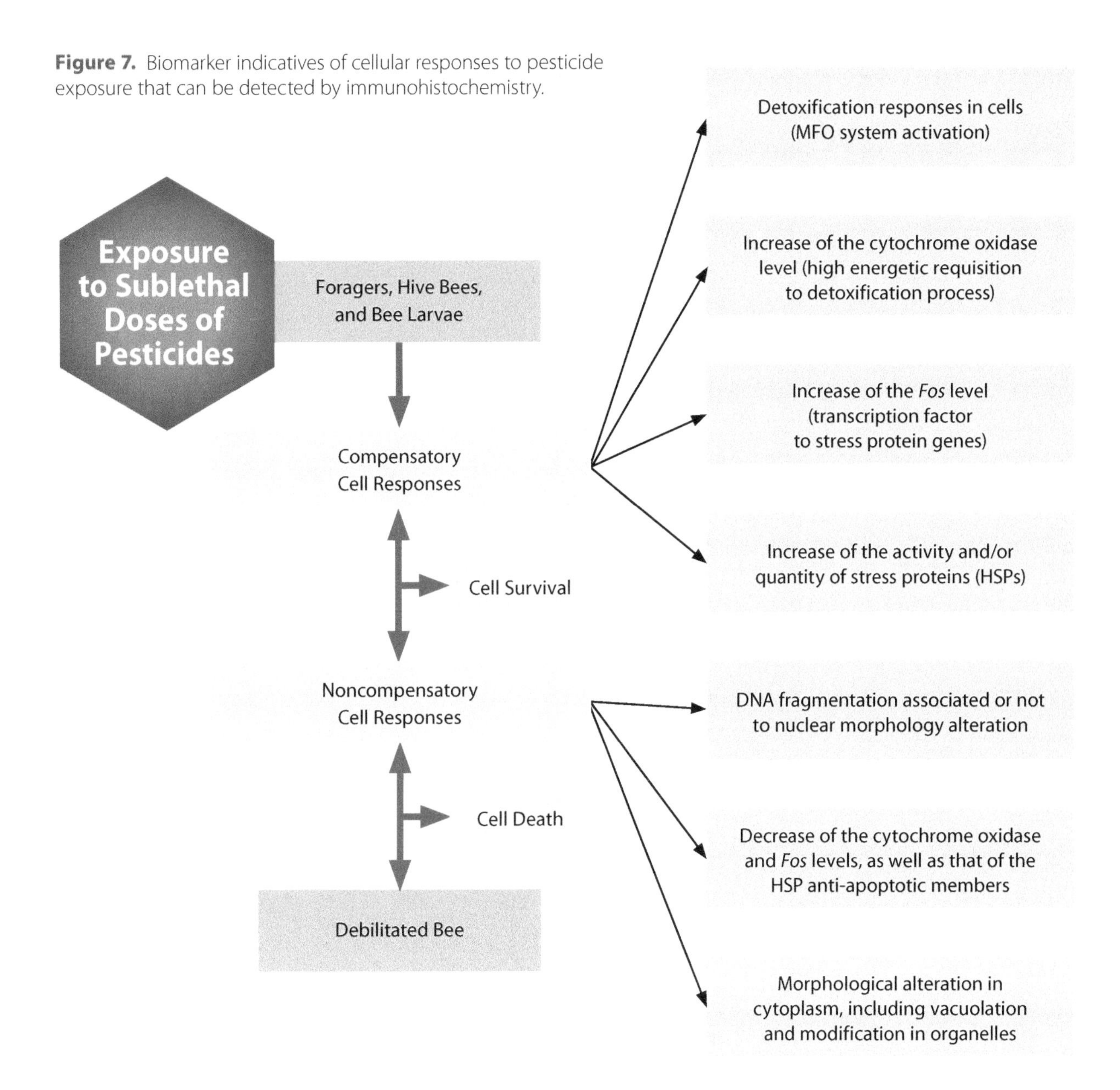

damage as evidenced by the recent decrease in the density of bees in the vicinity of agricultural fields in various parts of the world.

Scientists have identified a phenomenon called Colony Collapse Disorder (CCD) (Ellis et al., 2010) in the United States where beekeepers lost up to 90% of their hives. One of the symptoms of CCD is that forager bees do not return to the hives, thus leaving the colony with a low bee population. In spite of the increased use of pesticides in agriculture, the causes of high colony mortality are not yet well understood.

Chemical substances from the environment are an important source of stress to the bee colony, both the individual and at the cellular level. Disturbed cellular homeostasis can induce corresponding cellular restoration, and cellular functioning can be partially affected or can lead to cell death. These changes are primarily on the molecular level, proceeding to changes in the secondary structure and cellular activity, and are defined as sublethal changes. Furthermore, development of tolerance to the environmental stressors is a potential result of sublethal effects. Tolerance may have a genetic basis or may be acquired by physiological adaptations to stress and cannot be passed to the offspring.

Investigations of the role of pesticides on bees and understanding how organisms respond to sublethal doses, as well as identifying possible cellular markers that allow identification of the damage early in the process, can bring substantial benefits to agro-ecosystems. Cellular defense systems can contribute to reducing the effects of environmental stressors not only on the cellular level, but also on the level of the individual bee or the bee colony. Defense systems in the colony and on an organismal level are thus very complex. Proper colony management can contribute to the prevention of irreversible changes and can increase bee survival and reduce individual bee or colony mortality. Understanding the role of all the potential environmental pollutants, both within and outside the colony, that affect honey bees can help us further understand pollinators as environmental services of ecosystems, effective and necessary for the world agricultural production.

Differences Among Fungicides Targeting Beneficial Fungi Associated with Honey Bee Colonies

CHAPTER 16

Jay A. Yoder, Brian Z. Hedges, Derrick J. Heydinger,
Diana Sammataro, and Gloria DeGrandi-Hoffman

Abstract This chapter draws attention to individual fungi commonly found in stored pollen (bee bread) and how they may respond to different fungicides. We evaluated the effects of boscalid and pyraclostrobin (Pristine® BASF), propiconazole (OrbitTMTilt® Syngenta), and chlorothalonil (Bravo® Syngenta), on 12 fungi species isolated from bee bread. A key observation was that no two fungi cultured from bee bread respond the same with regard to percentage reduction in radial growth rate, mortality, and lowest effective concentration. Chlorothalonil was fungistatic (slowed growth without killing) and was least effective on *Aspergillus* spp., *Penicillium* sp., *Cladosporium* sp., and *Ascosphaera apis*. Boscalid and pyraclostrobin mixture was almost entirely fungicidal, especially against *Aspergillus* spp. and *Penicillium* sp. Consistently, *Rhizopus* sp. was the most sensitive to the fungicides and *A. apis* was the most tolerant. Parallel studies with antibiotics showed no effect on limiting growth of the 12 fungi. Thus, effectiveness of test fungicides is boscalid and pyraclostrobin > propiconazole > chlorothalonil. Our studies suggest that exposure to fungicides applied when plants are in bloom and collected by foraging bees could have a negative effect on colony health because they disrupt the mycoflora bees use to process and store their food.

Acknowledgments This work was funded, in part, by the California State Beekeepers Association and by the Almond Board of California.

Introduction

Fungicides are often applied to crops in bloom even when honey bees, *Apis mellifera*, are present for pollination because they are deemed safe to bees (United States Environmental Protection Agency, 2006). However, recent studies show fungicide residue contamination of the pollen and associated bee products as the result of this practice (Charlton and Jones, 2007; Smodiš Škerl et al., 2009). When studies have been conducted on the bees, it was discovered that certain fungicides can modify bee behavior by acting as repellents or causing bees to become disoriented during foraging. Other studies have found fungicides to be lethal to adult bees and larvae (Soloman and Hooker, 1989; Mussen et al., 2004; Ladurner et al., 2005, 2008; Alarcón et al., 2009; Kubik et al., 1999, 2000; Smodiš Škerl et al., 2009). While the application of fungicides in evening hours, when bees are not foraging, can reduce direct exposure, fungicides and their breakdown compounds nevertheless can be stored in the pollen and bee bread (Mullin et al., 2010).

Few experiments have investigated the impact of fungicides on the symbiotic and beneficial fungi that are present in the colony environment, especially during the processing of raw pollen into bee bread. Bee bread is an absolute dietary requirement for developing bee larvae (Gilliam, 1979; Gilliam and Vandenberg, 1997; Gilliam et al., 1989). The concern is that disrupting the mycoflora balance by altering the growth rate, or the permanent removal of a single fungus or select group of fungi, could allow undesirable fungi (including pathogens) to thrive. As a result, there could be pronounced changes in fungal composition and community reorganization that thwart the processing of pollen into bee bread (Gilliam et al., 1988, 1989; Yoder et al., 2008). Our interest was peaked by observations from commercial beekeepers of an increased incidence of chalkbrood mycoses symptoms in bee colonies that are regularly moved for pollination (D. Sammataro, unpublished observations, 2010). Emergence of chalkbrood may be the result of variable effects on the growth of fungi that suppress the growth of the chalkbrood pathogen (Gilliam et al., 1988; Gilliam and Vandenberg, 1997).

Our question is: what would be the effects of fungicides on the two major bee fungal pathogens *Ascosphaera apis*, causative agent of chalkbrood disease (Gilliam et al., 1997), and *Aspergillus flavus*, causative agent of stonebrood disease (Gliński and Buczek, 2003), as well as the array of beneficial fungi that bees need to process and store pollen? We have identified 12 fungi that comprise the majority of cultured isolates associated with stored pollen: *Absidia* sp., *Alternaria* sp., *A. apis*, *A. flavus*, *A. niger*, *Bipolaris* sp., *Cladosporium* sp., *Fusarium* sp., *Mucor* sp., *Penicillium* sp., *Rhizopus* sp., and *Trichoderma* sp. (Gilliam, 1997; Yoder et al., 2008). Most of these fungal species are easily recognized soil saprobes with high sporing (conidia) activity. These fungi are brought into colonies by the bees while they are collecting pollen (Gilliam et al., 1989; Osintseva and Chekryga, 2008). Conceivably, fungicides could be lethal to all fungi or perhaps not all fungi may be similarly affected.

The goal of our study was to determine if the growth and survival of all 12 species of fungi we identified from bee bread were similarly affected by the fungicides. Accordingly, we determined radial growth rates (K_r) of the 12 fungal species after exposure to three representative fungicides: chlorothalonil, propiconazole, and boscalid and pyraclostrobin. These fungicides are broad-spectrum and are routinely applied to various commercial crops, such as almonds and other stone fruits and berries. The fungicides were tested alone and in combination to determine if there were synergistic effects.

Another goal of this study was to examine the effects of antibiotics on the resident fungi in bee bread. Fumagillin is used for control of nosemosis, caused by *Nosema apis* and *N. ceranae* (Katznelson and Jamieson, 1952; Williams et al., 2008). Tylosin and oxytetracycline, which are used for the control of American foulbrood disease caused by *Paenibacillus larvae* (Hitchcock et al., 1970; Peng et al., 1996), were also tested. Common methods of delivery of these antibiotics to the colony include adding them to powdered sugar or to sugar syrup. Though these antibiotics are effective in controlling pathogenic microorganisms, they also may inhibit the growth of symbiotic fungi that play a key role in predigestion and fermentation of pollen and its conversion to bee bread (Gilliam, 1979; Gilliam et al., 1989). Antibiotics sometimes used by beekeepers to control the fungal-associated brood diseases chalkbrood and stonebrood may also affect the growth of beneficial microbes. Experiments conducted herein seek to examine this.

We emphasize that this chapter focuses on results of a laboratory study that evaluated representative fungicides and antibiotics, *in vitro*. The concentrations of fungicide solutions do not necessarily reflect what is found in the field, and strain variations in fungi of the same species (e.g., origin of collection, substrate) are expected to occur (Barnett and Hunter, 1998; Fisher and Cook, 1998), leading to a potentially different fungal response *in situ*. We also emphasize that our results apply only to Arizona bee colony fungi. Additionally, the experimental design that we used does not address effects of the test fungicides and antibiotics on bee bread fungal community that cannot be cultured or as a whole, but rather effects of these compounds on each of the 12 cultured fungi in isolation.

Characteristics and contribution of each fungus and how each compound affects each fungal component individually in bee bread can be separated. There is no doubt that fungicide residues are found in bee colonies (Charlton and Jones, 2007; Smodiš Škerl et al., 2009) and that they are having an impact on the fungi in bee bread (Yoder et al., 2011, this volume). Results presented in this chapter indicate that beneficial fungi in colonies are differentially affected, and this could lead to weakening the colony over time. The importance of this chapter is that it points out a necessity to evaluate fungicide effectiveness as a tool to determine which individual fungi are affected, and to what extent, so that more informed decisions can be made by beekeepers when considering fungicide use in orchards where their bees are located and antibiotic use in colonies.

Materials and Methods

Fungi and Test Compounds

Fungi were original isolates collected and isolated from European honey bee colonies and bee bread from the Carl Hayden Bee Research Center, (Tucson, Arizona); additional isolates were collected and stored using Benoit et al. (2004) and Yoder et al. (2008) methods. The majority of isolates were saprobic fungi: *Absidia* sp., *Alternaria* sp., *A. apis*, *A. flavus*, *A. niger*, *Bipolaris* sp., *Cladosporium* sp., *Fusarium* sp., *Mucor* sp., *Penicillium* sp., and *Rhizopus* sp. A *Trichoderma* sp. was the only mycoparasitic fungus (Fisher and Cook, 1998). Culturing was done in 100 × 15-mm Petri dishes (surface area 56.08 cm^2; Fisher Scientific, Pittsburgh, Pennsylvania). Growth media was potato dextrose agar (PDA; Fisher). An autoclave sterilized, fungicide-free bee bread supplemented nonnutritive agar (BBSA) was also used as a growth medium (coupled with incubation at 30°C and total darkness to mimic bee colony environment) to simulate the comb cell where pollen is converted into bee bread (modified from Folk et al., 2001; Hua and Feng, 2006). Analysis by GC/MS (QuEChERS extraction) confirmed that the bee bread used to prepare the BBSA contained no detectable pesticides (R. Simonds, USDA, National Science Laboratory, Gastonia, North Carolina, unpublished observations, 2010) that would interfere with this fungus growth rate experiment that involves fungicide evaluation.

Test fungicides were propiconazole, chlorothalonil, and boscalid and pyraclostrobin, and they were used directly from the packaging provided by the manufacturer. Chlorothalonil is a chlorinated benzonitrile fungicide, 54.0% chlorothalonil (tetrachloroisophthalonitrile). Propiconazole is a triazole derivative fungicide, 41.8% propiconazole (1-[[2-(2,4-dichlorophenyl)-4-propyl-1,3-dioxolan-2-yl]methyl]-1H-1,2,4-triazole). Boscalid and pyraclostrobin are a multi-site contact fungicide, 12.8% pyraclostrobin (carbamic acid, [2-[[[1-(4-chlorophenyl)-1H-pyrazol-3yl]oxy]methyl]phenyl]methoxy-, methyl ester) and 25.2% boscalid (3-pyridinecarboxamide, 2-chloro-N-(4'-chloro[1,1'-biphyenyl]-2-yl).

Antibiotics tested were fumagillin, tylosin, and oxytetracycline, which were obtained directly from the manufacturer. Glass micropipettes (± SE < 0.03%; Fisher) were used for fungicide and antibiotic solution applications, and all solution preparations and manipulations were done in glass.

Determination of Radial Growth Rate

Fungicide and antibiotic solutions (1.0%, 0.1%, 0.01%) were prepared in double-distilled deionized (DI) water with DI water as control, and 1 mL was spread over the surface of solidified agar and allowed to absorb for 30 minutes (Yoder et al., 2008). Next, three lines were drawn on the bottom of the dish radiating out from the center (Currah et al., 1987), and a 1-cm^3 block of fungus inoculum (edge of a 10-day-old mycelium from a pure culture) was placed on the agar surface in the center of the dish. Petri dishes were incubated at 30 ± 1°C and darkness (regular hive conditions; Chiesa et al., 1988) under 5% CO_2 and then transferred to aerobic conditions after a day (Gilliam et al., 1989). A total of five measurements were taken along each of the three lines as the mycelium spread over the agar surface. Measurements were fitted to the equation $K_r = (R_1 - R_0)/(t_1 - t_0)$, where K_r is the radial growth rate, R_0 and R_1 are colony radii at initial (t_0) and elapsed (t_1) times between start of linear, t_0, and stationary, t_1, growth phases (Baldrian and Gabriel, 2002). Each experiment was replicated three times using a different fungus mycelium as the source of inoculum for each replicate. Thus, each radial growth rate is mean ± SE and is based on a total of 45 measurements (replicates of 15 measurements/plate; N = 3). Data were compared using an analysis of covariance (ANCOVA), with Duncan multiple comparison test where applicable (Sokal and Rohlf, 1995; SPSS 14.0, Microsoft Excel and Minitab, Chicago, Illinois). Each fungicide was tested individually (one fungicide and one solution per plate) and then multiplied, with two different fungicides at a time and then three at a time, crossing the different fungicide concentration solutions reciprocally. Antibiotics were also tested singly (one concentration and one antibiotic per plate) and in combination to test for synergistic effects on fungal growth.

Different procedures were tested in a preliminary study to select the most suitable application technique of the fungicide and antibiotic solutions. In each procedure, a Petri dish was filled with 20 mL agar. Three different methods were used to incorporate the fungicide or antibiotics into the agar, and each technique was done 10 times for each dilution of fungicide for each of the 12 fungi and was replicated 3 times. In the first approach, 1 mL of fungicide or antibiotic solution was incorporated into the melted agar and swirled to mix, and then the agar was allowed to solidify. The swirling used in this technique did not allow for even distribution of the fungicide throughout the dish and produced mixed, inconsistent results because the mycelium did not grow in a concentric manner. As a second approach, a 15-mm i.d. well was punched into agar at the center and the 1-cm^3 block of fungal inoculum was placed into the well, and the remaining space was filled with 100 μL of the fungicide solution. This technique was not used because filling the punched-out well with fungicide solution resulted in dishes covered with small fungal colonies, eventually growing together, as a result of the fungicide solution carrying the conidia through the agar by diffusion; this obscured the growth of the mycelium from the center. In the third approach, 1 mL of fungicide solution was spread over and allowed to absorb into solidified agar (after adding the fungicide solution to the surface of the plate); the elapsed time before the fungal inoculums were added was 20 minutes. This approach yielded the most reproducible results (also did not have any of the other problems as observed with the other approaches) and was used for the reported tests.

Results

Control Growth Rates with Comparative Observations on Bee Bread-Supplemented Media

Radial growth rates of bee bread fungi on PDA at 30°C were distinctive, permitting separation of these fungi into classification groups (Table 1). The fastest growing were *Rhizopus* sp., *Mucor* sp., *Absidia* sp., and *Trichoderma* sp.; and the slowest growing were *A. apis*, *A. niger*, *A. flavus*, and *Cladosporium* sp. Growth of *Fusarium* sp., *Bipolaris* sp., *Penicillium* sp., and *Alternaria* sp. was moderate and between these extremes ($P < 0.05$). With the exception of *Absidia* sp. (0.274 mm/hour), *Pencillium* sp. (0.161 mm/hour) and *A. flavus* (0.063 mm/hour), all other bee bread fungi grew slower when plated on BBSA: *Rhizopus* sp. (0.377 mm/hour), *Mucor* sp. (0.285 mm/hour), *Trichoderma* sp. (0.212 mm/hour), *Fusarium* sp. (0.163 mm/hour), *Bipolaris* sp. (0.157 mm/hour), *Alternaria* sp. (0.119 mm/hour), *A. apis* (0.065 mm/hour), *A. niger* (0.052 mm/hour) and *Cladosporium* sp. (0.047 mm/hour) (± SE ≤ 0.031; $P < 0.05$). For controls used in the antibiotic experiment (Table 2), similar fast-, moderate-, and slow-growing groups were noted as in the fungicide treatment experiment (ranking radial growth rates statistically from fastest to slowest):

> P < 0.05): *Rhizopus* sp. = *Mucor* sp. > *Absidia* sp.
> > *Trichoderma* sp. >> *Alternaria* sp. > *Bipolaris* sp.,
> = *Fusarium* sp. > *Penicillium* sp. >> *A. niger* = *A. flavus*
> > *A. apis* = *Cladosporium* sp.

Consistently, radial growth rates of fungi were reduced by 7 to 23% when the various fungi were plated on BBSA compared to PDA ($P < 0.05$). Despite producing nearly identical trends with regard to fungus responsiveness toward the fungicides, the one feature that distinguished culturing on BBSA was a slight reduction in growth rate compared to PDA. The slower growth rate on BBSA that we found agrees with the reduced hyphal growth of molds germinating on pollen and bee bread observed by scanning electron microscopy analysis by Klungness and Peng (1983). Bee bread fungi grow at different rates and actual bee bread itself is suppressive against growth of chalkbrood agent (*A. apis*) and also favors the growth of *Pencillium* sp., which is a known chalkbrood inhibitor.

Effects of Chlorothalonil on Fungal Growth

Though growth rates were reduced for all of the 12 bee colony fungi, none were killed by chlorothalonil at the highest concentration (1.0%) (Table 1). However, the greatest reduction in radial growth rate of fungi occurred at the 1.0% fungicide concentration compared with 0.1% and 0.01% solutions. At the highest concentration, the most sensitive genera (gauged by the extent that radial growth rate was suppressed compared with control) were *Mucor* sp. (reduced by 85%), *Rhizopus* sp. and *Absidia* sp. (each with a 83% reduction), *Fusarium* sp. (76% reduction), *Penicillium* sp. (74% reduction), and *Cladosporium* sp. (72% reduction) ($P < 0.05$). Radial growth rate at the 1.0% fungicide concentration was about one-half the rate of controls for *Trichoderma* sp. (53% reduction), and *Bipolaris* sp. (52% reduction), *Alternaria* sp. (41% reduction), *A. niger* (39% reduction), and *A. apis* (45% reduction) ($P < 0.05$).

Table 1. Fungicides chlorothalonil, (Chl), propiconazole, and boscalid and pyraclostrobin effect on radial growth rate (K_r) of bee bread fungi on potato dextrose agar, 30°C.
Abs, *Absidia* sp.; Alt, *Alternaria* sp.; Asc, *Ascosphaera apis*; Asf, *Aspergillus flavus;*
Asn, *Aspergillus niger*; Bip, *Bipolaris* sp.; Cla, *Cladosporium* sp.; Fus, *Fusarium* sp.;
Muc, *Mucor* sp.; Pen, *Penicillium* sp.; Rhi, *Rhizopus* sp.; Tri, *Trichoderma* sp.;
killed, Kr = 0, no detectable measureable growth.

Radial Growth (mm/hour, 30°C), Potato Dextrose Agar (mean K_r ± SE ≤0.042)										
Control Fungus	Chl	Propiconazole			Boscalid and Pyraclostrobin					
		1.0%	0.1%	0.01%	1.0%	0.1%	0.01%	1.0%	0.1%	0.01%
Abs	0.277	0.088	0.132	0.334	0.012	0.010	0.148	0.045	0.053	0.083
Alt	0.146	0.082	0.119	0.126	0.041	0.049	0.147	Killed	0.008	0.102
Asc	0.084	0.042	0.039	0.073	Killed	0.041	0.068	Killed	0.011	0.062
Asf	0.063	0.057	0.052	0.073	Killed	0.039	0.049	Killed	0.013	0.030
Asn	0.072	0.051	0.079	0.087	0.045	0.051	0.065	Killed	Killed	0.024
Bip	0.169	0.109	0.121	0.170	Killed	0.071	0.092	0.031	0.032	0.097
Cla	0.058	0.015	0.027	0.051	Killed	Killed	0.027	Killed	0.046	0.051
Fus	0.182	0.049	0.110	0.174	Killed	0.049	0.093	0.036	0.027	0.065
Muc	0.301	0.072	0.229	0.361	0.008	0.239	0.294	0.128	0.112	0.132
Pen	0.152	0.032	0.042	0.119	Killed	Killed	0.054	Killed	Killed	0.037
Rhi	0.411	0.088	0.132	0.334	Killed	Killed	0.192	0.038	0.044	0.082
Tri	0.236	0.118	0.217	0.233	Killed	Killed	0.127	Killed	0.039	0.112

The least sensitive fungus to chlorothalonil was *A. flavus*, which experienced only a 17% reduction in radial growth rate compared with the control rate ($P < 0.05$). Sensitivity to chlorothalonil based on percentage reduction in radial growth rate and concentration was as follows (ranked statistically:

> $P < 0.05$): (most sensitive) *Rhizopus* sp. > *Mucor* sp.
> = *Absidia* sp. > *Fusarium* sp. = *Penicillium* sp.
> = *Bipolaris* sp. = *Cladosporium* sp. > *A. apis*
> = *A. niger* = *Trichoderma* sp. = *Alternaria* sp.
> > *A. flavus* (least sensitive)

At 1.0% concentration of chlorothalonil, radial growth rate was reduced by 79% for the fastest-growing fungus *Rhizopus* sp., 74% for the slowest-growing fungus *Cladosporium* sp., and 73% for *Fusarium* sp. with a radial growth rate that falls between these extremes (Table 1). Percentage reduction in radial growth rates is correlated with chlorothalonil concentration when all other fungi are compared (because all survived) at 0.01% fungicide concentration ($y = 11.9x$, $R = 0.83$; Figure 1). At other chlorothalonil concentrations, there is a lack of correlation between percentage reduction due to the effect of fungicide and control growth

Table 2. Antibiotics fumagillin (Fum), oxytetracycline, and tylosin exposure in relation to radial growth rate (K_r; potato dextrose agar, 30°C) of bee bread fungi: Abs, *Absidia* sp.; Alt, *Alternaria* sp.; Asc, *Ascosphaera apis*; Asf, *Aspergillus flavus* Asn, *Aspergillus niger*; Bip, *Bipolaris* sp.; Cla, *Cladosporium* sp.; Fus, *Fusarium* sp. Muc, *Mucor* sp.; Pen, *Penicillium* sp.; Rhi, *Rhizopus* sp.; Tri, *Trichoderma* sp.

Radial Growth (mm/hour, 30°C), Potato Dextrose Agar (mean K_r ± SE ≤0.071)										
Control Fungus	Fum	Oxytetracycline			Tylosin					
		1.0%	0.1%	0.01%	1.0%	0.1%	0.01%	1.0%	0.1%	0.01%
Abs	0.305	0.319	0.324	0.278	0.294	0.308	0.318	0.295	0.320	0.284
Alt	0.158	0.169	0.167	0.152	0.150	0.164	0.159	0.160	0.170	0.162
Asc	0.064	0.076	0.062	0.049	0.051	0.054	0.067	0.072	0.061	0.066
Asf	0.098	0.113	0.108	0.097	0.089	0.094	0.119	0.092	0.082	0.086
Asn	0.103	0.084	0.121	0.114	0.100	0.093	0.086	0.097	0.124	0.091
Bip	0.134	0.109	0.125	0.141	0.137	0.139	0.144	0.130	0.119	0.127
Cla	0.056	0.068	0.062	0.060	0.044	0.053	0.049	0.063	0.058	0.052
Fus	0.126	0.134	0.118	0.127	0.139	0.109	0.128	0.139	0.137	0.120
Muc	0.391	0.374	0.405	0.381	0.401	0.369	0.421	0.397	0.386	0.380
Pen	0.114	0.094	0.097	0.120	0.109	0.116	0.122	0.095	0.091	0.119
Rhi	0.424	0.404	0.437	0.413	0.409	0.426	0.433	0.419	0.428	0.411
Tri	0.263	0.268	0.245	0.250	0.271	0.266	0.247	0.275	0.253	0.269

rate (1.0%: $y = 18.0x$, R = 0.61; 0.1%: $y = 7.8x$, R = 0.30; $P > 0.05$). Therefore, chlorothalonil had no greater effect on faster-growing fungi than slower-growing fungi as concentration increased. Bee bread fungi were not equally sensitive to chlorothalonil; the least affected were inhibitory molds *Aspergillus spp.* and *Penicillium* sp. as well as the agent of chalkbrood *A. apis*.

Effects of Propiconazole on Fungal Growth

All 12 bee bread fungi we tested had suppressed growth in response to propiconazole exposure (Table 1). At the highest concentration (1.0% fungicide), eight of the 12 species were killed; $K_r = 0$, while growth in others was severely slowed (Table 1; $P < 0.05$). Fungi killed by propiconazole at the 1.0% concentration were *A. apis*, *A. flavus*, *Bipolaris* sp., *Cladosporium* sp., *Fusarium* sp., *Penicillium* sp., *Rhizopus* sp., and *Trichoderma* sp. The only fungi within this group that were killed with the 0.1% solution were *Cladosporium* sp., *Penicillium* sp., *Rhizopus* sp., and *Trichoderma* sp. A concentration of 0.01% fungicide was not lethal to any of the fungi, but there was a measurable reduction in radial growth of all 12 fungi. When the bee bread fungi are ranked statistically based on percentage reduction

Figure 1. Relationship between percentage growth reduction by chlorothalonil, propiconazole and boscalid and pyraclostrobin fungicides and ln radial growth rate (K_r) in 12 bee bread fungi at 30°C on potato dextrose agar: $y = 11.9x$, R = 0.83 (chlorothalonil), $y = 10.8x$, R = 0.51 (propiconazole), $y = 18.7x$, R = 0.57 (boscalid and pyraclostrobin). Data are shown for 0.01% fungicide as it is the only concentration where all fungi survived ($K_r \neq 0$) of the fungicides tested.

% Growth Reduction at [0.01%]

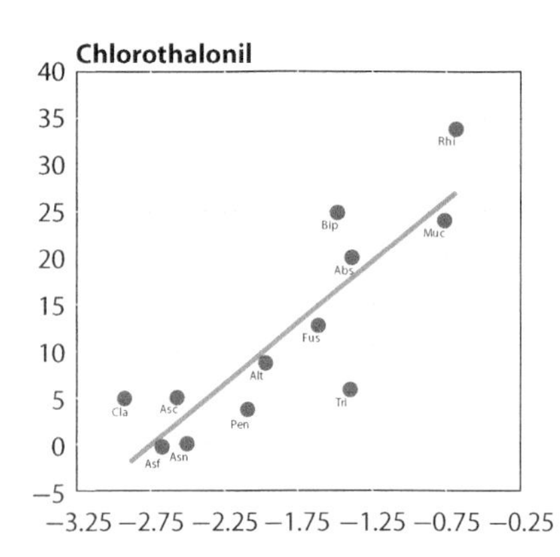

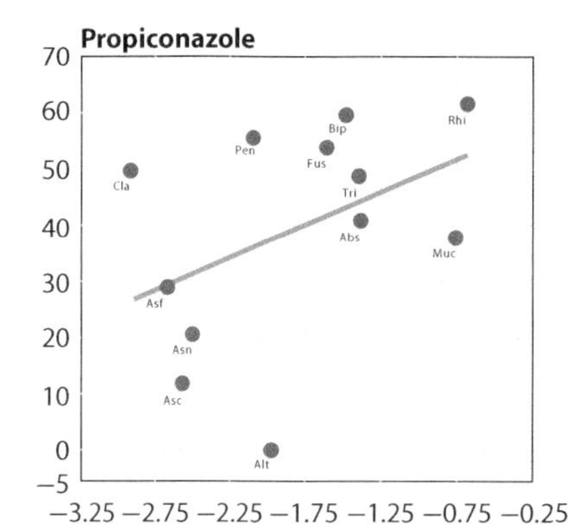

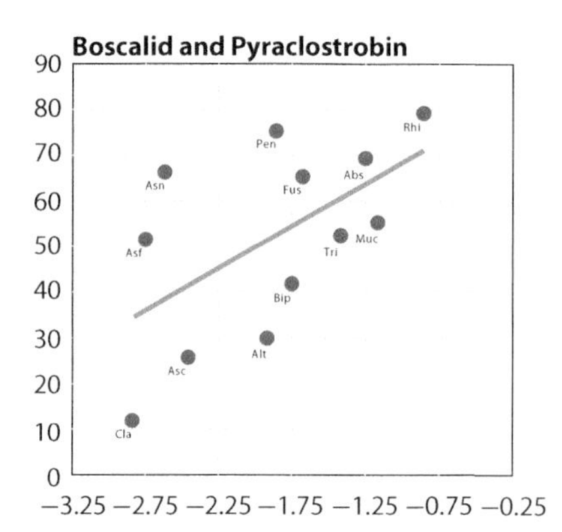

in radial growth rate and concentration of propiconazole (P < 0.05), the ranking is:

> (most sensitive) *Rhizopus* sp. > *Trichoderma* sp.
> = *Penicillium* sp.= *Fusarium* sp. = *Bipolaris* sp.
> = *Cladosporium* sp. > *Absidia* sp. = *Mucor* sp.
> > *A. flavus* > *A. niger* = *A. apis*
> > *Alternaria* sp. (least sensitive).

At concentration of 0.01% propiconazole, the percent reduction in radial growth was not a positive correlate of radial growth rate of untreated controls when all 12 bee bread fungi were compared (Figure 1; $y = 10.8x$, R = 0.51). A similar absence of correlation between percentage reduction and growth rate was found at fungicide concentrations of 1.0% ($y = 6.0x$, R = 0.26) and 0.1% ($y = 9.7x$, R = 0.29) (P > 0.05). At the lowest concentration (0.01%), *A. apis* experienced a 20% reduction in growth rate indicating that *A. apis* is somewhat resistant to these fungicides at lower concentrations compared to other bee bread fungi. We conclude that no two bee bread fungi respond the same to propiconazole exposure, and the effectiveness of any concentration does not depend upon the fungus being a slow or fast grower. Propiconazole is particularly effective in reducing the growth of *Penicillium* sp. (that inhibits the growth of chalkbrood), but not *Aspergillus* spp., and chalkbrood, *A. apis*, is not particularly affected.

Boscalid and Pyraclostrobin Effects on Fungal Growth

Exposure to a mixture of boscalid and pyraclostrobin fungicides was characterized by a pronounced lethal effect for the majority of the 12 bee bread fungi. Most of the fungi showed reductions in growth rate of approximately 80% or higher at fungicide concentrations of 0.1% fungicide compared to water-treated controls (Table 1; P < 0.05). Though no select group of fungi showed resistance to the fungicide, the least sensitive fungi (did not die over range of concentrations) were *Rhizopus* sp., *Mucor* sp., *Absidia* sp., and *Bipolaris* sp. The most vulnerable fungi were *Pencillium* sp. and *A. niger*, based on the large extent of radial growth rate suppression. *Rhizopus* sp. had the highest growth rate (K_r = 0.411 mm/hour) and was the most negatively affected (in terms of reduction in growth rate) by the fungicide (reduced K_r by 80% at 0.01% fungicide), whereas the fungus with the lowest growth rate, *Cladosporium* sp. (K_r = 0.058 mm/hour), was the least affected (12% reduction in K_r at 0.01% fungicide solution). However, there was no correlation between growth rate and percentage reduction when other fungi species were compared: $y = 18.7x$, R = 0.57 at 0.01% fungicide (Figure 1), $y = 8.2x$, R = 0.25 at 0.1% fungicide and $y = -11.9x$, R = 0.53 at 1.0% fungicide. Thus, there is no apparent relationship between fungicide concentration effect and radial growth rate. Over the range of concentrations tested, only two fungi were killed at 1.0% and 0.1% fungicide, and these were *Pencillium* sp. and *A. niger*. Ranking the fungi statistically ($P < 0.05$) based on sensitivity to a boscalid-pyraclostrobin mixture using lethality and extent of radial growth rate suppression combined over the different fungicide solutions:

(most tolerant) *Mucor* sp. > *Bipolaris* sp. = *Fusarium* sp.
= *Absidia* sp. > *Rhizopus* sp. >> *Cladosporium* sp.
> *A. apis* = *Alternaria* sp. > *A. flavus* = *Trichoderma* sp.
>> *A. niger* = *Penicillium* sp. (most sensitive)

Thus, these fungicides in combination cause nearly the same response by most of the bee bread fungi in that the majority of them are killed, and the response at lower concentrations is independent of the radial growth rate of the fungus. Noteworthy is that these fungicides are especially lethal to *Penicillium* sp. and *Aspergillus* spp.

Synergistic Effect of Chlorothalonil, Propiconazole and Boscalid and Pyraclostrobin Fungicides

When applied in combination with mixtures of two and three fungicides at a time, no pronounced increase in antifungal properties was observed. In all cases, the most concentrated fungicide in the combination (0.01%, 0.1%, 1.0%) prevails; that is, the results are most similar to those obtained at the concentration when that fungicide is applied alone. For example, in a dual application of 0.01% propiconazole and 0.1% chlorothalonil, radial growth rates for all 12 fungi were not significantly different compared to rates for fungi exposed to 0.1% chlorothalonil alone, and were as if 0.01% propiconazole had no effect (Table 1; $P > 0.05$). Similar results were obtained with boscalid and pyraclostrobin combinations: 0.01% boscalid and pyraclostrobin mixed with 0.1% propiconazole showed similar results at 0.1% propiconazole alone (Table 1; $P > 0.05$). When concentrations were the same, the response to fungicide was the one for the fungicide that is most potent, consistent with the greater effectiveness of boscalid and pyraclostrobin over propiconazole and propiconazole over chlorothalonil (i.e., boscalid and pyraclostrobin > propiconazole > chlorothalonil). For example, in a dual application of 0.01% boscalid and pyraclostrobin and 0.01% propiconazole, the radial growth rate of all 12 fungi was comparable to radial growth rates of fungi exposed to 0.01% boscalid and pyraclostrobin alone (Table 1; $P > 0.05$), showing that boscalid and pyraclostrobin is more potent.

To clarify this further with additional examples, in three-way combinations of 0.1% chlorothalonil, 1.0% propiconazole, 0.01% boscalid and pyraclostrobin, radial growth rates were most similar to those of fungi that were exposed to 1.0% propiconazole in that propiconazole exists in highest concentration among the three. Radial growth rates of all 12 fungi were most similar to 0.1% boscalid and pyraclostrobin alone (Table 1) in a three-way combination of 0.1% chlorothalonil, 0.1% propiconazole, 0.1% boscalid and pyraclostrobin ($P > 0.05$). Therefore, the effectiveness of chlorothalonil does not improve with the addition of another fungicide, which means there is no additive effect by fungicide combinations on suppressing the radial growth rate for any of the fungi tested. Our conclusion is that there is no increased antifungal activity when compared to single-fungicide applications, and these fungicides do not appear to synergize with each other.

Antibiotics (Fumagillin, Oxytetracycline, Tylosin) Effects on Fungal Growth

Like treatment with fungicides, the antibiotic exposure resulted in no changes in gross morphology (obverse/reverse pigmentation, colony, conidia, philiade characteristics by 40× microscopic observation) of any of the 12 fungi compared to water-treated controls, nor did they initiate production of teleomorphs. From Table 2, evidence of a lack of antibiotic sensitivity by exposure to fumagillin, oxytetracycline, and tylosin includes

1. Little deviation from control radial growth rates by treatment ($P > 0.05$), neither promoted nor suppressed growth, spanning all three antibiotics.
2. Failure to respond to the antibiotics across a broad range of 12 different fungal taxa.
3. None of the fungi were killed, $K_r \neq 0$.

Additionally, there were no consistent concentration trends; that is, no regular dose-response in that some correlate positively, some vary inversely (improbable if the antibiotic were having an effect), and some fail to correlate ($R \leq 0.67$). Slow-growing fungi *A. apis*, *A. niger*, *A. flavus*, and *Cladosporium* sp. exhibited no detrimental effects by being retained on antibiotic-treated surfaces longer and showed radial growth rates on treated surfaces that compared favorably to untreated controls ($P > 0.05$). When fungi were exposed to two or three different antibiotics simultaneously, with all two or three concentrations crossed reciprocally, all 12 fungi responded on treated surfaces as they did on untreated controls (i.e., no change in radial growth rate compared to controls; $P < 0.05$).

This indicates an absence of synergistic effects (data not shown) because the growth of bee bread fungi is not suppressed by exposure to fumagillin, tylosin, and oxytetracycline, alone or in combination. Neither are these antibiotics lethal to the fungi we could culture. Indeed, antibiotics are shown not to kill fungi (Jennings and Lysek, 1999), and this is the first direct evidence showing that this is the case for fungal strains in a honey bee colony. Relevant information for beekeepers is that the antibiotics fumagillin, tylosin, and oxytetracycline are not preventatives or treatments against stonebrood and chalkbrood. Because the antibiotics showed no effect on bee bread fungi, all further discussion will be on the fungicides.

Discussion

Chlorothalonil, propiconazole, and boscalid and pyraclostrobin are broad-spectrum fungicides and each has a different mode of action. Exposure to the fungicides had a negative impact on 12 bee bread fungi, either by showing pronounced reductions in radial growth rate (chlorothalonil, propiconazole, and a combination of boscalid and pyraclostrobin) or by either stopping growth or killing the fungus outright (propiconazole and a combination of boscalid and pyraclostrobin), regardless of whether they were cultured on BBSA or PDA. Slower radial growth that we measured on BBSA is corroborated by scanning electron microscopy that showed reduced hyphal size when fungi in bee bread are examined (Klungness and Peng, 1983). Slower growth could have a profound impact on the processing of pollen to bee bread, because of competitive interactions among the symbiotic and pathogenic microbes (Yoder et al., 2011, see Chapter 17, this edition).

In this study, the most effective fungicide against the 12 bee bread fungi was the boscalid and pyraclostrobin combination, followed by propiconazole and chlorothalonil. Chlorothalonil interferes with protein synthesis and is classified as a multisite contact chloronitrile, chlorinated benzonitrile. Propiconazole disrupts membrane synthesis by inhibiting demethylation of sterol biosynthesis and is classified as a sterol-inhibiting triazole-derivative fungicide. Finally, boscalid and pyraclostrobin prevent fungal respiration with two inhibitors in the mitochondrial electron transport chain. Boscalid inhibits succinate ubiquinone reductase (complex II) and pyraclostrobin blocks electron transfer in cytochromes b and c1 (ubiquinol oxidase at the Qo site) as the active ingredients (Tomlin, 2006). As anamorphs of ascomycetes, these 12 fungi associated with bee bread have high sporulating activity, indicating that those not killed could continue to spread and establish fungal colonies. Of the fungicides evaluated here, chlorothalonil is least detrimental to the growth of beneficial fungi (namely, *Penicillium* sp. and *Aspergillus* spp.; Gilliam et al., 1988) that make up the majority found in bee bread (Yoder et al., 2011, see Chapter 17, this edition).

Because of the differential activity toward individual fungi, an evaluation of each fungicide should be conducted for its effectiveness, with particular attention on activity toward *Aspergillus* spp. and *Penicillium* spp. as beneficial fungi that are inhibitory toward bee fungal pathogens (Gilliam et al., 1988). This study serves mainly as a point of public awareness, particularly for beekeepers. Before moving colonies into orchards sprayed with fungicides, beekeepers should ask for information related to fungicide effectiveness (fungicidal or fungistatic) toward specific, individual fungal species; in that way, more informed decisions can be made concerning which fungicide should be avoided.

Based on this study, we expect that different fungicides impact the diversity of fungi naturally occurring in an orchard. From our observations and discussions with growers, we found that many fungicides are used, and boscalid and pyraclostrobin applied last because of its superior effectiveness to kill plant fungi that are perceived to be resistant to the other fungicides used (D. Sammataro, personal communication, 2010). From the samples tested (California, United States), it does not appear that any of the fungal components in bee bread are strictly absent with regularity, but there is a pronounced reduction in overall fungi quantity

(Yoder et al., 2011, see Chapter 17, this edition). The key point is that the mycoflora of bee bread is modified both in the area where bees collect pollen as well as inside the colony. It suggests that fungicide exposure causes a reduction in quantity, but not quality, of fungi activity as a result of boscalid and pyraclostrobin use. This reduction in overall quantity of fungi in bee bread agrees with the *in vitro* effect of boscalid and pyraclostrobin that we now describe, that could have a negative effect on all isolated fungal components, while at the same time not killing any particular ones outright (field concentrations must not be sufficiently high to achieve this). Thus, results in the laboratory (*in vitro*; this study) reflect what is occurring in the field (*in vivo*; Yoder et al., 2010). Because of reduced fungus quantity, it is conceivable that key metabolites (as a result of mycoflora activity) needed by bees for food processing and preservation may be absent. The end result could be a chronic weakening of the colony from malnutrition.

There are reports that bee colonies in fungicide-sprayed areas exhibit chalkbrood symptoms, presumably the end result of low levels of inhibitory fungi *Aspergillus* spp. and *Penicillium* sp. (this study; Yoder et al., 2011, see Chapter 17, this edition). *A. apis* is an opportunistic pathogen and is abundant in most bee colonies (Gilliam et al., 1997; Gilliam and Vandenberg, 1997). Chalkbrood symptoms appear in colonies even though boscalid and pyraclostrobin limit the growth of *A. apis*, suggesting that the occurrence of *Aspergillus* spp. and *Pencillium* sp. is more important to controlling chalkbrood than the removal of *A. apis* as the etiological agent. From this fungicide evaluation, the relevant information for beekeepers is that different fungicides have different effects on the fungi present in bee bread *in vitro*. The major finding is that each fungicide has differential activity toward *Aspergillus* spp. and *Penicillium* sp. that fight against chalkbrood. Although the boscalid and pyraclostrobin combination is undoubtedly effective against all fungi and is superior for use in the field against crop diseases, chlorothalonil (not as effective as boscalid and pyraclostrobin, but still effective as a broad-spectrum fungicide) is a better option for beekeepers because of its lessened impact on inhibitory molds in the bee colony. One outcome of this study could be that beekeepers should request from the growers or the manufacturers information on what fungicide is being sprayed and its effect on these beneficial bee colony molds.

Fungicides Reduce Symbiotic Fungi in Bee Bread and the Beneficial Fungi in Colonies

CHAPTER

17

Jay A. Yoder, Derrick J. Heydinger, Brian Z. Hedges,
Diana Sammataro, Jennifer Finley, Gloria DeGrandi-Hoffman,
Travis J. Croxall, and Brady S. Christensen

Abstract *Aspergillus* spp. (primarily *A. niger* and secondarily *A. flavus*), *Penicillium* spp., *Cladosporium* spp., and *Rhizopus* spp. are the main fungi regularly found in bee bread. They function as a natural defense against the pathogenic fungal diseases chalkbrood and stonebrood, as inferred by *in vitro* fungal-fungal interaction bioassays. Ten other species are also present in bee bread, but as minor though necessary components for maintaining the proper balance of the bee bread mycoflora. Colonies in orchards that were sprayed with fungicides had low amounts of bee bread fungi. Agents known to suppress growth or kill bee bread fungi are fungicides, formic acid, oxalic acid miticides, and high fructose corn syrup. Communication between growers and beekeepers is encouraged to assess the timing of direct and nearby fungicide applications so that colonies could be moved if necessary.

Acknowledgments This work was funded in part by the California State Beekeepers Association and by the Almond Board of California. We gratefully acknowledge Eric Olson, Olson's Honey, Yakima, Washington, for the use of his colonies. We also wish to thank the three almond growers granting us access to their orchards in Turlock and Delhi, California.

Introduction

Honey bees (*Apis mellifera*) collect pollen as a protein source for rearing brood and feeding adult bees. The pollen is stored in the comb cells where it is fermented by the action of microorganisms (Gilliam, 1979; Gilliam et al., 1989; Vásquez and Oloffson, 2009). The fermented pollen is called *bee bread*. Bee bread differs from the raw pollen used to prepare it in both pH and chemical composition. When ingested by nurse bees, bee bread activates the hypopharyngeal or food glands, enabling the nurse bees to feed the immature larvae (Gilliam and Vandenberg, 1997) and the queen. Numerous species of fungi (which includes yeasts and molds) have been isolated and identified from bee bread using classic culture techniques (Gilliam et al., 1989; Gilliam, 1997); many more undoubtedly exist that cannot be cultured (West et al., 2007). Collectively, these fungi are responsible for the synthesis of vitamins, enzymes, sterols, and other compounds that are vital to bee health and aid in pollen digestion and preservation (Gilliam and Vandenberg, 1997).

The predominant, culturable fungi in bee bread include both fast-growing as well as slow/moderate-growing species. The fast-growing fungi include *Trichoderma* spp., *Fusarium* spp., *Bipolaris* spp., *Penicillium* spp., *Alternaria* spp., *Rhizopus* spp., *Mucor* spp., and *Absidia* spp. Slow- and moderate-growing fungi include *Ascosphaera apis* (an agent of chalkbrood, Gilliam et al., 1997), *Aspergillus niger*, *Aspergillus flavus* (an agent of stonebrood, Gliñski and Buczek, 2003) and *Cladosporium* spp. (Yoder et al., 2008). All of these fungi are common filamentous soil saprobes that function as agents of decay (Jennings and Lysek, 1999). All are also heavy spore- or conidia-producing genera, allowing for rapid spread and establishment in the bee colony and in bee bread. Bee bread also includes fungi that have been shown to inhibit the growth of other, non-conspecific fungi; most notable are *Penicillium* spp., *Aspergillus* spp., and *Rhizopus* spp. (Gilliam et al., 1988). *Trichoderma* spp. is mycoparasitic (Jennings and Lysek, 1999) and presumably plays a role in keeping levels of fungi from becoming too high. Key factors in maintaining the balance of fungal components in bee bread as well as in the colony environment as a whole are competitive interactions among these fungi for resources, the immune mechanisms of bees, the presence of beneficial microflora, and the near-constant high 30% to 35°C temperature of bee colony environment (Cooper, 1980; Gilliam et al., 1989; Gliñski and Buczek, 2003; Wilson-Rich et al., 2009).

Pollen that bees collect is essential for stimulating and maintaining the growth of symbiotic microbes in colonies; bees often come in contact with toxins such as fungicides while foraging for pollen. This is especially true when colonies are placed in orchards for pollination and transport the compounds back to the hive via the pollen (Alarcón et al., 2009; Škerl et al., 2009). Unlike other pesticides, fungicides are considered safe for bees because most do not show direct toxicity to adults or larvae (U.S. EPA, 2006). However, the effects of fungicides could be more subtle or unforeseen than toxic pesticides, such as reducing the number and diversity of beneficial fungi that bees require to digest and preserve their pollen stores. Indeed, our investigations demonstrated that fungicides can suppress, but not kill, the growth of fungi isolated from bee bread and grown in culture media *in vivo* with differential effects.

All fungi are not equally sensitive to a specific fungicide, and no two fungi respond the same (Yoder et al., 2011, Chapter 16, this edition), with the end result of modifying the mixture of fungal components in bee bread in the process. One goal of this study was to determine the impact of fungicides on the growth and diversity of fungi in bee bread collected from colonies in California almond orchards where fungicides were applied during bloom. Colonies were either in an orchard that was sprayed directly or indirectly (within 3.2 km bee flight range of sprayed areas), with fungicide containing boscalid and pyraclostrobin. This included an organic orchard that was originally our untreated control. We anticipated the fungal load to be higher there than in samples from sprayed areas.

Another goal was to examine how bee bread mycoflora differed geographically by analyzing samples from Arizona colonies for qualitative and quantitative differences in fungal composition compared to California samples. The last goal was an *in vitro* study to explore population dynamics among bee bread fungal species, focusing on the interspecific fungal interactions that may contribute to the balance of the mycofloral components and the prevention of brood diseases with an emphasis on fungal-to-fungal competition. Thus, we combined a field and laboratory experiment to demonstrate how the composition of bee bread fungi arises and its functional role toward prevention of fungal diseases.

Materials and Methods

In Vivo Analysis: Mycoflora Profile of Field-Collected Bee Bread

Source of the bee bread was from colonies of European honey bees that were placed in three commercial California almond orchards during bloom (February–March, 2009). The orchards (N = 3; two treated with fungicides) were located near Turlock, California. The organic control was in Delhi, California. Samples of bee bread (N = 11 colonies) were cut from the combs using a sterilized scalpel, after which the comb was wrapped in autoclaved foil, then plastic zip bags for shipment. All bee bread samples were refrigerated and shipped in coolers packed with icepacks to the Carl Hayden Honey Bee Research Center or to Wittenberg University (March, 2009) within 24 hours. Because there was no unsprayed pollen available in California (organic orchard excepted), samples of bee bread were collected from colonies at the Carl Hayden Bee Research Center (N = 10 colonies) in Tucson, Arizona, for comparison using the same methods as were used in California. These were shipped to Wittenberg University between December 2008 and January 2009. Initially, we were only investigating one fungicide, and agreed that one California orchard would be sprayed at night on 10 March 2009 following label application and directions. As a control for this field investigation, the organic orchard was going to serve as an untreated (unsprayed) control for comparison to fungicide-sprayed orchards because all these orchards are within the same agricultural habitat (thus exposed to similar fungi); fungicide spraying for the organic orchard is prohibited and thus is technically "unsprayed."

Table 1 provides a listing of the 21 colonies that were involved in this survey study. Samples have the following designations that key them back to the site of origin: "Direct spray" (CA) was treated with boscalid and pyraclostrobin (BP); "Non-target" (CA) orchard was separated from BP-sprayed orchards by more than 3.0 km; "Organic" (CA) was a certified organic almond orchard; and "No spray" (AZ) had no history of BP, or any fungicide, spraying for many years. However, we later learned from communication with local beekeepers and the orchard managers that all almond orchards in the California sampling sites were subjected to constant and multiple fungicide applications prior to and during our sampling period.

Collection of Bee Bread from Samples

Individual comb cells containing bee bread were excised and the bee bread was removed following a regular aseptic technique. The bee bread was left intact upon removal from the comb cell; the sample was weighed using a microbalance (precision of SD ± 0.2 μg, accuracy of ± 0.1 μg at 1 mg; CAHN, Ventron Co., Cerritos, California). Then the sample was placed into a sterile Petri dish (100 × 15 mm; Fisher Scientific, Pittsburgh, Pennsylvania), and separated into thirds that corresponded to the position of the bee bread in the comb cell: *top* (presumably more aerobic conditions); *middle* (moderately anaerobic conditions); or *bottom* (presumably more anaerobic conditions). Randomized block design was used to select samples from the same comb to test for variation between the cells. Each portion of bee bread (top, middle, or bottom third) was placed separately into 2 mL of sterile distilled double-deionized (DI) water and mixed thoroughly with a vortex. The bee bread-DI water extract was split such that 1 mL of the extract was used for quantitative analysis by enumeration and the remaining 1 mL was used for qualitative analysis for fungus identification. Samples of DI water were run in parallel as a negative control.

To conduct the fungus identification, 100 μL aliquots of the 1 mL of bee bread-DI water extract from each portion (top, middle or bottom third) were plated out on PDA and MMN in Petri dishes (Fisher), and incubated at 30°C in total darkness. Once fungal colonies began to appear, individual hyphal tips were excised from the agar with a scalpel in 1-cm^3 blocks (performed under 100× light microscopy) and transferred to a fresh plate of solidified agar. Subcultures were incubated at 30°C in darknss. Plates were examined daily under 100× magnification for the appearance of macroscopic (colony) and microscopic (e.g., philiades, conidia) culture characteristics that were suitable for identification. Identification was based on Barnett and Hunter's (1998) keys. As further confirmation of identity, isolates were compared to authentic cultures and microscopic preparations under high magnification (1000×). *Mycelia sterilia* was used to denote fungi that did not produce identifiable characteristics. Results were expressed as the identity of a fungus and corresponding number of isolates. Controls were DI water and samples of bee bread that had been autoclave-sterilized.

To carry out the enumeration, 1 mL of the bee bread-DI water extract from each portion (top, middle, or bottom third) was diluted serially (100 μL aliquots in 900 μL DI water) and 1 mL samples were plated out (Brown, 2007). Potato Dextrose agar (PDA) and modified Melin-Norkrans agar (MMN) served as culture media in Petri dishes (Fisher). Two agar growth media were used to maximize recovery of fungi that might be fastidious. Incubation conditions were 30 ± 1°C in total darkness to mimic bee colony conditions (Yoder et al., 2008). Counts of fungal colonies were made using an automatic colony counter (Bantex Co., Burlingame, California) at 24, 48, and 72 hours after incubation. Results were expressed as conidia/cm^3. DI water and autoclave-sterilized samples of bee bread served as controls.

A total of 10 samples of bee bread were analyzed from each of the 21 bee colony sources (Table 1). Data were pooled because the 10 samples were taken from the same treatment groups. Each intact bee bread sample was divided into thirds (top, middle, or bottom), and the experiment (enumeration/identification) on each third was replicated three times on each of the two agar media (PDA and MMN). An analysis of variance (ANOVA) was used to compare data, using an arcsin transformation in the case of percentages (SPSS 14.0 for Windows, IBM Corp., Armonk, New York; Excel, Microsoft Corp. Redmond, Washington; Minitab, Chicago, Illinois; Sokal and Rohlf, 1995). Data are mean ± SE.

In Vitro Analysis: Competitive Interactions of Bee Bread Fungi

The following bee bread fungal isolates were used: *Absidia* sp., *Alternaria* sp., *Ascosphaera apis*, *Aspergillus flavus*, *Aspergillus niger*, *Bipolaris* sp., *Cladosporium* sp., *Fusarium* sp., *Mucor* sp., *Penicillium* sp., *Rhizopus* sp., *Scopulariopsis* sp., and *Trichoderma* sp. To simulate the bee colony environment and to reflect interactions of what would occur in bee bread, incubation conditions were set at 30 ± 1°C (regular bee colony temperature; Chiesa et al., 1988) in total darkness using bee bread-supplemented nonnutritive agar (BBSA, Fisher Scientific, Pittsburgh, Pennsylvania), containing 40% bee bread (modified from Klungness and Peng, 1983; Folk et al., 2001; Hua and Feng, 2006) under 5% CO_2. All culturing and subsequent testing was done in disposable 100 × 15 mm-Petri plates.

After 24 hours, cultures were transferred to aerobic conditions as described by Gilliam et al. (1989). For the preparation of BBSA, the bee bread was taken from the combs of bee colonies that were confirmed to be free of fungicides by GC/MS (R. Simonds, USDA, National Science Laboratory, Gastonia, North Carolina, and D. Sammataro, USDA-ARS, Carl Hayden Bee Research Center, Tucson, Arizona, unpublished observations) and the bee bread was autoclave-sterilized before incorporation into the media (Yoder et al., 2008). Unless otherwise noted, the fungal inoculum

Fungicide Treatment	Fungicide	Colony: ID	Coll.	km from Application	Hours-Post	Day Foraging Post	Comments
California Direct Spray	CP+ BP	BP1	11 Mar 09	(0.0, 0.0)	16	0.5	—
	"	BP2	"	(0.0, 0.0)	16	0.5	—
	"	BP31	13 Mar 09	(0.0, 0.0)	68–70	2.5	Chalkbrood positive
	"	BP36	"	(0.0, 0.0)	68–70	2.5	—
	"	BP39	"	(0.0, 0.0)	68–70	2.5	Weak colony
	"	BP42	"	(0.0, 0.0)	68–70	2.5	Queenless ca. 1 week
	"	RH14	"	(0.0, 0.4)	68–70	2.5	—
	"	RH16	"	(0.0, 0.4)	68–70	2.5	Weak colony
	"	RH22	"	(0.0, 0.4)	68–70	2.5	—
Non-target	CP	KG28	13 Mar 09	(0.0, 5.9)	68–70	2.5	Chalkbrood positive
Organic	None	Organic	13 Mar 09	(30.0, 30.0)	68–70	2.5	surrounded by non-organic orchards
Arizona No Spray	None	AHB1	9 Dec 08	—	—	—	
		AHB2	"	—	—	—	—
		EHB	"	—	—	—	SDI
		EHB37	"	—	—	—	SDI
		EHB41	"	—	—	—	Inactive comb; SDI
		EHB60	"	—	—	—	MDI
		EHB62	"	—	—	—	MDI
		EHBH	"	—	—	—	Hygienic
		EHBR	"	—	—	—	Russian
		AHB3	29 Jan 09	—	—	—	—

Table 1. Colony source of bee bread used in this study from California and Arizona. All were active hives with good foraging based on presence of pollen in pollen traps unless otherwise noted; weak colonies were those having little or no foraging bees. "Direct spray" colonies were those in an orchard sprayed with boscalid and pyraclostrobin (BP). "Non-target" colonies were over 3.0 km away but within bee flight range, of BP-sprayed orchards. CP, cyprodinil applied on 26 February 09; P, BP applied on 10 March 09. AHB3 was not actively collecting pollen; SDI, single drone-inseminated queen; MDI, multidrone-inseminated queen.

("block") was a 1-cm^3 block excised from the edge of an established (>10 days old) mycelium, with each replicate (N = 3) utilizing a separate mycelium. Thus, a different mycelium was used as the source of inoculum during the study. Fungi were distinguished based on Barnett and Hunter's (1998) keys using both macroscopic (40/100×) and microscopic characteristics (e.g., conidia, phialides) at 1000× under oil. Three experiments were conducted: primary resource capture, secondary resource capture, and differential competition.

In order to analyze the antifungal compound production by the different fungi, primary resource capture (fungi's ability to inhabit unoccupied resources) was measured. A secondary aim of this experiment was to examine how the interaction of two fungi alters spore production. A modified version of the trisecting line method was used for radial growth rate (K_r) determination; two fungi were examined simultaneously in the same Petri dish (modified from Currah et al., 1987; Klepzig and Wilkens, 1997; Baldrian and Gabriel, 2002; Benoit et al., 2004).

A diameter line was drawn across the bottom of a Petri dish. At the end of each line from the edge of the dish, two additional lines were drawn at ±30° angles from the centerline. Two blocks of fungal mycelium were placed onto the agar surface directly opposite each other, over the top point of intersection of the three lines that radiated 5 mm from the edge of the plate. Daily measurements were taken along each of the three lines as the mycelium spread over the agar surface, before the fungi made hyphal contact with each other. The measurements were fit to Baldrian and Gabriel's (2002) growth rate equation $K_r = (R_1-R_0)/(t_1-t_0)$, where K_r is the radial growth rate, and R_0 and R_1 are colony radii between the beginning of the linear (t_0) and stationary (t_1) phases of growth, expressed as mm/hour.

Once the two fungi made hyphal contact, growth was allowed to continue for 2 days and then a block area of the zone on the agar plate where the two fungi overlapped was removed. This block was placed into 5 mL deionized, double distilled (DI) water in a 10-mL disposable test tube at 30°C and shaken overnight for 12 hours. Conidia were counted (based on 0.1% trypan blue exclusion with five counts with a hemocytometer, AO Spencer Bright-Line, St. Louis, Missouri) and adjusted to 1.0×10^7 conidia/mL with DI water and diluted serially (1.0×10^{-1}, 1.0×10^{-2}, 1.0×10^{-3}, and 1.0×10^{-4} conidia/mL) for enumeration (Brown, 2007). One milliliter of each serial dilution was plated on solidified agar. The number of colonies were counted with an automatic colony counter (Bantex Co., Burlingame, California) after 48 hours of incubation. Colonies were identified and the number of conidia of a particular fungus (as a proportion of initial 1.0×10^7 conidia/mL suspension) was expressed as number of conidia/1cm^2 mycelium. Two control groups were run: an isolated control (block of plain agar with no fungus placed across from a block of fungal inoculum) and a conspecific control (blocks of the same fungus were across from each other).

Secondary resource capture experiment (the capacity of a newly introduced fungus to establish on a substrate already occupied by another fungus) was also utilized to further analyze the growth competition of colony-associated fungi. A block of fungal inoculum was placed directly on top of an established, 1-week-old mycelium of another fungus that was growing on the plate. The experiment was conducted in a Petri dish where three lines that had been drawn radiating from the center midpoint of the plate (lines separated by ±120°) such that the radial growth rate (K_r) of the newly introduced fungus could be determined by the trisecting line method as described above. Daily measurements of the newly-introduced fungus were taken along three lines scored on the bottom of the Petri dish as the fungus spread over the existing mycelium. Radial growth rate was calculated using the growth rate equation on page 197, with absence of hyphae by microscopic observation (also evidenced by the white advancing edge of new mycelium) from the newly introduced block over the established mycelium, being defined as no growth. Two control groups were run: an isolated control where a block of fungal inoculum was placed onto solidified agar (not on top of an existing established mycelium) and a conspecific control where a block of fungus was placed on top of a mycelium of the same fungus species.

Differential competition determines which of two competing fungi grows faster. In order to test this, one fungus was designated as "A" and a second was designated as "B". Blocks of fungal inoculum (0.5-cm^3 block) were then removed from an established mycelium and placed directly onto the surface of agar in a Petri dish. Each dish contained 20 blocks of fungi (mixed combinations of A and B) spaced at least 1 cm apart using a randomized block design. The ratios of the 20 blocks of fungal inocula/Petri dish were as follows: 0.2:0.8, (4 blocks of fungus A with 16 blocks of fungus B); 0.4:0.6 (8 blocks of fungus A with 12 blocks of fungus B); 0.5:0.5 (10 blocks of fungus A with 10 blocks of fungus B); 0.6:0.4 (12 blocks of fungus A with 8 blocks of fungus B); 0.8:0.2 (16 blocks of fungus A with 4 blocks of fungus B). Control was 1:0 (20 blocks of fungus A with 0 blocks of fungus B) where all blocks were of the same fungus (modified from Klepzig and Wilkens, 1997; Benoit et al., 2004). The growth perimeter of the mycelium around each block of inoculum was measured with a planimeter (Professional Equipment Inc., Hauppauge, New York) after 5 days. Measurements of the area were used to calculate relative crowding coefficient (RCC) according to the following equation:

$$\text{RCC} = \frac{\left[\dfrac{(\text{area occupied by fungus A at 0.5:0.5})}{(\text{area occupied by fungus B at 0.5:0.5})}\right]}{\left[\dfrac{(\text{area occupied by fungus A at 1:0})}{(\text{area occupied by fungus B at 1:0})}\right]}$$

The equation indicates whether fungus A had the advantage over fungus B (if RCC > 1) or whether fungus B had the advantage over fungus A (RCC < 1) (modified from Novak et al., 1993; Klepzig and Wilkens, 1997; Benoit et al., 2004). Reversing the A and B assignments for the two members of the fungus pair resulted in the same conclusion.

Each experiment was replicated three times using ten plates per replicate, and each growth rate (primary resource capture experiment and secondary resource capture experiment) was calculated from a total N = 30 plates, 3 growth lines per plate. In total, 450 measurements per fungal species were made (each radial growth rate is the mean of 450 measurements). Determinations of conidial output in the primary resource capture experiment were made using eight blocks of mycelium from a zone where two fungi had overgrown and the enumeration was based on five different dilutions for a total of 40 Petri dishes per fungus pairing. In the secondary resource capture experiment, each pairing was replicated three times. In the differential competition experiment, each ratio combination of the different blocks of inocula was done in triplicate. In all cases, data are given as means ± SE. Data were compared by analysis of variance (ANOVA) using SPSS 14.0 for Windows, Microsoft Excel, and Minitab as described by Sokal and Rohlf (1995).

Results

In Vivo Analysis of Bee Bread Samples: Mycoflora Profile

Figure 1 illustrates a culture of bee bread that compares a sample from a fungal colony from Arizona (right) with one from California (left). Mycoflora profiles from select bee colonies from California and Arizona are shown in Tables 2 and 3, respectively. The percentages shown represent the diversity and amount of fungi found within a core of bee bread at any one time. In Tables 2 and 3, specific bee bread samples were chosen as representatives of the various fungi and their relative amounts; these compare favorably to mycoflora profiles of bee bread sampled from other colonies within the same yard. From Table 1, colonies selected for data presentation among the California samples (Table 2) were: BP1 as a representative of effects 16 hours after spraying with BP, and BP31 and BP36 and RH14 for effects 68 to 70 hours after spraying (BP31 also displayed symptoms of chalkbrood). Mycoflora data are presented for KG28, because this colony exhibited symptoms of chalkbrood infection and is located 5.5 km from "Direct spray" samples. For the organic orchard

Figure 1. Plated bee bread samples representative of a bee colony from Arizona with no history of fungicide-spraying (right; EHB37) and California from a colony in a fungicide-sprayed orchard (left; BP39). Scale is in cm. Incubation conditions were 30°C, darkness, 72 hours, Potato Dextrose agar, DF = 10. These plates were randomly chosen and are not ones that contained the most or least amount of fungi among the samples. (*Photo by J. Yoder.*)

conventional fungicide use was prohibited, but is within areas where fungicides are sprayed. For the Arizona samples (Table 3), these colonies are located in an urban area not subject to widespread fungicide applications: Consistently, bee bread from California contained less fungi than bee bread from Arizona. *Penicillium* spp. (17%) consistently made up the majority of the isolates in bee bread from colonies that had been exposed to fungicide (Table 2), with *Aspergillus niger* predominating (5 isolates, 12%) of the *Aspergillus* spp. isolates. For these samples, numbers of *Aspergillus* spp. isolates were nearly equivalent to the number of *Penicillium* spp. isolates ($P > 0.05$). Among the fungi identified, *Rhizopus* spp. was also present with regular frequency as well as *Cladosporium* spp. Frequency of occurrence of other fungi (*Alternaria* spp., *Collectotrichum* spp., *Mucor* spp., *Paecilomyces* spp., *Scopulariopsis* spp., *Stigmella* spp.) was less (1 to 4 isolates; 2% to 10%), although these minor components contributed to variability in mycoflora profile among the bee bread samples.

For the Arizona-based samples, a total of ten genera of fungi were also recovered, with slight variations from the California samples (Table 3). Based on frequency of isolates, the mycoflora from Arizona was dominated by *Aspergillus* spp. (mainly *A. niger* and *A. flavus*) and *Penicillium* spp. when abundances of the other fungi were compared and comprised 33% to 67% of the total isolates for *Aspergillus* spp. (41 isolates total combined for *A. flavus, A. niger*, and *Aspergillus* spp.) and 13% to 42% of the total isolates for *Pencillium* spp. (22 isolates) ($P < 0.05$). *Cladosporium* spp. (6% to 13% of total isolates; 8 isolates) and *Rhizopus* spp. (4% to 14% of total isolates; 6 isolates) were present to a lesser extent than *Aspergillus* spp. and *Penicillium* spp. ($P < 0.05$). Remaining components (*Aureobasidium* spp., *Bipolaris* spp., *Fusarium* spp., *Mucor* spp., *Paecilomyces* spp., and *Trichoderma* spp.) comprised only a small proportion (4% to 11%; 1 to 3 isolates) of the total number of isolates, although represented most of the variation in the mycoflora of bee bread when compared from one bee colony to another.

From the 21 bee colonies listed in Table 1, mycoflora profiles of bee bread samples were the most similar (based on percentage and identification of fungal isolates) among the following: RH16 and RH22 = KG28; BP2 = RH14; BP39 = Organic; and BP42 = BP1 from the California samples (Table 2). Among the Arizona samples, similarities existed between AHB2 and EHB37 = EHB60; EHB41 = EHBR; and EHB62 = EHB (Table 3) ($P > 0.05$). Samples from California contained higher proportions of *Aspergillus* spp., *Mucor* spp., *Paecilomyces* spp., and *Rhizopus* spp. than the ones from Arizona, whereas *A. flavus* and *A. niger* were more prevalent in the Arizona than the California samples (Table 2 and 3; $P < 0.05$). Relative amounts of *Cladosporium* spp. and *Penicillium* spp. between Arizona and California samples were approximately the same ($P > 0.05$). No detectable preference or occurrence was displayed by fungi for a particular position (top, middle, bottom) of the bee bread in the comb when comparing levels of fungi by stratification ($P > 0.05$), suggesting that they are uniformly distributed throughout the entire bee bread column. Species of fungi present in bee bread were similar between California and Arizona samples, but California samples contained less overall.

Total number of conidia in bee bread is presented in Table 4. Counts of fungal colonies after 48 and 72 hours incubation yielded nearly identical results, whereas counts after 24 hours of incubation yielded inconsistent results, because not all fungi display the same growth rate; only faster growers are prevalent at 24 hours. Bee bread from Arizona contained about 1.5- to 3.0-fold the amount of conidia compared to samples from California ($P < 0.05$). Among replicates there was no difference by enumeration in the total number of conidia present in all portions of the bee bread sample. There was also little, if any, variation in total conidia numbers between different bee bread samples coming from the same beeswax comb ($P > 0.05$). Therefore, conidia are uniformly spread throughout the bee bread column within the cell. Mass measurements of bee bread samples pulled from the comb were consistently similar (Table 4), thus the pronounced quantitative differences in conidia that we note between lower amounts in California and higher amounts in Arizona samples are not due to amounts of bee bread of varying sizes.

In Vitro Analysis of Bee Bread Fungi: Competitive Interactions

Of the 13 fungi examined, there was an 8-fold difference between the fungal species that grew the fastest (*Rhizopus* sp.) and those that grew slowest (*Cladosporium* sp.). Radial growth rates of the other fungal types fell between these extremes when grown alone in culture (isolated control; Table 5). All of these mesophilic fungi were capable of using bee bread as a nutritional source in the form of BBSA at 30°C; that is, $K_r \neq 0.00$ mm/hour. Radial growth rates varied between the 13 fungi, with the exception of *Scopulariopsis* sp. and *A. apis* that grew at similar rates (ranked statistically; $P < 0.05$), which permitted separation into fast, moderate, and slow-growing groups. The fast growers were *Rhizopus* sp. > *Mucor* sp. > *Absidia* sp. > *Trichoderma* sp. >> (moderate growers) *Fusarium* sp. > *Bipolaris* sp. > *Penicillium* sp. > *Alternaria* sp. >> (slow growers) *A. niger* > *Scopulariopsis* sp. = *A. apis* > *A. flavus* > *Cladosporium* sp. Radial growth rate correlated positively with conidia (spore) production (Table 6; number of viable conidia/1 cm^2 versus natural logarithm K_r; $y = 1832.5\times$, $R = 0.93$). When paired and cultured opposite a fungus of the same strain, both of the fungi displayed similar radial growth rates

Table 2. Select California samples of bee bread designated in Table 1 having known BP fungicide exposure either by direct spray ("Direct spray") or indirect spray ("Non-target") within flight range of sprayed orchards giving identity of fungi and number of isolates (%). Culturing on Potato Dextrose agar (shown) and modified Melin-Norkrans agar gave similar results. Organic orchard was adjacent to orchards sprayed with fungicides.

	No. Isolates (%) from Bee Bread Sample from California Colonies:							
	Direct Spray				Non-Target			
Fungi	**BP 1**	**BP31**	**BP36**	**RH14**	**KG28**	**Organic**	**Total**	**%**
Alternaria	1 (13)	0	0	1 (14)	0	0	2	5
Aspergillus flavus	0	0	0	1 (14)	1 (13)	0	2	5
Aspergillus niger	1 (13)	0	1 (14)	0	2 (25)	1 (20)	5	12
Aspergillus spp.	0	1 (17)	1 (14)	0	0	1 (20)	3	7
Cladosporium spp.	1 (13)	1 (17)	0	1 (14)	0	0	3	7
Collectotrichum	0	1 (17)	0	0	0	0	1	2
Mucor spp.	0	0	1 (14)	0	3 (38)	0	4	10
Mycelia sterilia	3 (38)	0	0	0	0	0	3	7
Paecilomyces spp.	0	0		0	1 (13)	1 (20)	2	5
Penicillium spp.	1 (13)	1 (17)	1 (14)	2 (29)	1 (13)	1 (20)	7	17
Scopulariopsis spp.	0	1 (17)	0	1 (14)	0	0	2	5
Stigmella spp.	0	0	1 (14)	0	0	0	1	2
Rhizopus spp.	1 (13)	1 (17)	2 (29)	1 (14)	0	1 (20)	6	14
Total	8	6	7	7	8	5	41	

Table 3. Select Arizona samples of bee bread having no history of fungicide exposure showing identification of fungi and number of isolates (%). Bee colony descriptions are in Table 1. Results were similar using Potato Dextrose agar (shown) and modified Melin-Norkrans agar.

	No. Isolates (%) from Bee Bread Sample from:							
Fungi	**AHB1**	**AHB2**	**EHB**	**EHB60**	**EHBH**	**EHBR**	**Total**	**%**
Aspergillus flavus	1 (7)	1 (8)	4 (22)	3 (20)	0	2 (9)	11	12
Aspergillus niger	5 (36)	3 (25)	2 (11)	6 (40)	3 (33)	8 (35)	27	30
Aspergillus spp.	1 (7)	0	0	1 (7)	1 (11)	0	3	3
Aureobasidium spp.	0	1 (8)	0	1 (7)	0	0	2	2
Bipolaris spp.	0	0	2 (11)	0	0	1 (4)	3	3
Cladosporium spp.	0	1 (8)	1 (6)	2 (13)	1 (11)	3 (13)	8	9
Fusarium spp.	0	0	1 (6)	0	0	0	1	1
Mucor spp.	1 (7)	0	0	0	0	1 (4)	2	2
Mycelia sterilia	0	0	0	0	1 (11)	2 (9)	3	3
Paecilomyces spp.	0	0	1 (6)	0	1 (11)	0	2	2
Penicillium spp.	4 (29)	5 (42)	4 (22)	2 (13)	2 (22)	5 (22)	22	24
Rhizopus spp.	2 (14)	1 (8)	(11)	0	0	1 (4)	6	7
Trichoderma spp.	0	0	1 (6)	0	0	0	1	1
Total	14	12	18	15	9	23	91	

(*Absidia* sp. = 0.289 mm/hour; *Bipolaris* sp. = 0.160 mm/hour; *Alternaria* sp. = 0.147 mm/hour; *Scopulariopsis* sp. = 0.084 mm/hour; *A. apis* = 0.067 mm/hour; *Cladosporium* sp. = 0.055 mm/hour; Table 5; P > 0.05). At no time was the gross morphology of the fungi (obverse/reverse pigmentation, colony, conidia, and philiade characteristics) altered in a way that would permit fungi to be separated from each other. Nor was there ever an inability to track hyphae through the agar by light microscopy. All of the fungi remained as anamorphs and did not produce structures assignable to their sexual counterparts; anamorph-teleomorph connections exist for some of the fungi used (Jennings and Lysek, 1999). Thus, differences we note are due to competitive interactions and not to large metabolic shifts associated with transition into a sexual form.

Table 5 shows the subset of the 13 fungi that displayed significant inhibitory activity toward another fungus by primary resource capture, by opposing dual culture prior to the fungi making contact in the Petri dish. Of the fast growers, *Rhizopus* sp. was inhibited by *Mucor* sp. by 17%, and *Mucor* sp. was inhibited by *Fusarium* sp. by 13%. No fungi reduced growth of *Absidia* sp. or *Trichoderma* sp. Growth of moderate growing fungi was reduced for *Fusarium* sp., 24% by *A. niger* and 38% by *Penicillium* sp.; *Bipolaris* sp. 35% by *Penicillium* sp. and 34% by *Trichoderma* sp.; for *Penicillium* sp. 12% by *Fusarium* sp. and 26% by *Trichoderma* sp.; and for *Alternaria* sp. 42% by *A. flavus*, 23% by *Penicillium* sp. and 29% by *Trichoderma* sp.

For slow-growing fungi, growth of *A. niger* was reduced 19% by *Trichoderma* sp.; *Scopulariopsis* sp. was reduced 16% by *A. flavus*, 17% by *Fusarium* sp. and 15% by *Trichoderma* sp. *A. apis* was suppressed 26% by *A. flavus*, 52% by *A. niger*, 22% by *Fusarium* sp., 13% by *Mucor* sp., 19% by *Penicillium* sp., 42% by *Rhizopus* sp., and 38% by *Trichoderma* sp.; *A. flavus* was reduced 26% by *A. niger*; and *Cladosporium* sp. was reduced 12% by *A. niger*, 33% by *Rhizopus* sp. and 24% by *Trichoderma* sp. The inhibitory fungi were not more (or less) effective against fungi that were slow, moderate, or fast growers (mean percentage reduction in K_r versus natural logarithm K_r; $y = -7.30x$, $R = 0.33$); the inhibitory fungi did not reduce growth when paired against the same species. In other words, faster-growing fungi did not produce more potent antifungal compounds. No inhibitory effect on the growth of the 13 fungi were observed by the presence of *Absidia* sp., *Bipolaris* sp., *Alternaria* sp., *Scopulariopsis* sp., *A. apis*, and *Cladosporium* sp. In other words, there was no significant difference in radial growth rate to isolated or conspecific controls (data not shown; P > 0.05), suggesting an absence of the production of antifungal substances by these species. We conclude that the effects of the antifungal compounds from the species we cultured vary independent of whether the fungus making the antifungal, or the target fungus, is a slow, moderate, or fast grower. The antifungal compounds suppressed fungal growth by 20% to 30% but in no instance inhibited it.

Bee Bread Sample	Mass (g)	No. Conidia/cm³
No spray (AZ)		
AHB1	0.13 ± 0.04[a]	114 ± 22[a]
AHB2	0.12 ± 0.06[a]	107 ± 18[a]
AHB3	0.11±0.04[a]	134 ± 11[b]
EHB	0.15 ± 0.04[a]	163 ± 16[c]
EHB37	0.16 ± 0.11[a]	159 ± 21[c]
EHB41	0.17 ± 0.08[a]	182 ± 18[d]
EHB60	0.11 ± 0.08[a]	151 ± 25[c]
EHB62	0.18 ± 0.07[a]	176 ± 14[d]
EHBH	0.13 ± 0.05[a]	87 ± 12[e]
EHBR	0.16 ± 0.02[a]	206 ±23[t]
Direct spray (CA)		
BP1	0.19 ± 0.03[a]	52 ± 8[g]
BP2	0.15 ± 0.06[a]	38 ± 13[h]
BP31	0.15 ± 0.07[a]	64 ± 17[i]
BP36	0.12 ± 0.05[a]	53 ± 12[g]
BP39	0.17 ± 0.08[a]	67 ± 8[l]
BP42	0.14 ± 0.05[a]	74 ± 11[j]
RH14	0.13 ± 0.04[a]	81 ± 10[e]
RH16	0.12 ± 0.07[a]	57 ± 14[l]
RH22	0.15 ± 0.10[a]	66 ± 7[i]
Non-target (CA)		
KG28	0.12 ± 0.04[a]	61 ± 10[l]
Organic	0.17 ± 0.06[a]	44 ± 8[h]

Table 4. Enumeration of fungi from bee bread samples after 48 hours incubation (30°C, darkness) from bee colony sources given in Table 1. No fungal colonies were present on autoclave sterilized bee bread or DI water used to prepare dilutions (controls). Data (mean ± SE) followed by same superscript letter within a column do not significantly differ (P < 0.05).

Figure 2. Representative plots showing colonizing ability of bee bread fungi against each other on bee bread supplemented nonnutritive agar, 30°C. Rhi, *Rhizopus* sp.; Muc, *Mucor* sp.; Alt, *Alternaria* sp.; Pen, *Penicillium* sp.; Asf, *Aspergillus flavus*; Cla, *Cladosporium* sp. Each point is mean ± SE ≤ 2.6.

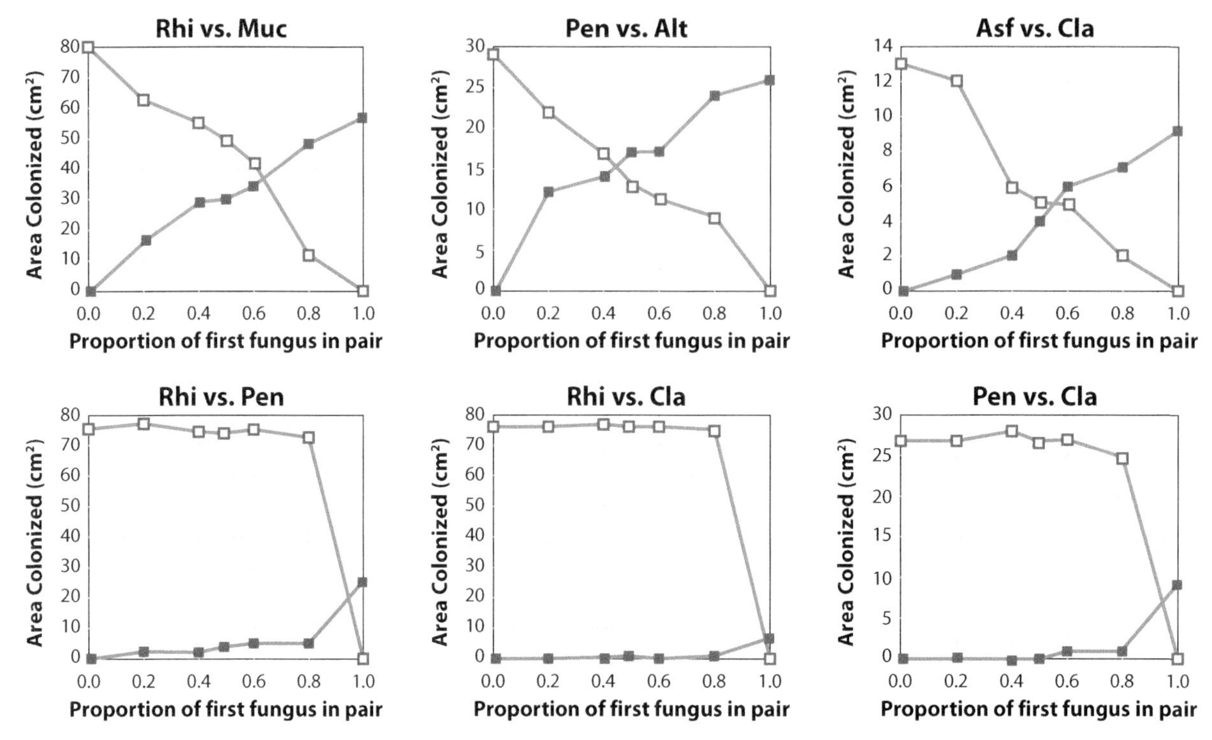

Any dual-culture situation had a marked impact on decreasing conidia output compared to isolated controls after the two fungi overgrew (Tables 6 to 8), including conspecific controls ($P < 0.05$). Fast-growing fungi *Rhizopus* sp. and *Absidia* sp. exerted the greatest reduction on *A. apis*, reducing the amount of *A. apis* conidia 30- and 9-fold, respectively (conidia output for other fungi was reduced by 2- to 9-fold by *Rhizopus* sp. and *Absidia* sp.). *Mucor* sp. had the greatest reduction (13-fold) on *Rhizopus* sp. (other fungi were reduced 2- to 11-fold by *Mucor* sp.). *Trichoderma* sp. had the greatest reduction (25-fold) on *Absidia* sp. (Table 6). No viable conidia were recovered by the majority of fungi in response to *Trichoderma* sp. (Table 6). Greatest reduction in conidial output caused by moderate growing fungi were observed for *Rhizopus* sp. (12-fold reduction by *Fusarium* sp.), *Mucor* sp. (13-fold reduction by *Bipolaris* sp.), *Bipolaris* sp. (13-fold reduction by *Pencillium* sp.), and *Trichoderma* sp. (reduced 18-fold by *Alternaria* sp.). Suppression in numbers of conidia ranged 2- to 12-fold for other fungi in response to competition with a moderate grower (Table 7). For slow growers, *A. niger*'s greatest reduction (7-fold) was on *Mucor* sp. The growth of *Scopulariopsis* was most greatly reduced (13-fold) on *Cladosporium* sp., *A. apis* on *A. niger* (10-fold), *A. flavus* on *Penicillium* sp. (6-fold), and *Cladosporium* sp. on *Rhizopus* sp. (8-fold) (other fungi showed reductions in response to a slow-growing fungus in conidia output of 2- to 6-fold; Table 8).

No viable conidia were recovered from *A. apis* when overtaken by *Mucor* sp., *Trichoderma* sp., *Penicillium* sp., *A. niger*, and *A. flavus*. Also, no growth (no viable conidia produced) was observed for *A. flavus* when overgrown by *A. niger*; *Bipolaris* sp. by *Trichoderma* sp.; *Cladosporium*

	Mean Radial GrowthRate, K_r (mm/hour ± SE ≤ 0.039) of Fungus in Presence of:							
		Competitor:						
Fungus	Control	Asf	Asn	Fus	Muc	Pen	Rhi	Tri
Abs	0.287	0.295	0.274	0.289	0.271	0.297	0.284	0.281
Alt	0.142	0.082*	0.156	0.149	0.151	0.109*	0.136	0.101*
Asc	0.069	0.051*	0.033*	0.054*	0.060*	0.056*	0.041*	0.043*
Asf	0.058	0.056	0.043*	0.064	0.064	0.059	0.062	0.065
Asn	0.083	0.079	0.084	0.077	0.087	0.083	0.085	0.067*
Bip	0.164	0.158	0.162	0.159	0.169	0.107*	0.173	0.108*
Cla	0.051	0.050	0.045*	0.052	0.055	0.047	0.034*	0.039*
Fus	0.176	0.181	0.133*	0.171	0.178	0.109*	0.169	0.182
Muc	0.312	0.301	0.317	0.273*	0.315	0.310	0.308	0.314
Pen	0.149	0.155	0.147	0.131*	0.153	0.144	0.150	0.110*
Rhi	0.411	0.414	0.417	0.396	0.341*	0.389	0.416	0.409
Sco	0.075	0.063*	0.078	0.062*	0.085	0.081	0.071	0.064*
Tri	0.237	0.246	0.233	0.242	0.239	0.240	0.236	0.229

Table 5. Growth rates (K_r) of bee bread fungi at 30°C on bee bread-supplemented nonnutritive agar in opposing dual cultures in Petri dishes prior to fungal contact. Data are shown for only those pairings that resulted in significant reduction from the rate of an isolated control ($P < 0.05$).
Abs, *Absidia* sp.;
Alt, *Alternaria* sp.;
Asc, *Ascosphaera apis*;
Asf, *Aspergillus flavus*;
Asn, *Aspergillus niger*;
Bip, *Bipolaris* sp.;
Cla, *Cladosporium* sp.;
Fus, *Fusarium* sp.;
Muc, *Mucor* sp.;
Pen, *Penicillium* sp.;
Rhi, *Rhizopus* sp.;
Sco, *Scopulariopsis* sp.;
Tri, *Trichoderma* sp.
* Denotes a significant difference in growth rate compared to isolated control ($P < 0.05$).

sp. by *Rhizopus* sp., *Mucor* sp., *Penicillium* sp. and *A. niger*; *Fusarium* sp. by *Trichoderma* sp.; *Mucor* sp. by *Trichoderma* sp.; *Penicillium* sp. by *Trichoderma* sp., *Fusarium* sp. and *A. niger*; and *Scopulariopsis* sp. by *Trichoderma* sp., *Penicillium* sp. and *A. niger*. Amount of decrease in conidia production did not depend upon whether the fungus was in dual competition with a fast, moderate, or slow grower (mean percentage reduction versus natural logarithm K_r; $y = -125.10\times$, $R = 0.41$); competition with a faster grower did not result in greater reductions in conidia production than when in competition with a slower grower, for example. The diminished grow rates and conidia outputs are the likely result of hoarding resources.

As shown by data collected from the secondary resource capture experiment, none of the 13 fungi were capable of growing ($K_r = 0.00$ mm/hour) when placed onto a mycelium of a conspecific fungus or when applied to a mycelium of *Absidia* sp., *Mucor* sp., or *Rhizopus* sp. (Table 9). Fungi that were capable of establishing on a previously existing mycelium were *A. flavus*, *A. niger*, *Mucor* sp., *Rhizopus* sp., and *Trichoderma* sp. All displayed reduced growth rates compared to fungi on uninhabited substrates. *A. flavus* grew on mycelia of *Bipolaris* sp. at 5-fold the regular rate and *Cladosporium* sp. at 2-fold. Growth on *Alternaria* sp., *Fusarium* sp., and *Scopulariopsis* sp. resulted in a 2-fold growth reduction of *A. niger*. Growth rate of *A. niger* was reduced 3-fold by *Cladosporium* sp. and 4-fold by *Trichoderma* sp. Growth rates of *Mucor* sp. were reduced 2- to 3-fold when placed on a mycelium of *A. apis*, *Cladosporium* sp. and *Penicillium* sp., and 5-fold by *Alternaria* sp. and *Bipolaris* sp. Placement on *Trichoderma* sp. resulted in a 6-fold decrease in the growth of extremely fast-growing *Rhizopus* sp., whereas *Rhizopus* sp. experienced a 2- to 3-fold decrease when placed on *A. apis*, *A. flavus*, *Cladosporium* sp. and *Scopulariopsis* sp. *Trichoderma* sp. was the most effective at growing

Table 6. Impact of fast-growing fungi on conidial output by bee bread fungi after fungus overgrowth at 30°C on bee bread supplemented nonnutritive agar in paired competition trials. Abbreviations of fungi as per Table 5; NG, no growth indicates an absence of colonies. All are significantly different compared to the isolated control ($P < 0.05$).

	No. Viable Conidia/1 cm² Mycelium (Mean ± SE ≤ 67.4) of Fungus in Presence of:				
		Fast-Growing Fungal Competitor			
Fungus	Control	Rhi	Muc	Abs	Tri
Abs	2997.3	449.2	261.1	661.3	118.7
Alt	1941.9	345.3	555.4	594.2	372.4
Asc	358.1	11.5	NG	39.1	NG
Asf	972.8	232.7	462.1	122.8	103.5
Asn	1416.6	532.7	428.4	612.3	240.9
Bip	1807.4	506.1	261.4	249.2	NG
Cla	233.5	NG	NG	95.1	13.7
Fus	1488.2	622.8	431.6	218.5	NG
Muc	3751.7	559.1	1349.3	642.9	NG
Pen	2649.3	731.2	552.5	1239.4	NG
Rhi	4021.4	1649.2	316.2	628.8	268.4
Sco	155.7	83.1	17.2	49.1	NG
Tri	3591.4	1297.9	993.4	705.1	1226.1

Table 7. Effect of moderate-growing fungus after direct contact with bee bread-associated fungi on conidial output (bee bread supplemented nonnutritive agar, 30°C). Abbreviations of fungi as per Table 5; NG, no growth indicates an absence of colonies. All are significantly different compared to the isolated control ($P < 0.05$).

	No. Viable Conidia/1 cm² Mycelium (Mean ± SE ≤ 44.6) of Fungus in Presence of:				
		Moderate-Growing Fungal Competitor			
Fungus	Control	Fus	Bip	Pen	Alt
Abs	2997.3	527.9	583.4	275.3	649.1
Alt	1941.9	506.1	297.5	555.7	249.2
Asc	358.1	54.3	193.5	NG	114.7
Asf	972.8	213.6	481.6	61.3	152.9
Asn	1416.6	507.3	253.4	842.4	646.3
Bip	1807.4	449.1	423.1	142.8	723.1
Cla	233.5	94.3	56.4	NG	88.4
Fus	1488.2	331.9	742.1	492.1	552.7
Muc	3751.7	1554.2	291.6	670.4	367.5
Pen	2649.3	NG	647.2	736.4	411.8
Rhi	4021.4	318.4	1935.7	399.6	550.4
Sco	155.7	44.2	19.5	NG	74.1
Tri	3591.4	1462.5	794.3	866.4	197.6

Table 8. Conidial output by bee bread-associated fungi after making contact with a slow-growing fungus on bee bread-supplemented nonnutritive agar, 30°C. Abbreviations of fungi as per Table 5; NG, no growth indicates an absence of colonies. All are significantly different compared to the isolated control ($P < 0.05$).

	No. Viable Conidia/1 cm² Mycelium (Mean ± SE ≤ 44.6) of Fungus in Presence of:					
		Slow-Growing Fungal Competitor				
Fungus	Control	Asn	Sco	Asc	Asf	Cla
Abs	2997.3	579.1	1258.3	406.9	799.2	1317.8
Alt	1941.9	822.5	701.3	1242.7	562.9	568.2
Asc	358.1	NG	125.2	84.7	NG	183.7
Asf	972.8	NG	211.5	346.9	67.9	527.6
Asn	1416.6	434.7	195.9	129.2	271.8	506.8
Bip	1807.4	1225.7	298.6	1674.5	857.2	1225.4
Cla	233.5	NG	17.2	107.6	64.7	49.7
Fus	1488.2	571.8	954.1	233.5	423.8	242.7
Muc	3751.7	555.7	413.6	1210.4	2118.3	1576.5
Pen	2649.3	NG	891.0	827.6	437.5	949.4
Rhi	4021.4	1664.2	671.3	2345.8	1824.2	487.2
Sco	155.7	NG	69.2	106.5	44.9	23.5
Tri	3591.4	2798.3	1505.7	1297.4	2416.2	1973.4

Substrate	Mean Radial Growth Rate, K_r (mm/hour ± SE ≤ 0.048)*: Asf	Asn	Muc	Rhi	Tri
Uninhabited (control)	0.083	0.058	0.312	0.411	0.237
Occupied by:					
Abs	NG	NG	NG	NG	NG
Alt	NG	0.026	0.067	NG	NG
Asc	NG	0.014	0.108	0.116	0.074
Asf	NG	NG	NG	0.241	0.061
Asn	NG	NG	NG	NG	0.108
Bip	0.017	NG	0.064	NG	NG
Cla	0.041	0.019	0.159	0.135	0.145
Fus	NG	0.033	NG	NG	0.071
Muc	NG	NG	NG	NG	NG
Pen	NG	NG	0.091	NG	NG
Rhi	NG	NG	NG	NG	NG
Sco	NG	0.035	NG	0.148	0.168
Tri	NG	NG	NG	0.062	NG

Table 9. Growth by bee colony fungi when placed on a previously existing mycelium (K_r, bee bread-supplemented nonnutritive agar, 30°C). Data are shown for only those combinations where a growth rate could be measured (NG, no growth). Abbreviations of fungi as per Table 5. All radial growth rates reported are significantly different from controls ($P < 0.05$).

*No growth (Kr = 0.00 mm/hour) for *Absidia* sp., *Alternaria* sp., *Ascosphaera apis*, *Bipolaris* sp., *Cladosporium* sp., *Fusarium* sp., *Penicillium* sp., and *Scopulariopsis* sp.

on an existing mycelium. Radial growth rate of *Trichoderma* sp. were reduced approximately 2-fold when applied to *A. niger*, *Cladosporium* sp., and *Scopulariopsis* sp. and 3- to 4-fold when placed on *A. apis*, *A. flavus*, and *Fusarium* sp. While the capacity to establish on an existing mycelium did not depend upon whether the secondary fungus was a fast, moderate, or slow grower, those fungi that *were* capable of secondary resource capture (all except *Fusarium* sp. and *Penicillium* sp.) displayed evidence of antifungal production (Table 5).

With regard to differential competition, *Rhizopus* sp. rapidly outcompeted *Penicillium* sp. and *Cladoporium* sp., and the same was observed when *Penicillium* sp. was plated with *Cladosporium* sp. (Figure 2). In combinations of *Rhizopus* sp. and *Penicillium* sp. (fast versus moderate grower), *Rhizopus* sp. and *Cladosporium* sp. (fast versus slow grower), and *Penicillium* sp. and *Cladosporium* sp. (moderate versus slow grower), the area that was occupied by the mycelium did not correlate positively with the amount of inoculum in an anticipated linear, dose-response manner ($R \leq 0.52$). This demon-strates an inundation effect by the faster-growing fungus of the pairing despite being present in lower levels of inoculum (Table 5). A linear dose-response between the area colonized and proportion of inoculum was observed when growth rates of the two fungi were compared: as in *Rhizopus* sp. and *Mucor* sp. (R = 0.93/0.89, respectively), two moderate growers, such as *Penicillium* sp. and *Alternaria* sp. (R = 0.87/0.91, respectively), or two slow growers, such as *A. flavus* and *Cladosporium* sp. (R = 0.92/0.90, respectively) (Figure 2).

Other fungi that exhibited a proportional trend (increasing area occupied with increasing inoculum; in all cases, $R \geq 0.85$; plots resembled *Rhizopus* sp./*Mucor* sp. in Figure 2) were: *Penicillium* sp. with *Fusarium* sp., *Bipolaris* sp., *Alternaria* sp., and *A. niger*; *Mucor* sp. with *Absidia* sp.

	Relative Crowding Coefficients (RCC) Fungus A												
Fungus B	Abs	Alt	Asc	Asf	Asn	Bip	Cla	Fus	Muc	Pen	Rhi	Sco	Tri
Abs	—												
Alt	6.91	—											
Asc	7.92	6.05	—										
Asf	5.80	0.57	0.64	—									
Asn	5.92	9.17	1.68	3.10	—								
Bip	7.26	1.43	8.32	11.79	5.51	—							
Cla	7.94	8.22	13.03	0.89	1.54	6.27	—						
Fus	24.92	2.26	4.83	5.72	3.79	1.94	21.60	—					
Muc	1.83	5.23	8.35	7.25	4.80	5.44	9.37	5.21	—				
Pen	6.36	0.69	7.64	8.11	4.28	0.73	30.3	0.61	12.49	—			
Rhi	1.46	4.06	6.33	6.71	5.82	6.19	18.30	2.74	1.14	5.97	—		
Sco	8.21	10.60	2.04	0.91	11.40	6.92	2.54	5.36	7.69	16.30	7.12	—	
Tri	2.47	6.91	7.22	8.08	6.53	5.39	7.17	3.79	4.04	6.72	3.18	7.48	—

Table 10. Relative crowding coefficients (RCC) of bee colony fungi when applied in equal amounts (30°C, bee bread-supplemented nonnutritive agar). Fungus A has the advantage when RCC > 1 and Fungus B has the advantage when RCC < 1. Data for reciprocal pairings (indicated by Table designations where data can be found) yield same conclusion. Abbreviations of fungi as per Table 5.
— Not determined (conspecific).

and *Trichoderma* sp.; *Fusarium* sp. with *Bipolaris* sp. and *Alternaria* sp.; *Absidia* sp. with *Trichoderma* sp.; *A. apis* with *A. flavus*; and *Scopulariopsis* sp. with *A. apis*, *Cladosporium* sp. and *A. niger*. This reflects their similarity in radial growth rate (Table 5). All other combinations of the 13 fungi tested resembled plots depicted by *Rhizopus* sp./*Penicillium* sp. (Figure 2). In all cases, $R \leq 0.64$ for these fungus pairings, and all were characterized by a large disparity of growth rate between both fungi in the pair. Corresponding relative crowding coefficient (RCC) values for fungus pairings are presented in Table 10 and, in all cases, reflect the fungus with the higher growth rate (Table 5) and greater ability to colonize with exception of *Penicillium* sp. > *Fusarium* sp; *Cladosporium* sp. > *A. flavus*; and *A. flavus* >*Alternaria* sp., *Scopulariopsis* sp., and *A. apis*.

Discussion

Field Analysis

In spite of differences in locality of origin, pollen source (almond in California, mixed in Arizona), level of fungicide exposure in the habitat, and positive symptoms of chalkbrood, bee bread samples display a relatively standard qualitative mycoflora dominated by *Aspergillus* spp. (*A. niger* > *A. flavus*), unidentified *Aspergillus* spp., *Penicillium* spp., *Rhizopus* spp., and *Cladosporium* spp. as the major isolates. Other fungal components contribute subtle variations, as has been reported on isolated colonies (Gilliam et al., 1989; Osintseva and Chekryga, 2008). Noteworthy among this group of fungi are *Aspergillus* spp. and *Pencillium* spp. that produce inhibitory compounds that limit the growth of chalkbrood, *A. apis* (Gilliam et al., 1988).

In our survey, two of the bee colonies (BP31 and KG28) displayed symptoms of chalkbrood and both had low levels of *Aspergillus* spp. and *Pencillium* spp. We suspect that other colonies displaying reduced levels of these inhibitory molds, namely the bulk of the California colonies, are at elevated risk for developing this disease due to the absence of inhibition provided by beneficial fungal associations. It is important to note that no *A. apis* was recovered from bee bread from these colonies, most likely because we used aerobic methods of culturing rather than anaerobic that is required for germination of *A. apis* spores (Gilliam et al., 1997). Another noteworthy fungus that was recovered in our census, occurring in high frequency and abundance, was *A. flavus*, agent of stonebrood disease, but none of the 21 colonies displayed overt symptoms of this disease, suggesting that *A. flavus* exists, probably like *A. apis* (Gilliam et al., 1988) as an opportunistic pathogen requiring a specific trigger to switch from saprobe to parasite.

We also emphasize that none of the bee bread samples showed evidence of spoiling or were spoiled, which occasionally can occur and thus reduce available food to bees. Fungi that can be responsible for this are *Fusarium* spp., *Penicillium* spp., *Rhizopus* spp., and *Aspergillus* spp. (Batra et al., 1973), but they typically fail to pose a threat to the colony because of behavioral and immunoregulatory mechanisms (Gliñski and Buczek, 2003). Taken together, evidence indicates that a heavy, diverse fungus load can be supported by bee bread without any noticeable detrimental effects or changes in bee bread fermentation.

The consistency in mycoflora and perpetuation of a specific subset of fungal components (*Aspergillus* spp., *Penicillium* spp., *Rhizopus* spp., *Cladosporium* spp.) reflects the fact that the bee colony is a consistent, stable, high-temperature, moisture-rich environment and that bee bread creates an ideal environment (30°C to 35°C, 60% to 80% RH; Chiesa et al., 1988) that favors growth and proliferation of a small number of mesophilic taxa, particularly anamorphs. Thus, it is not surprising that these are the dominant fungi in bee bread, especially when samples from different regions are compared. None of the fungi that were cultured are unusual.

Under our culture conditions, many more fungi are likely present in bee bread because our methods were not tailored for their recovery (Gilliam, 1979; Gilliam et al., 1989). The fungi that were recovered are easily recognizable soil saprobes that produce copious amounts of spores. None of them are fastidious, not selective for Potato Dextrose agar or modified Melin-Norkrans agar, which reflects their ubiquitous nature. During the course of this study, all fungi remained in their anamorphic state for 1 to 3 months after recovery even though some are known to have teleomorphs (Jennings and Lysek, 1999). Different bee bread samples from the same bee colony displayed nearly the same mycoflora profile, an indicator of widespread distribution and infiltration of conidia within the colony and spread by bee activity and sporulation by the fungi themselves. None of the fungi in bee bread showed a particular preference for location within the packed pollen in the comb cell, thus the fungi did not stratify according to their oxygen requirements; the recoverable fungi in bee bread are exposed to each other, equally competing for resources. A specific select group of fungi (*Aspergillus* spp., *Penicillium* spp., *Rhizopus* spp., *Cladosporium* spp.) is perpetuated as the end result. Thus, mycoflora composition of bee bread is expected to be proportionally similar wherever bee colonies are found as a product of an unchanging bee colony environment.

	Boscalid	Pyraclostrobin	Chlorothalonil	Cyprodinil	Fenbuconazole	Iprodione	Pyrimethanil
Organic	N.D.	15	N.D.	23	21	3290	76
Organic	N.D.	89	15	N.D.	28	25800	84
Direct	N.D.	N.D.	3	3690	17.9	799	10
Direct	N.D.	21	10	2770	142	3520	15
Direct	871	356	70	339	95	2810	6
Direct	2800	1130	70	245	38.7	1270	3
Non-target	N.D.	40	47	4230	47.9	1820	204
Non-target	N.D.	N.D.	2	124	1250	31	2
Non-target	N.D.	9	N.D.	17.9	97	4850	2
Totals	3671	1660	218	11439	1738	44472	403

Table 11. Amount of fungicides found in almond pollen (in ppb) and subsequently the bee bread, collected Feb–March, 2009, from the California orchards (unpublished data D. Sammataro and R. Simonds).

Differences between sites were in *Alternaria* spp., *Colletotrichum* spp., *Scopulariopsis* spp., and *Stigmella* spp. that were unique to bee bread from California colonies, and *Aureobasidium* spp., *Bipolaris* spp., *Fusarium* spp., and *Trichoderma* spp. (mycoparasitic) that were unique to Arizona colonies; however, all were minor components. Fungi as minor components that were in common between the Arizona and California samples were *Mucor* spp., *Paecilomyces* spp., and *Mycelia sterilia*. Frequent occurrence of *M. sterilia* in bee bread is not surprising considering that sterility is common in unchanging environmental conditions and darkness (Chapman, 1993), which are precisely the conditions that prevail in the bee colony environment.

Quantitatively, bee bread samples from California featured a reduced number of isolates of major fungal components (*A. niger, A. flavus, Aspergillus* spp., *Penicillium* spp., *Cladosporium* spp., *Rhizopus* spp.) than samples from Arizona colonies. This is corroborated by results of our enumeration study where the total number of conidia in bee bread from California was nearly 3- to 5-fold less than number of conidia in Arizona bee bread. The quantitative difference that we note in amount of fungi between California (low fungal load) and Arizona (high fungal load) necessarily implies differences in the amount of conidia being brought via pollen into the bee colony.

The effects of fungicides on reducing growth and proliferation of fungi result in the production of fewer conidia, their primary mode of spread (Jennings and Lysek, 1999). Thus, fungal levels are greatly reduced in bee bread because the amount of conidia in a treated habitat is appreciably less; that is, bees bring pollen back to the colony that is coated with fewer conidia in a fungicide-sprayed habitat. The source of the reduced number of conidia in California samples of bee bread compared to the Arizona samples is unknown. Because fungal loads were similar in quantity between bee bread from orchards that were directly sprayed with fungicide (even near the organic orchard), it seems reasonable to suggest that the lower amount of fungi in the California samples is due to lower amounts of fungi (hence, conidia) in the habitat as a result of general fungicide spraying in all the orchards within the sampling locations.

Kubik et al. (1999, 2000) and Škerl et al. (2009) have shown that fungicide contamination in bee colonies via pollen in treated habitats can occur, and fungicides can also be transported into the colony directly on the bees themselves (Charlton and Jones, 2007). We speculate that bee bread from the almond orchards shows low levels of fungi (reflected in the number of isolates and conidia) due to decreased levels of fungi in the environment where the bees were foraging. Therefore, the bee bread mycoflora are a direct reflection of the habitats that bees visit and the pollen brought back to the colony that is coated with fewer conidia as a result of its fungicide-sprayed habitat. If the reduced fungus load in bee bread from colonies in the organic orchard is attributed to fungicide contamination, then we would anticipate corresponding high levels of fungicide on pollen collected in pollen traps and subsequently processed bee bread. Indeed, analysis shows detectable quantities of BP, cyprodinil-containing fungicide, and iprodione-containing fungicide in the pollen (see Table 11). Qualitative changes to the bee bread mycoflora included a pronounced decrease in *Aspergillus* spp. and *Pencillium* spp., presumably making such colonies more prone to chalkbrood (Table 1).

In Vitro Analysis

Ecologically, fungi cannot distinguish between species and recognize any mycelium, including a conspecific, as a challenger. As vigorously growing hyphae from different fungi compete with one another for nutrients, slower growth rates and increasingly fewer conidia result—largely as an outcome of depleting resources. Eventually, the more antagonistic fungus overgrows the weaker ones, leading to their eventual downfall through metabolic breakdown. The end result is a substrate that is completely exploited by the aggressor. The substrate becomes fixed as an ecological niche, rendering it unavailable to invasion by other fungi. Priority for colonization is given to the pioneering fungus that arrives, proliferates, and establishes on an unoccupied substrate first, either directly (hoarding resources, nutrients, moisture, or oxygen) or indirectly (chemical mediation by production of inhibitory antifungal substances). These factors contribute to fixing an ecological niche and act as territorial defense (Scardaci and Webster, 1981; Klepzig and Wilkens, 1997; Jennings and Lysek, 1999; Benoit et al., 2004). Since the fungus that secures a substrate first is the one that predominates, a key survival element for fungi is the ability to capture the substrate. This characteristic favors more aggressive growers (Scardaci and Webster, 1981; Klepzig and Wilkens, 1997; Jennings and Lysek, 1999; Benoit et al., 2004). As such, there is a special advantage for those fungi that have greater exploitative ability to use a substrate that is already inhabited by another fungus, demonstrating enhanced and uniquely modified capacity for overgrowth and dominance over an existing fungus.

The fungi in bee bread capable of exploiting substrates already inhabited by other fungi are *A. flavus*, *A. niger*, *Mucor* sp., *Rhizopus* sp., and *Trichoderma* sp. Their ability to capitalize on inhabited substrate by secondary resource capture is further enhanced by attributes of high radial growth rate and accelerated conidia production. Efficiency of rapid colonization as revealed by RCC values include the production of growth-inhibitory antifungals to suppress the competition. Despite lacking the capacity for secondary resource capture by *Penicillium* sp. and *Fusarium* sp., *Fusarium* sp. also produces antifungal against *Penicillium* sp.

Pencillium sp. likely prevails over *Fusarium* sp. by having twice the sporing capacity (2,649 versus 1,488 conidia/1 cm^2, *Pencillium* sp. and *Fusarium* sp., respectively) and more rapid colonizing ability (RCC > 1 in *Fusarium* sp./*Penicillium* sp. pairs when *Fusarium* is designated as fungus A; refer to RCC equation on page 199). Thus, the balance of the interactions in bee bread favors dominance of *Rhizopus* sp., *Mucor* sp., *Trichoderma* sp., *A. niger*, *A. flavus,* and *Penicillium* sp., which all exhibit documented antifungal activity against *A. apis* (Gilliam et al., 1988; Al-Ghamdi et al., 2004) and the capacity for overgrowth of an existing *A. apis* mycelium (this study). *A. apis* offers little resistance to takeover by fungi that are selectively propagated through the competitive interaction of bee bread fungi. Although both are favored, *A. niger* is especially antagonistic toward *A. flavus* (this study; Phillips et al., 1979), as is *Cladosporium* sp. This suggests an intervention against stonebrood, perhaps preventing the switch from saprobe to parasite by interactions of bee bread fungi. Many of the fungi are unculturable by our methods (Gilliam et al., 1989), leaving potential fungal interactions unresolved. Thus, whether *A. flavus*, *A. niger*, *Mucor* sp., *Rhizopus* sp., and *Trichoderma* sp. play a critical role in the conversion of bee bread from pollen is speculative, but cannot be ruled out.

Of the 13 bee bread fungi investigated, there is evidence of reduced growth (prior to hyphal contact) of *A. apis* and other fungal species as a result of the production of growth inhibitory substances by *A. flavus*, *A. niger*, *Fusarium* sp., *Mucor* sp., *Penicillium* sp., *Rhizopus* sp., and *Trichoderma* sp. (Gilliam et al., 1989; Al-Ghamdi et al., 2004). The antifungal substances are apparently not broad in spectrum, as not all of the 13 fungi were negatively affected. The extent to which the radial growth rate was suppressed was independent of the fungus that produced the antifungal substance and whether the target fungus was a fast, moderate, or slow grower. Antifungal compounds produced by this select group merely suppressed fungal growth, not stopping it, as evidenced by varying amounts of conidia output reduction. There was no evidence that a given fungus species produced an antifungal against itself, which is common for antifungal substances (Jennings and Lysek, 1999). Lower radial growth for conspecifics was observed only after hyphal contact (not before) and is the result of resource hoarding as with any fungal-to-fungal contact (Benoit et al., 2004).

Our results show that *A. apis* was the fungus that was the most vulnerable of the bee colony fungi in that:

1. Most of the bee colony fungi produced growth inhibitory substances against *A. apis*, suppressing spread and conidial output, thus there is overlapping sensitivity to chemical mediation by a wide range of fungi.
2. The presence of other fungi suppressed radial growth (by aggressive *A. niger* and *Rhizopus* sp.) of *A. apis* by > 40% (the largest reduction in growth experienced by any of the fungi tested); and
3. A mycelium of *A. apis* was capable of being exploited and easily overgrown, with *A. apis* offering little resistance in the way of territorial defense once it establishes.

The most versatile fungus was *Trichoderma* sp. in that the presence of this fungus had controlling effects on nearly all of the 13 fungi. This agrees with its mycoparasitic (mycophagy) ecologic classification and rather broad spectrum capacity to deter the growth of most fungal taxa (Elad, 2000; Barbosa et al., 2001; Roco and Pérez, 2001; Al-Ghamdi et al., 2004; Mónaco et al., 2004; Abdel-Fattah et al., 2007).

From the time of pollen collection and packing, the composition of the mycoflora in bee bread changes as it ages. Shifts in makeup of the mycoflora, quality, and quantity reflect the interactions *in vitro* of individual fungal components in bee bread as a competitive environment. *Pencillium* spp., *Aspergillus* spp., and *Rhizopus* spp. are the most dominant fungi isolated in completely processed bee bread (Gilliam et al., 1989; Gilliam and Vandenberg, 1997; Osintseva and Chekryga, 2008). Other minor components of bee bread (e.g., *Absidia* sp., *Bipolaris* sp., *Fusarium* sp., *Scopulariopsis* sp., *Trichoderma* sp., *Alternaria* sp.) are typically lost as the bee bread ages.

Although rare, *Fusarium* sp., *Penicillium* sp., *Rhizopus* sp., and *Aspergillus* sp. have been known to overgrow and negatively impact the hive by spoiling the bee bread and lead to natural population declines due to a reduced food supply (Batra et al., 1973). Fungal growth is restricted to the surface area of the substrate (Jennings and Lysek, 1999), thus the growth of the fungi in bee bread increases only to a level that the pollen's surface area will allow. This study shows reduced radial growth rate and conidial output when any two fungi interact (including conspecific fungi against each another), suggesting that controlling fungal blooms to reduce spoilage and preserve bee bread is facilitated in that the fungi in bee bread are self-regulated and limited by the amount of pollen that is packed into the cell. Nearly stable temperatures (around 30°C), the actions of other microorganisms, and the honey covering over the bee bread may also be important in suppressing fungal levels and preventing spoilage of bee bread (Sammataro, 1998). We conclude that the overall quantity of fungi in bee bread is internally regulated and select fungi that appear to be important for controlling chalkbrood and stonebrood are perpetuated qualitatively. Redirections in fungal interactions as a result of removal/lowering levels (via differential growth suppression) of key fungal competitors by frequently applied compounds (miticides, high fructose corn syrup, fungicides; Yoder et al., 2008; Alarcón et al., 2009; *Trichoderma* sp. as a biological control agent; Al-Ghamdi et al., 2004) are anticipated to alter food quality and weaken the level of natural mycoprophylactic protection that is generated via close fungal interactions. Though *A. apis* and *A. flavus* are ubiquitous in honey bee colonies, the symptoms of chalkbrood and stonebrood disease, respectively, are not always seen. *Aspergillus niger*, *Rhizopus* spp., and *Mucor* spp. isolated from bee bread produce growth-inhibitory substances against *A. apis* (Gilliam et al., 1988). These symbiotic microbes present in bee bread are thus instrumental in controlling the pathogenic species by providing direct, or indirect, negative effects on mold growth.

General Interpretations

Fungicides sprayed in the environment are reducing the overall amount of fungi that are present in bee bread, which may have a negative impact on the natural resistance mechanisms in the bee colony. The ability of the colony to fight against opportunistic chalkbrood and stonebrood depends in a large part on the presence of *Aspergillus* spp. and *Penicillium* spp., which act as natural regulators. As such, these fungal diseases are a pathogenic consequence of a lowering of fungal concentration in bee bread. Exposure to fungicides brought in by the bees via fungicide-contaminated pollen, or collecting fewer fungi because of decreased levels of conidia in the environment, has the effect of decreasing the amount of *Aspergillus* spp. and *Penicillium* spp. Any colony that is within the bee flight range of a sprayed orchard is vulnerable to the effects of these fungicides.

We anticipate that identifying colonies at risk for developing chalkbrood can be accomplished by testing for the levels of inhibitory molds (*Penicillium* spp. and *Aspergillus* spp.) and using bee bread as a screening tool. Not only do fungicides have suppressive effects on fungi, but miticides and high fructose corn syrup feeding similarly have negative effects on bee colony fungal growth, particularly *Aspergillus* spp. and *Penicillium* spp. (Yoder et al., 2008, 2011). Other supplements, such as antibiotics, could also have a negative effect (Yoder et al., 2011, Chapter 16, this edition). Consistent with the defense function that is conveyed by these fungi, bee colonies that have a history of repeated and excessive miticide (formic acid, oxalic acid) and high fructose corn syrup use typically show a high incidence of chalkbrood. How this effect of lowering a colony's defense shield contributes to Colony Collapse Disorder is unknown, but likely represents a facet of the overall disorder. Our evidence only shows a causal link between use of fungicide, miticides and high fructose corn syrup, and a higher frequency of chalkbrood in such colonies exposed to these compounds. The close interplay and regulatory activity of bee bread fungi as revealed by our *in vitro* competition experiments offers another intriguing hypothesis about consequences of altering fungal dynamics within the colony. The enhanced capacity to restrict or kill the bulk of bee colony fungi by *Trichoderma* sp., including the beneficial inhibitory molds *Aspergillus* sp. and *Penicillium* sp., suggests that *Trichoderma* sp. as a biological control agent against chalkbrood (Al-Ghamdi et al., 2004) and stonebrood. It should also be applied carefully so as not to disrupt the balance of all components in the bee bread mycoflora.

Interactions between Risk Factors in Honey Bees

CHAPTER

18

Yves Le Conte, Jean-Luc Brunet, Cynthia McDonnell,
Claudia Dussaubat, and Cédric Alaux

Abstract In recent years, a noticeable decline in managed honey bee populations has been observed. Consequently, research into plausible causative factors has been conducted, identifying several detriments to colonies, although no single factor has been shown to be the sole cause of the decline. We hypothesize that the decline in the honey bee population may be due to interactions between multiple detrimental factors, both environmental and human-induced. In this chapter, these possible combinations of factors are described in depth and suggestions for their resolution are made.

Acknowledgments C. Alaux was supported by an INRA young researcher position (INRA SPE department); C. McDonnell by a BEEDOC grant (FP7, RTD REG/E.4(2009) D/5612 21); and C. Dussaubat by a CONICYT/French Embassy of Chile grant.

Introduction

Over the past decades, a significant decline in the number of managed honey bee (*Apis mellifera*) colonies has been observed in both the United States (Ellis et al., 2010) and Europe (Potts et al., 2010). One of the most striking examples of such decline is the case of Colony Collapse Disorder (CCD), characterized by a sudden loss of workers from the colony (Cox-Foster et al., 2007). Although such extensive losses are not unique and have been historically documented (reviewed by Neumann and Carreck, 2010), this most recent honey bee decline served as a bellwether, captivating the attention of both the general public and scientists. As a consequence, a thorough search for the causative factors of colony losses was conducted that identified an extensive list of detrimental factors, including pathogens, parasites, and pesticides (Cox-Foster et al., 2007; Johnson et al., 2009a).

However, despite honey bees being challenged by many biotic and abiotic factors, no single factor has been shown to be the sole cause of CCD (Oldroyd, 2007; Stokstad, 2007; Anderson and East, 2008), suggesting that honey bees are not endangered by a monocausal syndrome but rather by a combination of multiple factors (Moritz et al., 2010). This last hypothesis is well-illustrated by a recent epizootiological survey of CCD in affected and nonaffected colonies (vanEngelsdorp et al., 2009). The authors of the survey quantified and compared 61 variables, including bee physiology, pathogen loads, and pesticides levels and concluded that "None of these measures on its own could distinguish CCD from control colonies" (vanEngelsdorp et al., 2009). Therefore, studies analyzing the interactions between risk factors are currently emerging as a new way of investigating honey bee mortality (Moritz et al., 2010).

The underlying hypothesis is that a single factor might not be harmful to the colony but that in combination with a second factor, its negative impact or the impact of both could be enhanced. For example, healthy colonies might not suffer from exposure to pesticides but the chemicals could become rapidly lethal if the colonies are already weakened by disease (Thompson, 2003). Indeed, there is some evidence that pesticides can weaken the insect immune system (Delpuech et al., 1996; George and Ambrose, 2004). Several others factors, mostly involving human impact on landscape and beekeeping practice (Oldroyd, 2007), are suspected to make colonies more vulnerable on a long-term basis to others stressors. Over the years, honey bee colonies have been threatened by pesticides or a large variety of infectious agents and parasites. However, nowadays it might be more and more difficult for bees to cope with those stress factors due to the current colony weakening. Therefore, analyzing the multiple interactions among risk factors is crucial for better understanding current honey bee losses and developing sustainable strategies for their protection. In this chapter, we discuss the main combinations of risk factors that might impact bee health and review the recent progress that has been made in this direction.

Interactions between Pesticides

Since the development of modern agriculture, pesticides have long been suspected to be a major factor contributing to honey bee declines. Indeed, honey bees are without doubt in contact with pesticides including systemic or systemic-like molecules used to treat crops via spraying, soil granules, or seed dressing. Those chemicals are recovered in pollen and nectar at low concentrations (Bonmatin et al., 2003; Laurent and Rathahao, 2003; Charvet et al., 2004; Bonmatin et al., 2005; Elbert et al., 2008) and brought back to the hive. Therefore, many pesticides, including their metabolites, and miticides originating respectively from agricultural or beekeeping practices have been detected in adult honey bees and apicultural matrices such as brood, wax, honey, pollen, and beebread (Bogdanov, 2006; Chauzat et al., 2009; vanEngelsdorp et al., 2009; Johnson et al., 2010; Mullin et al., 2010). Recently, unprecedented levels and types of miticides and agricultural pesticides (121 different pesticides and metabolites) coexisting in different combinations, were detected in honey bee colonies from the United States and one Canadian province (Mullin et al., 2010). Likewise in Europe, pesticide residues were found in different apicultural matrices but at lower levels (Wallner, 1999; Chauzat et al., 2009).

The recent finding that compared to others insects, the honey bee genome is deficient in genes coding for detoxification enzymes (cytochrome P450s, glutathione-S-transferases, and carboxylesterases) (Claudianos et al., 2006), combined with the large variety of pesticide residues found in the hives, suggests that honey bee health is endangered by those chemicals. However, most of the studies aiming at characterizing the impact of pesticides on honey bee health were performed on only one chemical at a time (see Chapter 14 on pesticides for a review, this edition). Until now, besides sublethal effects, no single pesticide has been associated with colony collapse. On the contrary, a recent review of toxicological studies found that, despite the lack of genes coding for detoxification enzymes, honey bees are in general not more sensitive to, for example, insecticides (Hardstone and Scott, 2010). Now, however, the combined effects of multiple pesticides are suspected to significantly affect the bees. Indeed, the high frequency of pesticides in the hive environment with different modes of action represents an increased potential for compound interactions.

However, pesticide combinations pose major challenges because of the lack of data on many individual pesticides, pathways of chemicals interactions, and pertinent experimental design due to large numbers of molecules (Lydy et al., 2004; Monosson, 2005). In addition, different types of interactions can be considered. *Additivity* is observed when the effect of the combination of two or more chemicals is the sum of expected individual responses, whereas *synergism* is observed when the effect of the combination is higher than that of the sum of expected individual responses. Indeed, synergism occurs when chemicals interact in a way that generally enhances or magnifies the effect of each other. The worst case of pesticide interactions is *potentiation*, where a nontoxic molecule enhances the toxicity of another one, or when two nontoxic molecules induce toxicological effects. Finally, *antagonism* is observed when the combination of two or more chemicals decreases the individual responses induced by a chemical.

Despite the scarcity of studies performed on chemical interaction, there is clear evidence of chemical synergy in honey bees between pyrethroid insecticides and azole fungicides that inhibit ergosterol biosynthesis (EBI). The first studies of such an interaction demonstrated that sublethal exposures to deltamethrin (pyrethroid) and prochloraz (EBI fungicides) lead to a synergistic increase in bee mortality (Colin and Belzunces, 1992; Belzunces et al., 1993). Afterward, those results were extended to a large variety of EBI fungicides in combination with a pyrethroid (Pilling and Jepson, 1993; Pilling et al., 1995). For example, the pyrethroid combined with EBI fungicides can enhance its toxicity by a ratio ranging from 366- to 1,786-fold. Synergistic effects are not limited to mortality, but also to bee physiology, such as thermoregulation, which can also be affected (Vandame and Belzunces, 1998). In addition, seasonal variation should be taken into account in future studies, as summer bees are more susceptible to the synergistic action of prochloraz and deltamethrin than winter bees (Meled et al., 1998).

Others insecticides, like nicotinoids (acetamiprid and thiacloprid), can act jointly with fungicides on honey bee toxicity (Iwasa et al., 2004). However, fungicides do not elicit toxicity of numerous nicotinoids and their metabolites (Schmuck et al., 2003; Iwasa et al., 2004). The origin of these synergies observed between fungicides and pesticides relies mainly on the fungicide activity. Indeed, EBI fungicides disrupt ergosterol biosynthesis via the inhibition of cytochromes P450 involved in detoxification (Brattsten et al., 1994). Thus, fungicides decrease the capacity of the organism to detoxify others chemicals, which might explain their synergy with pyrethroids or some nicotinoids (Belzunces et al., 1993; Brattsten et al., 1994; Iwasa et al., 2004). Similar interaction at the level of P450 detoxification has been found between miticides and pyrethroids, and between miticides (*tau*-fluvalinate and coumaphos) (Johnson et al., 2006; Johnson et al., 2009b).

During each foraging trip, honey bees are usually exposed to only a few pesticides at a time. But when they bring back the contaminated food, each chemical is stored in the different matrices of the hive, creating a complex of pesticides and metabolites that will interact with beekeeping miticides. Honey bees are unintentionally concentrating the different pesticides from the environment into a single spot (the hive), which could rapidly become lethal for them. That is why synergistic interactions needs to be further studied.

Pathogen/Pesticide Interactions

In modern agriculture, the combination of pathogens and pesticides has been successfully used to manage insect crop pests, as part of an Integrated Pest Management (IPM) approach (Maredia et al., 2003). IPM can be the tandem applications of an infectious organism and insecticide that together significantly enhance the lethality of each control agent through synergistic interactions. Because this method is highly efficient in killing insect pests, it is easy to imagine that beneficial insects, like pollinators, could also fall victim to those pathogen/pesticide combinations. Early work by Ladas (1972) showed that honey bees artificially infected with the parasitic microsporidia *Nosema apis* were more susceptible to some organochlorine and organophosphate pesticides, suggesting a link between both factors and honey bee survival.

Currently, the combination of imidacloprid (neonicotinoid) and fungal spores is being successfully employed for controlling numerous insect pests (like cockroaches, termites, burrowing bugs, caterpillars, and leaf-cutter ants) (Kaakeh et al., 1997; Ramakrishnan et al., 1999; Jaramillo et al., 2005; Purwar and Sachan, 2006; Santos et al., 2007). Interestingly, a similar combination of imidacloprid and the pathogen Nosema can also weaken honey bee health (Alaux et al., 2010a). In combination with sublethal doses of imidacloprid (encountered in the environment), the honey bee pathogen Nosema caused a significantly higher rate of individual mortality than either agent alone (Alaux et al., 2010a). In addition, while the single or combined treatments showed no effect on individual immunity, a measure of colony-level immunity, namely glucose oxidase activity, was significantly decreased only by the combined treatments, emphasizing their synergistic effects. Glucose oxidase activity enables bees to secrete antiseptics in honey and brood food. This suggests a higher susceptibility to pathogens over the long-term duration of the hive.

Until now, no relationship between colony mortality and pesticide residues has been found either in the United States or in Europe (Chauzat et al., 2009; vanEngelsdorp et al., 2009). However, each of those pesticides found at sublethal doses in the hive (Bogdanov, 2006; Chauzat et al., 2009; vanEngelsdorp et al., 2009; Johnson et al., 2010; Mullin et al., 2010) represents a potential candidate that could interact with any of the wide range of pathogens commonly infecting bees. Two scenarios could be envisaged regarding their interactions. On one hand, pesticides could make bees more susceptible to pathogens due to their potential impact on insect immune function (Delpuech et al., 1996; George and Ambrose, 2004). On the other hand, pathogens could enhance the toxicity of pesticides. As an example, even though imidacloprid is usually encountered at sublethal doses in the hive, higher consumption of contaminated food caused by Nosema infection can expose the bees to lethal doses of the pesticide (Alaux et al., 2010a). Indeed, Nosema infection induces an energetic stress in bees and thus a higher food demand (Mayack and Naug, 2009; Alaux et al., 2010a).

Although there is a large body of evidence that hives are contaminated and infested by a collection of pesticides and pathogens, studies looking at their potential interactions are largely missing. Such studies would bridge the gap between research on the intentional control of agricultural insect pests and the unintended consequences for beneficial insects.

Pathogen/Pathogen Interaction

From virus to parasites, many pathogens have been described that act on honey bee health. As pathogens continue to be discovered in the honey bee, the most recent being multiple viruses (Bromenshenk et al., 2010; Cox-Foster et al., 2007) and a fungus (Fries, 2010), we cannot exclude the possibility that a single unidentified pathogenic agent could be the major cause of colony losses worldwide. Interactions between different pathogens can be dramatic, with one pathogen favoring the development and the virulence of the other, and are therefore believed to be involved in colony losses. Many parasites and pathogens have been described in honey bee colonies, creating a multitude of potential interactions to be studied. In general, interactions between pathogens have been described at two different levels: the *transmission level* with a primary pathogen (mites) having the potential to transmit and disperse secondary pathogens (viruses, fungi) and the *development level* with a pathogen enhancing the development and negative impact of a second pathogen, or both pathogens promoting the development of each other.

When we look at the transmission level, a number of viruses, including deformed wing virus (DWV), acute bee paralysis virus (ABPV), chronic bee paralysis virus (CBPV), slow bee paralysis virus (SPV), black queen cell virus (BQCV), Kashmir bee virus (KBV), cloudy wing virus (CWV), and sacbrood virus (SBV), have been shown to be associated to varying degrees with Varroa mite (*Varroa destructor*) infestation (see Le Conte et al., 2010; Rosenkranz et al., 2010, for reviews). *V. destructor* is also implicated as the vector of stonebrood disease in honey bees, for which, *Aspergillus flavus* is the fungal agent (Benoit et al., 2004). The transmission of viruses is not restricted to the Varroa mite, as recently the ectoparasitic mite *Tropilaelaps mercedesae* (Acari, Laelapidae) has been described as a novel vector of honey bee viruses, in particular, DWV in European honey bees (Dainat et al., 2009; Forsgren et al., 2009).

Interactions at the development level can occur as a result of synergetic effects between two pathogens. Downey et al. (2000) demonstrated that the two parasites, *V. destructor* and *Acarapis woodi*, have biologically synergistic interactions at the individual and colony levels that are detrimental to their host colonies (Downey et al., 2000; Downey and Winston, 2001). *V. destructor* weakens the honey bee immune system, which might trigger viral multiplication, thus leading to honey bee death (see Le Conte et al., 2010; Rozenkranz et al., 2010 for review). The fungus *Nosema* ssp. is also implicated in synergetic effects with other pathogens. For example, *N. apis*, which infects the epithelium of the honey bee midgut, can increase susceptibility of the alimentary tract to infection by BQCV (Bailey et al., 1983). Recently, an interaction between *Nosema* and an iridovirus has been reported to be involved in colony losses; however, the effects were only additive (Bromenshenk et al., 2010).

Finally, while just a handful of studies have reported the detrimental interaction between two pathogenic agents, further studies are needed on the effects of two or more agents in the hive, as well as on the way in which developing pathogens compete within the hive. As colony losses have appeared recently worldwide, one way to proceed would be to focus either on interactions between newly emergent pathogens, recently described or introduced, like *Nosema ceranae*, Israeli acute paralysis virus (IAPV), and other viruses, or between newly emergent pathogens and those already established in the colonies.

Genetic Diversity/Other Risk Factors

Within a colony, genetic diversity, provided by a multiply mated queen, confers increased fitness to the hive as well as increased tolerance of pathogens (Tarpy, 2003; Jones et al., 2004; Mattila and Seeley, 2007; Seeley and Tarpy, 2007; Mattila et al., 2008). Thus, maintaining genetic diversity within a managed colony may replicate this natural propensity of the queen to hedge her bets against the range of pathogens or parasites that may befall the colony throughout the years, by increasing the number of patrilines within her hive. Genetically diverse colonies show increased fitness and productivity over genetically uniform colonies, as evidenced by higher rates of swarming success as well as increased foraging rates, food storage, and drone production (Mattila and Seeley, 2007). In addition, genetically diverse colonies have better thermoregulation (Jones et al., 2004) and increased foraging-related communication that signal food discoveries farther from the nest (Mattila et al., 2008).

Genetically diverse colonies also better withstand the pathogens that cause American foulbrood (*Paenibacillus larvae*) (Seeley and Tarpy, 2007) and chalkbrood (*Ascophaera apis*) (Tarpy, 2003), and show different infection rates by American foulbrood between patrilines within the colony. This might be explained by allelic variation in immune gene expression between different patrilines (Evans, 2004). The higher the genetic variation (number of patrilines), the greater the chance that one or more of the patrilines, and therefore the colony, will be able to defend against pathogen challenges. Lower genetic variation in the hive would diminish the chance for natural resistance to abiotic or biotic stressors. Therefore, the health of honey bees cannot be fully understood without considering the effect of genetic diversity on the response to environmental challenges. Genetic diversity within the colony and within the population might determine whether any stress factors will simply stress a colony or lead to its demise. In a challenge-free environment, colonies with a low or high genetic diversity might both survive, but the emergence of pathogens or parasites might trigger the collapse of the former over the latter.

Selective breeding using pure stock is a common practice in beekeeping, to improve apicultural aspects of the colony, such as honey production and quietness of the bees; but one of the drawbacks is a loss of biodiversity. For example, it has been shown that honey bees from a managed population in Europe have a reduced genetic diversity when compared to wild bee populations in Africa (Moritz et al., 2007; Jaffe et al., 2010). Given the benefits of increased genetic diversity within the hive, the question arises as to the current diversity of honey bees available in the wild, and if that diversity is sufficient to sustain a genetically diverse colony. In Europe, nearly the entire population of honey bees sampled in the wild is from managed colonies (Jaffe et al., 2010). Although, due to its wide range of climatic and vegetative zones, 10 of the 26 described ecotypes and subspecies are native to Europe. There is a need to conserve these ecotypes that have adapted to different climates in order to maintain genetic diversity (Meixner et al., 2010). In the United States, on the other hand, where the importation of honey bees is limited, genetic diversity depends on feral populations, derived from bees imported in the 19th century (Schiff and Sheppard, 1995). Feral populations of honey bees sampled from the central and southern central United States differ genotypically from bees found in managed conditions (Magnus and

Szalanski, 2010). There, feral bee populations that have survived parasites and pathogens without treatment may offer additional genetic variation to bee breeding efforts (Magnus and Szalanski, 2010). Yet feral bees, when brought into managed conditions, may not necessarily be more resistant to parasites and pathogens, as Seeley (2007) suggested with Varroa mite infection.

Genetic variability in the colony strengthens the hive overall, which may be the first (or best?) line of defense against a range of insults, including pathogens, parasites, and pesticides, because increasing genetic variability in the hive seems to increase its resilience in the face of attack (Tarpy, 2003). If feral bee populations cannot provide genetic variation, then other sources are needed because creating genetic diversity within the colony (i.e., genetically diverse drones mating with the queen) depends on the level of genetic variation in the population of bees.

Environmental Resources/Other Risk Factors

The development and maintenance of the honey bee colony is intimately associated with the floral resources from which they harvest their nutrients (pollen and nectar) (Haydak, 1970; Keller et al., 2005; Brodschneider and Crailsheim, 2010). Floral nectar stored as honey is the main source of carbohydrates, the energetic fuel of honey bee workers, and pollen provides most of the proteins, amino acid, and lipid requirements for their physiological development (Brodschneider and Crailsheim, 2010). Therefore, honey bee populations and beekeeping activities are tightly linked to the environmental resources on hand. However, a direct consequence of the current intensification of land-use and agriculture is the reduction or loss of areas constituting the foraging resources of honey bees (Decourtye et al., 2010). The relationship between the decline in biodiversity of wild bees and intensive agriculture (Kremen et al., 2002; Biesmeijer et al., 2006) suggests that honey bees might also be affected by this loss of floral resources. Supporting this hypothesis, beekeepers rank poor nutrition and starvation as two of the main reasons for bee losses (vanEngelsdorp et al., 2008).

If the immediate consequence of nutritional stress is the decrease in the colony population (Keller et al., 2005), then one of the long-term effects may be a deficiency in the physiology and health of individuals. In turn, this physiological deficiency could affect the threshold resistance of bees to others risk factors, leading to a synergistic effect between nutritional deficiency and pathogens or pesticides (Naug, 2009).

Recently, more studies are providing evidence of such interactions by showing that poor nutrition can diminish the resistance to abiotic or biotic stressors. For example, to develop and reproduce, parasites depend highly on the host's energy, which generally induces an energetic and thus nutritional stress in the host. This was found in honey bees parasitized with the microsporidia *Nosema ceranae*, which have a higher need of carbohydrates and at a higher hunger level (Mayack and Naug, 2009; Naug and Gibbs, 2009; Alaux et al., 2010a). The higher nutritional stress and mortality caused by Nosema parasitism could be reversed in the shortterm by *ad libitum* carbohydrate supplies (Mayack and Naug, 2009). Similarly, the quantity and quality of pollen provided to bees can affect their sensitivity to some pesticides, with a higher quantity or a high-quality pollen (protein content) reducing the toxicity of the pesticides (Wahl and Ulm, 1983). In addition, because a deficiency in nutrition, especially in proteins and amino acids, can impair immune function in animals (Field et al., 2002; Li et al., 2007), pollen nutrition is expected to affect the development of pathogens. This was confirmed by studies demonstrating that pollen nutrition can help the bees fight against parasites and pathogens, like the microsporidia *N. apis* (Rinderer and Elliott, 1977), the bacterial species *Paenibacillus larvae* (American foulbrood agent) (Rinderer et al., 1974), deformed wing virus (Degrandi-Hoffman et al., 2010), and the mite *V. destructor* in foragers (Janmaat and Winston, 2000). Later, it was found that dietary protein quantity within the pollen can enhance the immune function in honey bees (Alaux et al., 2010b). Low pollen diversity in the diet might also make honey bees less able to fight against pathogens or withstand different stress factors, as adults fed with a diet of one type of pollen end up with a weaker immune system compared to bees supplied with a more diverse diet of pollen (Alaux et al., 2010b). Notably, the production of glucose oxidase, involved in the production of antiseptics (hydrogen peroxide) in the brood food and honey, was higher with a polyfloral diet compared to the monofloral regime. Similar results were found in bumble bees, where larvae fed with a mix of pollen types were heavier than larvae fed with monofloral pollen of equivalent or higher protein content (Tasei and Aupinel, 2008).

Exposure to stress factors is costly and requires adequate nutrition in order to cope. However, with current agriculture practices, honey bees that usually pollinate a large variety of flowers, are forced to feed on single crops (monocultures) or are confronted with decreasing availability of resources. This malnutrition could chronically weaken honey bees, by decreasing their threshold of resistance to infectious disease and/or pesticide toxicity. In addition to local human activity, global climate change may also affect the plant phenology and distribution and thus environmental ressources of the bees (Le Conte and Navajas, 2008). Therefore, initiatives aimed at developing floral resources within the agro-environment are recommended for better honey bee health (Decourtye et al., 2010).

Conclusions

The role of interactions between factors is becoming a central paradigm for explaining current colony losses (Moritz et al., 2010), even though instances occur in which a single factor alone can lead to colony death [(e.g., *V. destructor* (Le Conte et al., 2010)]. In terms of their causal effects on honey bee losses, risk factors may interact additively or synergistically, thereby augmenting, in the case of a positive interaction, the individual effects of each other. Because risk factors can occur in the short term (e.g., environmental pesticides and pathogens) or the long-term (e.g., loss of genetic diversity and environmental resources) with regard to the lifetime of the bee population, future studies of interactions will need to integrate the differing timescales of individual factors. In this review we provided examples of known interactions involving the main honey bee risk factors (pesticides, pathogens, genetic diversity, and malnutrition), but due to the myriad possible combinations of chemicals and pathogens found in the hives, coordinated efforts among scientists are needed if we want to better characterize those interactions and develop sustainable strategies to improve honey bee health.

Understanding the Impact of Honey Bee Disorders on Crop Pollination

CHAPTER

19

Keith S. Delaplane

Abstract In spite of growing awareness of health and sustainability problems with bees, there is little scientific knowledge on the direct impacts of bee disorders on crop pollination. In the case of honey bees, a eusocial pollinator used widely in agriculture, disorders can affect pollination at any of three levels: (1) by impairing pollination efficacy of compromised individuals, (2) by impairing efficacy of compromised colonies, or (3) by reducing the availability of pollinators, that is, killing bee colonies. The available evidence suggests that honey bee disorders impact pollination through the simple agency of killing colonies; however, hypotheses should be framed and tested with different model bee disorders, different pollinator species, and different model plants. Answers to these questions will help identify appropriate research and response priorities. Should remedial action steer toward practices that maximize sheer colony numbers? Or should the emphasis rest on a more classical IPM approach in which within-colony pest levels and action thresholds play a more prominent role?

Introduction

The opening years of this century have seen an unprecedented degree of public inquiry into the health and welfare of the western honey bee, *Apis mellifera* L., a cosmopolitan species found on every verdant continent and considered a mainstay of honey production and crop pollination. There appear to be good reasons for the alarm, as sustained honey bee declines have been documented in North America (vanEngelsdorp et al., 2008, 2010) and Europe (Potts et al., 2010). The problem is not universal, as managed bee numbers remain stable in parts of Europe (Potts et al., 2010) and across regions as vast as Africa, Australia, and South America (Neumann and Carreck, 2010). But even when interpreted in the most optimistic light, the number of managed beehives is not keeping pace with the global demand for the pollination services these bees provide (Aizen and Harder, 2009). The reasons for the sluggish increase, or negative increase, of managed beehives are many and include causes social, economic, and organic. Entire issues of scientific journals have been dedicated to the matter (*Journal of Apicultural Research*, January, 2010; *Journal of Invertebrate Pathology*, January, 2010; *Apidologie*, June, 2010).

Honey bees elicit a range of human responses, and at the private level every stakeholder has his or her own reasons for caring about honey bee health and welfare. But in the public conversation, there is really only one reason on the table—pollination. Virtually every competitive grant proposal, every political solicitation, every news story—essentially every public appeal on behalf of the honey bee—invokes pollination as the *raison d'être* for public funds and goodwill. This is appropriate, for it is with pollination, not honey production, that honey bees most profoundly affect human well-being. This benefit goes beyond the mere dollars and cents contributed by bees to agricultural production (estimated at >$200 billion worldwide (Gallai et al., 2009) to the point that bees are determiners of the diversity and quality of human diets. Excluding crops that are passively self-pollinated, wind-pollinated, or parthenocarpic, Klein et al. (2007) showed that among all crops traded on the world market and directly consumed by humans, 13 are obligately pollinator-dependent, 30 are highly-dependent, 27 moderately, 21 slightly, 7 nondependent, and 9 unknown. To the extent that it is honey bees acting in these agricultural

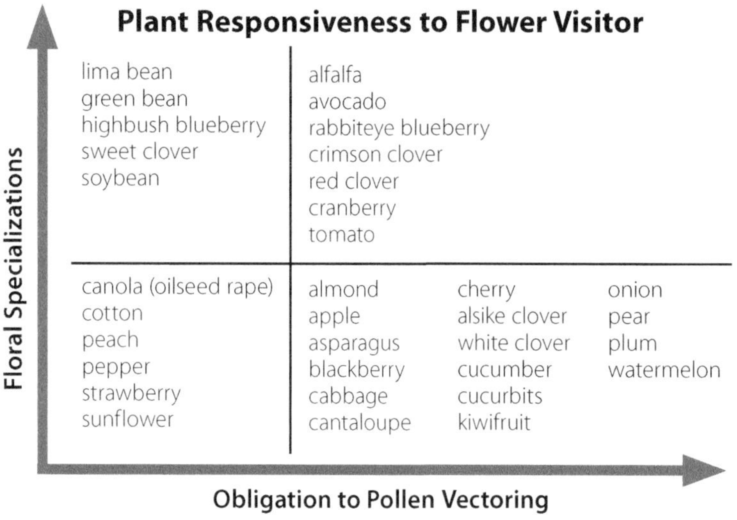
Plant Responsiveness to Flower Visitor
lima bean
green bean
highbush blueberry
sweet clover
soybean
alfalfa
avocado
rabbiteye blueberry
crimson clover
red clover
cranberry
tomato
canola (oilseed rape)
cotton
peach
pepper
strawberry
sunflower
almond
apple
asparagus
blackberry
cabbage
cantaloupe
cherry
alsike clover
white clover
cucumber
cucurbits
kiwifruit
onion
pear
plum
watermelon
Floral Specializations
Obligation to Pollen Vectoring

arenas—and owing to their manageability this is usually the case—honey bees are appropriate targets of public research investment. But the truth is, it's not so much honey bees that are the targets of research dollars, but rather honey bee disorders. Among the suspected causes of bee decline, the field is dominated by pathogens, parasites, and toxins. For proof, one need look no further than the contents of this book. It is therefore possible to reframe this conversation in new terms: to the extent that bee pathogens, parasites, and toxins impact pollination, they are appropriate targets of public research investment. It should come as a surprise, therefore, to learn that the direct effects of bee disorders on honey bee pollination is almost entirely untreated in the scientific literature.

Understanding a Pollinator's Impact

Before we discuss the direct effects of honey bee disorders on pollination, it is useful to consider some criteria by which we can appraise any flower visitor's impact on pollination. Foremost, the reproductive impact of a flower visitor is contextual in terms of the plant's reproductive responsiveness to flower visitors and the pollinating efficacy of the visitor. I propose the two-part model (below) as a tool for parsing out the relevant dynamics at work.

From the Plant's Perspective

The reproductive responsiveness of a flowering plant to a given flower visitor is the product of at least two interacting dynamics—the degree to which the plant expresses floral specializations that either invite or exclude the visitor, and the degree to which the flower is obligated to pollen vectoring in the first place. As we will see, the plant may require a pollen vector for at least two reasons—to mechanically transfer pollen from anthers to receptive stigmas or to facilitate genetic out-crossing.

Floral specialization can take the form of narrow flowering windows in time that preclude all visitors but those with the same active season. It can take the form of nectar constituents such as minerals or alkaloids that repel all but the most co-adapted pollinators. And it can take the form of complex flower morphologies that mechanically limit access to only those co-adapted pollinators possessing the necessary behaviors for reaching the flower's rewards. The y-arrow of the graph above reinforces that floral specialization occurs as a continuum, with greater or lesser degrees of complexity. The canola flower, for example, is morphologically simple and accessible to a broad array of visitors and occurs at the lower end of the spectrum; blueberry, on the other hand, is morphologically complex to the point that it must be actively vibrated by the flower visitor (a bee behavior called *sonication*) before it will release its pollen. The pool of candidate pollinators is therefore more narrow for blueberry.

A flower's obligation to pollen vectoring reaches its extreme form in dioecious plants, such as kiwifruit, in which only one sex of flower occurs on any one plant; it is easy in this case to understand the necessity for a vector to move pollen from male to female plants. A similar situation can occur in monoecious plants—ones in which both sexes of flower occur on the same plant, a good example of which are the cucurbits (squash and watermelon); again, it is easy to understand the need for a vector to move pollen from male to female flowers. Monoecious plants are also represented by many species, such as blueberry, apple, and pear, with hermaphroditic flowers in which male stamens and female stigmas occur in the same inflorescence. In these cases, mechanical pollen vectoring may be easy, owing to the fact that the male and female parts are close together. However, there is another factor that eclipses proximity as a predictor of the plant's responsiveness to the flower visitor, and that is the plant's obligation to genetic out-crossing. In monoecious plants, there is a wide range of genetic tolerance for self-pollen. Plants that are self-fruitful (such as soybean) have a high tolerance for self-pollen, and conversely, plants that are self-sterile (such as almond) have a high demand for out-crossing with pollen from another plant of the same species, or even a different variety of the same species.

The *x*-arrow of the graph on page 224 shows that obligation to pollen vectoring is a continuum. At the lower end we can invoke soybean and its near-total habit of selfing and at the higher end almond with its near-total self-incompatibility.

From this discussion we can conclude that a given flower visitor will have a greater chance of positively affecting reproduction in those plants with unspecialized flower morphology and high obligation to pollen vectoring. Plants with specialized flower morphology and high obligation to pollen vectoring include some important agricultural crops; however, their biology inherently narrows the field of effective candidate pollinators.

From the Pollinator's Perspective

The pollinating competence of a flower visitor can also be thought of as an interaction of two dynamics—the innate efficacy of the pollinator at achieving fruit-set and the pollinator's availability (see diagram below). The literature standard for reporting and comparing pollinator efficacy is percentage fruit-set per single flower visit. Fruit-set is the ratio of ripe fruit relative to initial number of available flowers. (This parameter can also be applied to seeds [seed-set], but for simplicity I'll refer hereafter to fruit-set.) This ratio is rarely 100%, owing to suboptimal pollination, normal levels of fruit abortion, herbivory, or cultural problems. In nature, fruit-set usually follows repeated flower visits by one or more species of pollinators; but when evaluating different pollinators, it is normal to compare fruit-set on the basis of single flower visits. The investigator will bag virgin flowers, un-bag them when they open, observe a single visitor, re-bag the flower, then follow the flower's progress for subsequent fruit development. The flower has a known history, and the efficacy of the specific pollinator can be

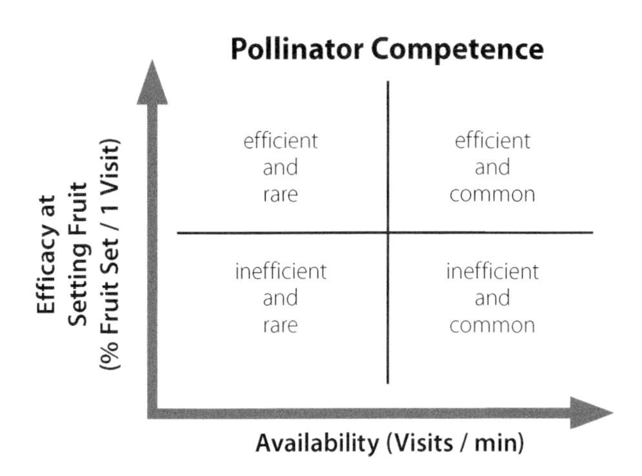

measured and compared with others. Table 1 shows some known single-visit fruit-set values for bees in rabbiteye blueberry, *Vaccinium ashei*. A general trend is that the Southeastern blueberry bee, *Habropoda laboriosa*, is an efficient pollinator, sometimes achieving in one visit a level of fruit-set comparable to that of open-visited flowers. Honey bees, in contrast, typically express low efficiencies at the single-visit level.

Single-visit fruit-set is the most basic and equitable measure of innate pollinator efficiency. It is a product of factors as diverse as the size and hairiness of the bee, the degree to which the bee contacts the flower's sexual surfaces, or specialized behaviors the bee may employ to access morphologically complex flowers. Highly complex and specialized behaviors for accessing difficult flowers are evidence of a co-evolutionary relationship between the bee and flower. These kinds of specialized—and efficient—pollinating behaviors are more common in solitary bee species, such as *H. laboriosa*, which tend to have a relatively brief active season. There is strong selection pressure for these bees to assure the survival—hence, the reproduction—of those plants upon which they depend for food and whose bloom phenology matches their own active season. Social bees like the honey bee, on the other hand, experience long, almost leisurely active seasons during which they visit the panoply of blooming plants available in their neighborhood. These bees are generalists, able to field a large forager force and make repeat flower visits but characteristically unspecialized as far as flower visitors go, experiencing little selection to be an efficient pollinator for any one plant.

However, there is evidence that this generalization may be simplistic. Recent work has shown in rabbiteye blueberry (Dedej and Delaplane, 2003; Ellis and Delaplane, 2008) that the comparative inefficiencies of honey bees are erased when fruit-set is measured not at the level of individual bee but of the colony. One honey bee colony can field a foraging force of tens of thousands, and the sum of many inefficient flower visits eventuates to a level of fruit-set comparable to open-visited flowers. Upon reflection this makes sense—even season-long generalist pollinators have an evolutionary motive to ensure survival of their host plants; but the agency of repeat flower visits arguably diminishes selection for efficient pollinating behavior at the level of individuals. On a practical level, it remains to be seen if the efficiency-enhancing benefit of repeat flower visits by social bees applies to other crops. If so, this may be a powerful argument for the continued importance of the honey bee as agriculture's most dependable pollinator.

But, important as repeat visits may be to the social and inefficient honey bee, it applies as well to any pollinator. It is safe to assume that many flower visits are better than fewer. This leads to the second dynamic of a pollinator's competence—its availability. Pollinator availability is most equitably and practically evaluated in the form of visitation rate—the number of flower visits per unit time. Ecologic theory predicts that foraging animals will invest the minimal energy necessary in their pursuit of resources; it follows, therefore, that high rates of bee flower visitation indicate the proximity of good bee nesting sites or, in the case of honey bees, managed hives. It is obvious that an absent pollinator will not pollinate; however, as with innate efficacy, pollinator availability occurs as a continuum with any given pollinator in any given locale occurring at higher or lower rates of visitation.

This is the reality behind the practice of commercial-scale movement of beehives for pollination. With one overnight delivery, the pollinator

Variety	Honey Bee	Bumble Bee Queen	South-Eastern Blueberry Bee	Orchard Bee	Open Visitation	Bees Excluded	Reference
'Triblue'	1	30	30		49	3	*
'Climax'	13		58		44	3	
'Delite'			20		51	5	
'Premier'	52		51	17		0	**
'Climax'	33						***

*Cane and Payne (1990), **Sampson and Cane (2000), ***Ellis and Delaplane (2008).

Table 1. Some single-visit fruit-set values (fruit set per 100 flowers) for bees in rabbiteye blueberry.

population of a farm can be increased by millions. This is also the reality behind frustrations voiced toward wild unmanaged pollinators: their availability is unpredictable, high or low depending on ecologic conditions outside the control of the farmer.

Understanding the Impact of Honey Bee Disorders

With this background we are better able to approach the over-arching question of the impact of bee disorders on crop pollination. It seems to me that disorders can affect pollination at any of three levels:

1. By impairing pollination efficacy of compromised individuals
2. By impairing efficacy of compromised colonies
3. By reducing availability of pollinators, that is, killing bee colonies

Although the pollination impacts of bee disorders have rarely been studied directly, there is a relevant literature on the impacts of bee disorders on honey bee flight and foraging. We know, for instance, that at the colony level, sacbrood virus and the microsporidian gut parasite *Nosema apis* decrease pollen foraging (Anderson and Giacon, 1992). The parasitic mite *Varroa destructor* at the colony level decreases the number of pollen foragers (Janmaat et al., 2000) and at the level of individual, the body weight and flight activity (Schmid-Hempel, 1998). The nest-scavenging beetle *Aethina tumida* reduces colony bee populations, brood area, and flight activity (Ellis et al., 2003). It is reasonable, therefore, to presume that negative effects such as these translate to reducing the honey bee's efficacy as a pollinator. However, this premise must be unpacked contextually and organized into a set of testable hypotheses.

My lab has begun making progress in this direction, most directly in the work of Ellis and Delaplane (2008). We studied the effects of Varroa on honey bee pollination of rabbiteye blueberry. Bee colonies were manipulated to achieve two levels of Varroa—"high" with an average 24-hour mite drop of 57 to 43 or "no invader control" with mite drop of 2.7 to 3.1. Then colonies were tented individually with plants and numerous measures taken of pollination efficacy. The most important finding was that pollination at the level of single-bee flower visits was impaired if the bee came from a Varroa-parasitized colony. However, this handicap disappeared when we examined pollination at the colony level; in other words, colonies with Varroa pollinated no less effectively than colonies without. We speculated that the eusocial colony life habit of *Apis mellifera* permits a degree of compensation for individual inefficiencies. To some extent, these large colonies field enough noncompromised foragers that average pollination efficacy is unaffected. If this is true, then it appears that the practical cost to pollination happens at level 3 (above) by killing colonies and reducing the availability of pollinators at the community level.

There is evidence for buffering or compensating effects of eusocial life in more than one context. In the context of adaptive benefits to the colony, buffering is apparent when bee colonies absorb large loads of Varroa mite without measurable cost to individual bee foraging profits (Ellis and Delaplane, 2009) or honey storage rates (Murilhas, 2002). In the community context of pollination, compensation is apparent when individually inefficient honey bees nevertheless achieve acceptable fruit-set, even in flowers that are adapted for specialist pollinators. It is arguable that compensatory action in pollination is also adaptive to the honey bee, albeit less directly than the examples on foraging. By being a competent pollinator, the perennially social honey bee ensures itself a season-long succession of blooming food plants. This compensation is credited to the ability of these colonial bees to field large forager populations and make repeat flower visits (Dedej and Delaplane, 2003).

A general finding by my group has been that the action of many inefficient individuals adds up to sufficient pollination at the colony level. In even plainer language, it seems that as far as honey bee pollination goes, the colony will pollinate provided it is alive and foraging. Intuitively, there must be a point of increasing colony Varroa density at which colony pollination deteriorates. After all, the foraging behaviors of individuals are expected to integrate with colony state (reviewed by Schmid-Hempel et al., 1993). However, such a continuum has not been characterized, and once it is, I anticipate it will constitute another example of the buffering benefits of eusocial life, not only to the bees but to the local plant community supported by the pollinating superorganism (Moritz and Southwick, 1992).

Although the limited evidence suggests that honey bee disorders impact pollination through the simple agency of killing colonies, I reinforce that the literature is thin on the pollination impacts of pollinator disorders. Hypotheses should be framed and tested with different model bee disorders, different pollinator species, and different model plants. Our earlier discussion on variability in plant response to flower visitors reminds us that the impact of bee disorders is likely to be asymmetrical—more critical for those crops with a narrower field of candidate pollinators.

To restate my central question: At what point do honey bee disorders exact a societal cost to pollination? The individually-parasitized forager? The parasitized colony? Or the bee-impoverished ecosystem? These kinds of questions are critical to our understanding of the agro-ecosystem impacts of honey bee disorders. Without this basic knowledge, we cannot accurately chart our research and management agenda. Is it in fact all-or-nothing? Living colonies pollinate; dead colonies don't? If the answer is yes, then applied research priorities should steer toward practices that maximize sheer colony numbers, perhaps with renewed attention to conserving feral bee populations, increasing beekeeper recruitment and profitability, or relaxing our borders to imported *Apis mellifera*. Or is it instead a more nuanced effect whereby the efficacy of individual foragers interacts to some degree with colony state? If the answer lies here, then this would steer applied research toward a more classical IPM approach in which within-colony mite levels and action thresholds play a more prominent role. These scenarios are not mutually exclusive, but they do represent a difference in focus and emphasis. In either case, the question and its answers strike at the heart of the worldwide conversation about pollinator decline.

Calculating and Reporting Managed Honey Bee Colony Losses

CHAPTER

20

Dennis vanEngelsdorp, Robert Brodschneider, Yves Brostaux, Romee van der Zee, Lennard Pisa, Robyn Underwood, Eugene J. Lengerich, Angela Spleen, Peter Neumann, Selwyn Wilkins, Giles E. Budge, Stéphane Pietravalle, Fabrice Allier, Julien Vallon, Hannelie Human, Mustafa Muz, Yves Le Conte, Dewey Caron, Kathy Baylis, Eric Haubruge, Stephen Pernal, Andony Melathopoulos, Claude Saegerman, Jeffery S. Pettis, and Bach Kim Nguyen

Abstract Quantifying colony losses is a two-part process. First, colony loss data needs to be collected by surveying beekeepers and then it needs to be calculated and reported in a standardized way. We propose using two different ways to calculate and communicate colony losses. The first we term the *total colony losses*, sometimes referred to as cumulative loss rate in other systems, which aggregates all losses suffered by all beekeepers surveyed. While the total loss calculation is straightforward, calculating a 95% CI for this metric is complicated by the need to account for the varying sizes of responding beekeeper operations and the nested nature of colony losses within those operations. The second reporting method, termed *average loss*, is the mean loss suffered by each responding beekeeper. The utility of these two reporting mechanisms differs, in that both are potentially biased by the demographics of the apicultural industry; total loss figures are more heavily influenced by the losses experienced by the few large operations, while average losses are more representative of the many small operations. Additionally, it is important to note that the results from this survey are representative of the responding population alone, and cannot be considered representative of the industry unless some means of identifying and adjusting for varying response is performed.

Introduction

Honey bees are commercial agriculture's "keystone" species because of the pollination services they provide. Considering this importance, it is not surprising that many nations have tabulated the number of colonies maintained within their country for well over half a century. The results of these efforts are stored in a freely accessible database maintained by the Food and Agriculture Organization (FAO 2009). A recent review of this database revealed that internationally managed honey bee colony populations have increased during this time period (Aizen and Harder 2009). Nevertheless, to interpret this trend as an indication of a healthy global apicultural industry is misleading (vanEngelsdorp and Meixner 2010) as gross trends based on annual colony tabulations do not take into account potentially dramatic fluctuations in colony numbers that can occur within a given year.

There have been systematic efforts to estimate the level of loss that occurs over the winter months since the reporting of unusually high colony mortality in North America during the winter of 2006–2007 (see Figure 1) (vanEngelsdorp et al. 2011) and subsequently in Europe (Neumann and Carreck 2010). Nevertheless, these losses did not translate into a reduction in the total number of managed colonies from year to year because beekeepers replaced their dead colonies using a variety of means including buying packages of bees from other beekeepers or by splitting their surviving colonies to make replacement colonies. Splitting involves making a new colony by moving frames of brood and adult bees from strong surviving colonies to empty hives. While the new colonies will raise their own queens, beekeepers commonly facilitate this process by introducing mature mated queens at the time of splitting. By splitting strong colonies, beekeepers can readily increase colony numbers.

Beekeepers experiencing losses of 30% or more can potentially replace these losses within one season by splitting surviving colonies. Nevertheless, splitting large colonies into smaller colonies has significant direct costs, including the price of labor and replacement queens as well as indirect costs such as lost productivity, which is dependent on overall colony size. Therefore, attempts to gauge the health of the apicultural industry must not be limited to simply counting colonies at a specific and regular time each year, but also needs to consider both the rates of colony loss and replacement (vanEngelsdorp and Meixner 2010). Estimation of colony loss and replacement rates is an essential step in developing an epidemiological approach to identifying the factors contributing to or mitigating losses (Nguyen et al. 2010a).

A necessary starting point for such an epidemiological approach is the accurate estimation of colony losses. Quantifying colony losses is a two-part process. First, a colony loss survey must be designed, validated and distributed, and the responses collected. The second part, the generation of meaningful and comparable loss data, involves validating, tabulating, and reporting the responses using a standardized, transparent, and accurate methodology. In recent years, neither the design of questions nor the selection of survey respondents was uniform among national winter loss survey efforts. These differences have often precluded the comparison of colony loss data among countries or regions or across time. While a detailed analysis of these various approaches is beyond the intent of this chapter, it is clear that a more standardized approach would be beneficial. To this end, a working group of an international COST (European Cooperation in Science and Technology) network of bee researchers with the acronym COLOSS (Prevention of Honeybee COlony LOSSes) has released a "model" winter loss questionnaire that includes a set of essential standardized winter loss questions. The international adoption of these questions should help make colony loss results more comparable.

The standardization of winter loss survey questions permits the establishment of standardized colony loss calculations and reporting. Providing guidelines on how loss numbers should be calculated and reported will facilitate transparency and the comparisons between different winter and/or other colony loss survey efforts. This chapter proposes a standardized colony loss calculation and reporting methodology.

Calculating Colony Losses

When reporting colony loss figures, there are two different calculations that we feel are meaningful. The first is total colony losses, which are an aggregate of all losses suffered by all respondents. The second is average loss, the mean of the losses suffered by each responding beekeeper. Each estimate is useful; however, each is potentially biased. This bias is particularly prevalent when surveys are conducted in regions where the apicultural industry is comprised of relatively few individuals (commercial beekeepers) who own the vast majority of colonies while the individuals who make up the vast majority of the industry (backyard beekeepers) own a small percentage of the total number of managed colonies. These skewed demographics of the beekeeping industry are particularly pronounced in countries such as the United States (Daberkow et al. 2009). As explained in detail below, total loss figures are more heavily influenced by the losses experienced by the few large operations, while average losses are more representative of all apicultural operations.

Figure 1. Since the winter of 2006–2007, high rates of colony mortality have been reported in North America and Europe. These losses were observed both in northern and southern locations. Depicted here is an apiary containing both living and dead colonies. The colonies in this apiary were owned by an east coast U.S. migratory beekeeper who lost over 66% percent of his operation in the 3 months since moving colonies from Pennsylvania to Florida in November 2006.
(*Photo by D. vanEngelsdorp.*)

Calculating Total Losses:

Total colony loss (or cumulative loss rate) figures are the percentage of all colonies lost in a defined group over a defined period of time. Total colony losses (%) are calculated using the same equation used by epidemiologists to calculate mortality; for example Equation 1, where TL = total losses over a period, td = total number of colonies that died over a given period, and tc = the total number of monitored colonies that are at risk of dying over the period (Koepsell and Weiss, 2003).

Equation 1:

$$TL = \left(\frac{td}{tc}\right) \times 100\%$$

TL could, for instance, be calculated from the response to two questions in the 2010 COLOSS standardized questionnaire:

1. How many colonies did you have on October 1, 2009?
2. How many colonies did you have on April 1, 2010?

To account for the beekeeper practice of either selling, giving away, buying, or making additional colonies over this time (winter) period, however, tc must be adjusted using information derived from two additional questions:

3. How many colonies did you sell or give away over the period between October 1st and April 1st, and
4. How many colonies did you make or buy over that period?

Thus, the total number of monitored colonies at risk of dying over the period (tc) is given by Equation 2, where the total number of colonies at the beginning of the survey period (tcb) is the number of colonies alive on October 1st minus any colonies sold or given away over the period (tcs) plus total colony number increases (tci), for example, colonies bought or made through splitting over the period.

Equation 2: $$tc = tcb - tcs + tci$$

The total number of colonies that died over the period (td; Equation 3), is then calculated as the difference between the total number of monitored colonies at risk of dying over the period (tc) and the total number of living colonies at the end of the period (te). For winter loss calculations in the Northern hemisphere, the period under consideration is October 1st to April 1st for any given year.

Equation 3: $$td = tc - te$$

In some previous reports, colony losses were not adjusted for changes in the colony counts that result from selling, buying, or increasing colony numbers by splitting (Brodschneider et al., 2010). Such an approach may have a minimal effect on total loss calculations in regions where colonies are not commonly replaced or sold over the time period in question (winter months) or among the group of individual respondents that would not likely perform such management, notably backyard beekeepers. For example, countries such as Canada, have a sharply defined temperate winter period over its major beekeeping regions during which few, if any, colonies are bought, sold, or split, making such corrections largely unnecessary ($tcs << tc$ and $tci << tc$). Such adjustments would also have minimal effect in regions where the total number of colonies sold is almost equal to the total increase in colony numbers ($tcs \approx tci$). For instance, in Belgium, 14.3% (n = 238) of responding beekeepers reported having sold or bought colonies over the 2009–2010 survey period. The total number of colonies these beekeepers bought and sold make up a relatively small portion of the total colonies monitored (8.2%). Furthermore, the difference in the number of colony decreases (colonies sold or given away) and the number of colony increases (splits or purchases) is relatively small ($n = 53$, or 1.7% of tc) (Nguyen et al., 2010b). Should the percentage of total loss calculations disregard the number of colonies bought or sold, the effect on total loss calculations would be minimal, changing the national total loss from 27.8% to 26.6%.

In contrast, disregarding the number of colonies bought, given, sold, or increased in regions where these practices are common can have pronounced effects on the total loss calculations. A significant example of this situation arose in the 2010 U.S. winter loss survey, when a minority of responding operations (16.5%, $n = 696$ of a total of 4,227) reported having bought, sold, or increased their operations during the survey period. While the number of colonies sold or given away in the period was minimal ($n = 8{,}086$ colonies or 1.4% of *tc*), the number of colonies bought or made during the period was substantial, totaling 135,837 colonies, or 23.7% of the total colonies monitored (*tc*) (vanEngelsdorp et al., 2010). Disregarding this information would have changed the total colony loss calculations from 34.7% to 14.0%, thus significantly underreporting losses.

Another approach to dealing with increases and decreases in colony numbers is to remove all data obtained from responding beekeepers who reported buying, giving, or selling colonies, or making increases. This approach, however, risks biasing survey results, especially if this behavior is related to colony losses. The degree of potential bias varies depending on the percent of beekeepers splitting, selling or buying colonies over the winter in different regions. As previously mentioned, only 14.3% of beekeepers in Belgium reported reducing or increasing their colony numbers through management or commerce for the 2009–2010 winter. Excluding these beekeepers from the calculation would have changed total loss figures for the country from 27.8% to 26.0%. In the United States, however, excluding beekeepers who bought, gave, sold, or made colony increases over the survey period would bias results by disproportionally excluding larger operations. Among commercial beekeepers (operating more than 500 colonies), 64% reported splitting, selling, or buying colonies compared to only 27.0% of sideline beekeepers (operating 51 to 500 colonies) and 9% of backyard beekeepers (1 to 50 colonies) who engaged in one or more of these practices. The outright exclusion of those beekeepers would reduce the number of colonies surveyed (*tc*) in the United States by 76% ($n = 572{,}641$), and would change the total colony loss figures from 34.7% to 31.6% for the observed period. Therefore, in the same way that national mortality figures include persons born in the period and exclude persons no longer residing within a region during a period, calculating total colony losses for a defined period (October 1st to April 1st) should exclude colonies removed from monitoring (colonies sold or given) and include increases in the number of colonies (made or bought) during the period.

In some cases it may be informative to report the losses of colonies observed for the entire surveyed period, that is, losses only among colonies observed at the start of the survey period (*tcb*). As previously mentioned, this calculation could be made by simply excluding operations that reported increasing or decreasing colony numbers. As before, this approach can seriously bias results in regions where a significant number of surveyed beekeepers sell or increase their colony numbers over the observation period. An alternative approach would be to report the potential range of loss experienced by the colonies included in *tcb*.

As outlined in Equation 4, the upper value of this range (TL_{min}) assumes that all the colonies that died over the period were those colonies included in the original population and kept over the winter ($tcb - tcs$), while the lower range value (TL_{max}) assumes that the number of dead colonies is larger than *tci* and that all *tci* colonies died. In cases where $td < tci$, TL_{min} should be reported as 0%, and in cases where $td \geq (tcb - tcs)$, the TL_{max} should be reported as 100%.

Equation 4:

$$TL_{max} = \frac{td}{tcb - tcs} \qquad TL_{min} = \frac{td - tc}{tcb - tcs}$$

Calculating 95% CI for Total Losses:

Confidence intervals (CI) are used to express the reliability of an estimate and are defined by two numbers, or confidence limits, that straddle a mean (Zar 1996). In principle, calculating the confidence limits for total losses uses the same approach used to calculate the confidence limits for any proportion (Wald Method, Equation 5).

Equation 5: Confidence limits for $TL = \widehat{TL} \pm Z_\alpha \times \text{s.e.}(\widehat{TL})$

$$\text{where s.e.}(\widehat{TL}) = \sqrt{\frac{\widehat{TL}\,(1 - \widehat{TL})}{n}}$$

To calculate a two-sided 95% CI, we use Equation 5 setting Z_α ($\alpha = 0.05/2$) to 1.96. The resulting confidence limits would encompass the surveyed population's true TL with 95% certainty. In other words, the mean TL resulting from other survey efforts of the same population would fall between the confidence limits 95% of the time. Calculating confidence limits using Equation 5 assumes a normal approximation of the binomial distribution, which is a safe assumption provided that td and tc – td (used to calculate TL in Equation 1) are greater or equal to 10 (Koepsell and Weiss, 2003). While it is hard to envision a total colony loss calculation where this would not be the case, should $td < 10$ or $tc - td < 10$, then confidence limits would be more accurate if based on the binomial distribution. Several alternative methods to calculate confidence limits using binomial distributions are available. These methods may also be useful if the estimate is close to 0% or 100%, at which point the confidence interval may become inappropriate, resulting, for example, in a lower confidence limit that is less than 0% or an upper confidence limit that is greater than 100% (Koepsell and Weiss, 2003, p. 85).

Another consideration that must be taken into account when using Equation 5 is selecting the appropriate value for "n" when calculating s.e. (standard error) for $\widehat{TL}$. Simply setting $n = tc$ clearly underestimates the real 95% CI (see Table 1; (Brown et al., 2001, 2002)). To avoid this problem, previous efforts that calculate 95% CI for total losses have opted to set n as the number of survey respondents (vanEngelsdorp et al., 2008; Brodschneider et al. 2010; vanEngelsdorp et al., 2010; Nguyen et al., 2010b). While expedient, this approach is also faulty, in that the resulting CI is overly wide (see Table 1) (Brown et al., 2001, 2002).

Underpinning the calculation of confidence limits in Equation 5 is an assumption that the data used to derive are independent. That is, each colony is equally likely to die over the period. However, colony losses are not independent, as there is a notable difference in the total loss experienced between different beekeeping operations. To accurately calculate 95% CI for colony loss data, we must take the nested nature of colony loss data into account (see Table 1). This approach requires calculating and using the s.e. of the intercept of a null Generalized Linear Model with quasi-binomial family error distribution and logit link, where the total colonies alive and dead at the end of the survey period are the independent variables (McCullagh and Nelder, 1989). The quasi-binomial family is introduced to take into account the increase in the standard error

Method to calculate s.e using Equation 5	95% CI for National Colony Losses 2009–2010 for Total National Colony Losses Belgium	U.S.	Result
$n = tc$	26.3–29.4	34.3–34.5	Conservative—underestimates true range
n = total respondents	22.1–33.5	32.9–35.9	Liberal—overestimates true range
s.e. of the Generalized Linear Model	23.9–32.2	33.7–35.1	Accurate

Table 1. A comparison of 95% CI of total colony loss figures calculated using various methods. Belgian and U.S. national loss figures for 2009–2010 are presented.

induced by dependence in the data, when compared to the standard Wald method. These calculations are easily conducted using the R statistical package, which is open source (R Development Core Team, v.2.10.0, 2009).

Calculating Average Losses:

Average losses are the mean of the total colony losses experienced by all responding beekeepers. The information needed to calculate average losses (AL; Equation 6) is identical to that needed to calculate TL; however, the TL for each responding operation (TL_O) is calculated and the number of responding beekeepers (nr) whose TL_O was calculated is required.

Equation 6:

$$AL = \frac{\sum TL_O}{nr}$$

Calculating 95% CI for Average Losses:

Calculating the 95% CI for average losses uses the same equation as presented in Equation 5, except that TL is replaced with AL, and

$$\text{s.e.} = \frac{\sigma TL_O}{\sqrt{nr}}$$

This approach assumes that the number of respondents used to calculate the AL was larger than 60, which permits us to assume that the sampling distribution of AL is normally distributed. In cases where nr is equal to or less than 60, a normal distribution of AL cannot be assumed. In such cases, the Z_α used in Equation 5 needs to be replaced with a value derived from a t-distribution, where the t-distribution chosen is based on the data's number of degrees of freedom ($nr - 1$). T-distribution values are easily obtained from statistical tables (Paoli et al., 2002). This estimation can be somewhat biased if the distribution of the number of colonies per respondent is asymmetric, as each respondent is given the same weight in the resulting AL, regardless of the number of colonies owned.

Reporting Losses:

A consistent system for reporting colony loss data will permit comparison of data across different regions and/or over time. As mentioned earlier, the most appropriate way to calculate losses may differ to accommodate regionally specific differences in apicultural practices, such as the frequency of splitting colonies over the survey period. Such differences need not negate the ability to make comparisons among regions if sufficient summary data are provided. To that end, in addition to total and average loss numbers, as well as their respective 95% CI, the total number of responding beekeepers, the total number of colonies at the beginning of the study (*tcb*), and total number of colonies that died (*td*) should be reported. Colony number increases (*tci*) and the number of colonies sold or given away (*tcs*) should also be reported to facilitate TL_{max} and TL_{min} calculations when those figures are not included in the losses report.

Limitations of Data Interpretation:

Consistent data collection, analysis, and reporting will allow for the comparison of losses across regions and time. Because these surveys are voluntary and respondents are not random, the survey results are representative only of the surveyed population. For results to be considered representative of a nation or region, steps must be made to identify and/or adjust for potential biases in the survey. To adjust for biases, there is a need for an accurate census of producers, ideally by characteristics such as the size of operation, in a given area. These producers can either be randomly sampled as part of the survey delivery design, or resulting data can be adjusted after the fact by weighing responses according to the known beekeeper and colony population structure. The more detailed the data on the colony and beekeeper distribution, the more accurate the weighting. For instance, in the United States where the number of beekeepers and colonies kept in the country are summarized every five years, colony loss results can be adjusted to ensure the responding population is representative of the actual population (both by operation size and location) as determined by the Agricultural Census. Unfortunately, for many countries, accurate colony numbers are not available. In these cases, pains must be made to stress the fact that the results are representative of the survey population only and not necessarily the entire apicultural industry.

Summary

Widespread reports of high colony losses over winter have resulted in numerous surveys that aim to quantify these losses. To permit comparisons between survey efforts, there is a need to standardize the way losses are calculated and reported. This chapter proposes and outlines methods by which total and average losses can be calculated and reported. The proposed approach takes into account several factors that may influence colony loss calculations, including the potentially dynamic nature of the number of colonies owned by a responding beekeeper over the observed season, the nested nature of loss data within operations, and incomplete survey responses. To accommodate regional differences in the way that loss data may be collected and calculated, it also proposes standard estimates that should be presented in future colony loss reports to permit comparisons between regions and over time.

Reflections: Conservation of Plant–Pollinator Mutualisms

CHAPTER

21

David W. Inouye

Abstract The conservation of plant–pollinator interactions is important to preserve the mutualism that provides much of the food that humans consume. This interaction is at risk due to factors including climate change, which may shift differentially the phenology of events like flowering and emergence of bee-introduced diseases and parasites that are detrimental to pollinators, and other environmental changes. Our understanding of the consequences of these threats is limited by our lack of knowledge of the distribution and abundance of pollinator populations, but both honey bees and bumble bees have declined in recent years. Data collection by both scientists and citizen-scientists is helping to solve this problem, for both urban and field studies.

Pollination is a mutualism, with plants benefitting from, and often requiring, the services of pollinators for sexual reproduction. Correspondingly, pollinators are typically dependent upon the floral resources of pollen and nectar for their own nutrition and reproduction. In order to conserve this mutualism, both plants and pollinators must be considered. This has become more challenging in recent years because of the effects of climate change, habitat changes, and introduction of diseases as pollinators are moved from continent to continent. The chapters in this book that focus on Colony Collapse Disorder in honey bees give some insight into the problems that this phenomenon alone has created.

An aspect of the plant–pollinator mutualism that has attracted increased interest in the past few years relates to phenology (the timing of seasonal events, such as flowering, or beginning of the flight period of bees), particularly in the context of climate change. Many studies have found that the phenology of flowering has been changing, typically to earlier blooming in response to earlier beginning of the growing season (Abu-Asab et al., 2001; Lambert et al., 2010), and/or increased temperatures. There is some evidence that as the growing season expands in length, species that were formerly closely sequential in flowering are being separated to the point that a gap in floral resources is developing (Miller-Rushing and Inouye, 2009; Aldridge et al., 2011). A modeling study found that under some scenarios, phenological shifts reduced the floral resources available to 17 to 50% of the pollinator community; up to half of the former period of activity no longer had floral resources, reducing the pollinators' diet breadth (Memmott et al., 2007). This kind of loss of seasonal synchrony could thus lead to extinction of plants, pollinators, and their interactions, and at a minimum will require simultaneous change in plants and pollinators, or the establishment of new interactions. The lack of data about pollinator phenology, particularly solitary species, hinders our ability to forecast consequences of changes in response to climate change.

The ecosystem service provided by pollinators, both native and introduced, is receiving increasing attention from a growing audience. Some of this new attention stems from our increased knowledge of the

Introduced species *Anthophora plumipes* (Apidae) working flowering quince blossoms (*Chaenomeles* spp.) (*Photos by D. Inouye.*)

biology of pollinator species and their susceptibility to anthropogenic influences such as habitat fragmentation, habitat loss, use of pesticides, and transmission of diseases (Kearns and Inouye, 1997; Kearns et al., 1998). There is also a greater appreciation of the economic value of this ecosystem service, reflecting the fact that about one out of every three bites of food we eat is the product of this service (Buchmann and Nabhan, 1996). Our understanding of the effects of global, regional, and local climate changes on pollinator populations is also growing. Groups such as the North American Pollinator Protection Campaign (www.nappc.org) and Xerces Society (www.xerces.org) are also helping to increase public awareness of the plight of many pollinator species.

There is still a surprising paucity of data about the abundance and distribution of pollinator species, and of the consequences of human activities such as introducing non-native pollinators and new diseases of native pollinators. For example, we have no information how the historic introduction of honey bees (*Apis mellifera*) affected the abundance and distribution of native bees in North America. It seems likely that the short-tongued bumble bee (*Bombus*) species may have been particularly affected, as workers of these species have the same proboscis length as honey bees (Inouye, 1977), which can reach much greater abundances. A recent study from Great Britain demonstrated an apparent effect of competition between honey bees and bumble bees, with bumble bees having smaller worker sizes in the presence of honey bees (Goulson and Sparrow, 2009). Previous studies have also demonstrated changes in foraging by bumble bees in the presence of honey bees (Walther-Hellwig et al., 2006). Colony reproductive success can also be affected by such competition (Thomson, 2006). Some solitary bee species are also likely competitors of honey bees (Paini and Roberts, 2005).

The National Academy of Science committee that produced a report about the status of pollinators in North America (Committee on the Status of Pollinators in North America, 2007) had difficulty making an assessment of pollinator populations because of the paucity of data for pollinators other than honey bees, and even for this species the data were not as abundant as hoped. The report called for increased collection of data on pollinator distributions and abundance, to provide a means of assessing future population changes.

The native pollinator group for which most is known about distribution and status is probably the bumble bees (*Bombus*), and the news is not good. A study of historically sampled localities in Illinois found that species richness declined substantially during the period 1940–1960, with four species locally extirpated and range contractions of four others (Grixti et al., 2008). In the United States, relative abundances of four species have

Left: A *Bombus rufocinctus* queen on a sneezeweed flower (*Dugaldia hoopsii*) in Colorado. **Right:** A *Bombus bifarius* queen on a dandelion flower (*Taraxacum officinale*) in Colorado. (*Photos by D. Inouye.*)

declined by nearly 96% and geographic ranges contracted by 23% to 87% (Cameron et al., 2011). Declines have also been reported elsewhere in North America, in Europe, and Asia (Williams et al., 2009; Williams and Osborne, 2009). At least in North America, some of these declines appear to be the consequence of increased infection levels of the microsporidian pathogen *Nosema bombi*, accompanied by lower genetic diversity of affected species compared with co-occurring populations of the stable (nondeclining) species (Cameron et al., 2011). Commerce in bumble bees for pollination in greenhouses is apparently responsible for the introduction of *Crithidia bombi*, a destructive protozoan pathogen commonly found in commercial *Bombus* (Otterstatter and Thomson, 2008).

An initial step in the conservation of an interaction is to recognize the threats that it faces. There is a growing appreciation of the problems that pollinators now encounter (Kearns and Inouye, 1997; Kearns et al., 1998). A subsequent need is to understand how these threats can be mitigated, for example through management practices. This phase too is now being addressed for pollinators. A large-scale study of agricultural practices in Europe, and how they can be modified to protect and encourage pollinator populations, is underway (Batáry et al., 2010). Other studies as well are providing insights about management practices that can benefit pollinators (Ekroos et al., 2007; Rundlöf et al., 2007). The recent problems that commercial honey bee operations have faced have triggered interest in alternative native pollinators in North America (Winfree et al., 2007; Winfree et al., 2008).

A honey bee working a raspberry flower (*Rubus* spp.). (*Photo by M. Frazier.*)

Although there are a few compendia of information about crop pollination, by Free (1970) and Roubik (1995), there is still very little quantitative information about how crop pollinators vary in time and space, over latitudinal or longitudinal gradients, or in their effectiveness as pollinators. Information is also needed about how best to encourage pollinator populations under different scenarios of agricultural ecosystems (e.g., organic or not).

Citizen science projects, which seek to involve nonscientists in research projects that can generate data of value for scientists, have expanded to include pollination projects. Monarch Watch (www.monarchwatch.org) is a cooperative network of students, teachers, volunteers, and researchers dedicated to the study of the Monarch butterfly, *Danaus plexippus*, and

Left: A *Bombus appositus* queen visiting *Corydalis caseana* (*Fumariaceae*) in Colorado. (*Photo by D. Inouye.*) **Center:** Honey bee gathering corn pollen (*Zea mays*). (*Photo by M. Frazier.*) **Right:** A *Bombus appositus* queen visiting *Delphinium nuttallianum* (*Ranunculaceae*) in Colorado. (*Photo by D. Inouye.*)

to increasing the abundance of its larval host plant, milkweeds. The efforts of these volunteers in tagging butterflies have provided tremendous new insights into their migration behavior and pathways. The Great Sunflower Project (www. greatsunflower.org) started with a focus on sunflowers, but has now expanded to include other species that are observed to quantify visitation rates of pollinators. Project Budburst (www. neoninc.org/budburst) and the USA National Phenology Network (www. usanpn.org) focus on the phenology of floral resources that support pollinator populations. All of these projects will generate many more data than could be collected by scientists alone.

Little is known about urban pollinators and pollination as well, although some recent research addresses that topic (e.g., Frankie et al., 2005; Hernandez et al., 2009; Werrell et al., 2009). The recent increase in interest in urban beekeeping, facilitated by legalization of apiaries in some major cities, may help to stimulate more research on this topic, although there can also be significant numbers of native pollinators in urban environments.

An indication of how much remains to be learned about pollinators is the report in 2011 of a new mammal pollination system; *Acomys subspinosus* (Cape spiny mouse) was demonstrated to be a regular visitor and pollinator of *Erica hanekomii* (Ericaceae) (Turner et al., 2011). There are undoubtedly many invertebrate pollinators that remain to be described by taxonomists, and much to be discovered about their activity as pollinators. As pointed out in other chapters in this book, honey bee pollinators face a variety of serious challenges, and they are only a tiny fraction of all the pollinator biodiversity. We can at least point, with a bit of optimism, to the fact that the pollinator community and the ecosystem service of pollination that they provide are now receiving greater attention from both scientists and nonscientists. This probably bodes well for their future.

References

Introduction

Allen, M., and B. Ball. 1996. The incidence and world distribution of honey bee viruses. *Bee World* 77: 141–162.

Aguilar, R., L. Ashworth, L. Galetto, and M. A. Aizen. 2006. Plant reproductive susceptibility to habitat fragmentation: Review and synthesis through a meta-analysis. *Ecol. Lett.* 9: 968–980.

Aizen, M. A., and L. D. Harder. 2009. The global stock of domesticated honey bees is growing slower than agricultural demand for pollination. *Curr. Biol.* 19: 1–4.

Alaux, C., F. Ducloz, D. Crauser, and Y. Le Conte. 2010a. Diet effects on honeybee immunocompetence. *Biol. Lett.* 6: 562–565.

Alaux, C., J. L. Brunet, C. Dussaubat, F. Mondet, S. Tchamitchan, M. Cousin, J. Brillard, A. Baldy, L. P. Belzunces, and Y. Le Conte. 2010b. Interactions between Nosema microspores and a neonicotinoid weaken honeybees (*Apis mellifera*). *Environ. Microbiol.* 12: 774–782.

Ashman, T. L., T. M. Knight, J. A. Steets, P. Amarasekare, M. Burd, D. R. Campbell, M. R. Dudash, M. O. Johnston, S. J. Mazer, R. J. Mitchell, M. T. Morgan, and W. G. Wilson. 2004. Pollen limitation of plant reproduction: Ecological and evolutionary causes and consequences. *Ecology* 85: 2408–2421.

Bailey, L., and B. V. Ball. 1991. *Honey Bee Pathology* (2nd ed.), Academic Press, London, UK.

Ball, B. V., and M. F. Allen. 1988. The prevalence of pathogens in honey bee (*Apis mellifera*) colonies infested with the parasitic mite *Varroa jacobsoni*. *Ann. Appl. Biol.* 113: 237–244.

Bowen-Walker, P. L., S. J. Martin, and A. Gunn. 1999. The transmission of deformed wing virus between honeybees (*Apis mellifera* L.) by the ectoparasite mite *Varroa jacobsoni* Oud. *J. Invertebr. Pathol.* 73: 101–106.

Bromenshenk, J. J., C. B. Henderson, C. H. Wick, M. F. Stanford, A. W. Zulich, R. E. Jabbour, S. V. Deshpande, P. E. McCubbin, R. A. Seccomb, P. M. Welch, T. Williams D. R. Firth, E. Skowronski, M. M. Lehmann, S. L. Bilimoria, J. Gress, K. W. Wanner, and R. A. Cramer Jr. 2010. Iridovirus and microsporidian linked to honey bee colony decline. *PLoS ONE* 5: E13181. doi: 10.1371/journal.pone.001318

Carreck, N. L., B. V. Ball, and S. J. Martin. 2010a. The epidemiology of cloudy wing virus infection in the honey bee colonies in the UK. *J. Apic. Res.* 49: 66–71.

Carreck, N. L., B. V. Ball, and S. J. Martin. 2010b. Honey bee colony collapse and changes in viral prevalence associated with *Varroa destructor*. *J. Apic. Res.* 49: 93–94.

Chauzat, M. P., M. Higes, R. Martín-Hernández. A. Meana, N. Cougoule, and J. P. Faucon. 2007. Presence of *Nosema ceranae* in French honeybee colonies. *J. Apic. Res.* 46: 127–128.

Chen, Y., J. D. Evans, I. B. Smith, and J. S. Pettis, 2008. *Nosema ceranae* is a long-present and widespread microsporidian infection of the European honey bee (*Apis mellifera*) in the United States. *J. Invertebr. Pathol.* 97: 186–188.

Chen, Y. P., and Z. Y. Huang. 2010. *Nosema ceranae*, a newly identified pathogen of *Apis mellifera* in the USA and Asia. *Apidologie* 41: 364–374.

Cox-Foster, D. L., S. Conlan, E. C. Holmes, G. Palacios, J. D. Evans, N. A. Moran, P.-L. Quan, T. Briese, M. Hornig, D. M. Geiser, V. Martinson, D. vanEngelsdorp, A. L. Kalkstein, A. Drysdale, J. Hui, J. Zhai, L. Cui, S. K. Hutchison, J. F. Simons, M. Egholm, J. S. Pettis, and W. I. Lipkin. 2007. A metagenomic survey of microbes in honey bee Colony Collapse Disorder. *Science* 318: 283–287. doi: 10.1126/science.1146498

DeGrandi-Hoffman, G., Y. Chen, E. Huang, and M. Huang. 2010. The effect of diet on protein concentration, hypopharyngeal gland development and virus load in worker honey bees (*Apis mellifera* L.). *J. Insect Physiol.* 56: 1184–1191.

De Jong, D., P. H. De Jong, and L. S. Gonçalves. 1982. Weight loss and other damage to developing worker honey bees from infestation with *V. jacobsoni*. *J. Apic. Res.* 21: 165–216.

Evans, J. D., K. Aronstein, Y. P. Chen, C. Hetru, J.-L. Imler, H. Jiang, M. Kanost, G. J. Thompson, Z. Zou, and D. Hultmark. 2006. Immune pathways and defense mechanisms in honey bees *Apis mellifera*. *Insect Mol. Biol.* 15: 645–656.

Evans, J. D., and M. Spivak. 2010. Socialized medicine: Individual and communal disease barriers in honey bees. *J. Invertebr. Pathol.* 103: S62–S72.

Fries, I. 1997. Protozoa. In *Honey Bee Pests, Predators and Diseases*. Morse, R. A., and K. Flottum (Eds.). A. I. Root Company, Medina, Ohio, U.S., pp. 59–76.

Fries, I., F. Feng, A. D. Silva, S. B. Slemenda, and N. J. Pieniazek. 1996. *Nosema ceranae n.* sp. (Microspora, Nosematidae), morphological and molecular characterization of a microsporidian parasite of the Asian honey bee *Apis cerana* (Hymenoptera, Apidae). *Eur. J. Protistol.* 32: 356–365.

Garedew, A., E. Schmolz, and I. Lamprecht. 2004. The energy and nutritional demand of the parasitic life of the mite *Varroa destructor*. *Apidologie* 35: 419–430.

Ghazoul, J. 2005. Buzziness as usual? Questioning the global pollination crisis. *Trends Ecol. Evol.* 20: 367–373.

Higes, M., R. Martín, and A. Meana. 2006. *Nosema ceranae*, a new microsporidian parasite in honey bees in Europe. *J. Invertebr. Pathol.* 92: 93–95.

Hölldobler, B., and E. O. Wilson. 1990. *The Ants*. The Belknap Press of Harvard University Press, Cambridge, Massachusetts, U.S.

Huang W. F., J. H. Jiang, Y. W. Chen, and C. H. Wang. 2007. A *Nosema ceranae* isolate from the honey bee *Apis mellifera*. *Apidologie* 38: 30–37.

Invernizzi, C., C. Abud, I. H. Tomasco, J. Harriet, G. Ramallo, J. Campá, H. Katz, G. Gardiol, and Y. Mendoza. 2009. Presence of *Nosema ceranae* in honey bees (*Apis mellifera*) in Uruguay. *J. Invertebr. Pathol.* 101: 150–153.

Karsai, I., and J. W. Wenzel. 1998. Productivity, individual-level and colony-level flexibility, and organization of work as consequences of colony size. *PNAS* 95: 8665–8669.

Kaspari, M., and E. L. Vargo. 1995. Colony size as a buffer against seasonality, Bergmann's rule in social insects. *Am. Nat.* 145: 610–632.

Klee, J., A. M. Besana, E. Genersch, S. Gisder, A. Nanetti, D. Q. Tam, T. X. Chinh, F. Puerta, J. M. Ruz, P. Kryger, D. Message, F. Hatjina, S. Korpela, I. Fries, and R. J. Paxton. 2007. Widespread dispersal of the microsporidian *Nosema ceranae*, an emergent pathogen of the western honey bee, *Apis mellifera*. *J. Invertebr. Pathol.* 96: 1–10.

Klein, A. M., B. E. Vaissiere, J. H. Cane, I. Steffan-Dewenter, S. A. Cunningham, C. Kremen, and T. Tscharntke. 2007. Importance of pollinators in changing landscapes for world crops. *Proc. R. Soc. B-Biol. Sci.* 274: 303–313.

Koch, W., and W. Ritter. 1991. Experimental examinations concerning the problem of deformed emerging bees after infestation with *Varroa jacobsoni*. *Zentralbl. Veterinaermed. B.* 38: 337–344.

Kovac, H., and K. Crailsheim. 1988. Lifespan of *Apis mellifera carnica* Pollm. infested by *Varroa jacobsoni* Oud. in relation to season and extent of infestation. *J. Apic. Res.* 27: 230–238.

Kralj, J., and S. Fuchs. 2006. Parasitic *Varroa destructor* mites influence flight duration and homing ability of infested *Apis mellifera* foragers. *Apidologie* 37: 577–587.

Larsson, R. 1986. Progress in protistology, ultrastructure, function, and classification of microsporidia. pp. 325–390. In *Progress in Protistology, Vol. 1*. Corliss J. O., and D. J. Patterson (Eds.). Vol. Biopress, Bristol, UK.

Lemaitre, B., and J. Hoffmann. 2007. The host defense of *Drosophila melanogaster*. *Annu. Rev. Immunol.* 25: 697–743.

Marcangeli, J., L. Monetti, and N. Fernandez. 1992. Malformations produced by *Varroa jacobsoni* on *Apis mellifera* in the province of Buenos Aires, Argentina. *Apidologie* 23: 399–402.

Martin, S. 1998. A population model for the ectoparasitic mite *Varroa jacobsoni* in honey bee (*Apis mellifera*) colonies. *Ecol. Model.* 109: 267–281.

Martin, S. J. 2001. The role of Varroa and viral pathogens in the collapse of honeybee colonies: A modelling approach. *J. Appl. Ecol.* 38: 1082–1093.

Martin, S. J., B. V. Ball, and N. L. Carreck. 2010. Prevalence and persistance of deformed wing virus (DWV) in untreated or acaricide-treated *Varroa destructor* infested honey bees (*Apis mellifera*) colonies. *J. Apic. Res.* 49: 72–79.

Matheson, A. 1993. World bee health update. *Bee World*. 74: 176–212.

McGregor, S. E. 1976. *Insect Pollination of Cultivated Crop Plants*. USDA Agriculture Handbook No. 496. U.S. Government Printing Office, Washington, DC. doi: http://afrsweb.usda.gov/sp2userfiles/place/53420300/onlinepollinationhandbook.pdf

Mullin, C. A., M. Frazier, J. L. Frazier, S. Ashcraft, R. Simonds, D. vanEngelsdorp, and J. S. Pettis. 2010. High levels of miticides and agrochemicals in North American apiaries: Implications for honey bee health. *PLoS ONE* 5(3): e9754. doi: 10.1371/journal.pone.0009754

Paxton, R. J., J. Klee, S. Korpela, and I. Fries. 2007. *Nosema ceranae* has infected *Apis mellifera* in Europe since at least 1998 and may be more virulent than *Nosema apis*. *Apidologie* 38: 558–565.

Potts, S. G., J. C. Biesmeijer, C. Kremen, P. Neumann, O. Schweiger, and W. E. Kunin. 2010. Global pollinator declines: Trends, impacts and drivers. *Trends Ecol. Evol.* 25: 345–353.

Ricketts, T. H., J. Regetz, I. Steffan-Dewenter, S. A. Cunningham, C. Kremen, and A. Bogdanski. 2008. Landscape effects on crop pollination services: Are there general patterns? *Ecol. Lett.* 11: 499–515.

Romero-Vera, C., and G. Otero-Colina. 2002. Effect of single and successive infestation of *Varroa destructor* and *Acarapis woodi* on the longevity of worker honey bees *Apis mellifera*. *Am. Bee J.* 142: 54–57.

Rueppell, O., C. Bachelier, M. K. Fondrik, and R. E. Page Jr. 2007. Regulation of life history determines lifespan of worker honey bees (*Apis mellifera* L.). *Exp. Gerontol.* 42: 120–132.

Schmid, M.R., A. Brockmann, C. W. W. Pirk, D. W. Stanley, and J. Tautz. 2008. Adult honey bees (*Apis mellifera* L.) abandon hemocytic, but not phenoloxidase-based immunity. *J. Insect Physiol.* 54: 439–444.

Schneider, P., and W. Drescher. 1987. The influence of *Varroa jacobsoni* Oud. on weight; development on weight and hypopharyngeal glands; and longevity of *Apis mellifera* L. *Apidologie* 18: 101–110.

Schultz, D. J., Z. Y. Huang, and G. E. Robinson. 1998. Effects of colony food shortage on behavioral development in honey bees. *Behav. Ecol. Sociobiol.* 42: 295–303.

Spivak, M., and G. S. Reuter. 2001. Resistance to American foulbrood disease by honey bee colonies *Apis mellifera* bred for hygienic behavior. *Apidologie* 32: 555–565.

Starks, P. T., C. A. Blackie, D. Thomas, and P. T. Seeley. 2000. Fever in honey bee colonies. *Naturwissenschaften* 87: 229–231.

Tentcheva, D., L. Gauthier, N. Zappulla, B. Dainat, F. Cousserans, M. E. Colin, and M. Bergoin. 2004. Prevalence and seasonal variations of six bee viruses in *Apis mellifera* L. and *Varroa destructor* mite populations in France. *Appl. Environ. Microbiol.* 70: 7185–7191.

Theopold, U., and M. S. Dushay. 2007. Mechanisms of Drosophila immunity—an innate immune system at work. *Current Immunol. Rev.* 3: 276–288.

Toth, A. L., and G. E. Robinson. 2005. Worker nutrition and division of labour in honey bees. *Animal Behav.* 69: 427–435.

Toth, A. L., S. Kantarovich, A. F. Meisel, and G. E. Robinson. 2005. Nutritional status influences socially regulated foraging ontogeny in honey bees. *J. Exp. Biol.* 208: 4641–4649.

Tsai, S. J., C. F. Lo, Y. Soichi, and C. H. Wang. 2003. The characterization of Microsporidian isolates (Nosematidae: Nosema) from five important lepidopteran pests in Taiwan. *J. Invertebr. Pathol.* 83: 51–59.

Wasson, K., and R. L. Peper. 2000. Mammalian microsporidiosis. *Vet. Pathol.* 37: 113–128.

Williams, G. R., A. B. A. Shafer, R. E. L. Rogers, D. Shutler, and D. T. Stewart. 2008. First detection of *Nosema ceranae*, a microsporidean parasite of European honey bees (*Apis mellifera*), in Canada and central U.S. *J. Invertebr. Pathol.* 97: 189–192.

Wilson, E. O. 1971. *The Insect Societies*. The Belknap Press of Harvard University Press, Cambridge, Massachusetts, U.S.

Wilson-Rich, N., S. T. Dres, and P. T. Starks. 2008. The ontogeny of immunity: Development of innate immune strength in the honey bee (*Apis mellifera*). *J. Insect Physiol.* 54: 1392–1399.

Yang, X., and D. Cox-Foster. 2007. Effects of parasitization by *Varroa destructor* on survivorship and physiological traits of *Apis mellifera* in correlation with viral incidence and microbial challenge. *Parasitology* 134: 405–412.

Yang, X. L., and D. L. Cox-Foster. 2005. Impact of an ectoparasite on the immunity and pathology of an invertebrate: Evidence for host immunosuppression and viral amplification. *PNAS* 102: 7470–7475.

Chapter 1

Honey Bee Health: The Potential Role of Microbes

Akira, S., S. Uematsu, and O. Takeuchi. 2006. Pathogen recognition and innate immunity. *Cell* 124: 783–801.

Asahara, T., K. Nomoto, K. Shimizu, M. Watanuki, and R. Tanaka. 2001. Increased resistance of mice to *Salmonella enterica serovar Typhimurium* infection by symbiotic administration of Bifidobacteria and transgalactosylated oligosaccharides. *J. Appl. Microbiol.* 91: 985–996.

Asahara, T., K. Shimizu, K. Nomoto, T. Hamabata, A. Ozawa, and Y. Takeda. 2004. Probiotic bifidobacteria protect mice from lethal infection with Shiga toxin-producing *Escherichia coli* O157: H7. *Infect. Immunol.* 72: 2240–2247.

Babendreier, D., D. Joller, J. Romeis, F. Bigler, and F. Widmer. 2007. Bacterial community structures in honey bee intestines and their response to two insecticidal proteins. *Federation of European Microbiological Societies (FEMS) Microbiol. Ecol.* 59: 600.

Baptist, J. N., M. Mandel, and R. L. Gherna. 1978. Comparative zone electrophoresis of enzymes in the genus *Bacillus*. *Int. J. Syst. Bacteriol.* 28: 229–244.

Burri, R. 1947. Die Beziehungen der Bakterien zum Lebenszyklus der Honigbiene. *Schweiz. Bienen-Ztg.* 70: 273–276.

Broderick, N. A., K. F. Raffa, R. M. Goodman, and J. Handelsman. 2004. Census of the bacterial community of the gypsy moth larval midgut by using culturing and culture independent methods. *Appl. Environ. Microbiol.* 70: 293–300.

Burnside, C. E. 1935. A disease of young bees caused by *Mucor. Am. Bee J.* 75: 75.

Cappelari, F. A., A. P. Turcatto, M. M. Morais, and D. DeJong. 2009. Africanized honey bees more efficiently convert protein diets into hemolymph protein than do Carniolan bee (*Apis mellifera carnica*). *Genet. Mol. Res.* 8: 1245–1249.

Casteel, D. B. 1912. *The behavior of the honey bee in pollen collecting*. Agr. Bulletin 121. USDA, Washington, DC, U.S.

Cherepov, V. T. 1966. The bacterial microflora of healthy honeybee colonies. *Tr. Nauch-Issled. Inst. Pchelovod.* 434–459 (in Russian)

Corthésy, B., H. R. Gaskins, and A. Mercenier. 2007. Cross-talk between probiotic bacteria and the host immune system. *J. Nutr.* 137: 781S–790S.

Cox-Foster, D. L., S. Conlan, E. C. Holmes, G. Palacios, J. D. Evans, N. A. Moran, P.-L. Quan, T. Briese, M. Hornig, D. M. Geiser, V. Martinson, D. vanEngelsdorp, A. L. Kalkstein, A. Drysdale, J. Hui, J. Zhai, L. Cui, S. K. Hutchison, J. F. Simons, M. Egholm, J. S. Pettis, and W. I. Lipkin. 2007. A metagenomic survey of microbes in honey bee Colony Collapse Disorder. *Science* 318: 283–287. doi: 10.1126/science.1146498

Czarnetzki, A. B., and C. C. Tebbe. 2004. Diversity of bacteria associated with Collembola—a cultivation-independent survey based on PCR-amplified 16S rRNA genes. *Federation of European Microbiological Societies (FEMS) Microbiol. Ecol.* 49: 217–227.

DeGrandi-Hoffman, G., G. Wardell, F. Ahumada-Secura, T. E. Rinderer, R. Danka, and J. Pettis. 2008. Comparisons of pollen substitute diets for honey bees: Consumption rates by colonies and effects on brood and adult populations. *J. Apic. Res.* 47: 265–270.

DeGrandi-Hoffman, G., Y. Chen, E. Huang, and M. Huang. 2010. The effect of diet on protein concentration, hypopharyngeal gland development and virus load in worker honey bees (*Apis mellifera* L.). *J. Insect Physiol.* 56: 1184–1191.

Dumas M. E., E. C. Maibaum, C. Teague, H. U. B. Zhou, J. C. Lindon, J. K. Nicholson, J. Stamler, P. Elliott, Q. Chan, and E. Holmes. 2006. Assessment of analytical reproducibility of 1H NMR spectroscopy-based metabonomics for large-scale epidemiological research: The INTERMAP Study. *Anal. Chem.* 78: 2199–2208.

Egert, M., B. Wagner, T. Lemke, A. Brune, and M. W. Friedrich. 2003. Microbial community structure in midgut and hindgut of the humus-feeding larva of *Pachnoda ephippiata* (Coleoptera: Scarabaeidae). *Appl. Environ. Microbiol.* 69: 6659–6668.

Evans, J. D., and T. N. Armstrong. 2006. Antagonistic interactions between honey bee bacterial symbionts and implications for disease. BMC *Ecology*: http://www.biomedcentral.com/1472–6785/6/4

Fogarty, W. M., P. J. Griffin, and A. M. Joyce. 1974a. Enzymes of *Bacillus* species part 1. *Process Biochem.* 9: 11–18, 24.

Fogarty, W. M., P. J. Griffin, and A. M. Joyce. 1974b. Enzymes of *Bacillus* species part 2. *Process Biochem.* 9: 27–29, 31, 33, 35.

Foote, H. L. 1957. Possible use of microorganisms in synthetic bee bread production. *Am. Bee J.* 97: 476–478.

Gill, S. R., M. Pop, R. DeBoy, P. B. Eckberg, P. J. Turnbaugh, B. S. Samuel, J. I. Gordon, D. A. Relman, C. M. Fraser-Liggett, and K. E. Nelson. 2006. Metagenomic analysis of the human distal gut microbiome. *Science* 312: 1355–1359.

Gilliam, M. 1971. Microbial sterility of the intestinal content of the immature honey bee, *Apis mellifera*. *Ann. Entomol. Soc. of Am.* 64: 315–316.

Gilliam, M. 1979a. Microbiology of pollen and bee bread: The yeasts. *Apidologie* 10: 43–53.

Gilliam, M. 1979b. Microbiology of pollen and bee bread: The genus *Bacillus*. *Apidologie* 10: 269–274.

Gilliam, M. 1997. Identification and roles of non-pathogenic microflora associated with honey bees. *Federation of European Microbiological Societies (FEMS) Microbiol. Lett.* 155: 1–10.

Gilliam, M., and D. K. Valentine. 1974. Enterobacteriaceae isolated from foraging worker honey bees. *J. Invertebr. Pathol.* 23: 38–41.

Gilliam, M., J. O. Moffett, and N. M. Kaufield. 1983. Examination of floral nectar of citrus, cotton, and Arizona desert plants for microbes. *Apidologie* 14: 299–302.

Gilliam, M., and D. B. Prest. 1987. Microbiology of feces of the larval honey bee, *Apis mellifera*. *J. Invertebr. Pathol.* 49: 70–75.

Gilliam, M., B. J. Lorenz, and G. V. Richardson. 1988. Digestive enzymes and micro-organisms in honey bees, *Apis mellifera*: Influence of streptomycin, age, season, and pollen. *Microbios* 55: 95–114.

Gilliam, M., D. B. Prest, and B. J. Lorenz. 1989. Microbiology of pollen and bee bread: Taxonomy and enzymology of molds. *Apidologie* 20: 53–68.

Gershwin, M. E., A. T. Borchers, and C. L. Keen. 2000. Phenotypic and functional considerations in the evaluation of immunity in nutritionally compromised hosts. *J. Infectious Diseases* 2000: 182 (Suppl 1): S108–14.

Handelsman, J. 2004. Metagenomics: Application of genomics to uncultured microorganisms. *Microbiol. Mol. Biol. Rev.* 68: 669–685.

Haydak, M. H., and A. E. Vivino. 1950. The changes in thiamine, riboflavin, niacin, and pantothenic acid content in the food of female honey bees during growth with a note on the vitamin K activity of royal jelly and bee bread. *Annu. Entomol. Soc. Am.* 43: 361–367.

Haydak, M. H. 1958. Pollen—pollen substitutes—bee bread. *Am. Bee J.* 98: 145–146.

Hetru, C., D. Hoffmann, and P. Bullet. 1998. Antimicrobial peptides from insects. In *Molecular Mechanisms of Immune Responses in Insects,* pp. 40–66. Brey, P. T., and D. Hultmark (Eds.). Chapman and Hall, London, UK.

Higes, H., R. Martin, and A. Meana. 2006. *Nosema ceranae,* a new microsporidian parasite in honeybees in Europe. *J. Invertebr. Pathol.* 92: 93–95.

Hongoh, Y., M. Ohkuma, and T. Kudo. 2003. Molecular analysis of bacterial microbiota in the gut of the termite *Reticulitermes speratus* (Isoptera; Rhinotermitidae). *Federation of European Microbiological Societies (FEMS) Microbiol. Ecol.* 44: 231–242.

Hubbell, S. P. 2006. Neutral theory and the evolution of ecological equivalence. *Ecology* 87: 1387–1398.

Human, H., and S. W. Nicolson. 2006. Nutritional content of fresh, bee-collected and stored pollen of *Aloe greatheadii* var. *davyana* (Asphodelaceae). *Phytochemistry* 67: 1486–1492.

Jeyaprakash, A., M. A. Hoy, and M. H. Allsopp. 2003. Bacterial diversity in worker adults of *Apis mellifera capensis* and *Apis mellifera scutellata* (Insecta: Hymenoptera) assessed using 16S rRNA sequences. *J. Invertebr. Pathol.* 84: 96–103.

Kaneda, T. 1977. Fatty acids of the genus *Bacillus*: An example of branched–chain preference. *Bacteriol. Rev.* 41: 391–418.

Katz, E., and A. L. Demain. 1977. The peptide antibiotics of *Bacillus*: Chemistry, biogenesis, and possible functions. *Bacteriol. Rev.* 41: 449–474.

Kelly D., T. King, and R. Aminov. 2007. Importance of microbial colonization of the gut in early life to the development of immunity. *Mutat. Res.* 622: 58–69.

Klaudiny, J., S. Albert, K. Bachanová, J. Kopernicky, and J. Simúth. 2005. Two structurally different defensin genes, one of them encoding a novel defensin isoform, are expressed in honey bee *Apis mellifera. Insect Biochem. Mol. Biol.* 35: 11–22.

Kluge, R. 1963. Untersuchungen ueber die Darmflora der Honigbiene, *Apis mellifera. Z. Bienenforsch.* 6: 141–169.

Klungness, L. M., and Y.-S. Peng. 1984. A histochemical study of pollen digestion in the alimentary canal of honey bees (*Apis mellifera* L.). *J. Insect Physiol.* 30: 511–521.

Kowalchuk, G. A., G. C. L. Speksnijder, K. Zhang, R. M. Goodman, and J. A. vanVeen. 2007. Finding the needles in the metagenome haystack. *Microbial Ecol.* 53: 475–485.

Lajudie, J., and V. C. Dumanoir. 1976. Recherche de l'activite pectinolytique chez le genre *Bacillus. Ann. Microbiol. Inst. Pasteur* 127A: 423–427.

Lamberty, M., S. Ades, S. Uttenwiler-Joseph, G. Brookharts, D. Bushey, J. A. Hoffman, and P. Bulet. 1999. Insect immunity: Isolation from the lepidopteran *Heliothis virescens* of a novel insect defensin with potent antifungal activity. *J. Biol. Chem.* 274: 9320–9326.

Ley, R. E., F. Bäckhed, P. Turnbaugh, C. A. Lozupone, R. D. Knight, and J. I. Gordon. 2005. Obesity alters gut microbial ecology. *PNAS* 102: 11070–11075.

Li, M., B. Wang, M. Zhang, M. Rantalainen, S. Wang, H. Zhou, Y. Zhang, J. Shen, X. Pang, M. Zhang, H. Wei, Y. Chen, H. Lu, J. Zuo, M. Su, Y. Qiu, J. Wei, C. Xiao, L. M. Smith, S. Yang, E. Holmes, H. Tang, G. Zhao, J. K. Nicholson, L. Li, and L. Zhao. 2008. Symbiotic gut microbes modulate human metabolic phenotypes. *PNAS* 105: 2117–2122.

Mattila, H. R., and G. W. Otis 2006. Influence of pollen diet in spring on the development of the honey bee (Hymenoptera: Apidae) colonies. *J. Econ. Entomol.* 99: 604–613.

Mitchell, J. A., M. J. Paul-Clark, G. W. Clarke, S. K. McMaster, and N. Cartwright. 2007. Critical role of toll-like receptors and nucleotide oligomerisation domain in the regulation of health and disease. *J. Endocrinol.* 193: 323–330.

Mohr, K. I., and C. C. Tebbe. 2006. Diversity and phylotype consistency of bacteria in the guts of three bee species (Apoidea) at an oilseed rape field. *Environ. Microbiol.* 8: 258–272.

Nicholson, J. K., E. Holmes, and I. D. Wilson. 2005. Gut microorganisms, mammalian metabolism and personalized health care. *Nat. Rev. Microbiol.* 3: 431–438.

Nout, M. J. R., and F. M. Rombouts. 1992. Fermentative preservation of plant foods. *J. Appl. Bacteriol. Symp. Suppl.* 73: 136S–147S.

Olofsson, T. C., and A. Vásquez. 2008. Detection and Identification of a novel lactic acid bacterial flora within the honey stomach of the honeybee *Apis mellifera*. *Curr. Microbiol.* 57: 356–363.

Olofsson, T. C., and A. Vásquez. 2009. The lactic acid bacteria involved in the production of bee pollen, and bee bread. *J. Apic. Res.* 48: 189–195.

Pain, J., and J. Maugenet. 1966. Recherches biochimiques et physiologiques sur le pollen emmagasine par les abeilles. *Annu. Abeille* 9: 209–236.

Rakoff-Nahoum, S., J. Paglino, F. Eslami-Varzaneh, S. Edberg, and R. Medzhitov. 2004. Recognition of commensal microflora by toll-like receptors is required for intestinal homeostasis. *Cell* 118: 229–241.

Reeson, A. F., T. Jankovic, M. L. Kasper, S. Rogers, and A. D. Austin. 2003. Application of 16S rDNA-DGGE to examine the microbial ecology associated with a social wasp *Vespula germanica*. *Insect Mol. Biol.* 12: 85–91.

Rennemeier, C., T. Frambach, and F. Hennicke. 2009. Microbial quorum-sensing moleculdes induce acrosome loss and cell death in human spermatozoa. *Infect. Immunol.* 77: 4990–4997.

Schmalenberger, A., and C. C. Tebbe. 2003. Bacterial diversity in maize rhizospheres: Conclusions on the use of genetic profiles based on PCR-amplified partial small subunit rRNA genes in ecological studies. *Mol. Ecol.* 12: 251–261.

Schneider, S. S., G. DeGrandi-Hoffman, and D. Smith. 2004. The African honeybee: Factors contributing to a successful biological invasion. *Annu. Rev. Entomol.* 49: 351–376.

Serzedello, A., and S. M. Tauk. 1974. Cellulase de bacterias isoladas de ninhos de *Atta laevigata* Smith. *Ciencia e Cultura* (Sao Paulo, Brazil) 26: 957–960.

Snodgrass, R. E. 1925. *Anatomy and Physiology of the Honeybee.* McGraw-Hill, New York, NY, U.S.

Standifer, L. N., W. F. McCaughey, S. E. Dixon, M. Gilliam, and G. M. Loper. 1980. Biochemistry and microbiology of pollen collected by honey bees (*Apis mellifera* L.) from almond, *Prunis dulcis*. II. Protein, amino acids and enzymes. *Apidologie* 11: 163–171.

Turnbaugh, P. J., R. E. Ley, M. A. Mahowald, V. Magrini, E. R. Mardis, and J. I. Gordon. 2006. An obesity associated gut microbiome with increased capacity for energy harvest. *Nature* 444: 1027–1031.

Turnbaugh, P. J., R. E. L. Ley, M. Hamady, C. M. Fraser-Liggett, R. Knight, and J. I. Gordon. 2007. The human microbiome project. *Nature* 449: 804–810.

Umenai, T., H. Hirai, N. Shime, T. Nakaya, T. Asahara, K. Nomoto, M. Kita, Y. Tanaka, and J. Imanishi. 2010. Eradication of the commensal intestinal microflora by oral antimicrobials interferes with the host response to lipopolysaccharide. *Eur. J. Clin. Microbiol. Infect. Dis.* 29: 633–641.

Woolford, M. K. 1978. Antimicrobial effects of mineral acids, organic acids, salts and sterilizing agents in relation to their potential as silage additives. *J. Br. Grassland Soc.* 33: 131–136.

Yamauchi, H. 2001. Two novel insect defensin from larvae of the Cupreous Chafer, *Anomala cupra*: Purification, amino acid sequences and antibacterial activity. *Insect Biochem. Mol. Biol.* 32: 75–78.

Zoetendal, E. G., A. D. L. Akkermans, W. M. Akkermans-vanVliet, J. A. G. M. deVisser, and W. M. deVos. 2001. The host genotype affects the bacterial community in the human gastrointestinal tract. *Microb. Ecol. Health Dis.* 13: 129–134.

Chapter 2

Seasonal Microflora, Especially Winter and Spring

Black, J. G. 2008. *Microbiology: Principles and Exploration* Seventh Edition. Wiley & Sons, New Jersey, U.S.

Cox-Foster, D. L., S. Conlan, E. C. Holmes, G. Palacios, J. D. Evans, N. A. Moran, P.-L. Quan, T. Briese, M. Hornig, D. M. Geiser, V. Martinson, D. vanEngelsdorp, A. L. Kalkstein, A. Drysdale, J. Hui, J. Zhai, L. Cui, S. K. Hutchison, J. F. Simons, M. Egholm, J. S. Pettis, and W. I. Lipkin. 2007. A metagenomic survey of microbes in honey bee Colony Collapse Disorder. *Science* 318: 283–287. doi: 10.1126/science.1146498

Fritzsch, W., and R. Bremer. 1975. *Bienengesundheitsdienst* (Bee Health Service), VEB Gustav Ficher Verlag, Jena Germany.

Gilliam, M. 1979. Microbiology of pollen and bee bread: The genus *Bacillus*. *Apidologie* 10: 269–274.

Gilliam, M., B. J. Lorenz, and G. V. Richardson. 1988. Digestive enzymes and micro-organisms in honey bees (*Apis mellifera*) influence of streptomycin, age, season and pollen. *Microbios* 55: 95–114.

Jeyaprakash, A., M. A. Hoy, and M. H. Allsopp. 2003. Bacterial diversity in worker adults of *Apis mellifera capensis* and *Apis mellifera scutellata* (Insecta: Hymenoptera) assessed using 16s rRNA sequencing. *J. Invertebr. Pathol.* 84: 96–103.

Kacaniova, M., R. Chlebo, M. Kopernicky, and A.Trakovicka. 2004. Microflora of the honey bee gastrointestinal tract. *Folia Microbiol.* 49(2): 169–171.

Kilwinski, J., M. Peters, A. Ashiralieva, and E. Genersch. 2004. Proposal to reclassify *Paenibacillus larvae* subsp. *pulvifaciens* DSM 3615 (ATCC49843) as *Paenibacillus larvae* subsp. *larvae*. Results from a comparative biochemical and genetic study. *Vet. Microbiol.* 104: 31–42.

Lyapunov, Ya. E., and R. Z. Kuzyaev, R. G. Khismatullin, and Bezgodova O. A., 2008. Intestinal enterobacteria of the hibernating *Apis mellifera mellifera* L. bees. *Microbiology* 77(3): 373–379.

Machova, M., and V. Rada, J. Huk, and F. Smekal. 1997. Development of probiotics. *Apiacta* 32:99–111.

Madigan, M. T., J. M. Martinko, P. V. Dunlap, and D. P. Clarck. 2009. *Brock Biology of Microorganisms* (12th ed.). Pearson Bejamin Cummings, New Jersey, U.S.

Manning, R. 2001. Fatty acids in pollen: A review of their importance for honey bees. *Bee World.* 82: 60–75.

Olofsson, T. C., and A. Vásquez. 2008. Detection and Identification of a novel lactic acid bacterial flora with in the honey stomach of the honey bee *Apis mellifera*. *Curr. Microbiol.* 57: 356–363.

Özkirim, A., and N. Keskin. 2002. Distribution of the major bacterial brood diseases diagnosed in Ankara and its surroundings. *Mellifera.* 2(4): 40–44.

Piccini, C., K. Antunez, and P. Zunino. 2004. An approach to the characterization of the honey bee hive bacterial flora. *J. Apic. Res.* 43(3): 101–104.

Sammataro, D., and J. Cicero. 2010. Functional morphology of the honey stomach wall of the European honey bee (Hymenoptera: Apidae). *Annals Entomol. Soc. Am. Section A* 103(6): 979–987.

Toumanoff, C. 1951. Flore bacterienne intestinale de l'abeille adulte. *Les Maladies Des Abeilles, La Revue Française d'apiculture,* no. 68–numero special, pp. 145–151. (in French)

Tysset, C., and M. Rousseau. 1967. Des germes phytopathogenes du genre *Erwinia commensaux* du tractus intestinal de l'abeille adulte (*Apis mellifera* L.). *B. Tech. Apic.* 2: 175–194.

Vásquez, A., and T. C. Olofsson. 2009. The lactic acid bacteria involved in the production of bee pollen and bee bread. *J. Apic. Res. and Bee World.* 48(3): 189–195.

Vassart, M., J. Thevenon, and F. Cotto. 1988. Étude sanitaire du rucher Neo-caledonien. *B. Tech. Apic.* 15(4): 195–202.

Chapter 3

Evaluation of Varroa Mite Tolerance in Honey Bees

Allsopp, M. 2006. Analysis of *Varroa destructor* infestation of southern African honey bee populations. MS thesis, University of Pretoria, Pretoria, South Africa.

Allsopp, M., V. Govan, and S. Davison. 1997. Bee health report: Varroa in South Africa. *Bee World* 78: 171–174.

Anderson, D. L., and J. W. H. Trueman. 2000. *Varroa jacobsoni* (Acari: Varroidae) is more than one species. *Exp. Appl. Acarol.* 24: 165–189.

Arathi, H. S., and M. Spivak. 2001. Influence of colony genotypic composition on the performance of hygienic behaviour in the honeybee, *Apis mellifera* L. *Anim. Behav.* 62: 57–66.

Boecking, O. 1993. Vergleich der Abwehrmechanismen von *Apis cerana* und *Apis mellifera* gegen die Varroamilbe. *Allg. Dtsch. Imkerztg.* 27: 23–27.

Büchler, R., S. Berg, and Y. Le Conte. 2010. Breeding for resistance to *Varroa destructor* in Europe. *Apidologie* 41: 393–408.

Calatayud, F., and M. J. Verdu. 1993. Hive debris counts in honey bee colonies: A method to estimate the size of small populations and rate of growth of the mite *Varroa jacobsoni* Oud. (Mesostigmata: Varroidae). *Exp. Appl. Acarol.* 17: 889–894.

Calatayud, F., and M. J. Verdu. 1995. Number of adult female mites *Varroa jacobsoni* Oud. on hive debris from honey bee colonies artificially infested to monitor mite population increase (Mesostigmata: Varroidae). *Exp. Appl. Acarol.* 19: 181–188.

Camazine, S. 1986. Differential reproduction of the mite *Varroa jacobsoni* (Mesostigmata, Varroidae), on Africanized and European honey bees (Hymenoptera, Apidae). *Ann. Entomol. Soc. Am.* 79: 801–803.

Camazine, S. 1988. Factors Affecting the Severity of *Varroa jacobsoni* Infestations on European and Africanized Bees. In *Africanized Honey Bees and Bee Mites*. Needham, G. R., and R. E. Page, Jr. (Eds.). Ellis Horwood Ltd., Chichester, UK, pp. 444–451.

Frazier, M., E. Muli, T. Conklin, D. Schmehl, B. Torto, J. Frazier, J. Tumlinson, J. D. Evans, and S. Raina. 2010. A scientific note on *Varroa destructor* found in East Africa; threat or opportunity? *Apidologie* 41: 463–465.

Fries, I., S. Camazine, and J. Sneyd. 1994. Population dynamics of *Varroa jacobsoni*: A model and a review. *Bee World*. 75: 5–28.

Fries, I., A. Imdorf, and P. Rosenkranz. 2006. Survival of mite (*Varroa destructor*) infested honey bee (*Apis mellifera*) colonies in a Nordic climate. *Apidologie* 37: 564–570.

Fries, I., and S. Raina. 2003. American foulbrood and African honey bees (Hymenoptera: Apidae). *J. Econ. Entomol.* 96: 1641–1646.

Fries, I., H. Wei, W. Shi, and S. J. Chen. 1996. Grooming behavior and damaged mites (*Varroa jacobsoni*) in *Apis cerana cerana* and *Apis mellifera ligustica*. *Apidologie* 27: 3–11.

Fuchs, S. 1990. Preference for drone brood cells by *Varroa jacobsoni* Oud. in colonies of *Apis mellifera carnica*. *Apidologie* 21: 193–199.

Harbo, J. R., and J. W. Harris. 2009. Responses to Varroa by honey bees with different levels of Varroa Sensitive Hygiene. *J. Apic. Res.* 48: 156–161.

Ibrahim, A., G. S. Reuter, and M. Spivak. 2007. Field trial of honey bee colonies bred for mechanisms of resistance against *Varroa destructor*. *Apidologie* 38: 67–76.

Ifantidis, M. D. 1984. Parameters of the population dynamics of the Varroa mite of honeybees. *J. Apic. Res.* 23: 227–233.

Korpela, S., A. Aarhus, I. Fries, and H. Hansen. 1993. *Varroa jacobsoni* Oud. in cold climates: Population growth, winter mortality and influence on survival of honey bee colonies. *J. Apic. Res.* 31: 157–164.

Le Conte, Y., G. De Vaublanc, D. Crauser, F. Jeanne, J. C. Rousselle, and J. M. Becard. 2007. Honey bee colonies that have survived *Varroa destructor*. *Apidologie* 38: 566–572.

Lee, K. V., R. D. Moon, E. C. Burkness, W. D. Hutchison, and M. Spivak. 2010. Practical sampling plans for *Varroa destructor* (Acari: Varroidae) in *Apis mellifera* (Hymenoptera: Apidae) colonies and apiaries. *J. Econ. Entomol.* 103: 1039–1050.

Martin, S. 1995. Ontogenesis of the mite *Varroa jacobsoni* Oud. in drone brood of the honey bee *Apis mellifera* L. under natural conditions. *Exp. Appl. Acarol.* 19: 199–210.

Medina, L. M., S. J. Martin, L. Espinosa-Montano, and F. L. W. Ratnieks. 2002. Reproduction of *Varroa destructor* in worker brood of Africanized honey bees (*Apis mellifera*). *Exp. Appl. Acarol.* 27: 79–88.

Mondragon, L., S. Martin, and R. Vandame. 2006. Mortality of mite offspring: A major component of *Varroa destructor* resistance in a population of Africanized bees. *Apidologie* 37: 67–74.

Peng, Y. S., Y. Fang, S. Xu, and L. Ge. 1987. The resistance mechanism of the Asian honey bee *Apis cerana* Fabr. to an ectoparasitic mite, *Varroa jacobsoni* Oudemans. *J. Invertebr. Pathol.* 49: 54–60.

Rath, W., and W. Drescher. 1990. Response of *Apis cerana* Fabr. towards brood infested with *Varroa jacobsoni* Oud. and infestation rate of colonies in Thailand. *Apidologie* 21: 311–321.

Rosenkranz, P., P. Aumeier, and B. Ziegelmann. 2010. Biology and control of *Varroa destructor*. *J. Invertebr. Pathol.* 103: S96–S119.

Seeley, T. D. 2007. Honey bees of the Arnot Forest: A population of feral colonies persisting with *Varroa destructor* in the northeastern United States. *Apidologie* 38(1): 19–29.

Spivak, M., and G. S. Reuter. 2001. *Varroa destructor* infestation in untreated honey bee (Hymenoptera: Apidae) colonies selected for hygienic behavior. *J. Econ. Entomol.* 94: 326–331.

Vandame, R., G. Otero-Colina, and M. Colin. 1995. Dinámica comparativa de las poplaciones de *Varroa jacobsoni* en colmenas de abejas europeas y africanizadas en Córdoba, Ver. pp. 61–62. In *IX Seminario Americano de Apicultura*. Colmina, Mexico.

Chapter 4

Status of Breeding Practices and Genetic Diversity in Domestic U.S. Honey Bees

Anonymous. 1859. Italian bees. *Am. Agriculturalist*. 18: 346.

Avetisyan, G. A. 1961. The relation between interior and exterior characteristics of the queen and fertility and productivity of the bee colony. In *XVIII International Beekeeping Congress,* Madrid, Spain, pp. 44–53.

Benton, F. 1905. Caucasian Bees. *Am. Bee. J.* 45: 60–61.

Bourgeois, L., W. S. Sheppard, H. A. Sylvester, and T. E. Rinderer. 2010. Genetic stock identification of Russian honey bees. *J. Econ. Entomol.* 103: 917–924.

Burgett, M., and C. Kitprasert. 1992. Tracheal mite infestation of queen honey-bees. *J. Apic. Res.* 31: 110–111.

Camazine, S., I. Çakmak, K. Cramp, J. Finley, J. Fisher, M. Frazier, and A. Rozo. 1998. How healthy are commercially-produced U.S. honey bee queens? *Am. Bee. J.* 138: 677–680.

Chen, Y., J. D. Evans, I. B. Smith, and J. S. Pettis. 2008. *Nosema ceranae* is a long-present and wide-spread microsporidian infection of the European honey bee (*Apis mellifera*) in the United States. *J. Invertebr. Pathol.* 97: 186–188.

Chen, Y. P., J. Evans, and M. F. Feldlaufer. 2006. Horizontal and vertical transmission of viruses in the honeybee, *Apis mellifera*. *J. Invertebr. Pathol.* 92: 152–159.

Chen, Y. P., J. S. Pettis, and M. F. Feldlaufer. 2005. Detection of multiple viruses in queens of the honey bee *Apis mellifera* L. *J. Invertebr. Pathol.* 90: 118–121.

Cobey, S. 2003. The extraordinary honey bee mating strategy and a simple field dissection of the Spermatheca—A three-part series. Part 1: Mating behavior. *Am. Bee. J.* 143: 67–69.

Collins, A. M. 2000. Relationship between semen quality and performance of instrumentally inseminated honey bee queens. *Apidologie* 31: 421–429.

Connor, L. J. 2008. *Bee Sex Essentials*. Wicwas Press, New Haven, Connecticut, U.S.

Dadant, C. 1877. Hardan Haines ventilated. *Am. Bee. J.* 13: 27.

Dedej, S., K. Hartfelder, P. Aumeier, P. Rosenkranz, and W. Engels. 1998. Caste determination is a sequential process: Effect of larval age at grafting on ovariole number, hind leg size and cephalic volatiles in the honey bee (*Apis mellifera carnica*). *J. Apic. Res.* 37: 183–190.

Delaney, D. A., M. D. Meixner, N. M. Schiff, and W. S. Sheppard. 2009. Genetic characterization of commercial honey bee (Hymenoptera: Apidae) populations in the United States by using mitochondrial and microsatellite markers. *Ann. Entomol. Soc. Am.* 102: 666–673.

Delaney, D. A., J. J. Keller, J. R. Caren, and D. R. Tarpy. 2011. The physical, insemination, and reproductive quality of honey bee queens (*Apis mellifera*). *Apidologie*, 42: 1–13.

Delaplane K. S., and D. F. Mayer. 2000. *Crop Pollination by Bees*. CABI Publishing, New York, NY, U.S.

Dodologlu, A., B. Emsen, and F. Gene. 2004. Comparison of some characteristics of queen honey bees (*Apis mellifera* L.) reared by using Doolittle method and natural queen cells. *J. Appl. Anim. Res.* 26: 113–115.

Eckert, J. E. 1934. Studies in the number of ovarioles in queen honeybees in relation to body size. *J. Econ. Entomol.* 27: 629–635.

Farrar, C. L. 1947. Nosema losses in package bees as related to queen supersedure and honey yields. *J. Econ. Entomol.* 40: 333–338.

Fischer, F., and V. Maul. 1991. Untersuchungen zu Aufzuchtbedingten Königinnenmerkmalen. (An inquiry into the characteristics of queens depending on queen rearing). *Apidologie* 22: 444–446.

Nelson, D. L., and N. E. Gary. 1983. Honey productivity of honey bee *Apis mellifera* colonies in relation to body weight attractiveness and fecundity of the queen. *J. Apic. Res.* 22: 209–213.

Norris, A. J. 1884. Carniolan Apiary. *Am. Bee. J.* 20: 139.

Oertel, E. 1976. Early records of honey bees in the Eastern United States. *Am. Bee. J.* 166: 70–71.

Palmer, K. A., and B. P. Oldroyd. 2000. Evolution of multiple mating in the genus *Apis*. *Apidologie* 31: 235–248.

Richard, F.-J., D. R. Tarpy, and C. M. Grozinger. 2007. Effects of insemination quantity on honey bee queen physiology. *PLoS ONE* 2: E980.

Ruttner F. 1975. Races of bees. pp. 19–38. In *The Hive and the Honey Bee*. C. P. Dadant (Ed.). Dadant and Sons, Hamilton, Illinois, U.S.

Ruttner, F. 1988. *Biogeography of the Honey Bee*. Springer-Verlag, Paris, France.

Schiff, N. M., and W. S. Sheppard. 1993. Mitochondrial DNA evidence for the 19th century introduction of African honey bees into the United States. *Experientia* 49: 530–532.

Schiff, N. M., W. S. Sheppard, G. M. Loper, and H. Shimanuki. 1994. Genetic diversity of feral honey bee (Hymenoptera: Apidae) populations in the southern United States. *Ann. Entomol. Soc. Am.* 87: 842–848.

Schiff, N. M., and W. S. Sheppard. 1995. Genetic analysis of commercial honey bees from the southeastern United States. *J. Econ. Entomol.* 88(5): 1216–1220.

Schiff, N. M., and W. S. Sheppard. Genetic differentiation in the queen breeding popualtion of the Western United States. *Apidologie* 27: 77–86.

Seeley, T. D. 2007. Honey bees of the Arnot Forest: A population of feral colonies persisting with *Varroa destructor* in the northeastern United States. *Apidologie* 38: 19–29.

Seeley, T. D., and D. R. Tarpy. 2007. Queen promiscuity lowers disease within honey bee colonies. *Proc. Royal Society of London, B*. 274: 67–72.

Shepherd, M. W. 1892. Punic (or Tunisian) bees. *Gleanings in Bee Culture* 20: 504.

Sheppard, W. S. 1988. Comparative study of enzyme polymorphism in US and European honey bee populations. *Ann. Entomol. Soc. Am.* 81: 886–889.

Sheppard, W. S. 1989. A history of the introduction of honey bee races into the United States, I and II. *Am. Bee J.* 129: 617–619, 664–667.

Sheppard, W. S., and M. D. Meixner. 2003. *Apis mellifera pomonella*, a new honey bee from the Tien Shan Mountains of Central Asia. *Apidologie* 34: 367–375.

Sherman, P. W., T. D. Seeley, and H. K. Reeve. 1988. Parasites, pathogens and polyandry in social hymenoptera. *Am. Nat.* 131: 602–610.

Tanaka, E. D., and K. Hartfelder. 2004. The initial stages of oogenesis and their relation to differential fertility in the honey bee (*Apis mellifera*) castes. *Arthropod Struct. Dev.* 33: 431–442.

Tarpy, D. R., and D. I. Nielsen. 2002. Sampling error, effective paternity, and estimating the genetic structure of honey bee colonies (Hymenoptera: Apidae). *Ann. Entomol. Soc. Am.* 95: 513–528.

Tarpy, D. R., and R. E. J. Page. 2002. Sex determination and the evolution of polyandry in honey bees. *Behav. Ecol. Sociobiology* 52: 143–150.

Tarpy, D. R., R. Nielsen, and D. I. Nielsen. 2004. A scientific note on the revised estimates of effective paternity frequency in *Apis*. *Insectes Sociaux* 51: 203–204.

Tarpy D. R., and T. D. Seeley. 2006. Lower disease infections in honeybee (*Apis mellifera*) colonies headed by polyandrous vs monandrous queens. *Naturwissenschaften* 93: 195–199.

Tefft, J. W. 1890. The keeping of bees and their improvement. *Am. Bee. J.* 26: 399.

USDA. 2007. Africanized Bees. honeybeenet.gsfc.nasa.gov/Honeybees/AHB.htm

vanEngelsdorp, D., J. Hayes, R. M. Underwood, and J. Pettis. 2008. A survey of honey bee colony losses in the U.S., fall 2007 to spring 2008. *PLoS ONE* 3: e4071. doi: 10.1371/journal.pone.0004071

vanEngelsdorp, D., J. D. Evans, C. Saegerman, C. Mullin, E. Haubruge, B. K. Nguyen, M. Frazier, J. Frazier, D. R. Tarpy, and J. S. Pettis. 2009. Colony Collapse Disorder: A Descriptive Study. *PLoS ONE*, 4: e6481. doi10.1371/journal.pone.0006481

Villa, J. D., and R. G. Danka. 2005. Caste, sex and strain of honey bees (*Apis mellifera*) affect infestation with tracheal mites (*Acarapis woodi*). *Exp. App. Acarol.* 37: 157–164.

Weaver, N. 1957. Effects of larval age on dimorphic differentiation of the female honey bee. *Ann. Entomol. Soc. Am.* 50: 283–294.

Whitfield, C. W., S. K. Behura, S. H. Berlocher, A. G. Clark, J. S. Johnson, W. S. Sheppard, D. R. Smith, A. V. Suarez, D. Weaver, and N. D. Tsutsui. 2006. Thrice out of Africa: Ancient and resent expansions of the honey bee, *Apis mellifera*. *Science* 314: 642–645.

Woyke, J. 1962. Natural and artificial insemination of queen honeybees. *Bee World* 43: 21–25.

Woyke, J. 1971. Correlations between the age at which honeybee brood was grafted, characteristics of the resultant queens, and results of insemination. *J. Apic. Res.* 10: 45–55.

Woyke, J. 1983. Dynamics of entry of spermatozoa into the spermatheca of instrumentally inseminated queen honeybees. *J. Apic. Res.* 22: 150–154.

Yang, X. L., and D. L. Cox-Foster. 2005. Impact of an ectoparasite on the immunity and pathology of an invertebrate: Evidence for host immunosuppression and viral amplification. *PNAS* 102: 7470–7475.

York, G. W. 1906. More testimony on Caucasian bees. *Am. Bee. J.* 46: 98.

Chapter 5
Global Status of Honey Bee Mites

Aggarwal, K., and R. P. Kapil. 1988. Observations on the effect of queen cell construction on *Euvarroa sinhai* infestation in drone brood of *Apis florea*. In *Africanized Honey Bees and Bee Mites*. G. R. Needham, R. E. Page, Jr., M. Delfinado-Baker, and C. E. Bowman (Eds.). Ellis Horwood Ltd., Chichester, UK, pp. 404–408.

Ali, M. A., M. D. Ellis, J. R. Coats, and J. Grodnitzky. 2002. Laboratory evaluation of 17 monoterpenoids and field evaluation of two monoterpenoids and two registered acaricides for the control of *Varroa destructor* Anderson and Trueman (Acari: Varroidae). *Am. Bee J.* 142(1): 50–53.

Aliano, N. P., and M. D. Ellis. 2005. A strategy for using powdered sugar to reduce varroa populations in honey bee colonies. *J. Apic. Res.* 44(2): 54–57.

Anderson, D. L., and M. J. Morgan. 2007. Genetic and morphological variation of bee-parasite *Tropilaelaps* mites (Acair: Laelapidae): New and re-defined species. *Exp. Appl. Acarol.* 43: 1–24.

Anderson, D. L., and J. W. H. Trueman. 2000. *Varroa jacobsoni* (Acari: Varroidae) is more than one species. *Exp. Appl. Acarol.* 24: 165–189.

Aumeier, P. 2001. Bioassay for grooming effectiveness towards *Varroa destructor* mites in Africanized and Carniolan honey bees. *Apidologie* 32: 81–90.

Baker, R. A. 2010. The parasitic mites of honeybees past, present and future. *J. Entomol. Res.* 1(4): 1–7.

Berry, J. A., W. B. Owens, and K. S. Delaplane. 2010. Small-cell comb foundation does not impede Varroa mite population growth in honey bee colonies. *Apidologie* 41(1): 40–44.

Berthoud, H., A. Imdorf, M. Haueter, S. Radloff, and P. Neumann. 2010. Virus infections and winter losses of honey bee colonies (*Apis mellifera*). *J. Apic. Res.* 49: 60–65. doi: 10.3896/IBRA.1.49.1.08

Boecking, O., and E. Genersch. 2008. Varroosis—the ongoing crisis in bee keeping. *J. Verbr. Lebensm.* [J. Consumer Protections and Food Safety], 3: 221–228.

Boncristiani, H. F., G. Di Prisco, J. S. Pettis, M. Hamilton, and Y. P. Chen. 2009. Molecular approaches to the analysis of deformed wing virus replication and pathogenesis in the honey bee, *Apis mellifera*. *Virology Journal* 6:221. doi: 10.1186/1743–422X–6–221

Booppha, B., S. Eittsayeam, K. Pengpat, and P. Chantawannakul. 2010. Development of bioactive ceramics to control mite and microbial diseases in bee farms. *Advanced Materials Res.* 93–94: 553–557.

Branco, M. R., N. A. C. Kidd, and R. S. Pickard. 2006. A comparative evaluation of sampling methods for *Varroa destructor* (Acari: Varroidae) population estimation. *Apidologie* 37: 452–461.

Büchler, R., S. Berg, and Y. Le Conte. 2010. Breeding for resistance to *Varroa destructor* in Europe. *Apidologie* 41(3): 393–408.

Çakmak, I. 2010. The over wintering survival of highly *Varroa destructor* infested honey bee colonies determined to be hygienic using the liquid nitrogen freeze killed brood assay. *J. Apic. Res.* 49(2): 197–201.

Çakmak, I., L. Aydin, S. Camazine, and H. Wells. 2002. Pollen traps and walnut-leaf smoke for Varroa control. *Am. Bee J.* 142(5): 367–370.

Çakmak, I., L. Aydin, and H. Wells. 2006. Walnut leaf smoke versus mint leaves in conjunction with pollen traps for control of *Varroa destructor*. *Bull. Veterinary Institute in Pulawy* 50(4): 477–479.

Calis, J. N. M., W. J. Boot, J. Beetsma, J. H. P. M. van den Eijnde, A. de Ruijter, and J. J. M. van der Steen. 1999. Effective biotechnical control of Varroa: Applying knowledge on brood cell invasion to trap honey bee parasites in drone brood. *J. Apic. Res.* 38(1/2): 49–61.

Campbell, E. M., G. E. Budge, and A. S. Bowman. 2010. Gene-knockdown in the honey bee mite *Varroa destructor* by a non-invasive approach: Studies on a glutathione S-transferase. *Parasites & Vectors* 3:73. http://www.parasitesandvectors.com/content/3/1/73

Chandler, D., K. D. Sunderland, B. V. Ball, and G. Davidson. 2001. Prospective biological control agents of *Varroa destructor* n. sp., an important pest of the European honeybee, *Apis mellifera*. *BioControl Sci. Technol.* 11(4): 429–448.

Charriere, J., A. Imdorf, B. Bachofen, and A. Tschan. 2003. The removal of capped drone brood: An effective means of reducing the infestation of Varroa in honey bee colonies. *Bee World* 84(3): 117–124.

Chen, Y. P., and R. Siede. 2007. Honey bee viruses. *Adv. in Virus Res.* 70: 33–80.

Cicero, J. M., and D. Sammataro. 2010. The salivary glands of adult female *Varroa destructor* (Acari: Varroidae), an ectoparasite of the honey bee, *Apis mellifera* (Hymenoptera: Apidae). *Int. J. Acarol.* 6(5): 377–386.

Corrêa-Marques, M. H., M. R. C. Issa, and D. de Jong. 2000. Classification and quantification of damaged *Varroa jacobsoni* found in the debris of honey bee colonies as criteria for selection? *Am. Bee J.* 140: 820–824.

Dainat, B., T. Ken, H. Berthoud, and P. Neumann. 2009. The ectoparasitic mite *Tropilaelaps mercedesae* (Acari: Laelapidae) as a vector of honeybee viruses. *Insect. Soc.* 56: 40–43.

Damiani, N., N. J. Fernández, L. M. Maldonado, A. R. Álvarez, M. J. Eguaras, and J. A. Marcangeli. 2010. Bioactivity of propolis from different geographical origins on *Varroa destructor* (Acari: Varroidae). *Parasitol. Res.* 107(1): 31–37.

de Guzman, L. I., and M. Delfinado-Baker. 1996. A new species of *Varroa* (Acari: Varroidae) associated with *Apis koschevnikovi* (Apidae: Hymenoptera) in Borneo. *Int. J. Acarol.* 22(1): 23–27.

de Guzman, L. I., D. M. Burgett, and T. E. Rinderer. 2001. Biology and life history of *Acarapis doraslis* and *Acarapis externus*. In *Mites of the Honey Bee*. T. C. Webster, and K. S. Delaplane (Eds). Dadant & Sons, Inc., Hamilton, Illinois, U.S., pp. 17–28.

de Guzman, L. I., T. E. Rinderer, and G. T. Delatte. 1998. Comparative resistance of four honey bee (Hymenoptera: Apidae) stocks to infestation by *Acarapis woodi* (Acari: Tarsonemidae). *J. Econ. Entomol.* 91(5): 1078–1083.

de Guzman, L. I., T. E. Rinderer, and J. A. Stelzer. 1999. Occurrence of two genotypes of *Varroa jacobsoni* Oud. in North America. *Apidologie* 30(1): 31–36.

de Miranda, J. R., G. Cordoni, and G. Budge. 2010. The Acute bee paralysis virus—Kashmir bee virus—Israeli acute paralysis virus complex. *J. Invertebr. Pathol.* 103 (Suppl. 1): S30–S47.

de Miranda, J. R., and E. Genersch. 2010. Deformed wing virus. *J. Invertebr. Pathol.* 103 (Suppl. 1): S48–S61.

Delaplane, K. S., J. A. Berry, J. A. Skinner, J. P. Parkman, and W. M. Hood. 2005. Integrated pest management against *Varroa destructor* reduces colony mite levels and delays treatment threshold. *J. Apic. Res.* 44: 157–162.

Delfinado-Baker, M., and E. W. Baker. 1982. Notes on honey bee mites of the genus *Acarapis* Hirst (Acari: Tarsonemidae). *Int. J. Acarol.* 8: 211–26.

Delfinado-Baker, M., and K. Aggarwal. 1987. A new Varroa (Acari: Varroidae) from the nest of *Apis cerana* (Apidae). *Int. J. Acarol.* 13: 233–237.

Desch, C. E. 2009. Human hair follicle mites and forensic acarology. *Exp. Appl. Acarol.* 49: 143–146. doi: 10.1007/s10493-009-9272-0

Devlin, S. M. 2001. Comparative analyses of sampling methods for Varroa mites (*Varroa destructor* Anderson and Trueman) on honey bees (*Apis mellifera* L.). Thesis submitted in partial fulfillment of the requirements for the degree of Masters of Pest Management in the Department of Biological Sciences, Simon Fraser University. 61 pp.

Dzierzawski, A., and W. Cybulski. 2010. Evaluation of the efficacy of Apiwarol AS [Ocena skuteczności Apiwarolu AS]. *Medycyna Weterynaryjna* 66(7): 475–479.

Eckert, J. E. 1961. Acarapis mites of the honey bee, *Apis mellifera* Linnaeus. *J. Insect Pathol.* 3: 409–25.

Eickwort, G. C. 1988. The origins of mites associated with honey bees. In *Africanized Honey Bees and Bee Mites.* G. R. Needham, R. E. Page, Jr., M. Delfinado-Baker, and C. E. Bowman (Eds.) Ellis Horwood Ltd., Chichester, UK, pp. 327–338.

Eischen, F. A., and T. Wilson. 1998. The effect of natural product smoke on *Varroa jacobsoni*: An update. *Am. Bee J.* 138: 293.

Ellis, A. M., G. W. Hayes, and J. D. Ellis. 2009a. The efficacy of dusting honey bee colonies with powdered sugar to reduce varroa mite populations. *J. Apicul. Res.* 48(1): 72–76.

Ellis, A. M., G. W. Hayes, and J. D. Ellis. 2009b. The efficacy of small cell foundation as a varroa mite (*Varroa destructor*) control. *Exp. Appl. Acarol.* 47(4): 311–316.

Ellis, Jr., J. D., K. S. Delaplane, and W. M. Hood. 2001. Efficacy of a bottom screen device, Apistan™, and Apilife VAR™, in controlling *Varroa destructor. Am. Bee J.* 141(11): 813–816.

Ellis, Jr., J. D., J. D. Evans, and J. Pettis. 2010. Colony losses, managed colony population decline, and Colony Collapse Disorder in the United States. *J. Apic. Res.* 49(1): 134–136.

Elzen, P. J., J. R. Baxter, F. A. Eischen, and W. T. Wilson. 1999b. Pesticide resistance in Varroa mites: Theory and practice. *Am. Bee J.* 139: 195–196.

Elzen, P. J., J. R. Baxter, M. Spivak, and W. T. Wilson. 1999c. Amitraz resistance in Varroa: New discovery in North America. *Am. Bee J.* 139: 362.

Elzen, P. J., J. R. Baxter, D. Westervelt, D. Causey, C. Randall, L. Cutts, and W. T. Wilson. 2001c. Acaricide rotation plan for control of Varroa. *Am. Bee J.* 141(6): 412.

Elzen, P. J., R. L. Cox, and W. A. Jones. 2004. Evaluation of food grade mineral oil treatment for varroa mite control. *Am. Bee J.* 144(12): 921–923.

Elzen, P. J., F. A. Eischen, J. R. Baxter, G. W. Elzen, and W. T. Wilson. 1999a. Detection of resistance in U.S. *Varroa jacobsoni* Oud. (Mesotigmata: Varroidae) to the acaricide fluvalinate. *Apidologie* 30: 13–18.

Elzen, P. J., F. A. Eischen, J. B. Baxter, J. Pettis, G. W. Elzen, and W. T. Wilson. 1998. Fluvalinate resistance in *Varroa jacobsoni* from several geographic locations. *Am. Bee J.* 138(9): 674–676.

Elzen, P. J., G. W. Elzen, and R. D. Stipanovic. 2001a. Biological activity of grapefruit leaf burning residue extract and isolated compounds on *Varroa jacobsoni. Am. Bee J.* 141(5): 369–371.

Elzen, P. J., R. D. Stipanovic, and R. Rivera. 2001b. Activity of two preparations of natural smoke products on the behavior of *Varroa jacobsoni* Oud. *Am. Bee J.* 141(4): 289–291.

Elzen, P. J., and D. Westervelt. 2002. Detection of coumaphos resistance in *Varroa destructor* in Florida. *Am. Bee J.* 142(4): 291–292.

Emsen, B., and A. Dodologlu. 2009. The effects of using different organic compounds against honey bee mite (*Varroa destructor* Anderson and Trueman) on colony developments of honey bee (*Apis mellifera* L.) and residue levels in honey. *J. Anim. Veterinary Advances* 8(5): 1004–1009.

Fakhimzadeh, K. 2000. Potential of super-fine ground, plain white sugar dusting as an ecological tool for the control of Varroasis in the honey bee (*Apis mellifera*). *Am. Bee J.* 140(6): 487–491.

Fassbinder, C., J. Grodnitzky, and J. Coats. 2002. Monoterpenoids as possible control agents for *Varroa destructor. J. Apic. Res.* 41(3–4): 83–88.

Fievet, J., D. Tentcheva, L. Gauthier, J. de Miranda, F. Cousserans, M. E. Colin, and M. Bergoin. 2006. Localization of deformed wing virus infection in queen and drone *Apis mellifera* L. *Virology Journal* 3: 16.

Forsgren, E., J. R. de Miranda, M. Isaksson, S. Wei, and I. Fries. 2009. Deformed wing virus associated with *Tropilaelaps mercedesae* infesting European honey bees (*Apis mellifera*). *Exp. Appl. Acarol.* 47(2): 87–97.

Fuchs, S. 1985. Quantitative diagnosis of the infestation of bee hives by *Varroa jacobsoni* Oud. and distribution of the parasitic mite within the hives. (Untersuchungen zur quantitativen Abschätzung des Befalls von Bienenvölkern mit *Varroa jacobsoni* Oudemans und zur Verteilung der Parasiten im Bienenvolk). *Apidologie* 16(4): 343–368.

Fuchs, S. 1990. Preference for drone brood cells by *Varroa jacobsoni* Oud in colonies of *Apis mellifera carnica*. *Apidologie* 21(3): 193–199.

García-Fernández, P., C. Santiago-Álvarez, and E. Quesada-Moraga. 2008. Pathogenicity and thermal biology of mitosporic fungi as potential microbial control agents of *Varroa destructor* (Acari: Mesostigmata), an ectoparasitic mite of honey bee, *Apis mellifera* (Hymenoptera: Apidae). *Apidologie* 39(6): 662–673.

Garedew, A., I. Lamprecht, E. Schmolz, and B. Schricker. 2002. The varroacidal action of propolis: A laboratory assay. *Apidologie* 33(1): 41–50.

Gashout, H. A., and E. Guzmán-Novoa. 2009. Acute toxicity of essential oils and other natural compounds to the parasitic mite, *Varroa destructor*, and to larval and adult worker honey bees (*Apis mellifera* L.). *J. Apic. Res.* 48(4): 263–269.

Glenn, G. M., A. P. Klamczynski, D. F. Woods, B. Chiou, W. J. Orts, and S. H. Imam. 2010. Encapsulation of plant oils in porous starch microspheres. *J. Agr. and Food Chem.* 58(7): 4180–4184.

González-Gómez, R., G. Otero-Colina, J. A. Villanueva-Jiménez, J. A. Pérez-Amaro, and R. M. Soto-Hernández. 2006. *Azadirachta indica* toxicity and repellence of *Varroa destructor* (Acari: Varroidae) [Toxicidad y repelencia de *Azadirachta indica* contra *Varroa destructor* (Acari: Varroidae)]. *Agrociencia* 40(6): 741–751.

Harbo, J. R., and J. W. Harris. 2004. Effect of screen floors on populations of honey bees and parasitic mites (*Varroa destructor*). *J. Apic. Res.* 43(3): 114–117.

Harbo, J. R.and J. W. Harris. 2001. Resistance to *Varroa destructor* (Mesostigmata: Varroidae) when mite-resistant queen honey bees (Hymenoptera: Apidae) were free-mated with unselected drones. *J. Econ. Entomol.* 94(6): 1319–1323.

Harris, J. W., R. G. Danka, and J. D. Villa. 2010. Honey bees (Hymenoptera: Apidae) with the trait of varroa sensitive hygiene remove brood with all reproductive stages of varroa mites (Mesostigmata: Varroidae). *Ann. Entomol. Soc. Am.* 103(2): 146–152.

Highfield, A. C., A. El Nagar, L. C. M. Mackinder, L. M.-L. J. Noël, M. J. Hall, S. J. Martin, and D. C. Schroeder. 2009. Deformed wing virus implicated in overwintering honeybee colony losses. *Appl. Environ. Microbiol.* 75(22): 7212–7220.

Huang, Z. 2001. Mite zapper—A new and effective method for Varroa mite control. *Am. Bee J.* 141(10): 730–732.

James, R. R., G. Hayes, and J. E. Leland. 2006. Field trials on the microbial control of varroa with the fungus *Metarhizium anisopliae*. *Am. Bee J.* 146(11): 968–972.

Johnson, R. M., M. D. Ellis, C. A. Mullin, and M. Frazier. 2010a. Pesticides and honey bee toxicity—USA. *Apidologie* 41(3): 312–331.

Johnson, R. M., Z. Y. Huang, and M. R. Berenbaum. 2010b. Role of detoxification in *Varroa destructor* (Acari: Parasitidae) tolerance of the miticide *tau*-fluvalinate. *Int. J. Acarol.* 36(1): 1–6.

Kanga, L. H. B., W. A. Jones, and R. R. James. 2003. Field trials using the fungal pathogen, *Metarhizium anisopliae* (Deuteromycetes: Hyphomycetes) to control the ectoparasitic mite, *Varroa destructor* (Acari: Varroidae) in honey bee, *Apis mellifera* (Hymenoptera: Apidae) colonies. *J. Econ. Entomol.* 96(4): 1091–1099.

Kanga, L. H. B., J. Adamczyk, J. Patt, C. Gracia, and J. Cascino. 2010. Development of a user-friendly delivery method for the fungus *Metarhizium anisopliae* to control the ectoparasitic mite *Varroa destructor* in honey bee, *Apis mellifera*, colonies. *Exp. Appl. Acarol.* 52(4): 327–342.

Kanga, L. H. B., R. R. James, and D. G. Boucias. 2002. *Hirsutella thompsonii* and *Metarhizium anisopliae* as potential microbial control agents of *Varroa destructor*, a honey bee parasite. *J. Invertebr. Pathol.* 81(3): 175–184.

Kanga, L. H. B., W. A. Jones, and C. Gracia. 2006. Efficacy of strips coated with *Metarhizium anisopliae* for control of *Varroa destructor* (Acari: Varroidae) in honey bee colonies in Texas and Florida. *Exp. Appl. Acarol.* 40(3–4): 249–258.

Krupka, I., and R. K. Straubinger. 2010. Lyme Borreliosis in dogs and cats: Background, diagnosis, treatment and prevention of infections with *Borrelia burgdorferi* sensu strict. *Veterinary Clinics of North America—Small Animal Practice* 40: 1103–1119. doi: 10.1016/j.cvsm.2010.07.011

Le Conte, Y., M. Ellis, and W. Ritter. 2010. Varroa mites and honey bee health: Can Varroa explain part of the colony losses? *Apidologie* 41(3): 353–363.

Lee, K. V., R. D. Moon, E. C. Burkness, W. D. Hutchison, and M. Spivak. 2010a. Practical sampling plans for *Varroa destructor* (Acari: Varroidae) in *Apis mellifera* (Hymenoptera: Apidae) colonies and apiaries. *J. Econ. Entomol.* 103(4): 1039–1050.

Lee, K. V., G. Reuter, and M. Spivak. 2010b. Standardized sampling plan to detect Varroa density in colonies and apiaries. *Am. Bee J.* 150: 1151–1155.

Leonivich, S. A. 2010. The lung mite *Pneumonyssus simicola* banks (Halarachidae) in lungs of the rhesus monkey *Macaca mulatta*. *Acarina* 18(1): 89–90.

Liu, X., Y. Zhang, X. Yan, and R. Han. 2010. Prevention of Chinese sacbrood virus infection in *Apis cerana* using RNA Interference. *Curr Microbiol.* 61: 422–428. doi: 10.1007/s00284-010-9633-2

Lodesani, M., M. Colombo, and M. Spreafico. 1995. Ineffectiveness of Apistan® treatment against the mite *Varroa jacobsoni* Oud. in several districts of Lombardy (Italy). *Apidologie* 26(1): 67–72.

Lodesani, M., C. Costa, G. Serra, R. Colombo, and A. G. Sabatini. 2008. Acaricide residues in beeswax after conversion to organic beekeeping methods. *Apidologie* 39(3): 324–333.

Macedo, P. A., J. Wu, and M. D. Ellis. 2002. Using inert dusts to detect and assess varroa infestations in honey bee colonies *J. Apic. Res.* 41(1–2): 3–7.

Maggi, M., N. Damiani, S. Ruffinengo, D. de Jong, J. Principal, and M. Eguaras. 2010a. Brood cell size of *Apis mellifera* modifies the reproductive behavior of *Varroa destructor*. *Exp. Appl. Acarol.* 50(3): 269–279.

Maggi, M. D., S. R. Ruffinengo, L. B. Gende, E. G. Sarlo, M. J. Eguaras, P. N. Bailac, and M. I. Ponzi. 2010b. Laboratory evaluations of *Syzygium aromaticum* (L.) Merr. et Perry essential oil against *Varroa destructor*. *J. Essential Oil Res.* 22(2): 119–122.

Maori, E., N. Paldi, S. Shafir, H. Kalev, E. Tsur, E. Glick, and I. Sela. 2009. IAPV, a bee-affecting virus associated with Colony Collapse Disorder can be silenced by dsRNA ingestion. *Insect Molec. Biol.* 18(1): 55–60.

Martin, S. J., B. V. Ball, and N. L. Carreck. 2010. Prevalence and persistence of deformed wing virus (DWV) in untreated or acaricide-treated *Varroa destructor* infested honey bee (*Apis mellifera*) colonies. *J. Apic. Res.* 49(1): 72–79.

Meikle, W. G., G. Mercadier, V. Girod, F. Derouané, and W. A. Jones. 2006. Evaluation of *Beauveria bassiana* (Balsamo) Vuillemin (Deuteromycota: Hyphomycetes) strains isolated from varroa mites in southern France. *J. Apic. Res.* 45(4): 219–220.

Meikle, W. G., G. Mercadier, N. Holst, and V. Girod. 2008a. Impact of two treatments of a formulation of *Beauveria bassiana* (Deuteromycota: Hyphomycetes) conidia on Varroa mites (Acari: Varroidae) and on honeybee (Hymenoptera: Apidae) colony health. *Exp. Appl. Acarol.* 46(1 4): 105–117.

Meikle, W. G., G. Mercadier, N. Holst, C. Nansen, and V. Girod. 2008b. Impact of a treatment of *Beauveria bassiana* (Deuteromycota: Hyphomycetes) on honeybee (*Apis mellifera*) colony health and on *Varroa destructor* mites (Acari: Varroidae). *Apidologie* 39(2): 247–259.

Meikle, W. G., G. Mercadier, N. Holst, C. Nansen, and V. Girod. 2007. Duration and spread of an entomopathogenic fungus, *Beauveria bassiana* (Deuteromycota: Hyphomycetes), used to treat varroa mites (Acari: Varroidae) in honey bee (Hymenoptera: Apidae) hives. *J. Econ. Entomol.* 100(1): 1–10.

Melathopoulos, A. P., M. L. Winston, R. Whittington, H. Higo, and M. le Doux. 2000. Field evaluation of neem and canola oil for the selective control of the honey bee (Hymenoptera: Apidae) mite parasites *Varroa jacobsoni* (Acari: Varroidae) and *Acarapis woodi* (Acari: Tarsonemidae). *J. Econ. Entomol.* 93(3): 559–567.

Milani, N. 1994. Possible presence of fluvalinate resistant strains of *Varroa jacobsoni* in northern Italy. In *New Perspectives on Varroa: Proc. Interntl. Meeting, Prague, 1993*. A. Matheson (Ed.). IBRA, Cardiff, UK, p.87.

Milani, N. 1995. The resistance of *Varroa jacobsoni* Oud. to pyrethroids: A laboratory assay. *Apidologie* 26: 415–429.

Milani, N. 1999. The resistance of *Varroa jacobsoni* Oud. to acaracies. *Apidologie* 30: 229–234.

Milani, N., G. Della Vedova, and M. Lodesani. 2009. Determination of the LC50 of chlorfenvinphos in *Varroa destructor*. *J. Apic. Res.* 48(2): 140–141.

Mullin, C. A., M. Frazier, J. L. Frazier, S. Ashcraft, R. Simonds, D. vanEngelsdorp, and J. S. Pettis. 2010. High levels of miticides and agrochemicals in North American apiaries: Implications for honey bee health. *PLoS ONE* 5(3): e9754. doi: 10.1371/journal.pone.0009754

Nadchatram, M. 2006. A review of endoparasitic acarines of Malaysia with special reference to novel endoparasitism of mites in amphibious sea snakes and supplementary notes on ecology of chiggers. *Tropical Biomed.* 23(1): 1–22.

Navajas, M., D. L. Anderson, L. I. de Guzman, Z. Y. Huang, J. Clement, T. Zhou, and Y. Le Conte. 2010. New Asian types of *Varroa destructor*: A potential new threat for world apiculture. *Apidologie* 41(2): 181–193.

Ogata, J. N., and A. Bevenue. 1973. Chlorinated pesticide residues in honey. *Bull. Environ. Contam. Toxicol.* 9(3): 143–147.

O'Meara, J. 2005. Walnut leaf smoke: A thrifty control of varroa mites. *Am. Bee J.* 145(1): 60–62.

Ostiguy, N., and D. Sammataro. 2000. A simplified technique for counting *Varroa jacobsoni* Oud. on sticky boards. *Apidologie* 31(6): 707–716.

Otis, G. W., and J. Kralj. 2001. Mites of economic importance not present in North America. In *Mites of the Honey Bee*. T. C. Webster and K. S. Delaplane (Eds). Dadant and Sons, Inc., Hamilton, Illinois, U.S., pp. 251–272.

Paldi, N., E. Glick, M. Oliva, Y. Zilberberg, L. Aubin, J. Pettis, Y. Chen, and J. D. Evans. 2010. Effective gene silencing in a microsporidian parasite associated with honeybee (*Apis mellifera*) colony declines. *App. Environ. Microbiol.* 76(17): 5960–5964.

Peng, C. Y. S., S. Trinh, J. E. Lopez, E. C. Mussen, A. Hung, and R. Chuang. 2000. The effects of azadirachtin on the parasitic mite, *Varroa jacobsoni* and its host honey bee (*Apis mellifera*). *J. Apic. Res.* 39(3–4): 159–168.

Peng, C. Y. S., X. Zhou, and H. K. Kaya. 2002. Virulence and site of infection of the fungus, *Hirsutella thompsonii*, to the honey bee ectoparasitic mite, *Varroa destructor*. *J. Invertebr. Pathol.* 81(3): 185–195.

Peng Y. S., Y. Fang, S. Xu, and L. Ge. 1987. The resistance mechanism of the Asian honey bee, *Apis cerana* Fabr. to an ectoparasitic mite *Varroa jacobsoni* Oudemans. *J. Invertebr. Pathol.* 49: 54–60.

Pettis, J. S. 2001. Biology and life history of tracheal mites. In *Mites of the Honey Bee*. T. C. Webster and K. S. Delaplane (Eds). Dadant & Sons, Inc., Hamilton, Illinois, U.S., pp. 29–41.

Phillips, E. F. 1923. The occurrence of disease of adult bees II. USDA Circular #287. Washington, DC, U.S.

Piccolo, F. D., R. Nazzi, G. D. Vedova, and N. Milani. 2010. Selection of *Apis mellifera* workers by the parasitic mite *Varroa destructor* using host cuticular hydrocarbons. *Parasitology* 137(6): 967–973.

Pooley, C. 2010. Electron and Confocal Microscope Unit, USDA-ARS. http://www.ars.usda.gov/pandp/docs.htm?docid=4807

Rath, W. 1992. The key to Varroa—the drones of *Apis cerana* and their call cap. *Am. Bee J.* 132(5): 329–331.

Richards, E. H., B. Jones, and A. Bowman. 2011. Salivary secretions from the honeybee mite, *Varroa destructor*: Effects on insect haemocytes and preliminary biochemical characterization. *Parasitology* 1: 1–7.

Rickli, M., P. M. Guerin, and P. A. Diehl. 1992. Palmitic acid released from honeybee worker larvae attracts the parasitic mite *Varroa jacobsoni* on a servosphere. *Naturwissenschaften* 79: 320–322.

Rinderer, T. E., L. I. de Guzman, V. A. Lancaster, G. T. Delatte, and J. A. Stelzer. 1999. Varroa in the mating yard: I. The effects of *Varroa jacobsoni* and Apistan® on drone honey bees. *Am. Bee J.* 139(2): 134–139.

Rinderer, T. E., L. I. de Guzman, G. T. Delatte, J. A. Stelzer, V. A. Lancaster, V. Kuznetsov, and L. Beaman. 2001. Resistance to the parasitic mite *Varroa destructor* in honey bees from far-eastern Russia. *Apidologie*, 32(4): 381–394.

Rinderer, T. E., J. W. Harris, G. J. Hunt, and L. I. de Guzman. 2010. Breeding for resistance to *Varroa destructor* in North America. *Apidologie* 41(3): 409–424.

Rodríguez, M., M. Gerding, and A. France. 2009a. Selection of entomopathogenic fungi to control *Varroa destructor* (Acari: Varroidae). [Selección de hongos entomopatógenos para el control de *Varroa destructor* (Acari: Varroidae)]. *Chilean J. Agricult. Res.* 69(4): 534–540.

Rodríguez, M., M. Gerding, A. France, and R. Ceballos. 2009b. Evaluation of *Metarhizium anisopliae* var. *anisopliae* Qu-M845 isolate to control *Varroa destructor* (Acari: Varroidae) in laboratory and field trials. [Evaluación del aislamiento Qu-M845 de *metarhizium anisopliae* var. *anisopliae* para el control de *Varroa destructor* (Acari: Varroidae) en ensayos de laboratorio y terreno]. *Chilean J. Agricult. Res.* 69(4): 541–547.

Romeh, A. A. 2009. Control of Varroa mite (*Varroa destructor*) on honey bees by Sycamore leaves (*Ficus Sycomorus*). *J. Appl. Sci. Res.* 5(2): 151–157.

Rosenkranz, P. 1999. Honey bee (*Apis mellifera* L.) tolerance to *Varroa jacobsoni* Oud. in South America. *Apidologie* 30: 159–172.

Rosenkranz, P., P. Aumeier, and B. Ziegelmann. 2010. Biology and control of *Varroa destructor*. *J. Invertebr. Pathol.* 103 (Suppl. 1): S96–S119.

Ruffinengo, S. R., M. Maggi, S. Fuselli, I. Floris, G. Clemente, N. H. Firpo, P. N. Bailac, and M. I. Ponzi. 2006. Laboratory evaluation of *Heterothalamus alienus* essential oil against different pests of *Apis mellifera*. *J. Essential Oil Research* 18(6): 704–707.

Salvy, M., C. Martin, A. G. Bagnères, É. Provost, M. Roux, Y. Le Conte, and J. L. Clément. 2001. Modifications of the cuticular hydrocarbon profile of *Apis mellifera* worker bees in the presence of the ectoparasitic mite *Varroa jacobsoni* in brood cells. *Parasitology* 122: 145–159.

Sammataro, D., and A. Avitabile. 2011. *Beekeeper's Handbook*. 4th ed. Cornell University Press, Ithaca, New York, U.S.

Sammataro, D., S. Cobey, B. H. Smith, and G. R. Needham. 1994. Controlling tracheal mites (Acari: Tarsonemidae) in honey bees (Hymenoptera: Apidae) with vegetable oil. *J. Econ. Entomol.* 57(4): 910–916.

Sammataro, D. 2006. An easy dissection technique for finding tracheal mites (Acari: Tarsonemidae) in honey bees. *Int. J. Acarol.* 32(4):339–343.

Sammataro, D., J. Finley, B. Leblanc, G. Wardell, F. Ahumada-Segura, and M. J. Carroll. 2009. Feeding essential oils and 2-heptanone in sugar syrup and liquid protein diets to honey bees (*Apis mellifera* L.) as potential varroa mite (*Varroa destructor*) controls. *J. Apic. Res.* 48(4): 256–262.

Sammataro, D., U. Gerson, and G. R. Needham. 2000. Parasitic mites of honey bees: Life history, implications and impact. *Annu. Rev. Entomol.* 45: 519–548.

Sammataro, D., B. Leblanc, M. J. Carroll, J. Finley, and M. T. Torabi. 2010. Antioxidants in wax cappings of honey bee brood. *J. Apic. Res.* 49(4): 293–301.

Sammataro, D., N. Ostiguy, and M. Frazier. 2002. How to use an IPM sticky board to monitor Varroa levels in honey bee colonies. *Am. Bee J.* 142: 363–366.

Santillán-Galicia, M. T., B. V. Ball, S. J. Clark, and P. G. Alderson. 2010. Transmission of deformed wing virus and slow paralysis virus to adult bees (*Apis mellifera* L.) by *Varroa destructor*. *J. Apic. Res.* 49(2): 141–148.

Scaife, S. H. 1952. The yellow-banded carpenter bee, *Mesotrichia caffra* Linn., and its symbiotic mites, *Dinogamasus brausni* Vitzthun. *J. Entomol. Soc. South Africa* 15: 63–76.

Schäfer, M. O., W. Ritter, J. S. Pettis, and P. Neumann. 2010. Winter losses of honeybee colonies (Hymenoptera: Pidae): The role of infestations with *Aethina tumida* (Coleoptera: Nitidulidae) and *Varroa destructor* (Parasitiformes: Varroidae). *J. Econ. Entomol.* 103(1): 10–16.

Schlüns, H., and R. H. Crozier. 2007. Relish regulates expression of antimicrobial peptide genes in the honeybee, *Apis mellifera*, shown by RNA interference. *Insect Molecular Biology* 16(6): 753–759.

Schulz, A. E. 1984. Reproduction and population dynamics of the parasitic mite *Varroa jacobsoni* Oud. in correlation with the brood cycle of *Apis mellifera*. *Apidologie* 5:401–419.

Shaddel-Telli, A.-A., N. Maheri-Sis, A. Aghajanzadeh-Golshani, A. Asadi-Dizaji, H. Cheragi, and M. Mousavi. 2008. Using medicinal plants for controlling varroa mite in honey bee colonies. *J. Anim. Veterinary Advances* 7(3): 328–330.

Shaw, K. E., G. Davidson, S. J. Clark, B. V. Ball, J. K. Pell, D. Chandler, and K. D. Sunderland. 2002. Laboratory bioassays to assess the pathogenicity of mitosporic fungi to *Varroa destructor* (Acari: Mesostigmata), an ectoparasitic mite of the honeybee, *Apis mellifera*. *Biological Control* 24(3): 266–276.

Spivak, M. 1996. Honey bee hygienic behavior and defense against *Varroa jacobsoni*. *Apidologie* 27: 245–260.

Spivak, M., and G. S. Reuter. 1998. Performance of hygienic honey bee colonies in a commercial apiary. *Apidologie* 29: 291–302.

Stanimirovič, Z., S. Jevrosima, A. Nevenka, and V. Stojíc. 2010. Heritability of grooming behaviour in grey honey bees (*Apis mellifera carnica*). *Acta Veterinaria* 60(2–3): 313–323.

Steenberg, T., P. Kryger, and N. Holst. 2010. A scientific note on the fungus *Beauveria bassiana* infecting *Varroa destructor* in worker brood cells in honey bee hives. *Apidologie* 41(1): 127–128.

Steere, A. C. 2001. Lyme disease. *New Eng. J. Med.* 345: 115–125. doi: 10.1056/NEJM200107123450207

Strange, J. P., and W. S. Sheppard. 2001. Optimum timing of miticide applications for control of *Varroa destructor* (Acari: Varroidae) in *Apis mellifera* (Hymenoptera: Apidae) in Washington State, U.S. *J. Econ. Entomol.* 94: 1324–1333.

Tabor, K. L., and J. T. Ambrose. 2001. The use of heat treatment for control of the honey bee mite, *Varroa destructor*. *Am. Bee J.* 141(10): 733–736.

Tu, S., X. Qiu, L. Cao, R. Han, Y. Zhang, and X. Liu. 2010. Expression and characterization of the chitinases from *Serratia marcescens* GEI strain for the control of *Varroa destructor*, a honey bee parasite. *J. Invertebr. Pathol.* 104 (2): 75–82.

United States Department of Agriculture. 2010. *http://www.aphis.usda.gov/plant_health/plant_pest_ info/honey_bees/downloads/survey_project_plan_2010.pdf*

vanEngelsdorp, D., J. D. Evans, C. Saegerman, C. Mullin, E. Haubruge, B. K. Nguyen, M. Frazier, J. Frazier, D. Cox-Foster, Y. Chen, R. Underwood, D. R. Tarpy, and J. S. Pettis. 2009. Colony Collapse Disorder: A Descriptive Study. *PLoS ONE* 4(8): e6481.

Villa, J. 2006. Autogrooming and bee age influence migration of tracheal mites to Russian and susceptible worker honey bees (*Apis mellifera* L). *J. Apic. Res.* 45(2): 28–31. doi: 10.3896/IBRA.1.45.2.06

Walter, D. E., G. Krantz, and E. Lindquist. 1996. Acari. The Mites. In *The Tree of Life Web Project*, http://tolweb.org/Acari/2554/1996.12.13

Wantuch, H. A., and D. R. Tarpy. 2009. Removal of drone brood from *Apis mellifera* (Hymenoptera: Apidae) colonies to control *Varroa destructor* (Acari: Varroidae) and retain adult drones. *J. Econ. Entomol.* 102(6): 2033–2040.

Warrit, N., D. R. Smith, and C. Lekprayoon. 2006. Genetic subpopulations of Varroa mites and their *Apis cerana* hosts in Thailand. *Apidologie* 37: 19–30.

Wilkinson, D., and G. C. Smith. 2002. Modeling the efficiency of sampling and trapping *Varroa destructor* in the drone brood of honey bees (*Apis mellifera*). *Am. Bee J.* 142(3): 209–212.

Wilson, W. T., J. S. Pettis, C. E. Henderson, and R. A. Morse. 1997. Tracheal Mites. In *Honey Bee Pests, Predators and Diseases*. Morse, R. A., and K. Flottum (Eds.). A. I. Root Company, Medina, Ohio, U.S., pp. 253–277.

Zhang, Q., J. R. Ongus, W. J. Boot, J. Calis, J.-M. Bonmatin, E. Bengsch, and D. Peters. 2007. Detection and localisation of picorna-like virus particles in tissues of *Varroa destructor*, an ectoparasite of the honey bee, *Apis mellifera*. *J. Invertebr. Pathol.* 96: 97–105.

Chapter 6

Biological Control of Honey Bee Pests

Blumberg, D., and S. M. Ferkovich. 1994. Development and encapsulation of the endoparasitoid, *Microplitis croceipes* (Hym.: Braconidae), in six candidate host species (Lep.). *Entomophaga* 39: 293–302.

Bugeme, D. M., M. Knapp, H. I. Boga, A. K. Wanjoya, and N. K. Maniania. 2010. Influence of temperature on virulence of fungal isolates of *Metarhizium anisopliae* and *Beauveria bassiana* to the two-spotted spider mite *Tetranychus urticae*. *Mycologia* 167: 221–227. doi: 10.1007/s11046-008-9164-6.

Burges, H. D., and L. Bailey. 1968. Control of the greater and lesser wax moths (*Galleria mellonella* and *Achroia grisella*) with *Bacillus thuringiensis*. *J. Invertebr. Pathol.* 11(2): 184–195.

Cabanillas, H. E., and P. J. Elzen. 2006. Infectivity of entomopathogenic nematodes (Steinernematidae and Heterorhabditidae) against the small hive beetle *Aethina tumida* (Coleoptera: Nitidulidae). *J. Apic. Res.* 45: 49–50.

Chandler, D., G. Davidson, J. K. Pell, B. V. Ball, K. Shaw, and K. D. Sunderland. 2000. Fungal biocontrol of Acari. *BioControl Sci. Technol.* 10: 357–384.

Chandler, D., K. D. Sunderland, B. V. Ball, and G. Davidson. 2001. Prospective biological control agents for *Varroa destructor* n. sp., an important pest of the European honey bee, *Apis mellifera*. *BioControl Sci. Technol.* 11: 429–448.

Chandler, D., D. Hay, and A. P. Reid. 1997. Sampling and occurrence of entomopathogenic fungi and nematodes in UK. soils. *Appl. Soil Ecology* 5(2): 133–141.

Charrière, J .D., and A. Imdorf. 1999. Protection of honey combs from wax moth damage. *Am. Bee J.* 139: 627–630.

Cherry, A., N. Jenkins, G. Heviefo, R. P. Bateman, and C. Lomer. 1999. A West African pilot scale production plant for aerial conidia of *Metarhizium* sp. for use as a mycoinsecticide against locusts and grasshoppers. *Biocontrol Sci. Technol.* 9: 35–51.

Chernov, K. S. 1981. Transmission of mycoses, an aspect of Varroa infestations. *Byulletin Vsesoyuznogo Instituta Eksperimental noi Veternarii* 41: 59–60. (in Russian)

Davidson, G., K. Phelps, K. D. Sunderland, J. K. Pell, B. V. Ball, K. E. Shaw, and D. Chandler. 2003. Study of temperature-growth interactions of entomopathogenic fungi with potential for control of *Varroa destructor* (Acari: Mesostigmata) using a nonlinear model of poikilotherm development. *J. Appl. Microbiol.* 94: 816–825.

Donovan, B. J., and F. Paul. 2005. Pseudoscorpions: The forgotten beneficials inside beehives and their potential for management for control of varroa and other arthropod pests. *Bee World* 86(4): 83–87.

Ellis, A. M., and G. W. Hayes. 2009. Assessing the efficacy of a product containing *Bacillus thuringiensis* applied to honey bee (Hymenoptera: Apidae) foundation as a control for *Galleria mellonella* (Lepidoptera: Pyralidae). *J. Entomol. Sci.* 44(2): 1–6.

Ellis, J. D., P. Neumann, R. Hepburn, and P. J. Elzen. 2002. Longevity and reproductive success of *Aethina tumida* (Coleoptera: Nitidulidae) fed different natural diets. *J. Econ. Entomol.* 95: 902–907.

Ellis, J. D., H. I. Rong, M. P. Hill, H. R. Hepburn, and P. J. Elzen. 2004. The susceptibility of small hive beetle (*Aethina tumida* Murray) pupae to fungal pathogens. *Am. Bee J.* 144(6): 486–488.

Ellis, J. D., S. Spiewok, K. S. Delaplane, S. Buchholz, P. Neumann, and W. L. Tedders. 2010. Susceptibility of *Aethina tumida* (Coleoptera: Nitidulidae) larvae and pupae to entomopathogenic nematodes. *J. Econ. Entomol.* 103(1): 1–9.

Farenhorst, M., J. C. Mouatcho, C. K. Kikankie, B. D. Brooke, R. H. Hunt, M. B. Thomas, L. L. Koekemoer, B. G. J. Knols, and M. Coetzee. 2009. Fungal infection counters insecticide resistance in African malaria mosquitoes. *Proc. Natl. Acad. Sci. USA* 106: 17443–17447.

García-Fernández, P., C. Santiago-Álvarez, and E. Quesada-Moraga. 2008. Pathogenicity and thermal biology of mitosporic fungi as potential microbial control agents of

Varroa destructor (Acari: Mesostigmata), an ectoparasitic mite of honey bee, *Apis mellifera* (Hymenoptera: Apidae). *Apidologie* 39: 662–673.

Goettel, M. S., and G. D. Inglis. 1997. Fungi: Hyphomycetes, pp. 213–249. In *Manual of Techniques in Insect Pathology*. Lacey, L. (Ed.). Academic Press, San Diego, California, U.S.

Gutierrez, A. P., L. Caltagirone, and W. Meikle. 1999. Evaluation of results: Economics of biological control. In *Handbook of Biological Control*. Bellows, T. S., and T. W. Fisher (Eds.). Academic Press, San Diego, California, U.S., pp. 243–252.

Harvey, J. A., and L. E. M. Vet. 1997. *Venturia canescens* parasitizing *Galleria mellonella* and *Anagasta kuehniella*: Differing suitability of two hosts with highly variable growth potential. *Entomologia Experimentalis et Applicata* 84: 93–100.

Herbert, E. W. Jr. 1992. Chapter 6. Honey bee nutrition. In *The Hive and the Honey Bee*. Graham, J. M. (Ed.). Dadant and Sons, Hamilton, Illinois, U.S., pp. 197–233.

Islam, M. T., S. J. Castle, and S. Ren. 2010. Compatibility of the insect pathogenic fungus *Beauveria bassiana* with neem against sweetpotato whitefly, *Bemisia tabaci*, on eggplant. *Entomologia Experimentalis et Applicata* 134: 28–34.

James, R. R., G. Hayes, and J. E. Leland. 2006. Field trials on the microbial control of varroa with the fungus *Metarhizium anisopliae*. *Am. Bee J.* 146(11): 968–972.

Jaronski, S. T. 2010. Ecological factors in the inundative use of fungal entomopathogens. *BioControl* 55: 159–185.

Johnson, K. N., J.-L. Zeddam, and L. A. Ball. 2000. Characterization and construction of functional cDNA clones of pariacoto virus, the first Alphanodavirus isolated outside Australasia. *J. Virology* 74(11): 5123–5132.

Kanga, L. H. B., J. Adamczyk, J. Patt, C. Gracia, and J. Cascino. 2010. Development of a user-friendly delivery method for the fungus *Metarhizium anisopliae* to control the ectoparasitic mite *Varroa destructor* in honey bee, *Apis mellifera*, colonies. *Exp. Appl. Acarol.* 52: 327–342.

Kanga, L. H. B., R. R. James, and D. G. Boucias. 2002. *Hirsutella thompsonii* and *Metarhizium anisopliae* as potential microbial control agents of *Varroa destructor*, a honey bee parasite. *J. Invertebr. Pathol.* 81: 175–184.

Kanga, L. H. B., W. A. Jones, and R. R. James. 2003. Field trials using the fungal pathogen, *Metarhizium anisopliae* (Deuteromycetes: Hyphomycetes) to control the ectoparasitic mite, *Varroa destructor* (Acari: Varroidae) in honey bee, *Apis mellifera* (Hymenoptera: Apidae) colonies. *J. Econ. Entomol.* 96: 1091–1099.

Kleespies, R. G., J. Radtke, and K. Bienefeld. 2000. Virus-like particles found in the ectoparasitic bee mite *Varroa jacobsoni* Oudemans. *J. Invertebr. Pathol.* 75(1): 87–90.

Koppenhöffer, E. M., and E. M. Fuzy. 2008. Attraction of four entomopathogenic nematodes to four white grub species. *J. Invertebr. Pathol.* 99(2): 227–234.

MAAREC (Mid-Atlantic Apicultureal Researech & Extension Consortium) 2000. Publication 4.5, Wax Moth. https://agdev.anr.udel.edu/maarec/wp-content/uploads/2010/03/Wax_Moth_pm.pdf

Meikle, W. G., G. Mercadier, F. Annas, and N. Holst. 2009. Effects of multiple applications of a Beauveria-based biopesticide on *Varroa destructor* (Acari: Varroidae) densities in honey bee (Hymenoptera: Apidae) colonies. *J. Apic. Res.* 48: 220–222. doi: 10.3896/IBRA.1.48.3.13

Meikle, W. G., G. Mercadier, V. Girod, F. Derouané, and W. A. Jones. 2006. Evaluation of *Beauveria bassiana* (Balsamo) Vuillemin (Deuteromycota: Hyphomycetes) strains isolated from varroa mites in southern France. *J. Apic. Res.* 45: 219–220.

Meikle, W. G., G. Mercadier, N. Holst, and V. Girod. 2008a. Impact of two treatments of a formulation of *Beauveria bassiana* (Deuteromycota: Hyphomycetes) conidia on Varroa mites (Acari: Varroidae) and on honeybee (Hymenoptera: Apidae) colony health. *Exp. Appl. Acarol.* 46: 105–117. doi: 10.1007/s10493-008-9160-z

Meikle, W. G., G. Mercadier, N. Holst, C. Nansen, and V. Girod. 2007. Duration and spread of an entomopathogenic fungus, *Beauveria bassiana* (Deuteromycota: Hyphomycetes), used to treat varroa mites, *Varroa destructor* (Acari: Varroidae), in honeybee hives (Hymenoptera: Apidae). *J. Econ. Entomol.* 100: 1–10

Meikle, W. G., G. Mercadier, N. Holst, C. Nansen, and V. Girod. 2008b. Impact of a treatment of *Beauveria bassiana* (Deuteromycota: Hyphomycetes) on honeybee (Hymenoptera: Apidae) colony health and on varroa mites (Acari: Varroidae). *Apidologie* 39: 1–13. doi: 10.1051/apido:2007057.

Meikle, W. G., and J. M. Patt. 2011. Temperature, diet and other factors on development, survivorship and oviposition of the small hive beetle, *Aethina tumida* Murray (Col.: Nitidulidae). *J. Econ. Entomol.* 104(3):753-763. doi: 10.1603/EC10364

Muerrle, T. M., P. Neumann, J. F. Dames, H. R. Hepburn, and M. P. Hill. 2006. Susceptibility of adult *Aethina tumida* (Coleoptera: Nitidulidae) to entomopathogenic fungi. *J. Econ. Entomol.* 99(1): 1–6.

Nielsen, R. A., and C. D. Bister. 1979. Greater wax moth: Behavior of larvae. *Ann. Entomol. Soc. Am.* 72: 811–815.

Niu, G., R. M. Johnson, and M. R. Berenbaum. 2010. Toxicity of mycotoxins to honeybees and its amelioration by propolis. *Apidologie* 42(1): 79–87. doi: 10.1051/apido/2010039

Perkins, J. H., and R. Garcia. 1999. Social and economic factors affecting research and implementation of biological control. In *Handbook of Biological Control*. Bellows, T. S., and T. W. Fisher (Eds.). Academic Press, San Diego, California, U.S., pp. 993–1009.

Perrard, A., J. Haxaire, A. Rortais, and C. Villemant. 2009. Observations on the colony activity of the Asian hornet *Vespa velutina* Lepeletiet 1836 (Hymenoptera: Vespidae: Vespinae) in France. *Annales de la Société Entomologique de France* 45(1): 119–127.

Richards, C. S., M. P. Hill, and J. F. Dames. 2005. The susceptibility of small hive beetle (*Aethina tumida* Murray) pupae to *Aspergillus niger* (van Tieghem) and *A. flavus* (Link: Grey). *Am. Bee J.* 145(9): 748–751.

Rodríguez, M., M. Gerding, and A. France. 2009a. Selection of entomopathogenic fungi to control *Varroa destructor* (Acari: Varroidae). *Chilean J. Agr. Res.* 69: 534–540.

Rodríguez, M., M. Gerding, A. France, and R. Ceballos. 2009b. Evaluation of *Metarhizium anisopliae* var. *anisopliae* Qu-M845 isolate to control *Varroa destructor* (Acari: Varroidae) in laboratory and field trials. *Chilean J. Agr. Res.* 69(4): 541–547.

Schöller M., and S. Prozell. 2001. The braconid wasp *Habrobracon hebetor* (Hymenoptera: Braconidae): A natural enemy of moths infesting stored products [Die Mehlmottenschlupfwespe *Habrobracon hebetor* (Hymenoptera: Braconidae) als Antagonist vorratsschädlicher Motten]. *Gesunde Pflanzen* 53(3): 82–89. (in German)

Shapiro, M. 2000. Enhancement in activity of homologous and heterologous baculoviruses infectious to beet armyworm (Lepidoptera: Noctuidae) by an optical brightener. *J. Econ. Entomol.* 93(3): 572–576.

Shapiro-Ilan, D. I., W. A. Gardner, J. R. Fuxa, B. W. Wood, K. B. Nguyen, B. J. Adams, R. A. Humber, and M. J. Hall. 2003. Survey of entomopathogenic nematodes and fungi endemic to pecan orchards of the Southeastern United States and their virulence to the pecan weevil (Coleoptera: Curculionidae). *Environ. Entomol.* 32(1): 187–195.

Shapiro-Ilan, D. I., J. A. Morales-Ramos, M. G. Rojas, and W. L. Tedders. 2010. Effects of novel entomopathogenic nematode-infected host formulation on cadaver integrity, nematode yield, and suppression of *Diaprepes abbreviates* and *Aethina tumida*. *J. Invertebr. Pathol.* 103(2): 103–108.

Shaw, K. E., G. Davidson, S. J. Clark, B. V. Ball, J. K. Pell, D. Chandler, and K. D. Sunderland. 2002. Laboratory bioassays to assess the pathogenicity of mitosporic fungi to *Varroa destructor* (Acari: Mesostigmata), an ectoparasitic mite of the honeybee, *Apis mellifera*. *Biol. Control* 24: 266–276.

Spence, K. O., G. N. Stevens, H. Arimoto, J. Ruiz-Vega, H. K. Kaya, and E. E. Lewis. 2011. Effect of insect cadaver desiccation and soil water potential during rehydration on entomopathogenic nematode (Rhabditida: Steinernematidae and Heterorhabditidae) production and virulence. *J. Invertebr. Pathol.* 106(2): 268–273.

Stafford, K. C. III, and S. A. Allan. 2011. Field applications of entomopathogenic fungi *Beauveria bassiana* and *Metarhizium anisopliae* F52 (Hypocreales: Clavicipitaceae) for the control of *Ixodes scapularis* (Acari: Ixodidae). *J. Med. Entomol.* 47: 1107–1115.

Steenberg, T., P. Kryger, and N. Holst. 2010. A scientific note on the fungus *Beauveria bassiana* infecting *Varroa destructor* in worker brood cells in honey bee hives. *Apidologie* 41: 127–128.

Su, N.-Y., Z. Hillis-Starr, P. M. Ban, and R. H. Scheffrahn. 2003. Protecting historic properties from subterranean termites: A case study with Fort Christiansvaern, Christiansted National Historic Site, U.S. Virgin Islands. *Am. Entomol.* 49: 20–32.

Tanada, Y., and H. K. Kaya. 1993. *Insect Pathology*, Academic Press, San Diego, California, U.S.

Tsagou, V., A. Lianou, D. Lazarakis, N. Emmanouel, and G. Aggelis. 2004. Newly isolated bacterial strains belonging to Bacillaceae (*Bacillus* sp.) and *Micrococcaceae accelerate* death of the honey bee mite, *Varroa destructor* (*V. jacobsoni*), in laboratory assays. *Biotechnol. Lett.* 26(6): 529–532.

Tseng, Y. K., Y. W. Tsai, M. S. Wu, and R. F. Hou. 2008. Inhibition of phagocytic activity and nodulation in *Galleria mellonella* by the entomopathogenic fungus *Nomuraea rileyi*. *Entomologia Experimentalis et Applicata* 129(3): 243–250.

Tu, S., X. Qiu, L. Cao, R. Han, Y. Zhang, and X. Liu. 2010. Expression and characterization of the chitinases from *Serratia marcescens* GEI strain for the control of *Varroa destructor*, a honey bee parasite. *J. Invertebr. Pathol.* 104(2): 75–82.

Uçkan, F., E. Ergin, and F. Ayaz. 2004. Modelling age- and density-structured reproductive biology and seasonal survival of *Apanteles galleriae* Wilkinson (Hym., Braconidae). *J. Appl. Entomol.* 128(6): 407–413. doi: 10.1111/j.1439-0418.2004.00864

Vandenberg, J. D., and H. Shimanuki. 1990. Viability of *Bacillus thuringiensis* and its efficacy for larvae of the greater wax moth (Lepidoptera: Pyralidae) following storage of treated combs. *J. Econ. Entomol.* 83(3): 760–765.

Whitfield, J. B., S. A. Cameron, S. R. Ramirez, K. Roesch, S. Messinger, O. M. Taylor, and D. Cole. 2001. Review of the *Apanteles* species (Hymenoptera: Braconidae) attacking lepidoptera in *Bombus* (Fervidobombus) (Hymenoptera: Apidae) colonies in the new world, with description of a new species from South Africa. *Ann. Entomol. Soc. Am.* 94: 851–857.

Zimmerman, G. 1986. The "Galleria bait method" for detection of entomopathogenic fungi in soil. *J. Appl. Entomol*. 102: 213–215.

Chapter 7
Molecular Forensics for Honey Bee Colonies

Chen, Y., J. Evans, M. Hamilton, and M. Feldlaufer. 2007. The influence of RNA integrity on the detection of honey bee viruses: Molecular assessment of different sample storage methods. *J. Apic. Res.* 46: 81–87.

Chen, Y., Y. Zhao, Y. Zhao, J. Hammond, H.-T. Hsu, J. Evans, and M. Feldlaufer. 2004. Multiple virus infections in the honey bee and genome divergence of honey bee viruses. *J. Invertebr. Pathol.* 87: 84–93.

Chomczynski, P., and N. Sacchi. 1987. Signal-step method of RNA isolation by acid guanidinium thiocyanate-phenol-chloroform extraction. *Ann. Biochem.* 162: 156–159.

Cox-Foster, D. L., S. Conlan, E. C. Holmes, G. Palacios, J. D. Evans, N. A. Moran, P.-L. Quan, T. Briese, M. Hornig, D. M. Geiser, V. Martinson, D. vanEngelsdorp, A. L. Kalkstein, A. Drysdale, J. Hui, J. Zhai, L. Cui, S. K. Hutchison, J. F. Simons, M. Egholm, J. S. Pettis, and W. I. Lipkin. 2007. A metagenomic survey of microbes in honey bee Colony Collapse Disorder. *Science* 318: 283–287. doi: 10.1126/science.1146498

De Graaf, D. C., A. M. Alippi, M. Brown, J. D. Evans, M. Feldlaufer, A. Gregorc, M. Hornitzky, S. F. Pernal, D. M. Schuch, D. Titěra, V. Tomkies, and W. Ritter. 2006. Diagnosis of American foulbrood in honey bees: A synthesis and proposed analytical protocols. *Lett. Appl. Microbiol.* 43: 583–590.

de Miranda J. R., and I. Fries. 2008. Venereal and vertical transmission of deformed wing virus in honeybees (*Apis mellifera* L.). *J. Invertebr. Pathol.* 98: 184–189.

Evans, J. D. 2004. Transcriptional immune responses by honey bee larvae during invasion by the bacterial pathogen, *Paenibacillus larvae*. *J. Invertebr. Pathol.* 85: 105–111.

Evans, J. D. 2006. Beepath: An ordered quantitative-PCR array for exploring honey bee immunity and disease. *Handbook of Environmental Chemistry* 5(93): 135–139.

Evans J. D., Y. P. Chen, G. Di Prisco, J. Pettis, and V. Williams. 2009. Bee cups: Single-use cages for honey bee experiments. *J. Apic. Res.* 48: 300–302.

Genersch, E., W. Von Der Ohe, H. Kaatz, A. Schroeder, C. Otten, R. Büchler, S. Berg, W. Ritter, W. Mühlen, S. Gisder, M. Meixner, G. Liebig, and P. Rosenkranz. 2010. The German bee monitoring project: A long term study to understand periodically high winter losses of honey bee colonies. *Apidologie* 41: 332–352.

Grabensteiner E., T. Bakonyi, W. Ritter, H. Pechhacker, and N. Nowotny. 2007. Development of a multiplex RT-PCR for the simultaneous detection of three viruses of the honeybee (*Apis mellifera* L.): Acute bee paralysis virus, Black queen cell virus and Sacbrood virus. *J. Invertebr. Pathol.* 94: 222–225.

Honey Bee Genome Sequencing Consortium. 2006. Insights into social insects from the genome of the honeybee *Apis mellifera*. *Nature* 443: 931–949.

Johnson, R. M., J. D. Evans, G. E. Robinson, and M. R. Berenbaum. 2009. Changes in transcript abundance relating to Colony Collapse Disorder in honey bees (*Apis mellifera*). *PNAS* 106: 14790–14795.

Kukielka, D., F. Esperon, M. Higes, and J. M. Sanchez-Vizcaino. 2008. A sensitive one-step real-time RT-PCR method for detection of deformed wing virus and black queen cell virus in honeybee *Apis mellifera*. *J. Virol. Methods* 147: 275–281.

Lourenço A. P., A. Mackert, A. D. S. Cristino, and Z. L. P. Simões. 2008. Validation of reference genes for gene expression studies in the honey bee, *Apis mellifera*, by quantitative real-time RT-PCR. *Apidologie* 39: 372–385.

Nguyen, B. K., C. Saegerman, and E. Haubruge. 2009. Study on the contamination by *Paenibacillus larvae* of honey from the south part of Belgium (Walloon Region) and relation with the clinical expression of American foulbrood in honey bee colonies. *Annales Med. Veterinaire* 153: 219–223.

Pfaffl, M. W. 2001. A new mathematical model for relative quantification in real-time RT-PCR. *Nucleic Acids Res.* 29: 2002–2007.

Pinto, F. L., A. Thapper, W. Sontheim, and P. Lindblad. 2009. Analysis of current and alternative phenol based RNA extraction methodologies for cyanobacteria. *BMC Mol. Biol.* 10: 79.

Roetschi, A., H. Berthoud, R. Kuhn, and A. Imdorf. 2008. Infection rate based on quantitative real-time PCR of *Melissococcus plutonius*, the causal agent of European foulbrood, in honeybee colonies before and after apiary sanitation. *Apidologie* 39: 362–371.

Scharlaken, B., D. C. De Graaf, K. Goossens, M. Brunain, L. J. Peelman, and F. J. Jacobs. 2008. Reference gene selection for insect expression studies using quantitative real-time PCR: The head of the honeybee, *Apis mellifera*, after a bacterial challenge. *J. Insect Sci.* 8(33), available online: insectscience.org/8.33

Siede R, M. König, R. Büchler, K. Failing, and H. J. Thiel 2008. A real-time PCR based survey on acute bee paralysis virus in German bee colonies. *Apidologie* 39: 650–661.

Teixeira, E. W., Y. Chen, D. Message, J. Pettis, and J. D. Evans. 2008. Virus infections in Brazilian honey bees. *J. Invertebr. Pathol.* 99: 117–119.

vanEngelsdorp, D., J. D. Evans, C. Saegerman, C. Mullin, E. Haubruge, B. K. Nguyen, M. Frazier, J. Frazier, D. Cox-Foster, Y. Chen, R. Underwood, D. R. Tarpy, and J. S. Pettis. 2009. Colony Collapse Disorder: A Descriptive Study. *PLoS ONE* 4(8): e6481.

Ward, L., M. Brown, P. Neumann, S. Wilkins, J. Pettis, and N. Boonham. 2007. A DNA method for screening hive debris for the presence of small hive beetle (*Aethina tumida*). *Apidologie* 38: 272–280.

Chapter 8

Honey Bee Viruses and Their Effect on Bee and Colony Health

Allen, M., and B. V. Ball. 1996. The incidence and world distribution of the honey bee viruses. *Bee World* 77: 141–162.

Anderson, D. L., and A. J. Gibbs. 1988. Inapparent virus infections and their interactions in pupae of the honey bee (*Apis mellifera* L.) in Australia. *J. Gen. Virol.* 69: 1617–1625.

Anderson, D. L., and J. W. H. Trueman. 2000. *Varroa jacobsoni* (*Acari: Varroidae*) is more than one species. *Exp. Appl. Acarol.* 24: 165–189.

Bailey, L. 1965. The occurrence of chronic and acute bee paralysis viruses in bees outside Britain. *J. Invertebr. Pathol.* 7: 167–169.

Bailey, L., A. J. Gibbs, and R. D. Woods. 1964. Sacbrood virus of the larval honey bee (*Apis mellifera* Linnaeus). *Virology* 23: 425–429.

Bailey, L., and B. V. Ball. 1991. *Honey Bee Pathology* (2nd ed.). Academic Press, London, UK.

Bailey L., B. V. Ball, J. M. Carpenter, and R. D. Woods. 1980. Small virus-like particles in honey bees associated with chronic paralysis virus and with a previously undescribed disease. *J. Gen. Virol.* 46: 149–155.

Bailey, L., B. V. Ball, and J. N. Perry. 1981. The prevalence of viruses of honey bees in Britain. *Ann. Appl. Biol.* 97: 109–118.

Bailey, L., B. V. Ball, and J. N. Perry. 1983a. Honeybee paralysis: Its natural spread and its diminished incidence in England and Wales. *J. Apic. Res.* 22: 191–195.

Bailey, L., B. V. Ball, and J. N. Perry. 1983b. Association of viruses with two protozoal pathogens of the honey bee. *Ann. Appl. Biol.* 103: 13–20.

Bailey, L., J. M. Carpenter, and R. D. Woods. 1981. Properties of filamentous virus of the honey bee. *Virology* 114: 1–7.

Bailey, L., and E. F. W. Fernando. 1972. Effects of sacbrood virus on adult honey bees. *Ann. Appl. Biol.* 72: 27–35.

Bailey, L., A. J. Gibbs, and R. D. Woods. 1963. Two viruses from adult honey bees (*Apis mellifera* Linnaeus). *Virology* 21: 390–395.

Bailey, L., and R. D. Woods. 1974. Three previously undescribed viruses from the honeybee. *J. Gen. Virol.* 25: 175–186.

Bailey, L., and R. D. Woods. 1977. Two more small RNA viruses from honey bees and further observations on sacbrood and acute bee-paralysis viruses. *J. Gen. Virol.* 37: 175–182.

Ball, B. V. 1989. *Varroa jacobsoni* as a virus vector. In *Present Status of Varroatosis in Europe and Progress in the Varroa Mite Control*. Proc. Meeting, Udine, Italy, 1988. Cavalloro, E. (Ed.). EC-Experts Group, Luxembourg, pp. 241–244.

Ball, B. V., and M. F. Allen. 1988. The prevalence of pathogens in honey bee colonies infested with the parasitic mite *Varroa jacobsoni*. *Ann. Appl. Biol.* 113: 237–244.

Ball, B. V., and L. Bailey. 1997. Viruses. In *Honey Bee Pest, Predators, and Diseases*. Morse, R.A., and K. Flottum (Eds.). The A. I. Root Company, Medina, Ohio, U.S., pp. 11–31.

Berenyi, O., T. Bakonyi, I. Derakhshifar, H. Koglberger, and N. Nowotny. 2006. Occurrence of six honeybee viruses in diseased Austrian apiaries. *Appl. Environ. Microbiol.* 72: 2414–2420.

Berenyi, O., T. Bakonyi, I. Derakhshifar, H. Koglberger, G. Topolska, W. Ritter, H. Pechhacker, and N. Nowotny. 2007. Phylogenetic analysis of deformed wing virus genotypes from diverse geographic origins indicates recent global distribution of the virus. *Appl. Environ. Microbiol.* 73: 3605–3611.

Berthoud, H., A. Imdorf, M. Haueter, S. Radloff, and P. Neumann. 2010. Virus infections and winter losses of honey bee colonies (*Apis mellifera*). *J. Apic. Res.* 49: 60–65.

Blanchard, P., M. Ribière, O. Celle, P. Lallemand, F. Schurr, V. Olivier, A. L. Iscache, and J. P. Faucon. 2007. Evaluation of a real-time two step RT-PCR assay for quantitation of chronic bee paralysis virus (CBPV) genome in experimentally-infected bee tissues and in life stages of a symptomatic colony. *J. Virol. Methods* 141: 7–13.

Bromenshenk, J. J., C. B. Henderson, C. H. Wick, M. F. Stanford, A. W. Zulich, R. E. Jabbour, S. V. Deshpande, P. E. McCubbin, R. A. Seccomb, P. M. Welch, T. Williamsn, D. R. Firth, E. Skowronski, M. M. Lehmann, S. L. Bilimoria, J. Gress, K. W. Wanner, and R.A. Cramer. 2010. Iridovirus and microsporidian linked to honey bee colony decline. *PLoS-ONE* 5: e13181.

Bowen-Walker, P. L., S. J. Martin, and A. Gunn. 1999. The transmission of deformed wing virus between honey bees (*Apis mellifera* L.) by the ectoparasitic mite *Varroa jacobsoni* Oud. *J. Invertebr. Pathol.* 73: 101–106.

Burnside, C. E. 1945. The cause of paralysis of honeybees. *Am. Bee J.* 85: 354–363.

Camazine, S. M., and T. P. Liu. 1998. A putative Iridovirus from the honey bee mite, *Varroa jacobsoni* Oudemans. *J. Invertebr. Pathol.* 71: 177–178.

Carreck, N. L., B. V. Bell, and S. J. Martin. 2010. Honey bee colony collapse and changes in viral prevalence associated with *Varroa destructor*. *J. Apic. Res.* 49: 93–94.

Celle, O., P. Blanchard, F. Schurr, V. Olivier, N. Cougoule, J. P. Faucon, and M. Ribière. 2008. Detection of Chronic Bee Paralysis Virus (CBPV) genome and RNA replication in various hosts: Possible ways of spread. *Virus Res.* 133: 280–284.

Chen, Y. P., J. S. Pettis, A. Collins, and M. F. Feldlaufer. 2006. Prevalence and transmission of honey bee viruses. *Appl. Environ. Microbiol.* 72: 606–611.

Chen, Y. P., J. A. Higgins, and M. F. Feldlaufer. 2005. Quantitative real-time reverse transcription-PCR analysis of deformed wing virus infection in the honeybee (*Apis mellifera* L.). *Appl. Environ. Microbiol.* 71: 436–441.

Chen, Y. P., J. S. Pettis, J. D. Evans, M. Kramer, and M. F. Feldlaufer. 2004. Molecular evidence for transmission of Kashmir bee virus in honey bee colonies by ectoparasitic mite, *Varroa destructor*. *Apidologie* 35: 441–448.

Chen, Y. P., and R. Siede. 2007. Honey bee viruses. *Adv. Virus Res.* 70: 33–80.

Clark, T. B. 1978. A filamentous virus of the honey bee. *J. Invertebr. Pathol.* 32: 332–340.

Cox-Foster, D. L., S. Conlan, E. C. Holmes, G. Palacios, J. D. Evans, N. A. Moran, P.-L. Quan, T. Briese, M. Hornig, D. M. Geiser, V. Martinson, D. vanEngelsdorp, A. L. Kalkstein, A. Drysdale, J. Hui, J. Zhai, L. Cui, S. K. Hutchison, J. F. Simons, M. Egholm, J. S. Pettis, and W. I. Lipkin. 2007. A metagenomic survey of microbes in honey bee Colony Collapse Disorder. *Science* 318: 283–287. doi: 10.1126/science.1146498

Dainat, B., T. Ken, H. Berthoud, and P. Neumann. 2008. The ectoparasitic mite *Tropilaelaps mercedesae* (*Acari, Laelapidae*) as a vector of honeybee viruses. *Insect. Soc.* 56: 40–43.

DeJong, D., P. H. DeJong, and L. S. Goncalves. 1982. Weight loss and other damage to developing worker honey bees from infestation with *Varroa jacobsoni*. *J. Apic. Res.* 21: 165–167.

de Miranda, J. R. 2008. Diagnostoc techniques for virus detection in honey bees. In *Virology and the Honey Bee*. Aubert, M. F. A., B. V. Ball, I. Fries, R. F. A. Morritz, N. Milani, and I. Bernardinelli (Eds.). EEC Publications, Luxembourg, pp. 121–232.

de Miranda, J. R., and I. Fries. 2008. Venereal and vertical transmission of deformed wing virus in honeybees (*Apis mellifera* L.). *J. Invertebr. Pathol.* 98: 184–189.

de Miranda, J. R., and E. Genersch. 2010. Deformed wing virus. *J. Invertebr. Pathol.* 103: S48–S61.

de Miranda, J. R., G. Cordoni, and G. Budge. 2010a. The acute bee paralysis virus—Kashmir bee virus—Israeli acute paralysis virus complex. *J. Invertebr. Pathol.* 103: S30–S47.

de Miranda, J. R., B. Dainat, B. Locke, G. Cordoni, H. Berthoud, L. Gauthier, P. Neumann, G. E. Budge, B. V. Ball, and D. B. Stoltz. 2010b. Genetic characterisation of slow paralysis virus of the honeybee (*Apis mellifera* L.). *J. Gen. Virol.* 91: 2524–2530.

Denholm, C. H. 1999. Inducible honey bee viruses associated with *Varroa jacobsoni*. pp. 1–225. Keele University, UK. PhD thesis.

Ellis, J. D., and P. A. Munn. 2005. The worldwide health status of honey bees. *Bee World*. 86: 88–101.

Eyer, M., Y. P. Chen, M. O. Schäfer, J. Pettis, and P. Neumann. 2009. Small hive beetle, *Aethina tumida*, as a potential biological vector of honeybee viruses. *Apidologie* 40: 419–428.

Federici, B. A., D. K. Bideshi, Y. Tan, T. Spears, and Y. Bigot. 2009. Ascoviruses: Superb manipulators of apoptosis for viral replication and transmission. *Curr. Top. Microbiol. Immunol.* 328: 171–196.

Fievet, J., D. Tentcheva, L. Gauthier, J. R. de Miranda, F. Cousserans, M. E. Colin, and M. Bergoin. 2006. Localization of deformed wing virus infection in queen and drone *Apis mellifera* L. *Virol. J.* 3: e16.

Forgach, P., T. Bakonyi, Z. Tapaszti, N. Nowotny, and Rusvai, M. 2008. Prevalence of pathogenic bee viruses in Hungarian apiaries: Situation before joining the European Union. *J. Invertebr. Pathol.* 98: 235–238.

Forsgren, E., J. R. de Miranda, M. Isaksson, S. Wei, and I. Fries. 2009. Deformed wing virus associated with *Tropilaelaps mercedesae* infesting European honey bees (*Apis mellifera*). *Exp. Appl. Acarol.* 47: 87–97.

Fujiyuki, T., E. Matsuzaka, T. Nakaoka, H. Takeuchi, A. Wakamoto, S. Ohka, K. Sekimizu, A. Nomoto, and T. Kubo. 2009. Distribution of Kakugo virus and its effects on the gene expression profile in the brain of the worker honeybee *Apis mellifera* L. *J. Virol.* 83: 11560–11568.

Fujiyuki, T., S. Ohka, H. Takeuchi, M. Ono, A. Nomoto, and T. Kubo. 2006. Prevalence and phylogeny of Kakugo virus, a novel insect picorna-like virus that infects the honeybee (*Apis mellifera L.*), under various colony conditions. *J. Virol.* 80: 11528–11538.

Fujiyuki, T., H. Takeuchi, M. Ono, S. Ohka, T. Sasaki, A. Nomoto, and T. Kubo. 2004. Novel insect picorna-like virus identified in the brains of aggressive worker honeybees. *J. Virol.* 78: 1093–1100.

Gauthier, L., D. Tentcheva, M. Tournaire, B. Dainat, F. Cousserans, M. E. Colin, and M. Bergoin. 2007. Viral load estimation in asymptomatic honey bee colonies using the quantitative RT-PCR technique. *Apidologie* 38: 426–436.

Gauthier, L., M. Ravallec, M. Tournaire, F. Cousserans, M. Bergoin, B. Dainat, and J. R. de Miranda. 2011. Viruses associated with ovarian degeneration in *Apis mellifera* L. queens. *PLoS ONE* 6(1): e16217. doi: 10.137/journal.pone.0016217

Genersch, E. 2010. Honey bee pathology: Current threats to honey bees and beekeeping. *Appl. Microbiol. Biotechnol.* 87: 87–97.

Genersch, E., W. von der Ohe, H. Kaatz, A. Schroeder, C. Otten, R. Büchler, S. Berg, W. Ritter, W. Mühlen, S. Gisder, M. Meixner, G. Liebig, and P. Rosenkranz. 2010. The German bee monitoring project: A long term study to understand periodically high winter losses of honey bee colonies. *Apidologie* 41: 332–352.

Gisder, S., P. Aumeier, and E. Genersch. 2009. Deformed wing virus (DWV): Viral load and replication in mites (*Varroa destructor*). *J. Gen. Virol.* 90: 463–467.

Hails, R. S., B. V. Ball, and E. Genersch. 2008. Infection strategies of insect viruses. In *Virology and the Honey Bee*. Aubert, M. F. A., B. V. Ball, I. Fries, R. F. A. Morritz, N. Milani, and I. Bernardinelli (Eds.). EEC Publications, Luxembourg, pp. 255–275.

Higes, M., R. Martín-Hernandez, C. Botias, E. G. Bailon, A. Gonzales-Porto, L. Barrios, M. J. del Nozal, P. G. Palencia, and A. Meana. 2008. How natural infection by *Nosema ceranae* causes honeybee colony collapse. *Environ. Microbiol.* 10: 2659–2669.

Highfield, A. C., A. El Nagar, L. C. M. Mackinder, L. M. L. J. Noel, M. J. Hall, S. J. Martin, and D. C. Schroeder. 2009. Deformed wing virus implicated in overwintering honeybee colony losses. *Appl. Environ. Microbiol.* 75: 7212–7220.

Hitchcock, J. D. 1966. Transmission of sacbrood disease to individual honey bee larvae. *J. Econ. Entom.* 59: 1154–1156.

Iqbal, J., and U. Müller. 2007. Virus infection causes specific learning deficits in honeybee foragers. *Proc. Royal Soc. London-B* 274: 1517–1521.

Johnson, R. M., J. D. Evans, G. E. Robinson, and M. R. Berenbaum. 2009. Changes in transcript abundance relating to Colony Collapse Disorder in honey bees (*Apis mellifera*). *PNAS* 106: 14790–14795.

Katsuma, S., S. Tanaka, N. Omuro, L. Takabuchi, T. Daimon, S. Imanishi, S. Yamashita, M. Iwanaga, K. Mita, S. Maeda, M. Kobayashi, and T. Shimada. 2005. Novel macula-like virus identified in *Bombyx mori* cultured cells. *J. Virol.* 79: 5577–5584.

Kovac, H., and K. Crailsheim. 1988. Lifespan of *Apis mellifera carnica* Pollm. infested by *Varroa jacobsoni* Oud. in relation to season and extent of infestation. *J. Apic. Res.* 27: 230–238.

Kukielka, D., and J. M. Sánchez-Vizcaino. 2009. One-step real-time quantitative PCR assays for the detection and field study of sacbrood honeybee and acute bee paralysis viruses. *J. Virol. Methods* 161: 240–246.

Kulincevic, J., B. V. Ball, and V. Mladjan. 1990. Viruses in honey bee colonies infested with *Varroa jacobsoni*: First findings in Yugoslavia. *Acta Vet.* 40: 37–42.

Lanzi, G., J. R. de Miranda, M. B. Boniotti, C. E. Cameron, A. Lavazza, L. Capucci, S. M. Camazine, and C. Rossi. 2006. Molecular and biological characterization of deformed wing virus of honeybees (*Apis mellifera L.*). *J. Virol.* 80: 4998–5009.

Le Gall, O., P. Christian, C. M. Fauquet, A. M. Q. King, N. J. Knowles, N. Nakashima, G. Stanway, and A. E. Gorbalenya. 2008. *Picornavirales*, a proposed order of positive-sense single-stranded RNA viruses with a pseudo-T=3 virion architecture. *Arch. Virol.* 153: 715–727.

Lommel, S. A., T. J. Morris, and D. E. Pinnock. 1985. Characterization of nucleic acids associated with Arkansas bee virus. *Intervirology* 23: 199–207.

Maori, E., S. Lavi, R. Mozes-Koch, Y. Gantman, Y. Peretz, O. Edelbaum, E. Tanne, and I. Sela. 2007. Isolation and characterization of Israeli acute paralysis virus, a dicistrovirus affecting honeybees in Israel: Evidence for diversity due to intra- and inter-species recombination. *J. Gen. Virol.* 88: 3428–3438.

Martin S. J., B. V. Ball, and N. L. Carreck. 2010. Prevalence and persistance of deformed wing virus (DWV) in untreated or acaricide-treated *Varroa destructor* infested honey bees (*Apis mellifera*) colonies. *J. Apic. Res.* 49: 72–79.

Moore, J., A. Jironkin, D. Chandler, N. Burroughs, J. D. Evans, and E. V. Ryabov. 2010. Recombinants between deformed wing virus and *Varroa destructor* virus-1 may prevail in *Varroa destructor*-infested honeybee colonies. *J. Gen. Virol.* 92: 156–161.

Nielsen, S. L., M. Nicolaisen, and P. Kryger. 2008. Incidence of acute bee paralysis virus, black queen cell virus, chronic bee paralysis virus, deformed wing virus, Kashmir bee virus and sacbrood virus in honey bees (*Apis mellifera*) in Denmark. *Apidologie* 39: 310–314.

Nordström, S. 2000. Virus infections and *Varroa* mite infestations in honey bee colonies. pp. 1–65. Agraria 209, Acta Universitatis Agriculturae Sueciae. PhD thesis.

Nordström, S. 2003. Distribution of deformed wing virus within honey bee (*Apis mellifera*) brood cells infested with the ectoparasitic mite *Varroa destructor*. *Exp. Appl. Acarol.* 29: 293–302.

Nordström, S., I. Fries, A. Aarhus, H. Hansen, and S. Korpela. 1999. Virus infections in Nordic honey bee colonies with no, low or severe *Varroa jacobsoni* infestations. *Apidologie* 30: 475–484.

Olivier, V., P. Blanchard, S. Chaouch, P. Lallemand, F. Schurr, O. Celle, E. Dubois, N. Tordo, R. Thiery, R. Houlgatte, and M. Ribière. 2008a. Molecular characterization and phylogenetic analysis of chronic bee paralysis virus, a honey bee virus. *Virus Res.* 132: 59–68.

Olivier, V., I. Massou, O. Celle, P. Blanchard, F. Schurr, M. Ribière, and M. Gauthier. 2008b. *In situ* hybridization assays for localization of the chronic bee paralysis virus in the honey bee (*Apis mellifera*) brain. *J. Virol. Methods* 153: 232–237.

Ongus, J. R., D. Peters, J. M. Bonmatin, E. Bengsch, J. M. Vlak, and M. M. van Oers. 2004. Complete sequence of a picorna-like virus of the genus Iflavirus replicating in the mite *Varroa destructor*. *J. Gen. Virol.* 85: 3747–3755.

Overton, H. A., K. W. Buck, L. Bailey, and B. V. Ball. 1982. Relationships between the RNA components of chronic bee-paralysis virus and those of chronic bee-paralysis virus associate. *J. Gen. Virol.* 63: 171–179.

Ribiere, M., V. Olivier, and P. Blanchard. 2010. Chronic bee paralysis virus. A disease and a virus like no other? *J. Invertebr. Pathol.* 103: S120–S131.

Ribière, M. B. V. Ball, and M. F. A. Aubert. 2008. Natural history and geographic distribution of honey bee viruses. In *Virology and the Honey Bee*. Aubert, M. F. A., B. V. Ball, I. Fries, R. F. A. Morritz, N. Milani, and I. Bernardinelli (Eds.). EEC Publications, Luxembourg, pp. 15–84.

Rortais, A., D. Tentcheva, A. Papachristoforou, L. Gauthier, G. Arnold, M. E. Colin, and M. Bergoin. 2006. Deformed wing virus is not related to honey bees' aggressiveness. *Virol. J.* 3: 61. doi: 10.1186/1743-422X-3-61

Rosenkranz, P., P. Aumeier, and B. Ziegelmann. 2010. Biology and control of *Varroa destructor*. *J. Invertebr. Pathol.* 103: S96–S119.

Santillán-Galicia, M. T., B. V. Ball, S. J. Clark, and P. G. Alderson. 2010. Transmission of deformed wing virus and slow paralysis virus to adult bees (*Apis mellifera* L.) by *Varroa destructor*. *J. Apic. Res.* 49: 141–148.

Schneemann, A. 2006. The structural and functional role of RNA in icosahderal virus assembly. *Annu. Rev. Microbiol.* 60: 51–67.

Shen, M. Q., X. L. Yang, D. Cox-Foster, and L. W. Cui. 2005. The role of *Varroa* mites in infections of Kashmir bee virus (KBV) and deformed wing virus (DWV) in honey bees. *Virology* 342: 141–149.

Shimanuki, H., N. W. Calderone, and D. A. Knox. 1994. Parasitic mite syndrome: The symptoms. *Am. Bee J.* 134: 827–828.

Siede, R., and R. Büchler. 2003. Symptomatischer Befall von Drohnenbrut mit dem Black Queen Cell Virus auf hessischen Bienenständen [Symptomatic black queen cell virus infection of drone brood in Hessian apiaries]. *Berl. Münch. Tierärztl. Wschr.* 116: 130–133. (in German)

Sitaropoulou, N., E. P. Neophytou, and G. N. Thomopoulos. 1989. Structure of the nucleocapsid of a filamentous virus of the honey bee (*Apis mellifera*). *J. Invertebr. Pathol.* 53: 354–357.

Sumpter, D. J. T., and S. J. Martin. 2004. The dynamics of virus epidemics in *Varroa*-infested honey bee colonies. *J. Anim. Ecol.* 73: 51–63.

Tentcheva, D., L. Gauthier, L. Bagny, J. Fievet, B. Dainat, F. Cousserans, M. E. Colin, and M. Bergoin. 2006. Comparative analysis of deformed wing virus (DWV) RNA in *Apis mellifera* and *Varroa destructor*. *Apidologie* 37: 41–50.

Tentcheva, D., L. Gauthier, N. Zappulla, B. Dainat, F. Cousserans, M. E. Colin, and M. Bergoin. 2004. Prevalence and seasonal variations of six bee viruses in *Apis mellifera* L. and *Varroa destructor* mite populations in France. *Appl. Environ. Microbiol.* 70: 7185–7191.

Terio, V., V. Martella, M. Camero, N. Decaro, G. Testini, E. Bonerba, G. Tantillo, and C. Buonavoglia. 2008. Detection of a honeybee iflavirus with intermediate characteristics between kakugo virus and deformed wing virus. *New Microbiol.* 31: 439–444.

Todd, J. H., J. R. de Miranda, and B. V. Ball. 2007. Incidence and molecular characterization of viruses found in dying New Zealand honey bee (*Apis mellifera*) colonies infested with *Varroa destructor*. *Apidologie* 38: 354–367.

vanEngelsdorp, D., J. D. Evans, C. Saegerman, C. Mullin, E. Haubruge, B. K. Nguyen, M. Frazier, J. Frazier, D. Cox-Foster, Y. Chen, R. Underwood, D. R. Tarpy, and J. S. Pettis. 2009. Colony Collapse Disorder: A Descriptive Study. *PLoS ONE* 4(8): e6481.

vanEngelsdorp, D., and M. D. Meixner. 2010. A historical review of managed honey bee populations in Europe and the United States are the factors that may affect them. *J. Invertebr. Pathol.* 103: 580–595.

Venter, J. C., K. Remington, J. F. Heidelberg, A. L. Halpern, D. Rusch, J. A. Eisen, D. Wu, I. Paulsen, K. E. Nelson, W. Nelson, D. E. Fouts, S. Levy, A. H. Knap, M. W. Lomas, K. Nealson, O. White, J. Peterson, J. Hoffman, R. Parsons, H. Baden-Tillson, C. Pfannkoch, Y. Rogers, and H. O. Smith. 2004. Environmental genome shotgun sequencing of the Sargasso Sea. *Science* 304: 66–74.

Webby, R., and J. Kalmakoff. 1998. Sequence comparison of the major capsid protein from 18 diverse iridoviruses. *Arch. Virol.* 143: 1949–1966.

White, G. F. 1913. *Sacbrood, A Disease of Bees. Circ.* USDA, Washington, DC, U.S., pp. 1–5.

Yang, X., and D. L. Cox-Foster. 2005. Impact of an ectoparasite on the immunity and pathology of an invertebrate: Evidence for host immunosuppression and viral amplification. *Proc. Natl. Acad. Sci. USA* 102: 7470–7475.

Yue, C., and E. Genersch. 2005. RT-PCR analysis of deformed wing virus in honeybees (*Apis mellifera*) and mites (*Varroa destructor*). *J. Gen. Virol.* 86: 3419–3424.

Yue, C., M. Schröder, K. Bienefeld, and E. Genersch. 2006. Detection of viral sequences in semen of honeybees (*Apis mellifera*): Evidence for vertical transmission of viruses through drones. *J. Invertebr. Pathol.* 92: 105–108.

Yue, C., M. Schröder, S. Gisder, and E. Genersch. 2007. Vertical-transmission routes for deformed wing virus of honeybees (*Apis mellifera*). *J. Gen. Virol.* 88: 2329–2336.

Zioni, N., V. Soroker, and N. Chejanovski. 2011. Involvement of deformed wing virus (DWV) and *Varroa destructor* virus 1 (VaDV-1) in the deformed wing syndrome of the honey bee. *J. Gen. Virol.* (in review).

Chapter 9

PCR for the Analysis of Nosema in Honey Bees

Bailey, L. 1972. *Nosema apis* in drone honeybees. *J. Apic. Res.* 11: 171–174.

Bailey, L. 1981. *Honey Bee Pathology*. Academic Press, London, UK.

Becnel, J. J., and T. G. Andreadis. 1999. Microsporidia in insects. In *The Microporidia and Microsporidiosis*. M. Wittner and L. M. Weiss (Eds.). ASM Press, Washington, DC, U.S., pp. 447–501.

Bourgeois, A. L., T. E. Rinderer, L. D. Beaman, and R. G. Danka. 2010. Genetic detection and quantification of *Nosema apis* and *N. ceranae* in the honey bee. *J. Invertebr. Pathol.* 103: 53–58.

Burges, H. D., E. U. Canning, and I. K. Hullis. 1974. Ultrastructure of *Nosema oryzaephili* and the taxonomic value of the polar filament. *J. Invertebr. Pathol.* 23: 135–139.

Chamberlain, J. S., and J. R. Chamberlain. 1994. Optimization of multiplex PCRs. In *The Polymerase Chain Reaction*. Mullis, K. B., F. Ferre, and R. A. Gibbs (Eds.). Birkhauser, Boston, Massachusetts, U.S., pp. 38–64.

Chen, Y., J. D. Evans, I. B. Smith, and J. S. Pettis. 2008. *Nosema ceranae* is a long-present and wide-spread microsporidian infection of the European honey bee (*Apis mellifera*) in the United States. *J. Invertebr. Pathol.* 97: 186–188.

Chen, Y., J. D. Evans, L. Zhou, H. Boncristiani, K. Kimura, T. Xiao, A. M. Litkowski, and J. S. Pettis. 2009a. Asymmetrical coexistence of *Nosema ceranae* and *Nosema apis* in honey bees. *J. Invertebr. Pathol.* 101: 204–209.

Chen, Y. P., J. D. Evans, C. Murphy, R. Gutell, M. Zunker, D. Gundensen-Rindal, and J. S. Pettis. 2009b. Morphological, molecular, and phylogenetic characterization of *Nosema ceranae*, a microsporidian parasite isolated from the European honey bee, *Apis mellifera*. *J. Eukaryot. Microbiol.* 56: 142–147.

Chen, Y. P., J. A. Higgins, and M. F. Feldlaufer. 2005. Quantitative real-time reverse transcription-PCR analysis of deformed wing virus infection in the honeybee (*Apis mellifera* L.). *Appl. Environ. Microbiol.* 71: 436–441.

Chen, Y. P., and Z. Y. Huang. 2010. *Nosema ceranae*, a newly identified pathogen of *Apis mellifera* in the USA and Asia. *Apidologie* 41: 364–374.

Cox-Foster, D. L., S. Conlan, E. C. Holmes, G. Palacios, J. D. Evans, N. A. Moran, P.-L. Quan, T. Briese, M. Hornig, D. M. Geiser, V. Martinson, D. vanEngelsdorp, A. L. Kalkstein, A. Drysdale, J. Hui, J. Zhai, L. Cui, S. K. Hutchison, J. F. Simons, M. Egholm, J. S. Pettis, and W. I. Lipkin. 2007. A metagenomic survey of microbes in honey bee Colony Collapse Disorder. *Science* 318: 283–287. doi: 10.1126/science.1146498

Edlind, T. D., J. Li., G. S. Visvesvara, M. H. Vodkin, G. L. McLaughlin, and S. K. Katiyar. 1996. Phylogenetic analysis of beta-tublin sequences from amitrochondrial protozoa. *Mol. Phylgen. Evol.* 5: 359–367.

Farrar, C. L. 1954. Fumagillin for nosema control in package bees. *Am. Bee J.* 94: 52–53, 60.

Fries, I. 1989. Observations on the development and transmission of *Nosema apis* Z. in the ventriculus of the honeybee. *J. Apic. Res.* 28: 107–117.

Fries, I. 1993. *Nosema apis*—A parasite in the honey bee colony. *Bee World* 74: 5–19.

Fries, I. 2010. *Nosema ceranae* in European honey bees (*Apis mellifera*). *J. Invertebr. Pathol.* 103: S73–S79.

Fries, I., F. Feng, A. da Silva, S. B. Slemenda, and N. J. Pieniazek. 1996. *Nosema ceranae* n. sp. (Microspora, Nosematidae), morphological and molecular characterization of a microsporidian parasite of the Asian honey bee *Apis cerana* (Hymenoptera, Apidae). *Eur. J. Protistol.* 32: 356–365.

Gatehouse, H. S., and L. A. Malone. 1998. The ribosomal RNA gene region of *Nosema apis* (Microspora): DNA sequence for small and large subunit rRNA genes and evidence of a large tandem repeat unit size. *J. Invertebr. Pathol.* 71: 97–105.

Germot, A., H. Philippe, and H. Le Guyader. 1997. Evidence for loss of mitochondria in Microsporidia from a mitochondrial-type *Hsp70* in *Nosema locustae*. *Mol. Biochem. Parasitol.* 87: 159–168.

Gochnauer, T. A. 1953. Chemical control of American foulbrood and nosema diseases. *Am. Bee J.* 93: 410–411.

Heid, C. A., J. Stevens, K. J. Livak, and P. M. Williams. 1996. Real time quantitative PCR. *Genome Res.* 6: 986–994.

Higes, M., R. Matín-Hernández, E. Garrido-Bailón, P. García-Palencia, and A. Meana. 2008. Detection of infective *Nosema ceranae* (Microsporidia) spores in corbicular pollen of forager honeybees. *J. Invertebr. Pathol.* 97: 76–78.

Higes, M., R. Matín-Hernández, and A. Meana. 2010. *Nosema ceranae* in Europe: An emergent type C nosemosis. *Apidologie* 41: 375–392.

Higes, M., R. Matín, and A. Meana. 2006. *Nosema ceranae*, a new microsporidian parasite in honeybees in Europe. *J. Invertebr. Pathol.* 92: 93–95.

Hirt, R. P., B. Healy, C. R. Vossbrinck, E. U. Canning, and T. M. Embley. 1997. A mitchondrial HSP70 orthologue in *Vairimorpha necatrix*: Molecular evidence that microsporidia once contained mitrochondria. *Curr. Biol.* 7: 995–998.

Hirt, R. P., J. M. Logsdon, B. Healy, M. W. Dorey, W. F. Doolittle, and T. M. Embrey. 1999. Microsporidia are related to fungi: Evidence from the largest subunit of RNA polymerase II and other proteins. *PNAS* 96: 580–585.

Huang, W.-F., J.-H. Jiang, Y.-W. Chen, and C.-H. Wang. 2007. A *Nosema ceranae* isolate from the honeybee *Apis mellifera*. *Apidologie* 38: 30–37.

Katznelson, H., and C. A. Jamieson. 1952. Control of nosema disease of honeybees with fumagillin. *Science* 115: 70–71.

Keeling, P., and W. Doolittle. 1996. Alpha-tubulin from early-diverging eukaryotic lineages and the evolution of the tubulin family. *Mol. Biol. Evol.* 13: 1297–1305.

Klee, J., A. M. Besana, E. Genersch, S. Gisder, A. Nanetti, D. Q. Tam, T. X. Chinh, F. Puerta, J. M. Ruz, P. Kryger, D. Message, F. Hatjina, S. Korpela, I. Fries, and R. J. Paxton. 2007. Widespread dispersal of the microsporidian *Nosema ceranae*, an emergent pathogen of the western honey bee, *Apis mellifera*. *J. Invertebr. Pathol.* 96: 1–10.

Liu, T. P. 1984. Ultrastructure of the midgut of the worker honey bee *Apis mellifera* heavily infected with *Nosema apis*. *J. Invertebr. Pathol.* 44: 282–291.

Matín-Hernández, R., A. Meana, L. Prieto, A. M. Salvador, E. Garrido-Bailón, and M. Higes. 2007. The outcome of the colonization of *Apis mellifera* by *Nosema ceranae*. *Appl. Environ. Microbiol.* AEM. 00270–00207.

Matheson, A. 1993. World bee health report. *Bee World* 74: 176–212.

Moffett, J. O., J. J. Lackett, and J. D. Hitchcock. 1969. Compounds tested for control of *Nosema* in honey bees. *J. Econ. Entomol.* 62: 866–889.

Mullis, K. B., and F. A. Faloona. 1987. Specific synthesis of DNA *in vitro* via a polymerase-catalyzed chain reaction, In *Methods Enzymol*, Ray, W. (Ed.). Academic Press, London, UK, pp. 335–350.

OIE 2008. Chapter 22.4 Nosemosis of honey bees. In *OIE Terrestrial Manual*, ed. World Organisation of Animal Health, Paris, France, pp. 410–414.

Shimanuki, H., and D. A. Knox. 2000. Diagnosis of honey bee diseases. Agri. Handbook No. AH-690. USDA, Washington, DC, U.S. doi: http://www.ars.usda.gov/is/np/honeybeediseases/honeybeediseases.pdf

Tapaszti, Z., P. Forgách, C. Kövágó, L. Békési, T. Bakonyi, and M. Rusvai. 2009. First detection and dominance of *Nosema ceranae* in Hungarian honeybee colonies. *Acta Veterinaria Hungarica* 57: 383–388.

Tay, W. T., E. M. O'Mahony, and R. J. Paxton. 2005. Complete rRNA gene sequences reveal that the microsporidium *Nosema bombi* infects diverse bumblebee (*Bombus* spp.) hosts and contains multiple polymorphic sites. *J. Eukaryot. Microbiol.* 52: 505–513.

VanGuilder, H. D., K. E. Vrana, and W. M. Freeman. 2008. Twenty-five years of quantitative PCR for gene expression analysis. *BioTechniques* 44: 619–626.

Viljoen, G. J., L. H. Nel, and J. R. Crowther. 2005. *Molecular Diagnostic PCR Handbook*. Springer, Dordrecht, The Netherlands.

Weiss, L. M., and C. R. Vossbrinck. 1999. Molecular biology, molecular phylogeny, and molecular diagnostic approaches to the Microsporidia. In *The Microporidia and Microsporidiosis*. Wittner, M., and L. M. Weiss (Eds.). ASM Press, Washington, DC, U.S., pp. 129–171.

Chapter 10

Nosema ceranae Detection by Microscopy and Antibody Tests

Chen, Y., J. D. Evans, I. B. Smith, and J. S. Pettis. 2008. *Nosema ceranae* is a long-present and wide-spread microsporidian infection of the European honey bee (*Apis mellifera*) in the United States. *J. Invertebr. Pathol.* 97(2): 186–188.

Chen, Y. P., J. D. Evans, C. Murphy, R. Gutell, M. Zuker, D. Gundensen-Rindal, and J. S. Pettis. 2009. Morphological, molecular, and phylogenetic characterization of *Nosema ceranae*, a microsporidian parasite isolated from the European honey bee, *Apis mellifera*. *J. Eukaryot. Microbiol.* 56: 142–147

Fries, I., F. Feng, A. da Silva, S. B. Slemenda, and N. J. Pieniazek. 1996. *Nosema ceranae* sp. (Microspora, Nosematidae), morphological and molecular characterization of a microsporidian parasite of the Asian honey bee *Apis cerana* (Hymenoptera, Apidae). *Eur. J. Protistol.* 32: 356–365.

Gisder, S., K. Hedtke, N. Möckel, M.-C. Frielitz, A. Linde, and E. Genersch. 2010. Five-year cohort study of *Nosema* spp. in Germany: Does climate shape virulence and assertiveness of *Nosema ceranae*? *App. Environ. Microbiol.* 76(9): 3032–3038.

Higes, M., R. Martin, and A. Meana. 2006. *Nosema ceranae*, a new microsporidian parasite in honeybees in Europe. *J. Invertebr. Pathol.* 92: 93–95.

Huang, W.-F., J. Jing-Hao, C. Yue-Wen, and W. Chung-Hsiung. 2007. A *Nosema ceranae* isolate from the honeybee *Apis mellifera*. *Apidologie* 38(1): 30–37.

Klee, J., A. Besana, E. Genersch, S. Gisder, A. Nanetti, D. Q. Tam, T. X. Chinh, F. Puerta, J. M. Ruz, P. Kryger, D. Message, F. Hatjina, S. Korpela, I. Fries, and R. J. Paxton. 2007. Widespread dispersal of the microsporidian *Nosema ceranae*, an emergent pathogen of the Western honey bee, *Apis mellifera*. *J. Invertebr. Pathol.* 96: 1–10

Olsen, P. E., W. A. Rice, and T. P. Liu 1986. *In vitro* germination of *Nosema apis* spores under conditions favorable for the generation and maintenance of sporoplasms. *J. Invertebr. Pathol.* 47: 65–73.

Chapter 11

Chalkbrood Re-Examined

Anderson, D. L., and N. L. Gibson. 1998. New species and isolates of spore-cyst fungi (Plectomycetes: *Ascosphaera*les) from Australia. *Aust. Syst. Bot.* 11: 53–72.

Anderson, D. L., H. Giacon, and N. Gibson. 1997. Detection and thermal destruction of the chalkbrood fungus (*Ascosphaera apis*) in honey. *J. Apic. Res.* 36: 163–168.

Anderson, D. L., A. J. Gibbs, and N. L. Gibson. 1998. Identification and phylogeny of sporecyst fungi (*Ascosphaera* spp.) using ribosomal DNA sequences. *Mycol. Res.* 102: 541–547.

Arbia, A., and B. Babbay. 2011. Management strategies of honey bee diseases *J. Entomol.* 8: 1–15.

Aronstein, K., and R. H. ffrench-Constant. 1995. PCR based monitoring of specific *Drosophila* (Diptera: Drosophilidae) cyclodiene resistance alleles in the presence and absence of selection. *Bull. Entomol. Res.* 85: 5–9.

Aronstein, K., and G. Hayes. 2004. Antimicrobial activity of allicin against honeybee pathogens. *J. Apic. Res.* 43: 57–59.

Aronstein, K., and E. Saldivar. 2005. Characterization of a honey bee Toll related receptor gene Am18w and its potential involvement in antimicrobial immune defense. *Apidologie* 36: 3–14.

Aronstein, K., and K. D. Murray. 2010. Chalkbrood disease in honey bees. *J. Invertebr. Pathol.* 103: 20–29.

Aronstein, K. A., K. D. Murray, and E. Saldivar. 2010. Transcriptional responses in honey bee larvae infected with chalkbrood fungus. *BMC Genomics* 11(391).

Aronstein, K. A., K. D. Murray, J. de Leon, X. Qin, and G. Weinstock. 2007. High mobility group (HMG-box) genes in the honey bee fungal pathogen *Ascosphaera apis. Mycologia* 99(4): 553–561.

Bailey, L. 1981. *Honey Bee Pathology*. Academic Press, London, UK.

Bailey, L., and B. V. Ball. 1991. *Honey Bee Pathology*. (2nd ed.) Academic Press, London, UK.

Bissett, J. 1988. Contribution toward a monograph of the genus *Ascosphaera*. *Can. J. Bot.* 66: 2541–2560.

Bogdanov, S. 2006. Contaminants of bee products. *Apidologie* 37: 1–18.

Borum, A. E., and M. Ulgen. 2008. Chalkbrood (*Ascosphaera apis*) infection and fungal agents of honey bees in north-west Turkey. *J. Apic. Res.* 47(2): 170–171.

Chorbinski, P. 2004. Identification of *Ascosphaera apis* strains using ribosomal DNA sequences. *Vet. Med.* 60(2): 190–192.

Chorbinski, P., and K. Rypula. 2003. Studies on the morphology of strains *Ascosphaera apis* isolated from chalkbrood disease of the honey bees. *Vet. Med.* 6(2): 1–12.

Christensen, M., and M. Gilliam. 1983. Note on the *Ascosphaera* species inciting chalkbrood in honey bees. *Apidologie* 14(4): 291–297.

Coppin, E., R. Debuchy, S. Arnaise, and M. Picard. 1997. Mating types and sexual development in filamentous ascomycetes. *Microbiol. Mol. Biol. Rev.* 61(4): 411–428.

Evans, J. D., and M. Spivak. 2010. Socialized medicine individual and communal disease barriers in honey bees. *J. Invertebr. Pathol.* 103: S62–S72.

Evans, J. D., K. Aronstein, Y. P. Chen, C. Hetru, J.-L. Imler, H. Jiang, M. Kanost, G. J. Thompson, Z. Zou, and D. Hultmark. 2006. Immune pathways and defense mechanisms in honey bees *Apis mellifera*. *Insect Mol. Biol.* 15: 645–656.

Flores, J., M. Spivak, and I. Gutierrez. 2005. Spores of *Ascosphaera apis* contained in wax foundation can infect honeybee brood. *Vet. Microbiol.* 108: 141–144.

Flores, J. M., I. Gutierrez, and R. Espejo. 2005a. The role of pollen in chalkbrood disease in *Apis mellifera*: Transmission and predisposing conditions. *Mycologia* 97: 1171–1176.

Gilliam, M. 1978. Fungi. In *Honey Bee Pests, Predators, and Diseases*. Morse, R. A. (Ed.). Cornell University Press, Ithaca, New York, U.S., pp. 78–101.

Gilliam, M. 1986. Infectivity and survival of the chalkbrood pathogen, *Ascosphaera apis*, in colonies of honey bees, *Apis mellifera*. *Apidologie* 17(2): 93–100.

Gilliam, M., and S. Taber. 1991. Diseases, pests, and normal microflora of honeybees, *Apis mellifera*, from feral colonies. *J. Invertebr. Pathol.* 58(2): 286–289.

Gilliam, M., and J. D. Vandenberg. 1997. In *Honey Bee Pests, Predators, and Diseases*. Morse, R. A., and K. Flottum (Eds.). Cornell University Press, Ithaca, New York, U.S., pp. 81–110.

Gilliam, M., S. Taber, B. J. Lorenz, and D. B. Prest. 1988. Factors affecting development of chalkbrood disease in colonies of honey bee, *Apis mellifera*, fed pollen contaminated with *Ascosphaera apis*. *J. Invertebr. Pathol.* 52: 314–325.

Heath, L. A. F. 1982. Chalkbrood pathogens: A review. *Bee World* 63: 130–135.

Heath, L. A. F. 1985. Occurrence and distribution of chalkbrood disease of honeybees. *Bee World* 66: 9–15.

Heath, L. A. F., and B. M. Gaze. 1987. Carbon dioxide activation of spores of the chalkbrood fungus *Ascosphaera apis*. *J. Apic. Res.* 26(4): 243–246.

Hornitzky, M. 2001. *Literature review of chalkbrood*. A report for the RIRDC 842. Publication No. 01/150, Kingston, New South Wales, Australia. doi: www.hgsc.bcm.tmc.edu/projects/microbial/documents/review_Chalkb.pdf

James, R. R., and J. S. Skinner. 2005. PCR diagnostic methods for *Ascosphaera* infections in bees. *J. Invertebr. Pathol.* 90(2): 98–103.

Jensen, A. B., R. R. James, and J. Eilenberg. 2009. Long-term storage of *Ascosphaera aggregata* and *A. apis*, pathogens of the leafcutting bee (*Megachile rotundata*) and the honey bee (*Apis mellifera*). *J. Invertebr. Pathol.* 101(2): 157–160.

Johnson, R. N., M. T. Zaman, M. M. Decelle, A. J. Siegel, D. R. Tarpy, E. C. Siegel, and P. T. Starks. 2005. Multiple micro-organisms in chalkbrood mummies: Evidence and implications. *J. Apic. Res.* 44(1): 29–32.

Kim, I., S. H. Kim, Y. S. Lee, E. K. Yun, H. S. Lee, J. W. Kim, K. S. Ryu, P. D. Kang, and I. H. Lee. 2004. Immune stimulation in the silkworm, *Bombyx mori* L., by CpG oligodeoxynucleotides. *Arch. Insect Biochem. Physiol.* 55(1): 43–48.

Lapidge, K., B. Oldroyd, and M. Spivak. 2002. Seven suggestive quantitative trait loci influence hygienic behavior of honey bees. *Naturwissenschaften*. 89: 565–568.

Maassen, A. 1913. Weitere Mitteilungen ueber die seuchenhaften Brutkrankheiten der Bienen [Further communication on the epidemic brood diseases of bees]. *Mitteilungen aus der Kaiserlichen Biologischen Anstalt fuer Land- und Forstwirtschaft* 14: 48–58. (in German)

Murray, K. D., K. A. Aronstein, and W. A. Jones. 2005. A molecular diagnostic method for selected *Ascosphaera* species using PCR amplification of internal transcribed spacer regions of rDNA. *J. Apic. Res.* 44: 61–64.

Nugent, K. L., E. Sangvichen, P. Sihanonth, N. Ruchikachorn, and A. J. S. Whalley. 2006. A revised method for the observation of conidiogenous structures in fungi. *Mycologist* 20: 111–114.

Orantes-Bermejo, F. J., A. G. Pajuelo, M. M. Megías, and C. T. Fernández-Píñar. 2010. Pesticide residues in beeswax and beebread samples collected from honey bee colonies (*Apis mellifera* L.) in Spain: Possible implications for bee losses. *J. Apic. Res.* 49(3): 243–250.

Poggeler, S. 2001. Mating-type genes for classical strain improvements of Ascomycetes. *Appl. Microbiol. Biotechnol.* 56: 589–601.

Reynaldi, F. J., A. C. Lopez, G. N. Albo, and A. M. Alippi. 2003. Genomic fingerprinting. *J. Apic. Res.* 42(4): 68–76.

Sadd, B. M., and P. Schmid-Hempel. 2006. Insect immunity shows specificity in protection upon secondary pathogen exposure. *Curr. Biol.* 16(12): 1206–1210.

Schmid-Hempel, P. 2005. Evolutionary ecology of insect immune defenses. *Annu. Rev. Entomol.* 50: 529–551.

Sheridan, R., B. Policastro, S. Thomas, and D. Rice. 2008. Analysis and occurrence of 14 sulfonamide antibacterials and chloramphenicol in honey by solid-phase extraction followed by LC/MS/MS analysis. *J. Agric. Food. Chem.* 56(10): 3509–3516.

Skou, J. P. 1972. *Ascosphaera*les. *Friesia* 10(1): 1–24.

Skou, J. P. 1988. More details in support of the class Ascosphaeromycetes. *Mycotaxson.* XXXI(1): 191–198.

Spiltoir, C. F. 1955. Life cycle of *Ascosphaera apis*. *Am. J. Bot.* 42: 501–518.

Spivak, M., and M. Gilliam. 1993. Facultative expression of hygienic behavior of honeybees in relation to disease resistance. *J. Apic. Res.* 32: 147–157.

Spivak, M., and G. S. Reuter. 1998. Honey bee hygienic behavior. *Am. Bee J.* 138: 283–286.

Spivak, M., and D. L. Downey. 1998. Field assays for hygienic behavior in honey bees (Hymenoptera: Apidae). *J. Econ. Entomol.* 91: 64–70.

Spivak, M., and G. S. Reuter. 2001. Resistance to American foulbrood disease by honey bee colonies, *Apis mellifera*, bred for hygienic behavior. *Apidologie* 32: 555–565.

Starks, P. T., C. A. Blackie, and T. D. Seeley. 2000. Fever in honeybee colonies. *Naturwissenschaften* 87: 229–231.

Tarpy, D. R. 2003. Genetic diversity within honeybee colonies prevents severe infections and promotes colony growth. *Proc. R. Soc. Ser. B* (270): 99–103.

Theantana, T., and P. Chantawannakul. 2008. Protease and β-N-acetylglucosaminidase of honey bee chalkbrood pathogen *Ascosphaera apis*. *J. Apic. Res.* 47(1): 68–76.

Chapter 12

Critical Transition Temperature (CTT) of Chalkbrood Fungi and Its Significance for Disease Incidence

Bamford, S., and L. A. F. Heath. 1989. The effects of temperature and pH on the germination of spores of the chalkbrood fungus, *Ascosphaera apis*. *J. Apic. Res.* 28: 36–40.

Christensen, M., and M. Gilliam. 1983. Notes on the *Ascosphaera* species inciting chalkbrood in honey bees. *Apidologie* 14: 291–297.

Cooper, B. 1980. Fluctuating broodnest temperature rhythm. *Br. Isles Bee Breeders News* 18: 12–16.

Flores, J. M., J. A. Ruiz, J. M. Ruz, F. Puerta, M. Bustos, F. Padilla, and F. Campano. 1996. Effect of temperature and humidity of sealed brood on chalkbrood development under controlled conditions. *Apidologie* 27: 185–192.

Gilliam, M., S. Taber, and J. B. Rose. 1978. Chalkbrood disease of honeybees, *Apis mellifera* L.: A progress report. *Apidologie* 9: 75–89.

Hale, P. J., and D. M. Menapace. 1980. Effect of time and temperature on the viability of *Ascosphaera apis* [chalkbrood]. *J. Invertebr. Pathol.* 36: 429–430.

James, R. R. 2005. Temperature and chalkbrood development in the alfalfa leafcutting bee, *Megachile rotundata*. *Apidologie* 36: 15–23.

Johnson, C. G. 1940. The maintenance of high atmospheric humidities for entomological work with glycerol-water mixtures. *Ann. Appl. Biol.* 27: 295–299.

Johnson, R. N., M. T. Zaman, M. M. Decelle, A. J. Siefel, D. R. Tarpy, E. C. Siegel, and P. T. Starks. 2005. Multiple micro-organisms in chalkbrood mummies: Evidence and implications. *J. Apic. Res.* 44: 29–32.

Sokal, R. R., and F. J. Rohlf. 1995. *Biometry: The Principles and Practice of Statistics in Biological Research*. W. H. Freeman, San Francisco, California, U.S.

Thammavongs, B., J.-M. Panoff, and M. Guéguen. 2000. Phenotypic adaptation to freeze-thaw stress of yeast-like fungus *Geotrichum candidum*. *Int. J. Food Microbiol.* 6: 99–105.

Yoder, J. A., B. S. Christensen, T. J. Croxall, J. L. Tank, and D. Sammataro. 2008. Suppresion of growth rate of colony-associated fungi by high fructose corn syrup feeding supplement, formic acid, and oxalic acid. *J. Apic. Res.* 47: 126–130.

Chapter 13

Small Hive Beetle (*Aethina tumida*) Contributions to Colony Losses

Ambrose, J. T., M. S. Stanghellini, and D. I. Hopkins. 2000. A scientific note on the threat of small hive beetles (*Aethina tumida* Murray) to bumble bee (*Bombus* spp.) colonies in the United States. *Apidologie* 31(3): 455–456.

Arbogast, R. T., B. Torto, S. Willms, and P. E. A. Teal. 2009. Trophic habits of *Aethina tumida* (Coleoptera: Nitidulidae): Their adaptive significance and relevance to dispersal. *Environ. Entomol.* 38(3): 561–568.

Arbogast, R. T., B. Torto, and P. E. A. Teal. 2010. Potential for population growth of the small hive beetle *Aethina tumida* (Coleoptera: Nitidulidae) on diets of pollen dough and oranges. *Fla. Entomol.* 93(2): 224–230.

Buchholz, S., M. O. Schafer, S. Spiewok, J. S. Pettis, M. Duncan, W. Ritter, R. Spooner-Hart, and P. Neumann. 2008. Alternative food sources of *Aethina tumida*. (Coleoptera: Nitidulidae). *J. Apic. Res.* 47(3): 202–209.

Cabanillas, H. E., and P. J. Elzen. 2006. Infectivity of entomopathogenic nematodes (Steinernematidae and Heterorhabditidae) against the small hive beetle *Aethina tumida* (Coleoptera: Nitidulidae). *J. Apic. Res.* 45: 49–50.

de Guzman, L. I., and A. M. Frake. 2007. Temperature affects *Aethina tumida* (Coleoptera: Nitidulidae) development. *J. Apic. Res.* 46(2): 88–93.

de Guzman, L. I., A. M. Frake, and T. E. Rinderer. 2008. Detection and removal of brood infested with eggs and larvae of small hive beetles (*Aethina tumida* Murray) by Russian honey bees. *J. Apic. Res.* 47(3): 216–221.

Delaplane, K. S., J. D. Ellis, and W. M. Hood. 2010. A test for interactions between *Varroa destructor* (Acari: Varroidae) and *Aethina tumida* (Coleoptera: Nitidulidae) in colonies of honey bees (Hymenoptera: Apidae). *Ann. Entomol. Soc. Am.* 103(5): 711–715.

Ellis, J. D. 2005a. Progress towards controlling small hive beetles with IPM: Knowing our options—Part I of two parts. *Am. Bee J.* 145(2): 115–119.

Ellis, J. D. 2005b. Progress towards controlling small hive beetles with IPM: Integrating current treatments—Part II of two parts. *Am. Bee J.* 145(3): 207–210.

Ellis, J. D. 2005c. Reviewing the confinement of small hive beetles (*Aethina tumida*) by western honey bees (*Apis mellifera*). *Bee World* 86(3): 56–62.

Ellis, J. D., and K. S. Delaplane. 2008. Small hive beetle (*Aethina tumida*) oviposition behavior in sealed brood cells with notes on the removal of the cell contents by European honey bees (*Apis mellifera*). *J. Apic. Res.* 47(3): 210–215.

Ellis, J. D., and A. M. Ellis. 2008. Small hive beetle, *Aethina tumida* Murray (Nitidulidae: Coleoptera). In *Encyclopedia of Entomology*, Vol 4. Capinera, J. L. (Ed.). Kluwer Academic Publishers, Dordrecht, The Netherlands, pp. 3415–3418.

Ellis, J. D., and H. R. Hepburn. 2006. An ecological digest of the small hive beetle (*Aethina tumida*), a symbiont in honey bee colonies (*Apis mellifera*). *Insect. Soc.* 53(1): 8–19.

Ellis, J. D., and P. Munn. 2005. The worldwide health status of honey bees. *Bee World* 86(4): 88–101.

Ellis, J. D., C. W. W. Pirk, H. R. Hepburn, G. Kastberger, and P. J. Elzen. 2002a. Small hive beetles survive in honeybee prisons by behavioural mimicry. *Naturwissenschaften* 89: 326–328.

Ellis, J. D., P. Neumann, R. Hepburn, and P. J. Elzen. 2002b. Longevity and reproductive success of *Aethina tumida* (Coleoptera: Nitidulidae) fed different natural diets. *J. Econ. Entomol.* 95(5): 902–907.

Ellis, J. D., H. R. Hepburn, K. S. Delaplane, and P. J. Elzen. 2003a. A scientific note on small hive beetle (*Aethina tumida*) oviposition and behavior during European (*Apis mellifera*) honey bee clustering and absconding events. *J. Apic. Res.* 42(1–2): 47–48.

Ellis, J. D., H. R. Hepburn, K. S. Delaplane, P. Neumann, and P. J. Elzen. 2003b. The effects of adult small hive beetles, *Aethina tumida* (Coleoptera: Nitidulidae), on nests and flight activity of Cape and European honey bees (*Apis mellifera*). *Apidologie* 34: 399–408.

Ellis, J. D., A. J. Holland, R. Hepburn, P. Neumann, and P. J. Elzen. 2003c. Cape (*Apis mellifera capensis*) and European (*Apis mellifera*) honey bee guard age and duration of guarding small hive beetles (*Aethina tumida*). *J. Apic. Res.* 42(3): 32–34.

Ellis, J. D., C. S. Richards, H. R. Hepburn, and P. J. Elzen. 2003d. Oviposition by small hive beetles elicits hygienic responses from Cape honeybees. *Naturwissenschaften* 90: 532–535.

Ellis, J. D., K. S. Delaplane, C. S. Richards, R. Hepburn, J. A. Berry, and P. J. Elzen. 2004a. Hygienic behavior of Cape and European *Apis mellifera* (Hymenoptera: Apidae) toward *Aethina tumida* (Coleoptera: Nitidulidae) eggs oviposited in sealed bee brood. *Ann. Entomol. Soc. Am.* 97(4): 860–864.

Ellis, J. D., H. R. Hepburn, and P. J. Elzen. 2004b. Confinement of small hive beetles (*Aethina tumida*) by Cape honeybees (*Apis mellifera capensis*). *Apidologie* 35: 389–396.

Ellis, J. D., H. R. Hepburn, B. Luckman, and P. J. Elzen. 2004c. Effects of soil type, moisture, and density on pupation success of *Aethina tumida* (Coleoptera: Nitidulidae). *Environ. Entomol.* 33(4): 794–798.

Ellis, J. D., H. I. Rong, M. P. Hill, H. R. Hepburn, and P. J. Elzen. 2004d. The susceptibility of small hive beetle (*Aethina tumida* Murray) pupae to fungal pathogens. *Am. Bee J.* 144(6): 486–488.

Ellis, J. D., K. S. Delaplane, A. Cline, and J. V. McHugh. 2008. The association of multiple sap beetle species (Coleoptera: Nitidulidae) with western honey bees (*Apis mellifera*) colonies in North America. *J. Apic. Res.* 47(3): 188–189.

Ellis, J. D., S. Spiewok, K. S. Delaplane, S. Buchholz, P. Neumann, and W. L. Tedders. 2010. Susceptibility of *Aethina tumida* (Coleoptera: Nitidulidae) larvae and pupae to entomopathogenic nematodes. *J. Econ. Entomol.* 103 (1): 1–9.

Elzen, P. J., J. R. Baxter, D. Westervelt, C. Randall, K. S. Delaplane, L. Cutts, and W. T. Wilson. 1999. Field control and biology studies of a new pest species, *Aethina tumida* Murray (Coleoptera, Nitidulidae), attacking European honey bees in the Western Hemisphere. *Apidologie* 30 (5): 361–366.

Elzen, P. J., J. R. Baxter, P. Neumann, A. Solbrig, C. Pirk, H. R. Hepburn, D. Westervelt, and C. Randall. 2001. Behaviour of African and European subspecies of *Apis mellifera* toward the small hive beetle, *Aethina tumida*. *J. Apic. Res.* 40(1): 40–41.

Evans, J. D., J. S. Pettis, W. M. Hood, and H. Shimanuki. 2007. Tracking an invasive honey bee pest: Mitochondrial DNA variation in North American small hive beetles. *Apidologie* 34(2): 103–109.

Evans, J. D., S. Spiewok, E. W. Teixeira, and P. Neumann. 2008. Microsatellite loci for the small hive beetle, *Aethina tumida*, a nest parasite of honey bees. *Mol. Ecol. Resour.* 8(3): 698–700.

Eyer, M., Y. P. Chen, M. O. Schäfer, J. S. Pettis, and P. Neumann. 2009a. Honey bee sacbrood virus infects adult small hive beetles, *Aethina tumida* (Coleoptera: Nitidulidae). *J. Apic. Res.* 48: 296–297.

Eyer, M., Y. P. Chen, M. O. Schäfer, J. S. Pettis, and P. Neumann. 2009b. Small hive beetle, *Aethina tumida*, as a potential biological vector of honeybee viruses. *Apidologie* 40: 419–428.

Greco, M. K., D. Hoffmann, A. Dollin, M. Duncan, R. Spooner-Hart, and P. Neumann. 2010. The alternative Pharaoh approach: Stingless bees mummify beetle parasites alive. *Naturwissenschaften* 97(3): 319–323.

Hepburn, H. R., and S. Radloff. 1998. *Honey Bees of Africa*. Springer-Verlag, Berlin, Germany. 370 pp.

Hoffman, D., J. S. Pettis, and P. Neumann. 2008. Potential host shift of the small hive beetle (*Aethina tumida*) to bumblebee colonies (*Bombus impatiens*). *Insectes Soc.* 55(2): 153–162.

Hood, W. M. 2000. Overview of the small hive beetle, *Aethina tumida*, in North America. *Bee World* 81(3): 129–137.

Hood, W. M. 2004. The small hive beetle, *Aethina tumida*: A review. *Bee World* 85(3): 51–59.

Hood, W. M. 2006. Evaluation of two small hive beetle traps in honey bee colonies. *Am. Bee J.* 146(10): 873–876.

Hood, W. M., and G. A. Miller. 2003. Trapping small hive beetles. (Coleoptera: Nitidulidae) inside colonies of honey bees (Hymenoptera: Apidae). *Am. Bee J.* 143(5): 405–409.

Lounsberry, Z., S. Spiewok, S. F. Pernal, T. S. Sonstegard, W. M. Hood, J. Pettis, P. Neumann, and J. D. Evans. 2010. Worldwide diaspora of *Aethina tumida* (Coleoptera: Nitidulidae), a nest parasite of honey bees. *Ann. Entomol. Soc. Am.* 103(4): 671–677.

Lundie, A. E. 1940. The small hive beetle, *Aethina tumida*. Union of South Africa Department of Agriculture and Forestry, Entomological Series 2. *Science Bulletin* 220.

Muerrle, T. M., P. Neumann, J. F. Dames, H. R. Hepburn, and M. P. Hill. 2006. Susceptibility of adult *Aethina tumida* (Coleoptera: Nitidulidae) to entomopathogenic fungi. *J. Econ. Entomol.* 99(1): 1–6.

Neumann, P., and J. D. Ellis. 2008. The small hive beetle (*Aethina tumida* Murray, Coleoptera: Nitidulidae): Distribution, biology and control of an invasive species. *J. Apic. Res.* 47(3): 181–183.

Neumann, P., and P. J. Elzen. 2004. The biology of the small hive beetle (*Aethina tumida*, Coleoptera: Nitidulidae): Gaps in our knowledge of an invasive species. *Apidologie* 35(3): 229–247.

Neumann, P., and S. Härtel. 2004. Removal of small hive beetle (*Aethina tumida*) eggs and larvae by African honey bee colonies (*Apis mellifera scutellata*). *Apidologie* 35: 31–36.

Neumann, P., and W. Ritter. 2004. A scientific note on the association of *Cychramus luteus* (Coleoptera: Nitidulidae) with honey bee (*Apis mellifera*) colonies. *Apidologie* 35: 665–666.

Nolan, M. R., and W. M. Hood. 2008. Comparison of two attractants to small hive beetles, *Aethina tumida,* in honey bee colonies. *J. Apic. Res.* 47(3): 229–233.

Pettis, J. S., and H. Shimanuki. 2000. Observations on the small hive beetle, *Aethina tumida* Murray, in the United States. *Am. Bee. J.* 140: 152–155.

Richards, C. S., M. P. Hill, and J. F. Dames. 2005. The susceptibility of small hive beetle (*Aethina tumida* Murray) pupae to *Aspergillus niger* (van Tieghem) and *A. flavus* (Link: Grey). *Am. Bee J.* 145(9): 748–751.

Schäfer, M. O., W. Ritter, J. S. Pettis, and P. Neumann. 2010. Small hive beetles, *Aethina tumida*, are vectors of *Paenibacillus larvae*. *Apidologie* 41: 14–20.

Schmolke, M. D. 1974. Study of *Aethina tumida*: The small hive beetle. University of Rhodesia, Certificate in Field Ecology Project Report.

Shapiro-Ilan, D. I., J. A. Morales-Ramos, M. G. Rojas, and W. L. Tedders. 2010. Effects of novel entomopathogenic nematode-infected host formulation on cadaver integrity, nematode yield, and suppression of *Diaprepes abbreviates* and *Aethina tumida*. *J. Invertebr. Pathol.* 103(2): 103–108.

Spiewok, S., and P. Neumann. 2006a. Cryptic low-level reproduction of small hive beetles in honeybee colonies. *J. Apic. Res.* 45: 47–48.

Spiewok, S., and P. Neumann. 2006b. Infestation of commercial bumblebee (*Bombus impatiens*) field colonies by small hive beetles (*Aethina tumida*). *Ecol. Entomol.* 31: 623–628.

Stanghellini, M. S., J. T. Ambrose, and D. I. Hopkins. 2000. Bumble bee colonies as potential alternative hosts for the small hive beetle. *Am. Bee J.* 140(1): 71–75.

Suazo, A., B. Torto, P. E. A. Teal, and J. H. Tumlinson. 2003. Response of the small hive beetle (*Aethina tumida*) to honey bee (*Apis mellifera*) and beehive-produced volatiles. *Apidologie* 34(6): 525–533.

Swart, D. J., J. F. Johannsmeier, G. D. Tribe, and P. Kryger. 2001. Disease and pests of honeybees. In *Beekeeping in South Africa* (3rd ed.), Plant Protection Research Institute Handbook No. 14, Johannsmeier, M. F. (Ed.). Agricultural Research Council, Pretoria, South Africa, pp. 199–222.

Torto, B., R. T. Arbogast, H. Alborn, A. Suazo, D. vanEngelsdorp, D. Boucias, J. H. Tumlinson, and P. E. A. Teal. 2007a. Composition of volatiles from fermenting pollen dough and attractiveness to the small hive beetle *Aethina tumida,* a parasite of the honeybee *Apis mellifera*. *Apidologie* 38(4): 380–389.

Torto, B., D. Boucias, R. T. Arbogast, J. H. Tumlinson, and P. E. A. Teal. 2007b. Multitrophic interaction facilitates parasite-host relationship between an invasive beetle and the honey bee. *PNAS* 104(20): 8474–8378.

Torto, B., A. Suazo, H. Alborn, J. H. Tumlinson, and P. E. A. Teal. 2005. Response of the small hive beetle (*Aethina tumida*) to a blend of chemicals identified from honeybee (*Apis mellifera*) volatiles. *Apidologie* 36(4): 523–532.

vanEngelsdorp, D., J. Hayes, R. M. Underwood, and J. Pettis. 2008. A survey of honey bee colony losses in the U.S., Fall 2007 to Spring 2008. *PLoS ONE* 3(12): e4071. doi: 10.1371/journal.pone.0004071

vanEngelsdorp, D., J. Hayes Jr., R. M. Underwood, and J. S. Pettis. 2010. A survey of honey bee colony losses in the United States, Fall 2008 to Spring 2009. *J. Apicult. Res.* 49(1): 7–14.

West, J. 2004. The new West small hive beetle trap. *Am. Bee J.* 144(2): 89.

Chapter 14

Pesticides and Honey Bee Toxicity in the United States

Adey, M., P. Walker, and P. T. Walker. 1986. *Pest Control Safe for Bees*. IBRA, London, UK.

Ailouane, Y., M. Lambin, C. Armengaud, A. El Hassani, and V. Gary. 2009. Subchronic exposure of honeybees to sublethal doses of pesticides: Effects on behavior. *Environ. Toxicol. Chem.* 28: 113–122.

Alaux, C., J. Brunet, C. Dussaubat, F. Mondet, S. Tchamitchan, M. Cousin, J. Brillard, A. Baldy, L. P. Belzunces, and Y. Le Conte. 2009. Interactions between Nosema microspores and a neonicotinoid weaken honeybees (*Apis mellifera*). *Environ. Microbiol.* 12(3): 774–782. doi: 10.1111/j.14622920.2009.02123.x

Albero, B., C. Sanchez-Brunete, and J. L. Tadeo. 2004. Analysis of pesticides in honey by solid-phase extraction and gas chromatography—mass spectrometry. *J. Agric. Food Chem.* 52: 5828–5835.

Alder, L., K. Greulich, G. Kempe, and B. Vieth. 2006. Residue analysis of 500 high priority pesticides: Better by GC-MS or LC-MS/MS? *Mass Spectrometry Reviews* 25: 838–865.

Aliano, N. P., and M. D. Ellis. 2008. Bee-to-bee contact drives oxalic acid distribution in honey bee colonies. *Apidologie* 39: 481–487.

Anderson, J. F., and W. A. Wojtas M.A. 1986. Honey bees (Hymenoptera, Apidae) contaminated with pesticides and polychlorinated biphenyls. *J. Econ. Entomol.* 79: 1200–1205.

Arpaia, S. 1996. Ecological impact of Bt-transgenic plants: 1. Asessing possible effects of CrtIIIB toxin on honey bee (*Apis mellifera*) colonies. *J. Genet. Breed.* 50: 315–319.

Atkins, E. L. 1992 Injury to honey bees by poisoning. In *The Hive and the Honey Bee (*Rev.). Graham, J. M. (Ed.), Dadant and Sons, Inc., Hamilton, Illinois, U.S., p. 1324.

Aupinel, P., D. Fortini, H. Dufour, J. N. Taséi, and B. Michaud. 2005. Improvement of artificial feeding in a standard *in vitro* method for rearing *Apis mellifera* larvae. *Bull. Insectol.* 58: 107–111.

Babendreier, D., D. Joller, J. Romeis, F. Bigler, and F. Widmer. 2007. Bacterial community structures in honeybee intestines and their response to two insecticidal proteins. *Federation of European Microbiological Societies (FEMS) Microbiol. Ecol.* 59: 600–610.

Babendreier, D., N. M. Kalberer, J. Romeis, P. Fluri, E. Mulligan, and F. Bigler. 2005. Influence of Bt-transgenic pollen, Bt-toxin and protease inhibitor (SBTI) ingestion on development of the hypopharyngeal glands in honeybees. *Apidologie* 36: 585–594. doi: 10.1051/apido:2005049

Babendreier, D., J. Romeis, J. Bigler, and P. Fluri. 2006. *Neue Erkenntnisse zu möglichen Auswirkungen von transgenem Bt-Mais auf Bienen*. Forschungsanstalt Agroscope Liebefeld-Posieux ALP (Swiss Bee Research Centre), Switzerland. (in German) doi: http://www.agroscope.admin.ch/imkerei/00302/00305/index.html

Bailey, J., C. Scott Dupree, R. Harris, J. Tolman, and B. Harris. 2005. Contact and oral toxicity to honey bees in Ontario, Canada. *Apidologie* 36: 623–633.

Balayiannis, G., and P. Balayiannis. 2008. Bee honey as an environmental bioindicator of pesticides' occurrence in six agricultural areas of Greece. *Arch. Environ. Contam. Toxicol.* 55: 462–470.

Barron, A. B., R. Maleszka, R. K. Vander Meer, and G. E. Robinson. 2007. Octopamine modulates honey bee dance behavior. *Proc. Natl. Acad. Sci. USA* 104: 1703–1707.

Bernal, J. L., J. J. Jimenez, M. J. del Nozal, M. Higes, and J. Llorente. 2000. Gas chromatographic determination of acrinathrine and 3-phenoxybenzaldehyde residues in honey. *J. Chromatogr. A* 882: 239–243.

Berry, J. 2009. Pesticides, bees and wax: An unhealthy, untidy mix. *Bee Culture* 137: 33–35.

Blasco, C., M. Fernandez, A. Pena, C. Lino, M. I. Silveira, G. Font, and Y. Pico. 2003. Assessment of pesticide residues in honey samples from Portugal and Spain. *J. Agric. Food Chem.* 51: 8132–8138.

Blasco, C., G. Font, and Y. Pico. 2008. Solid-phase microextraction-liquid chromatography-mass spectrometry applied to the analysis of insecticides in honey. *Food Additives and Contaminants* 25: 59–69.

Boecking, O., and E. Genersch. 2008, Varroosis -the ongoing crisi in bee keeping. *Journal fuer Verbraucherschutz und Lebensmittelsicherheit* [Journal of Consumer Protection and Food Safety] 3: 221–228.

Bogdanov, S. 2004. Beeswax: Quality issues today. *Bee World* 85: 46–50.

Bogdanov, S. 2006. Contaminants of bee products. *Apidologie* 37: 1–18.

Bogdanov, S., V. Kilchenmann, and A. Imdorf. 1997. Acaricide residues in beeswax and honey. *Apiacta* 32: 72–80.

Bogdanov, S., V. Kilchenmann, and A. Imdorf. 1998. Acaricide residues in some bee products. *J. Apic. Res.* 37: 57–67.

Bogdanov, S., V. Kilchenmann, K. Seiler, H. Pfefferli, T. Frey, B. Roux, P. Wenk, and J. Noser. 2004. Residues of para-dichlorobenzene in honey and beeswax. *J. Apic. Res.* 43: 14–16.

Bonmatin, J. M., P. A. Marchand, R. Charvet, I. Moineau, E. R. Bengsch, and M. E. Colin. 2005. Quantification of imidacloprid uptake in maize crops. *J. Agric. Food Chem.* 53: 5336–5341.

Burley, L., R. Fell, and R. Saacke. 2008. Survival of honey bee (Hymenoptera: Apidae) spermatozoa incubated at room temperature from drones exposed to miticides. *J. Econ. Entomol.* 101: 1081–1087.

Canadian Honey Council. 2005. Conditions of use for oxalic acid dihydrate. *Hivelights 18* (Special Suppl.).

Cao, L. C., T. W. Honeyman, R. Cooney, L. Kennington, C. R. Scheid, and J. A. Jonassen. 2004. Mitochondrial dysfunction is a primary event in renal cell oxalate toxicity. *Kidney Int.* 66: 1890–1900.

CDPR (California Department of Pesticide Regulation). 2006. Top 100 pesticides by acres treated in 2005. CADPR, Sacramento, California, U.S. doi: www.cdpr.ca.gov/docs/pur/pur05rep/top100_ais.pdf (accessed by author 24 August 2009)

CDPR (California Department of Pesticide Regulation). 2008. Authorization for Section 18 use of Hivastan (Powerpoint document. CADPR, Sacramento, California, U.S. doi: www.cdpr.ca.gov/docs/dept/prec/2010/20100318_section18.pdf (accessed by author 28 August 2009)

Charrière, J. D., and A. Imdorf. 2002. Oxalic acid treatment by trickling against *Varroa destructor*: Recommendations for use in central Europe and under temperate climate conditions. *Bee World* 82: 51–60.

Chauzat, M. P., and J. P. Faucon. 2007. Pesticide residues in beeswax samples collected from honey bee colonies (*Apis mellifera* L.) in France. *Pest Manage. Sci.* 63: 1100–1106.

Chauzat, M. P., P. Carpentier, A. C. Martel, S. Bougeard, N. Cougoule, P. Porta, J. Lachaize, F. Madec, M. Aubert, and J. P. Faucon.(2009. Influence of pesticide residues on honey bee (Hymenoptera: Apidae) colony health in France. *Environ. Entomol.* 38: 514–523.

Chauzat, M. P., J. P. Faucon, A. C. Martel, J. Lachaize, N. Cougoule, and M. Aubert. 2006. A survey of pesticide residues in pollen loads collected by honey bees in France. *J. Econ. Entomol.* 99: 253–262.

Choudhary, A., and D. C. Sharma. 2008a. Pesticide residues in honey samples from Himachal Pradesh (India). *Bull. Environ. Contam. Toxicol.* 80: 417–422.

Choudhary, A, and D. C. Sharma. 2008b. Dynamics of pesticide residues in nectar and pollen of mustard (*Brassica juncea* (L.) Czern.) grown in Himachal Pradesh (India). *Environ. Monit. Assess.* 144: 143–150.

Claudianos, C., H. Ranson, R. M. Johnson, S. Biswas, M. A. Schuler, M. R. Berenbaum, R. Feyereisen, and J. G. Oakeshott. 2006. A deficit of detoxification enzymes: Pesticide sensitivity and environmental response in the honeybee. *Insect Mol. Biol.* 15: 615–636.

Colin, M. E., J. M. Bonmatin, I. Moineau, C. Gaimon, S. Brun, and J. P. Vermandere. 2004. A method to quantify and analyze the foraging activity of honey bees: Relevance to the sublethal effects induced by systemic insecticides. *Arch. Environ. Contam. Toxicol.* 47: 387–395.

Collins, A. M., J. S. Pettis, R. Wilbanks, and M. F. Feldlaufer. 2004. Performance of honey bee (*Apis mellifera*) queens reared in beeswax cells impregnated with coumaphos. *J. Apic. Res.* 43: 128–134.

Committee on the Status of Pollinators in North America, National Research Council. 2007. *Status of Pollinators in North America*. National Academy Press, Washington, DC, U.S. doi: www.nap.edu (entire report online)

Crane, E., and P. Walker. 1983. *The Impact of Pest Management on Bees and Pollination*. IBRA, London, UK.

Croft, B. A. 1990. *Arthropod Biological Control Agents and Pesticides*. Wiley, New York, NY, U.S.

Cutler, G. C., and C. D. Scott-Dupree. 2007. Exposure to clothianidin seed-treated canola has no long-term impact on honey bees. *J. Econ. Entomol.* 100: 765–772.

Davies, T. G. E., L. M. Field, P. N. R. Usherwood, and M. S. Williamson. 2007. DDT, pyrethrins, pyrethroids and insect sodium channels. *IUBMB Life* 59: 151–162.

Decourtye, A., J. Devillers, S. Cluzeau, M. Charreton, and M. H. Pham-Delègue. 2004. Effects of imidacloprid and deltamethrin on associative learning in honeybees under semi-field and laboratory conditions. *Ecotoxicol. Environ. Safety* 57: 410–419.

Decourtye, A., J Devillers, E. Genecque, K. Le Menach, and H. Budzinski. 2005. Comparative sublethal toxicity of nine pesticides on olfactory learning performances of honey bees. *Apis mellifera*. *Arch. Environ. Toxicol. Chem.* 28: 113–122.

Decourtye, A., E. Mader, and N. Desneux. 2010. Landscape enhancement of floral resources for honey bees in agro-ecosystems. *Apidologie* 41, 264–277.

Decourtye, A., and M. H. Pham-Delègue. 2002. The proboscis extension response: Assessing the sublethal effects of pesticides on the honey bee. In *Honey Bees: Estimating the Environmental Impact of Chemicals*. Devillers, J., and M. H. Pham-Delègue (Eds.), Taylor and Francis, London, UK.

Desneux, N., A. Decourtye, and J. M. Delpuech. 2007. The sublethal effects of pesticides on beneficial arthropods. *Annu. Rev. Entomol.* 52, 81–106.

Devillers, J., A. Decourtye, H. Budzinski., M. H. Pham-Delègue, S. Cluzeau, and C. Maurin. 2003. Comparative toxicity and hazards of pesticides to *Apis* and non-*Apis* bees: A chemometrical study. *SAR and QSAR in Environ. Res.* 14: 389–403.

Duan, J. J., M. Marvier, J. Huesing, G. Dively, and Z. Y. Huang. 2008. A meta-analysis of effects of Bt crops on honey bees (Hymenoptera: Apidae). *PLoS ONE* 3: e1415.

Eischen, F. A., and R. H. Graham. 2008. Feeding overwintering honey bee colonies infected with *Nosema ceranae*. *In* Proceedings of the American Bee Research Conference. *Am. Bee J.* 148: 555.

Ellis, M. D., and F. P. Baxendale. 1997. Toxicity of seven monoterpenoids to tracheal mites (Acari: Tarsonemidae) and their honey bee (Hymenoptera: Apidae) hosts when applied as fumigants. *J. Econ. Entomol.* 90: 1087–1091.

Ellis, M. D., F. P. Baxendale, and D. L. Keith. 1998. Protecting bees when using insecticides. NebGuide G98-1347, University of Nebraska Cooperative Extension, Lincoln, Nebraska, U.S.

Ellis, M. D., R. Nelson, and C. Simonds. 1988. A comparison of the fluvalinate and ether roll methods of sampling for Varroa mites in honey bee colonies. *Am. Bee J.* 128: 262–263.

Elzen, P. J., and D. Westervelt. 2002. Detection of coumaphos resistance in *Varroa destructor* in Florida. *Am. Bee J.* 142: 291–292.

Elzen, P. J., J. R. Baxter, M. Spivak, and W. T. Wilson. 2000. Control of *Varroa jacobsoni* Oud. resistant to fluvalinate and amitraz using coumaphos. *Apidologie* 31: 437–441.

Elzen, P. J., F. A. Eischen, J. B. Baxter, G. W. Elzen, and W. T. Wilson. 1999. Detection of resistance in US *Varroa jacobsoni* Oud. (Mesostigmata: Varroidae) to the acaricide fluvalinate. *Apidologie* 30: 13–17.

Enan, E. 2001. Insecticidal activity of essential oils: Octopaminergic sites of action. *Comparative Biochemistry and Physiology Part C, Toxicol. Pharmacol.* 130, 325–337.

EPA (Environmental Protection Agency). 2009a. *Federal Insecticide Fungicide and Rodenticide Act.* http://www.epa.gov/pesticides/regulating/laws.htm (accessed 24 August 2009).

EPA (Environmental Protection Agency). 2009b. *Pesticide Registration Review.* http://www.epa.gov/oppsrrd1/registration_review/highlights.htm (accessed 24 August 2009).

Estep, C. B, G. N. Menon, H. E. Williams, and A. C. Cole. 1977. Chlorinated hydrocarbon insecticide residues in Tennessee honey and beeswax. *Bull. Environ. Contam. Toxicol.* 17: 168–174.

Evans, P. D., and J. D. Gee. 1980. Action of formamidine pesticides on octopamine receptors. *Nature* 287: 60–62.

Faucon, J. P., L. Mathieu, M. Ribiere, A. C. Martel, P. Drajnudel, S. Zeggane, C. Aurieres, and M. F. A. Aubert. 2002. Honey bee winter mortality in France in 1999 and 2000. *Bee World* 83: 14–23.

Federal Register. 2000. Coumaphos: Pesticide tolerance for emergency action. U.S. Environmental Protection Agency, Washington, DC, U.S.

Fernandez-Muino, M. A., M. T. Sancho, S. Muniategui, J. F. Huidobro, and J. Simallozano. 1995. Nonacaricide pesticide residues in honey—Analytical methods and levels found. *J. Food Protect.* 58: 1271–1274.

Fernandez, M., Y. Pico, and J. Manes. 2002. Analytical methods for pesticide residue determination in bee products. *J. Food Protect.* 65: 1502–1511.

Floris, I., A. Satta, P. Cabras, V. L. Garau, and A. Angioni. 2004. Comparison between two thymol formulations in the control of *Varroa destructor*: Effectiveness, persistence, and residues. *J. Econ. Entomol.* 97: 187–191.

Frazier, M., C. Mullin, J. Frazier, and S. Ashcraft. 2008. What have pesticides got to do with it? *Am. Bee J.* 148: 521–523.

Fries, I. 1991 Treatment of sealed honey bee brood with formic acid for control of *Varroa jacobsoni*. *Am. Bee J.* 131: 313–314.

Ghini, S., M. Fernandez, Y. Pico, R. Marin, F. Fini, J. Manes, and S. Girotti. 2004. Occurrence and distribution of pesticides in the province of Bologna, Italy, using honeybees as bioindicators. *Arch. Environ. Contam. Toxicol.* 47: 479–488.

Girolami, V., M. Greatti, A. Di Bernardo, A. Tapparo, C. Giorio, A. Squartini, L. Mazzon, M. Mazaro, and N. Mori. 2009. Translocation of neonicotinoid insecticides from coated seeds to seedling guttation drops: A novel way of intoxication for bees. *J. Econ. Entomol.* 102: 1808–1815.

Gregorc, A., and I. Bowen. 2000. Histochemical characterization of cell death in honeybee larvae midgut after treatment with *Paenibacillus larvae*, amitraz and oxytetracycline. *Cell Biol. Int.* 24: 319–324.

Gregorc, A., and M. I. Smodiš-Škerl. 2007. Toxicological and immunohistochemical testing of honeybees after oxalic acid and rotenone treatments. *Apidologie* 38: 296–305.

Gunasekara, A., T. Truong, K. Goh, F. Spurlock, and R. Tjeerdema. 2007. Environmental fate and toxicology of fipronil. *J. Pestic. Sci.* 32: 189–199.

Haarmann, T., M. Spivak, D. Weaver, B. Weaver, and T. Glenn. 2002. Effects of fluvalinate and coumaphos on queen honey bees (Hymenoptera: Apidae) in two commercial queen rearing operations. *J. Econ. Entomol.* 95: 28–35.

Halm, M. P., A. Rortais, G. Arnold, J. N. Taséi, and S. Rault. 2006. New risk assessment approach for systemic insecticides: The case of honey bees and imidacloprid (Gaucho). *Environ. Sci. Technol.* 40: 2448–2454.

Higes. M., A. Meana, M. Suarez, and J. Llorente. 1999. Negative long-term effects on bee colonies treated with oxalic acid against *Varroa jacobsoni* Oud. *Apidologie* 30: 289–292.

ISB (Inf. Syst. Biotechnol.). 2007–2010. *Field test release applications in the U.S.* Database provided by APHIS. USDA, USDA Animal and Plant Health Inspection Service (APHIS) Biotechnology Regulatory Services. http://www.isb.vt.edu/search.aspx?CommandName=search&searchterm=field+test+release+application+in+the+u..+database

Isman, M. B. 2006. Botanical insecticides, deterrents, and repellents in modern agriculture and an increasingly regulated world. *Annu. Rev. Entomol.* 51: 45–66.

Iwasa, T., N. Motoyama, J. T. Ambrose, and R. M. Roe. 2003 Mechanism for the differential toxicity of neonicotinoid insecticides in the honey bee, *Apis mellifera*. *Crop Prot.* 23: 371–378.

Jeschke, P., and R. Nauen. 2008. Neonicotinoids—from zero to hero in insecticide chemistry. *Pest Manage. Sci.* 64: 1084–1098.

Jimenez, J. J., J. L. Bernal, M. J. del Nozal, and M. T. Martin. 2005. Residues of organic contaminants in beeswax. *Eur. J. Lipid Sci. Technol.* 107: 896–902.

Johansen, C. A. 1977. Pesticides and pollinators. *Annu. Rev. Entomol.* 22: 177–192.

Johansen, C. A., and D. F. Mayer. 1990. *Pollinator Protection: A Bee and Pesticide Handbook.* Wicwas Press, Cheshire, Connecticut, U.S.

Johnson, R. M., H. S. Pollock, and M. R. Berenbaum. 2009. Synergistic interactions between in-hive miticides in *Apis mellifera*. *J. Econ. Entomol.* 102: 474–479.

Johnson, R. M., Z. Wen, M. A. Schuler, and M. R. Berenbaum. 2006. Mediation of pyrethroid insecticide toxicity to honey bees (Hymenoptera: Apidae) by cytochrome P450 monooxygenases. *J. Econ. Entomol.* 99: 1046–1050.

Keyhani, J., and E. Keyhani. 1980. EPR study of the effect of formate on cytochrome oxidase. *Biochem. Biophys. Res. Commun.* 92: 327–333.

Kim, Y. J., S. H. Lee, S. W. Lee, and Y. J. Ahn. 2004. Fenpyroximate resistance in *Tetranychus urticae* (Acari: Tetranychidae): Cross-resistance and biochemical resistance mechanisms. *Pest Manage. Sci.* 60: 1001–1006.

Kubik, M., J. Nowacki, A. Pidek, Z. Warakomska, L. Michalczuk, and W. Goszczynski. 1999. Pesticide residues in bee products collected from cherry trees protected during blooming period with contact and systemic fungicides. *Apidologie* 30: 521–532.

Kubik, M., J. Nowacki, A. Pidek, Z. Warakomska, L. Michalczuk, W. Goszczynski, and B. Dwuznik. 2000. Residues of captan (contact) and difenoconazole (systemic) fungicides in bee products from an apple orchard. *Apidologie* 31: 531–541.

Laurent, F. M., and E. Rathahao. 2003. Distribution of [^{14}C]-imidacloprid in sunflowers (*Helianthus annuus* L.) following seed treatment. *J. Agric. Food Chem.* 51: 8005–8010.

Lemaux, P. G. 2008. Genetically engineered plants and foods: A scientist's analysis of the issues. *Annu. Rev. Plant Biol.* 59: 771–812.

Li, A. Y., R. B. Davey, R. J. Miller, and J. E. George, 2004. Detection and characterization of amitraz resistance in the southern cattle tick, *Boophilus microplus* (Acari: Ixodidae). J. Med. Entomol. 41: 193–200.

Li, A. Y., J. H. Pruett, R. B. Davey, and J. E. George. 2005. Toxicological and biochemical characterization of coumaphos resistance in the San Roman strain of *Boophilus microplus* (Acari: Ixodidae). *Pestic. Biochem. Physiol.* 81: 145–153.

Lodesani, M., and C. Costa. 2005. Limits of chemotherapy in beekeeping: Development of resistance and the problem of residues. *Bee World* 86: 102–109.

Lodesani, M., M. Colombo, and M. Spreafico. 1995. Ineffectiveness of Apistan® treatment against the mite *Varroa jacobsoni* Oud. in several districts of Lombardy (Italy). *Apidologie* 26: 67–72.

Lodesani, M., C. Costa, G. Serra, R. Colombo, and A. G. Sabatini. 2008. Acaricide residues in beeswax after conversion to organic beekeeping method. *Apidologie* 39: 324–333.

Lodesani, M., A. Pellacani, S. Bergomi, E. Carpana, T. Rabitti, and P. Lasagni. 1992. Residue determination for some products used against Varroa infestation in bees. *Apidologie* 23: 257–272.

Macedo, P. A., M. D. Ellis, and B. D. Siegfried. 2002. Detection and quantification of fluvalinate resistance in Varroa mites in Nebraska. *Am. Bee J.* 147: 523–526.

Marchetti, S., and R. Barbattini. 1984. Comparative effectiveness of treatments used to control *Varroa jacobsoni* Oud. *Apidologie* 15: 363–378.

Martel, A. C., S. Zeggane, C. Aurieres, P. Drajnudel, J. P. Faucon, and M. Aubert. 2007. Acaricide residues in honey and wax after treatment of honey bee colonies with Apivar® or Asuntol® 50. *Apidologie* 38: 534–544.

Meimaridou, E., J. Jacobson, A. M. Seddon, A. A. Noronha-Dutra, W. G. Robertson, and J. S. Hothersal. 2005. Crystal and microparticle effects on MDCK cell superoxide production: Oxalate-specific mitochondrial membrane potential changes. *Free Radical Biology and Medicine* 38: 1553–1564.

Mineau, P., K. M. Harding, M. Whiteside, M. R. Fletcher, D. Garthwaite, and L. D. Knoppers. 2008. Using reports of bee mortality in the field to calibrate laboratory-derived pesticide risk indices. *Environ. Entomol.* 37: 546–554.

Motoba, K., H. Nishizawa, T. Suzuki, H. Hamaguchi, M. Uchida, and S. Funayama. 2000. Species-specific detoxification metabolism of fenpyroximate, a potent acaricide. *Pestic. Biochem. Physiol.* 67: 73–84.

Motoba, K., T. Suzuki, and M. Uchida. 1992. Effect of a new acaricide, fenpyroximate, on energy metabolism and mitochondrial morphology in adult female *Tetranychus urticae* (two-spotted spider mite). *Pestic. Biochem. Physiol.* 43: 37–44.

Mullin, C. A., M. Frazier, J. L. Frazier, S. Ashcraft, R. Simonds, D. vanEngelsdorp, and J. S. Pettis. 2010. High levels of miticides and agrochemicals in North American apiaries: Implications for honey bee health. *PLoS ONE* 5(3): e9754. doi: 10.1371/journal.pone.0009754 (accessed 19 March 2010)

Nguyen, B. K., C. Saegerman, C. Pirard, J. Mignon, J. Widart, B. Tuirionet, F. J. Verheggen, D. Berkvens, E. De Pauw, and E. Haubruge. 2009. Does imidacloprid seed-treated maize have an impact on honey bee mortality? *J. Econ. Entomol.* 102: 616– 623.

NOD Apiary Products. Directions for use. www. miteaway. com/MAII_flyer.pdf (accessed 20 August 2009)

OEPP/EPPO. 1992. Guideline on test methods for evaluating the side effects of plant protection products on honeybees. *Bull. OEPP/EPPO* 22: 201–215.

PAN (Pesticide Action Network). http://www.pesticideinfo.org (accessed 12 August 2009)

Pettis, J. S., A. M. Collins, R. Wilbanks, and M. F. Feldlaufer. 2004. Effects of coumaphos on queen rearing in the honey bee, *Apis mellifera*. *Apidologie* 35, 605–610.

Pilling, E. D., and P. C. Jepson. 1993. Synergism between EBI fungicides and a pyrethroid insecticide in the honey bee (*Apis mellifera*). *Pestic. Sci.* 39: 293–297.

Pilling, E. D., K. A. C. Bromley-Challenor, C. H. Walker, and P. C. Jepson. 1995. Mechanism of synergism between the pyrethroid insecticide λ-cyhalothrin and the imidazole fungicide orochloraz, in the honeybee (*Apis mellifera* L.). *Pestic. Biochem. Physiol.* 51: 1–11.

Plapp, F. W. Jr. 1979. Synergism of pyrethroid insecticides by formamidines against Heliothis pests of cotton. *J. Econ. Entomol.* 72: 667–670.

PMRA—Canada's Pest Management Regulatory Agency. 2009. PMRA approves emergency use of Apivar in Canada. http://cba.stonehavenlife.com/2009/07/ apivar-varroa-emergency-use-in-canada/ (accessed 10 November 2009)

Priestley, C. M., E. M. Williamson, K. A. Wafford, and D. B. Sattelle. 2003. Thymol, a constituent of thyme essential oil, is a positive allosteric modulator of human GABA (A) receptors and a homooligomeric GABA receptor from *Drosophila melanogaster*. *British J. Pharmacol.* 140: 1363–1372.

Quarles, W. 1996. EPA exempts least-toxic pesticides. *IPM Practitioner* 18: 16–17.

Quarles, W. 2008. Pesticides and honey bee Colony Collapse Disorder. *IPM Practitioner* 30: 1–10.

Rademacher, E., and M. Harz. 2006. Oxalic acid for the control of varroosis in honey bee colonies—a review. *Apidologie* 37: 98–120.

Ramirez-Romero, R., J. Chaufaux, and M. H. Pham-Delègue. 2005. Effects of Cru1Ab protoxin, deltamethrin and imidacloprid on the foraging activity and the learning performances of the honeybee, *Apis mellifera*, a comparative approach. *Apidologie* 36: 601–611.

Rhodes, H. A., W. T. Wilson, P. E. Sonnet, and A. Stoner. 1979. Exposure of *Apis mellifera* (Hymenoptera, Apidae) to bee-collected pollen containing residues of microencapsulated methyl parathion. *Environ. Entomol.* 8: 944–948.

Rinderer, T. E., L. I. de Guzman, V. A. Lancaster, G. T. Delatte, and J. A. Stelzer. 1999. Varroa in the mating yard: I. The effects of *Varroa jacobsoni* and Apistan® on drone honey bees. *Am. Bee J.* 139: 134–139.

Rissato, S. R., M. S. Galhiane, M. V. de Almeida, M. Gerenutti, and B. M. Apon. 2007. Multiresidue determination of pesticides in honey samples by gas chromatography-mass spectrometry and application in environmental contamination. *Food Chem.* 101: 1719–1726.

Rissato, S. R., M. S. Galhiane, F. R. N. Knoll, and B. M. Apon. 2004. Supercritical fluid extraction for pesticide multiresidue analysis in honey: Determination by gas chromatography with electron-capture and mass spectrometry detection. *J. Chromatogr. A* 1048: 153–159.

Robertson, J. L., R. M. Russell, H. K. Preisler, and N. E. Savin, 2007. *Bioassays with Arthropods*. CRC Press, Boca Raton Florida, U.S.

Rortais, A., G. Arnold, M. P. Halm, and F. Touffet-Briens. 2005. Modes of honeybees exposure to systemic insecticides: Estimated amounts of contaminated pollen and nectar consumed by different categories of bees. *Apidologie* 36: 71–83.

Russell, D., R. Meyer, and J. Bukowski. 1998. Potential impact of microencapsulated pesticides on New Jersey apiaries. *Am. Bee J.* 138: 207–210.

Sammataro, D., P. Untalan. F. Guerrero, and J. Finley. 2005. The resistance of Varroa mites (Acari: Varroidae) to acaricides and the presence of esterase. *Int. J. Acarol.* 31: 67–74.

Schmidt, L. S., J. O. Schmidt, H. Rao, W. Wang, and X. Ligen. 1995. Feeding preferences and survival of young worker honey bees (Hymenoptera: Apidae) fed rape, sesame and sunflower pollen. *J. Econ. Entomol.* 88: 1591–1595.

Schulz, D. J., and G. E. Robinson. 2001. Octopamine influences division of labor in honey bee colonies. *J. Compar. Physiol. A* 187: 53–61.

Sherer, T. B., J. R. Richardson, C. M. Testa, B. B. Seo, A. V. Panov, T. Yagi, A. Matsuno-Magi, G. W. Miller, and J. T. Greenamyre. 2007. Mechanism of toxicity of pesticides acting at complex I: Relevance to environmental etiologies of Parkinson's disease. *J. Neurochem.* 100: 1469–1479.

Smith, R. K., and M. M. Wilcox. 1990. Chemical residues in bees, honey and beeswax. *Am. Bee J.* 130, 188–192.

Soberon, M., S. Gill, and A. Bravo A. 2009. Signaling versus punching hole: How do *Bacillus thuringiensis* toxins kill insect midgut cells? *Cell. Mol. Life Sci.* 66: 1337–1349.

Song, C., and M. E. Scharf. 2008. Formic acid: A neurologically active, hydrolyzed metabolite of insecticidal formate esters. *Pestic. Biochem. Physiol.* 92: 77–82.

Stoner, A., W. T. Wilson, and J. Harvey. 1985. Honey bee exposure to beeswax foundation impregnated with fenvalerate or carbaryl. *Am. Bee J.* 125: 513–516.

Suchail, S., D. Guez, and L. P. Belzunces. 2000. Characteristics of imidacloprid toxicity in two *Apis mellifera* subspecies. *Environ. Toxicol. Chem.* 19: 1901–1905.

Suchail, S., D. Guez, and L. P. Belzunces. 2001. Discrepancy between acute and chronic toxicity induced by imidacloprid and its metabolites in *Apis mellifera. Environ. Toxicol. Chem.* 20: 2482–2486.

Sylvester, H. A., R. P. Watts, L. I. de Guzman, J. A. Stelzer, and T. E. Rinderer. 1999. Varroa in the mating yard: II. The effects of Varroa and fluvalinate on drone mating competitiveness. *Am. Bee J.* 139: 225–227.

Taséi, J. N. 2003. Impact of agrochemicals on non-*Apis* bees, In *Honey Bees: Estimating the Environmental Impact of Chemicals.* Devillers, J., and M. H. Pham-Delègue (Eds.). Taylor and Francis, London, UK, pp. 101–131.

Taylor, M. A., R. M. Goodwin, H. M. McBrydie, and H. M. Cox. 2007. Destroying managed and feral honey bee (*Apis mellifera*) colonies to eradicate honey bee pests. *N. Z. J. Crop Hort. Sci.* 35: 313–323.

Thompson, H. M. 2003. Behavioral effects of pesticides in bees: Their potential for use in risk assessment. *Ecotoxicology* 12: 317–330.

Thrasyvoulou, A. T., and N. Pappas. 1988. Contamination of honey and wax with malathion and coumaphos used against the Varroa mite. *J. Apic. Res.* 27: 55–61.

Trapp, S., and L. Pussemier. 1991. Model calculations and measurements of uptake and translocation of carbamates by bean plants. *Chemosphere* 22: 327–339.

Tremolada, P., I. Bernardinelli, M. Colombo, M. Spreafico, and M. Vighi. 2004. Coumaphos distribution in the hive ecosystem: Case study for modeling applications. *Ecotoxicology* 13: 589–601.

Underwood, R., and R. Currie. 2003. The effects of temperature and dose of formic acid on treatment efficacy against *Varroa destructor* (Acari: Varroidae), a parasite of *Apis mellifera* (Hymenoptera: Apidae). *Exp. Appl. Acarol.* 29, 303–313.

USDA-AMS. 2009. Pesticide data program, annual summary, calendar year 2008. http://www.ams. usda.gov/pdp (accessed 20 January 2010)

USDA-Biotech Crop Data. 2009. Adoption of genetically engineered crops in the U.S. http://www.ers.usda.gov/Data/BiotechCrops/#2009-7-1 (accessed 20 August 2009)

USDA-PDP (Pesticide Data Program). 2008. Pesticide testing history 1991–2008. http://www.ams.usda.gov/AMSv1.0/pdp (accessed 20 August 2009)

vanEngelsdorp, D., J. D. Evans, L. Donovall, C. Mullin, M. Frazier, J. Frazier, D. R. Tarpy, J. Hayes, and J. S. Pettis. 2009a. "Entombed Pollen": A new condition in honey bee colonies associated with increased risk of colony mortality. *J. Invertebr. Pathol.* 101: 147–149.

vanEngelsdorp, D., J. D. Evans, C. Saegerman, C. Mullin, E. Haubruge, B. K. Nguyen, M. Frazier, J. Frazier, D. Cox-Foster, Y. Chen, R. Underwood, D. R. Tarpy, and J. S. Pettis. 2009b. Colony Collapse Disorder: A descriptive study. *PLoS ONE* 4: 1–17, e6481.

Varrox Vaporizer production information. 2007. *Hivelights* 20: 16.

Vita Europe Ltd. 2009. ApiGuard® production information. http://www.vita-europe.com (accessed 15 August 2009)

Wallner, K. 1995. The use of varroacides and their influence on the qualify of bee products. *Am. Bee J.* 135: 817–821.

Wallner, K. 1999. Varroacides and their residues in bee products. *Apidologie* 30: 235–238.

Walorczyk, S., and B. Gnusowski. 2009. Development and validation of a multi-residue method for the determination of pesticides in honeybees using acetonitrile-based extraction and gas chromatography-tandem quadrupole mass spectrometry. *J. Chromatogr. A* 1216: 6522–6531.

Wang, R. W., Z. Q. Liu., K. Dong, P. J. Elzen, J. Pettis, and Z. Y. Huang. 2002. Association of novel mutations in a sodium channel gene with fluvalinate resistance in the mite, *Varroa destructor. J. Apic. Res.* 41: 17–25.

Wellmark International. 2009. Hivistan® production information. http://www.centralapiary.com/products.htm#hivastan (accessed 30 August 2009)

Whittington, R., M. L. Winston, A. P. Melathopoulos, and H. A. Higo. 2000. Evaluation of the botanical oils neem, thymol, and canola sprayed to control *Varroa jacobsoni* Oud. (Acari:

Varroidae) and *Acarapsis woodi* (Acari: Tarsonemidae) in colonies of honey bees (*Apis mellifera* L., Hymenoptera: Apidae). *Am. Bee J.* 140: 565–572.

Chapter 15

Cellular Response in Honey Bees to Non-Pathogenic Effects of Pesticides

Ait-Aisa, S. 2000. Activation of the HSP 70 promoter by environmental inorganic and organic chemicals: Relationships with citotoxity and liphophilicity. *Toxicology* 145: 147–157.

Aliouane, Y., K. Adessalam, E. L. Hassani, V. Gary, C. Armengaud, M. Lambin, and M. Gauthier. 2009. Subchronic exposure of honeybees to sublethal doses of pesticides: Effects on behavior. *Environ. Toxicol. Chem.* 28: 113–122.

Armengaud, C, J. Ait-Oubah, N. Causse, and M. Gauthier. 2001. Nicotinic acetylcholine receptor ligands differently affect cytochrome oxidase in the honeybee brain. *Neurosci. Lett.* 304: 97–101.

Ashburner, M. 1982. The effects of heat shock and other stress on gene activity: An introduction. In *Heat Shock Proteins: From Bacteria to Human*, Schlesinger, M. J., M. Ashburner, and A. Tissiéres (Eds.). Spring Harbor Laboratory Press, New York, NY, U.S., pp.1–9.

Attencia, V. M., M. C. C. Ruvolo-Takasusuki, and V. A. A. DeToledo. 2005. Esterase activity in *Apis mellifera* after exposure to organophosphate insecticides (Hymenoptera: Apidae). *Sociobiology* 45: 587–595.

Barycki, J. J., and R. F. Colman. 1997. Identification of the nonsubstrate steroid binding site of the rat liver glutathione S-transferase, isoenzyme 1-1, by the steroid affinity label, 3β-(iodoacetoxy)dehydroisoandrosterone. *Arch. Biochem. Biophys.* 345: 16–31.

Becker, J., and E. A. Craig. 1994. Heat-shock proteins as molecular chaperones. *Eur. J. Biochem.* 219: 11–23.

Beckmann, R. P., M. Lovett, and W. J. Welch. 1992. Examining the function and regulation of HSP70 in cells subjected to metabolic stress. *J. Cell Bio.* 117: 1137–1150.

Beliën, T., J. Kellers, K. Heylen, W. Keulemans, J. Billen, L. Arckens, R. Huybrechts, and B. Gobin. 2009. Effects of sublethal doses of crop protection agents on honey bee (*Apis mellifera*) global colony vitality and its potential link with aberrant foraging activity. *Commun. Agric. Appl. Biol. Sci.* 74: 245–253.

Bell, H. S., I. R. Whittle, M. Walker, H. A. Leaver, and S. B. Wharton. 2001. The development of necrosis and apoptosis in glioma: Experimental findings using spheroid culture systems. *Neuropathol. Appl. Neurobiol.* 27: 291–304.

Berridge, M. J., and J. L. Oschmann. 1972. *Transporting Epithelia*. Academic Press, New York, NY, U.S.

Bidmon, H. J, N. A. Granger, and W. E. Stumpf. 1991. Co-localization of ecdysteroids receptors and c-*fos*-like protein in the brain of *Manduca sexta* larvae. *Dev. Genes Evol.* 200: 149–155.

Bierkens, J. G. E. A. 2000. Applicatios and pitfalls of stress-proteins in biomonitoring. *Toxicology* 153: 61–72.

Bortolli, L., R. Montanari, J. Marcelino, P. Mendrzychi, S. Maini, and C. Porrini. 2003. Effects of sub-lethal imidacloprid doses on the homing rate and foraging activity of honey bees. *B. Insectol.* 56: 63–67.

Bowen, I. D., K. Mullarkey, and S. M. Morgan. 1996. Programmed cell death during metamorphosis in the blow-fly *Calliphora vomitoria*. *Microsc. Res. Tech.* 34: 202–217.

Bradley, T. J. 1985. The excretory system: Structure and physiology. In *Comprehensive Insect Physiology, Biochemestry and Pharmacology.* Kerkut, G. A., and L. I. Gilbert (Eds). Pergamon Press, London, UK, pp. 421–465.

Busbee, D. L., J. Guyden, T. Kingston, F. L. Rose, and E. T. Cantrell. 1978. Metabolism of benzo(a)pyrene in animals with high aryl hydrocarbon hydroxylase levels and high rates of spontaneous cancer. *Cancer Lett.* 4: 61–68.

Buzzard, K. A. A. J. Giaccia, M. Killender, and R. L. Anderson. 1998. Heat shock protein 72 modulates pathways of stress-induced apoptosis. *J. Biol. Chem.* 273: 17147–17153.

Carvalho, S. M., G. A. Carvalho, C. F. Carvalho, J. S. S. Bueno Filho, and A. P. M. Batista. 2009. Toxicidade de acaricidas/inseticidas empregados na citricultura para a abelha africanizada *Apis mellifera* L., 1758 (Hymenoptera: Apidae). *Arquivos do Instituto Biológico,* 76(4): 597–606.

Cavalcante, V. M., and C. Cruz-Landim. 1999. Types of cells present in the midgut of the insects: A review. *Naturalia* 24: 19–40.

Chapman, R. F. 1998. *The Insects: Structure and Function* (4th ed.). Cambridge University Press, Cambridge, UK.

Chauzat, M. P., J. P. Faucon, A. C. Martel, J. Lachaize, N. Cougoule, and M. Aubert. 2006. A survey of pesticide residues in pollen loads by honeybees in France. *J. Econ. Entomol.* 99: 253–262.

Chiang, H., S. R. Terlecky, C. P. Plant, and J. F. Dice. 1989. A role for a 70-kilodalton heat shock protein in lysosomal degradation of intracellular proteins. *Science* 246: 382–385.

Cochran, D. G. 1994. *Toxic Effects of Boric Acid on the German Cockroach*. Virginia Polytechnic Institute and State University, Blacksburg, Virginia, U.S.

Cruz, A., E. C. M. Silva-Zacarin, O. C. Bueno, and O. Malaspina. 2010. Morphological alterations induced by boric acid and fipronil in the midgut of worker honeybee (*Apis mellifera* L.) larvae. *Cell Biol. Toxicol.* 26: 165–176.

Cruz-Landim, C. 1998. Specializations of the Malpighian tubules cells in a stingless bee, *Melipona quadrifasciata anthidioides* Lep. (Hymenoptera, Apidae). *Acta Microscopica* 7: 26–33.

Cruz-Landim, C. 2009. *Abelhas: Morfologia e Função Dos Sistemas.* Editora UNESP, São Paulo, Brazil.

Cruz-Landim, C., and R. A. Melo. 1981. Desenvolvimento e envelhecimento de larvas e adultos de *Scaptotrigona postica latreille* (Hymenoptera, Apidae): Aspectos histológicos e histoquímicos. ACIESP, São Paulo, Brazil.

Cuervo, A. M. M., and Dice, J. F. 1996. A receptor for the selective uptake and degradation of proteins by lysosomes. *Science* 273: 5101–5103.

Curran, T., G. Peters, C. Van Beveren, N. M. Teich, and M. Verna. 1982. The FBJ murine osteosarcoma virus: Identification and molecular cloning of biologicall active proviral DNA. *J. Virol.* 44: 674–682.

Curran, T., W. P. MacCornell, F. Van Sraaten, and I. M. Verna. 1983. Structure of the FBJ murine osteosarcoma virus genome: Molecular cloning of its associated helper virus and the cellular homolog of the v-*fos* gene from mouse and human cells. *Mol. Cell Bio.* 3: 914–921.

Cymborowski, B. 1996. Expression of C-*fos*-like protein in the brain and prothoracic gland of *Galleria mellonella* larvae during chilling stress. *J. Insect Physiol.* 42: 367–371.

Cymborowski, B., and V. King. 1996. Circadian regulation of *Fos*-like expression in brain of the blow fly *Calliphora vicina*. *Comp. Biochem. Phys.* C. 115: 239–246.

Decombel, L., L. Tirry, and G. Smagghe. 2005. Action of 24-epibrassinolide on a cell line of the beet armyworm, *Spodoptera exigua*. *Arch. Insect Biochem. Physiol.* 58: 145–156.

Decourtye, A., J. Devillers, E. Genecque, K. Le Menach, H. Budzinski, S. Cluseau, and M. H. Pham-Delegue. 2005. Comparative sublethal toxicity of nine pesticides on olfactory learning performances of the honeybees *Apis melifera*. *Arch. Environ. Contam. Toxicol.* 48: 242–250.

Decourtye, A., M. Armengaud, M. Renou, J. Devillers, S. Cluseau, M. Gauthier, and M. Pham-Delegue. 2004. Imidacloprid impairs memory and brain metabolism in the honeybee (*Apis mellifera* L.). *Pesticide Biochem. Physiol.* 78: 83–92.

Déglise, P., M. Dacher, E. Dion, M. Gauthier, and C. Armengaud. 2003. Regional brain variations of cytochrome oxidase staining during olfactory learning in the honey bee. *Behav. Neurosci.* 117: 540–547.

Diao, Q., K. Yuan, P. Liang, and X. Gao. 2006. Tissue distribution and properties of glutathione S-transferases in *Apis cerana* Fabricius and *Apis mellifera ligustica* Spinola. *J. Apic. Res.* 45: 145–152.

Eckwet, E., G. Alberti, and H. R. Köhler. 1997. The induction of stress proteins (HSP) in *Oniscux asellus* (Isopoda) as a molecular marker of multiple heavy metal exposure: I. Principles and toxicological assessment. *Ecotoxicology* 6: 249–262.

Edwards, R., D. P. Dixon, and V. Walbot. 2000. Plant glutathione S-transferases: Enzymes with multiple functions in sickness and health. *Trends Plant Sci.* 5: 193–198.

Ellis, J. D., and P. A. Munn. 2005. The worldwide health status of honey bees. *Bee World* 86: 88–101.

Ellis, J. D., J. D. Evans, and J. Pettis. 2010. Colony losses, managed colony population decline, and Colony Collapse Disorder in the United States. *J. Apic. Res.* 49: 134–16.

Ellis, R. E., J. Y. Yuan, and H. R. Horvitz. 1991. Mechanisms and functions of cell death. *Annu. Rev. Cell. Biol.* 7: 663–698.

EPPO. 1993. Decision-making schemes for the environmental risk assessment of plant protection products honey bees. *OEPP-EPPO Bull.* 23: 151–165. doi: 10.1111/j.1365-2338.1993.tb01040.x

Farris, S. M., G. E. Robinson, R. L. Davis, and S. E. Fahrbach. 1999. Larval and pupal development of the mushroom bodies in the honey bee, *Apis mellifera*. *J. Comp. Neurol.* 4: 97–113.

Feder, M. E., D. A. Parsell, and S. L. Lindquist. 1995. The stress response and stress proteins. In *Cell Biology of Trauma*. Lemasters, J. J., and C. Oliver (Eds.). CRC Press, Boca Raton, Florida, U.S, pp. 177–191.

Fonta, C., J. Gascuel, and C. Masson. 1995. Brain *fos*-like expression in developing and adults honeybees. *Neuro. Report* 6: 745–749.

Fontanetti, C. S., M. I. Camargo-Mathias, and F. H. Caetano. 2001. Apocrine secretion in the midgut of *Plusioporus setiger* (Brolemann, 1901) (Diplopoda, Spirostreptidae). *Naturalia* 26: 35–42.

Francis, F., E. Haubruge, and P. Dierickx. 2002. Glutathione S-transferase isoenzymes in the two-spot ladybird, *Adalia bipunctata* (Coleoptera: Coccinellidae). *Arch. Insect Biochem. Physiol.* 49: 158–166.

Fuchs, D., G. Baier-Bitterlich, I. Wede, and H. Wachter. 1997. Reactive oxygen and apoptosis. In *Oxidative Stress and the Molecular Biology of Antioxidant Defenses*. Scandalios, J. G. (Ed.). Cold Spring Harbor Press, Woodbury, New York, NY, U.S.

Garrido, C., S. Gurbuxani, L. Ravagnan, and G. Kroemer. 2001. Heat shock proteins: Endogenous modulators of apoptotic cell death. Biochem. *Biophys. Res. Commun.* 286: 433–442.

Georgopoulos, C., and W. J. Welch. 1993. Role of the major heat shock proteins as molecular chaperones. *Ann. Rev. Cell Biol.* 9: 601–634.

Giesen, K., U. Lammel, D. Langehans, K. Krukkert, I. Bunse, and C. Klämbt. 2003. Regulation of glial cell number and differentiation by ecdysone and Fos signaling. *Mech. Dev.* 120: 401–413.

Gilbert, M. D., and C. F. Wilkinson. 1974. Microsomal oxidases in the honey bee, *Apis mellifera* (L.). Pesticide Biochem. *Physiol.* 4: 56–66.

Gilbert, M. D., and C. F. Wilkinson. 1975. An inhibitor of microsomal oxidation from gut tissues of the honey bee (*Apis mellifera*). *Comp. Biochem. Physiol.* 50B: 613–619.

Gracey, A. Y., J. V. Troll, and G. N. Somero. 2001. Hypoxia-induced gene expression profiling in the euryoxic fish *Gillichthys mirabilis. PNAS* 98: 1993–1998.

Gregorc, A., and M. I. Smodiš-Škerl. 2007. Toxicological and immunohistochemical testing of honeybees after oxalic acid and rotenone treatments. *Apidologie* 38: 296–305.

Gregorc, A., A. Pogacnik, and I. D. Bowen. 2004. Cell death in honeybee larvae treated with oxalic or formic acid. *Apidologie* 35: 453–460.

Gregorc, A., and I. D. Bowen. 1997. Programmed cell death in the honeybee (*Apis mellifera* L.) larvae midgut. *Cell Biol. Int.* 21: 151–158.

Gregorc, A., and I. D. Bowen. 1998. Histopathological and histochemical changes in honeybee larvae (*Apis mellifera* L.) after infection with *Bacillus larvae*, the causative agent of American foulbrood disease. *Cell Biol. Int.* 22: 137–144.

Gregorc, A., and I. D. Bowen. 1999. *In situ* localization of heat-shock and histone proteins in honeybee (*Apis mellifera* L.) larvae infected with *Paenibacillus larvae. Cell Biol. Int.* 23: 211–218.

Gregorc, A., and I. D. Bowen. 2000. Histochemical characterization of cell death in honeybee larvae midgut after treatment with *Paenibacillus larvae*, amitraz and oxytetracycline. *Cell Biol. Int.* 24: 319–324.

Guengerich, F. P. 1990. Enzymatic oxidation of xenobiotic chemicals. *CRC Crit. Rev. Biochem. Mol. Biol.* 25: 97–153.

Hansson, B. S., and S. Anton. 2000. Function and morphology of the antennal lobe: New developments. *Annu. Rev. Entomol.* 45: 203–231.

Hendrick, J. P., and F. U. Hartl. 1993. Molecular chaperone functions of heat-shock proteins. *Annu. Rev. Biochem.* 62: 349–384.

Herrera, D. G., and H. A. Robertson. 1996. Activation of c-*fos* in the brain. *Progress in Neurobiology* 50: 83–107.

Hevener, R. F., and M. T. T. Wong. 1991. Neuronal expression of nuclear and mitochondrial genes for cytochrome oxidase (CO) subunits analyzed by *in situ* hybidrization: Comparison with CO activity and protein. *J. Neurosci.* 11: 1942–1958.

Hodgson, E. 1985. Microsomal mono-oxygenases. In *Comprehensive Insect Physiology, Biochemistry and Pharmacology* (11th ed.). Kerkut, G. A., and L. I. Gilbert (Eds.). Pergamon Press, Oxford, UK.

Hoffmann, G., and D. Lyo. 2002. Anatomical markers of activity in neuroendocrine systems: Are we all "*fos*-ed out"? *J. Neuroendocrinol.* 14: 259–268.

ICPBR Discussion and Recommendations of the Fifth Meeting. 1993. In *Proceedings of the Fifth International Symposium on the Hazards of Pesticides to Bees*. Harrison, E. G. (Ed.). Wageningen, The Netherlands, pp. 10–14.

Jakob, U., and J. Buchner. 1994. Assisting spontaneity: The role of HSP90 and small HSPs as molecular chaperones. *Reviews TIBS* 19: 205–211.

Jesus, D., O. Malaspina, and E. C. M. Silva-Zacarin. 2005. Histological studies in the midgut and Malpighian tubules of *Apis mellifera* workers treated with boric acid. *Braz. J. Morphol. Sci.* 22 (Suppl): 168.

Jimenez, D. R., and M. Gilliam. 1988. Cytochemistry of peroxisomal enzymes in microbodies of the midgut of the honey bee, *Apis mellifera. Comp. Biochem. Phys. B.* 90: 757–766.

Jimenez, D. R., and M. Gilliam. 1990. Ultrastructure of the ventriculus of the honey bee, *Apis mellifera* (L.): Cytochemical localization of acid phosphatase, alkaline phosphatase, and nonspecific esterase. *Cell Tissue Res.* 261: 431–443.

Johnson, R. M., H. S. Pollock, and M. R. Berenbaum. 2009. Synergistic interactions between in-hive miticides in *Apis mellifera. J. Econ. Entomol.* 102: 474–479.

Johnson, R. M., Z. Wen, M. A. Schuler, and M. R. Berenbaum. 2006. Mediation of pyrethroid insecticide toxicity to honey bees (Hymenoptera: Apidae) by cytochrome P450 monooxygenases. *J. Econ. Entomol.* 99: 1046–1050.

Kar Chowdhuri, D., D. K. Saxena, and P. N. Viswanathan. 1999. Effect of hexachlorocyclohexane (HCH) its isomers, and metabolites on HSP 70 expression in transgenic *Drosophila mellanogaster*. *Pestic. Biochem. Physiol.* 63: 15–25.

Kevan, P. G. 1999. Pollinators as bioindicators of the state of the environment: Species activity and diversity. *Agricult. Ecosyst. Environ.* 74: 373–393.

Kezic, N., D. Sulimanovic, and D. Lucic. 1989. Induction of monooxygenase activity in honey bees as a bioassay for detection of environmental xenobiotics in honey. *32nd Int. Congr. Apic., Rio de Janeiro*. Editions Apimondia, Bucharest, Romania.

Kezic, N., S. Britvic, M. Protic, J. E. Simmons, M. Rijavec, R. K. Zahn, and B. Kurelec. 1983. Activity of benzo(a)pyrene monooxygenase in fish from Sava river: Correlation with pollution. *Sci. Total. Environ.* 27: 59–69.

Kockel, L., J. G. Homsy, and D. Bohmann. 2001. Drosophila AP-1: Lessons from an invertebrate. *Oncogene* 20: 2347–2364.

Köhler. H. R., and R. Triesbskorn. 1998. Assessment of the cytotoxic impact of heavy metals on soil invertebrates using a protocol integrating qualitative and quantitative components. *Biomarkers* 3: 109–127.

Köhler, H. R., R. Triebskorn, W. Stöcker, P. M. Kloetzel, and G. Alberti. 1992. The 70 kD heat shock protein (HSP70) in soil invertebrates: A possible tool for monitoring environmental toxicants. *Arch. Environm. Toxicol.* 22: 334–338.

Korsloot, A., C. A. M. Van Gestel, and N. M. Van Straalen. 2004. *Environmental Stress and Cellular Response in Arthropods*. CRC Press, Boca Raton, Florida, U.S.

Kretzschmar, D., and G. O. Pflugfelder. 2002. Glia in development, function, and degeneration of the adult insect brain. *Brain Res. B.* 57: 121–131.

Lambin, M., C. Armengaud, S. Raymond, and M. Gauthier. 2001. Imidacloprid-induced facilitation of the proboscis extension reflex habituation in the honeybee. *Arch. Insect Biochem. Physiol.* 48: 129–134.

Langer, T., C. Lu, H. Echols, J. Flanagan, M. K. Hayer, and F. U. Hartl, 1992. Successive action of DnaK, DnaJ and GroEL along the pathway of chaperone-mediated protein folding. *Nature* 356: 683–689.

Levy, R., M. Benchaib, H. Cordonier, C. Souchier, and J. F. Guerin. 1998. Annexin V labelling and terminal transferase-mediated DNA end labelling (TUNEL) assay in human arrested embryos. *Mol. Hum. Reprod.* 4: 775–783.

Lewis, S., R. D. Handy, U. B. Cordi, Z. Billinghurst, and M. H. Depledge. 1999. Stress proteins (HSP's): Methods of detection and their use as an environmental biomarker. *Ecotoxicology* 8: 351–368.

Lindquist, S., and E. A. Craig. 1988. The heat shock proteins. *Annu. Rev. Genet.* 22: 631–677.

Liu, T. P. 1984. Ultrastructure of the midgut of the worker honey bee *Apis mellifera* heavy infected with *Nosema apis*. *J. Invertebr. Pathol.* 44: 282–291.

Malaspina, O. 1979. Estudo genético da resistência ao DDT e relação com outros caracteres em *Apis mellifera* (Hymenoptera, Apidae). [Genetic study of resistance to DDT and relationship with other characters in *Apis mellifera* (Hymenoptera, Apidae).] Dissertation, (Mestrado em Zoologia de Invertebrados), Instituto de Biociências, UNESP, Rio Clara, Brazil. (in Portuguese)

Malaspina, O., and E. C. M. Silva-Zacarin. 2006. Cell markers for ecotoxicological studies in target organs of bees. *Braz. J. Morphol. Sci.* 23: 129–136.

Matylevitch, N. P., S. T. Schuschereba, J. R. Mata, G. R. Gilligan, D. F. Lawlor, C. W. Goodwin, and P. D. Bowman. 1998. Apoptosis and accidental cell death in cultured human keratinocytes after thermal injury. *Am. J. Pathol.* 153: 567–577.

Medrzycki, P., R. Montanari, L. Bortolotti, A. G. Sabatini, S. Maini, and C. Porrini. 2003. Effects of imidacloprid administered in sub-lethal doses on honey bee behaviour. Laboratory tests. *B. Insectol.* 56: 59–62.

Moraes, R. L. M. S., and I. D. Bowen. 2000. Modes of cell death in the hypopharyngeal gland of the honey bee (*Apis mellifera* L.). *Cell Biol. Int.* 24: 734–737.

Morimoto, R. I., A. Tissiéres, and C. Georgopoulous. 1994. *The Biology of Heat Shock Proteins and Molecular Chaperones*. Cold Spring Harbor Laboratory Press, New York, NY, U.S.

Mukhopadhyay, I., A. Nazir, K. Mahmood, D. K. Saxena, M. Das, S. K. Khanna, and D. K. Chowdhuri. 2002. Toxicity of argemone oil: Effect on HSP70 expression and tissue damage in transgenic *Drosophila melanogaster* (HSP70-lacZ) Bg[9]. *Cell Biol. Toxicol.* 18: 1–11.

Mukhopadhyay, I., H. R. Siddique, V. K. Bajpai, D. K. Saxena, and D. K. Chowdhuri. 2006. Synthetic pyrethroid Cypermetrhrin induced cellular damage in reproductive tissues of *Drosophila melanogaster*: HSP70 as a marker of cellular damage. *Arch. Environ. Contam. Toxicol.* 51: 673–680.

Nadeau, S. I., and J. Landry. 2007. Mechanisms of activation and regulation of the heat shock-sensitive signaling pathways. *Adv. Exp. Med. Biol.* 594: 100–113.

Nazir, A., D. K. Saxena, and D. Kar Chowdhuri. 2003a. Induction of HSP70 in transgenic *Drosophila*: Biomarker of exposure against phthalimide group of chemicals. *Biochem. Biophys. Acta* 1621: 218–225.

Nazir, A., Mukhopadhyay, I. D. K. Saxena, M. S. Siddiqui, and D. K. Chowdhuri. 2003b. Evaluation of toxic potential of captan: Induction of HSP70 and tissue damage in transgenic *Drosophila melanogaster* (HSP70-lacZ) Bg^9. *J. Biochem. Mol. Toxicol.* 17: 98–107.

Neuhaus-Steinmetz, U., and L. Rensing. 1997. Heat shock protein induction by certain chemical stressors is correlated with their cytotoxicity, lipophilicity and protein-denaturing capacity. *Toxicology* 123: 185–195.

Nims, R. W., A. D. P. Shoemaker, M. A. Bauernschub, L. J. Rec, and J. W. Harbell. 1998. Sensitivity of isoenzyme analysis for the detection of interspecies cell line cross-contamination. *In Vitro Cell. Dev. Biol.* 34A: 35–39.

Nocelli, R. C. F., T. R. Roat, E. C. M. Silva-Zacarin, M. S. Palma, and O. Malaspina. 2010. Social insects: morphophysiology of the nervous system. In *Social Insects: Structure, Function, and Behavior.* Stewart, E. M. (Ed.). Nova Publishers, Hauppauge, New York, U.S.

OEDC. 1998. Guidelines for the Testing of Chemicals: Honeybees, Acute Oral and Contact Toxicity Test. Organisation for Economic Co-operation and Development (OECD), Paris, France, pp. 213–214.

Pajot, S. 2001. Dossier gaucho. *Abeilles et Fleurs* 616: 160–165.

Pannabecker, T. 1995. Physiology of the Malpighian tubule. Annu. Rev. Entomol. 40: 493–510.

Papaefthimiou, C., V. Pavlidou, A. Gregorc, and G. Theophilidis. 2002. The action of 2,4-dichlorophenoxyacetic acid on the isolated heart of insect and amphibian. *Environ. Toxicol. Pharmacol.* 11: 127–140.

Perkins, K. K., G. M. Dailey, and R. Tjian. 1988. Novel *jun-* and *fos*-related proteins in *Drosophila* are functionally homologues to enhancer factor AP-1. *EMBO J.* 7: 4265–4273.

Pettis, J. S., A. M. Collins, R. Wilbanks, and M. F. Feldlaufer. 2004. Effects of coumaphos on queen rearing in the honeybee, *Apis mellifera*. *Apidologie* 35: 605–610.

Pham-Delégue, M.-H., A. Decourtye, L. Kaisr, and J. Devillers. 2002. Behavioural methods to assess the effects of pesticides on honey bees. *Apidologie* 33: 425–432.

Rachinsky, A., C. Strambi, A. Strambi, and K. Hartfelder. 1990. Caste and metamorphosis: Hemolymph titers of juvenile hormone and ecdysteroids in last instar honeybee larvae. *Gen. Comp. Endocrinol.* 79: 31–38.

Ranson, H., L. Prapanthadara, and J. Hemingway. 1997. Cloning and characterization of two glutathione S-transferases from a DDT-resistant strain of *Anopheles gambiae*. *Biochem. J.* 324: 97–102.

Rassow, J., W. Voos, and N. Pfanner. 1995. Partner proteins determine multiple functions of HSP70. *Trends in Cell Biol.* 5: 207–212.

Renucci, M., A. Tirard, P. Charpin, R. Augier, and A. Strambi. 2000. c-*Fos* related antigens in central nervous system of an insect *Acheta domesticus*. *Arch. Insect Biochem. Physiol.* 45: 139–148.

Ribi, W., T. J. Senden, A. Sakellariou, A. Limaye, and S. Zhang. 2008. Imaging honey bee brain anatomy with micro-X-ray-computed tomography. *J. Neurosci. Methods* 171: 93–97.

Richter, C., and M. Schweizer. 1997. Oxidative stress in mitochondria. In *Oxidative Stress and the Molecular Biology of Antioxidant Defenses,* Scandalios, J. G. (Ed). Cold Spring Harbor Laboratory, New York, NY, U.S., pp. 169–200.

Roat, T. C. 2010. Efeitos toxicológicos do inseticida fipronil em operárias e rainhas de *Apis mellifera* (Hymenoptera: Apidae): Atividade neural e proteínas de desintoxicação. [Toxicological effects of the insecticide fipronil in workers and queens of *Apis mellifera* (Hymenoptera: apidae): neural activity and detoxification proteins.] *Relatório FAPESP Proc.* 2008/05018-7. (in Portuguese)

Robinson, G. E. 1992. Regulation of division of labor in insect societies. *Annu. Rev. Entomol.* 37: 637–665.

Rodpradit, P., S. Boonsuepsakul, T. Chareonviriphap, M. J. Bangs, and P. Rongnoparut. 2005. Cytochrome P450 genes: Molecular cloning and overexpression in a pyrethroid-resistant strain of *Anopheles minimus* mosquito. *J. Am. Mosq. Control Assoc.* 21: 71–79.

Rousseau, E., and E. S. Goldstein. 2001. The gene structure of the *Drosophila melanogaster* homolog of the human proto-oncogene *fos*. *Gene* 272: 315–322.

Sanders, B. M. 1993. Stress proteins in aquatic organisms: An environmental perspective. *Crit. Rev. Toxicol.* 23: 49–75.

Schuler, F., and J. E. Casida. 2001. The insecticide target in the PSST subunit of complex I. *Pest Manag. Sci.* 57: 932–940.

Sgonc, R., and J. Gruber. 1998. Apoptosis detection: An overview. *Exp. Gerontol.* 33: 525–533.

Sheehan, D., G. Meade, V. M. Foley, and C. A. Dowd. 2001. Structure, function, and evolution of glutathione transferases: Implications for classification of non-mammalian members of an ancient enzyme superfamily. *Biochem. J.* 360: 1–16.

Silva, E. C. M. 2002. Glândulas salivares larvais das abelhas. In *Glândulas Exócrinas Das Abelhas*. Cruz-Landim, C., and F. C. Abdalla (Eds.). FUNPEC, Ribeirão Preto, Brazil, pp. 21–49.

Silva-Zacarin, E. C. M. 2007. Authophagy and apoptosis coordinate physiological cell death in larval salivary glands of *Apis mellifera* (Hymenoptera: Apidae). *Autophagy* 3: 516–518.

Silva-Zacarin, E. C. M., A. Gregorc, and R. L. M. Silva de Moraes. 2006. *In situ* localisation of heat-shock proteins and cell death labeling in the salivary gland of acaricide-treated honeybee larvae. *Apidologie* 37: 507–516.

Silva-Zacarin, E. C. M., G. A. Tomaino, M. R. B. Brocheto-Braga, S. R. Taboga, and R. L. M. Silva de Moraes. 2007. Programmed cell death in the larval salivary glands of *Apis mellifera* (Hymenoptera: Apidae). *J. Biosci.* 32: 309–328.

Silva-Zacarin, E. C. M., R. Ferreira, R. C. F. Nocelli, T. C. Roat, M. S. Palma, and O. Malaspina. 2010. Structure and function of the intestine and Malpighian tubules: From bee biology to cell marker development for toxicological analysis. In *Social Insects: Structure, Function, and Behavior.* Stewart, E. M. (Ed.). Nova Publishers, Hauppauge, New York, U.S.

Silva-Zacarin, E. C. M., R. L. M. Silva de Moraes, and S. R. Taboga. 2003. Silk formation mechanism in the larval salivary glands of *Apis mellifera* (Hymenoptera: Apidae). *J. Biosci.* 28: 753–764.

Silva-Zacarin, E. C. M., S. R. Taboga, and R. L. M. Silva de Moraes. 2008. Nuclear alterations associated to programmed cell death in larval salivary glands. *Micron* 39: 117–127.

Smirle, M. J. 1990. The influence of detoxifying enzymes on insecticide tolerance in honey bee colonies (Hymenoptera: Apidae). *J. Econ. Entomol.* 83: 715–720.

Smirle, M. J., and M. L. Winston. 1987. Intercolony variation in pesticide detoxification by the honey bee (Hymenoptera: Apidae). *J. Econ. Entomol.* 80: 5–8.

Smirle, M. J., and M. L. Winston. 1988. Detoxifying enzyme activity in worker honey bees: An adaptation for foraging in contaminated ecosystems. *Can. J. Zool.* 66: 1938–1942.

Smodiš-Škerl, M. I., and A. Gregorc. 2010. Heat shock proteins and cell death *in situ* localisation in hypopharyngeal glands of honeybee (*Apis mellifera carnica*) workers after imidacloprid or coumaphos treatment. *Apidologie* 41: 73–86.

Snodgrass, R. E. 1956. *Anatomy and Physiology of the Honeybees*. Comstock Publishing, New York, NY, U.S.

Sorour, J. 2001. Ultrastructural variations in *Lethocerus niloticum* (Insecta: Hemiptera) caused by pollution in Lake Mariut, Alexandria, Egypt. *Ecotoxicol. Environ. Safety* 48: 268–274.

Souid, S., and C. Yanicostas. 2003. Differential expression of the two Drosophila *fos/kayak* transcripts during oogenesis and embryogenesis. *Dev. Dynamics* 227: 150–154.

Sperandio, S., I. De Belle, and D. E. Bredesen. 2000. An alternative, nonapoptotic form of programmed cell death, *PNAS* 97: 14376–14381.

Strausfeld, N. J., U. Homburg, and P. Kloppenberg. 2000 Parallel organization in honey bee mushroom bodies by peptidergic kenyon cells. *J. Comp. Neurol.* 424(1): 179–195.

Sumida, S., E. C. M. Silva-Zacarin, P. Décio, O. Malaspina, F. C. Bueno, and O. C. Bueno. 2010. Morphological changes in organs of *Atta sexdens rubropilosa* (Formicidae: Hymenoptera) treated with boric acid in toxicological bioassays. *J. Econ. Entomol.* 103: 676–690.

The Honeybee Genome Sequencing Consortium. 2006. Insights into social insects from the genome of the honeybee *Apis mellifera*. *Nature* 443: 931–349.

Thompson, H. M. 2003. Behavioural effects of pesticides in bees—Their potential for use in risk assessment. *Ecotoxicology* 12: 317–300.

vanEngelsdorp, D., J. D. Evans, L. Donovall, C. Mullin, M. Frazier, J. Frazier, D. R. Tarpy, J. Hayes, and J. S. Pettis. 2009. "Entombed pollen": A new condition in honey bee colonies associated with increased risk of colony mortality. *J. Invertebr. Pathol.* 101: 147–149.

Vandame, R., M. Meled, M. E. Colin, and L. P. Belzunces. 1995. Alteration of the homing-flight in the honey bee *Apis mellifera* exposed to sublethal dose of deltamethrin. *Environ. Toxicol. Chem.* 14: 855–860.

Weick, J., and R. S. Thorn. 2002. Effects of acute sublethal exposure to coumaphos or diazinon on acquisition and discrimination of odor stimuli in the honey bee (Hymenoptera: Apidae). *J. Econ. Entomol.* 95: 227–236.

Wigglesworth, V. B. 1974. *The Principles of Insect Physiology*. Chapman & Hall, London, UK.

Wong, H. R., I. Y. Menendez, M. A. Ryan, A. G. Denenberg, and J. R. Wispe. 1998. Increased expression of heat shock protein-70 protects A549 cells against hyperoxia. *Am. J. Physiol.* 275: 836–841.

Yokoyama, N., M. Hirata, K. Ohtsuka, Y. Nishiyama, K. Fujii, M. Fujita, K. Kuzushima, T. Kiyono, and T. Tsurumi. 2000. Co-expression of human chaperone HSP70 and HSDJ or

HSP40 co-factor increases solubility of overexpressed target proteins in insect cells. *Biochem. Biophys. Acta* 1493: 119–124.

Yu, S. J., F. A. Robinson, and J. L. Nation. 1984. Detoxication capacity in the honey bee, *Apis mellifera* L. *Pestic. Biochem. Physiol.* 22: 360–368.

Chapter 16

Differences Among Fungicides Targeting the Beneficial Fungi Associated with Honey Bee Colonies

Alarcón, R., G. DeGrandi-Hoffman, and G. Wardell. 2009. Fungicides can reduce, hinder pollination potential of honey bees. *Western Farm Press* 31: 17–21.

Baldrian P., and J. Gabriel. 2002. Intraspecific variability in growth response to cadmium of the wood-rotting fungus *Piptoporus betulinus*. *Mycologia* 94: 428–436.

Barnett H. L., and B. B. Hunter. 1998. *Illustrated Genera of Imperfect Fungi* (4th ed.). American Phytopathological Society Press, St. Paul, Minnesota, U.S.

Benoit, J. B., J. A. Yoder, D. Sammataro, and L. W. Zettler. 2004. Mycoflora and fungal vector capacity of the parasitic mite *Varroa destructor* (Mesostigmata: Varroidae) in honey bee (Hymenoptera: Apidae) colonies. *Int. J. Acarol.* 30: 103–106.

Benoit, J. B., J. A. Yoder, J. T. Ark, and E. J. Rellinger. 2005. Growth response to squalene, a tick allomonal component, by fungi commonly associated with the American dog tick, *Dermacentor variabilis* (Say). *Int. J. Acarol.* 31: 269–275.

Brennan, J. M., B. Fagan, A. vanMaanen, B. M. Cooke, and F. M. Doohan. 2003. Studies on *in vitro* growth and pathogenicity of European *Fusarium* fungi. *Eur. J. Plant Pathol.* 109: 577–587.

Charlton, A. J. A., and A. Jones. 2007. Determination of imidazole and triazole fungicide residues in honeybees using gas chromatography-mass spectrometry. *J. Chromatography A* 1141: 117–122.

Chiesa, F., N. Milani, and M. D'Agaro. 1989. Observations of the reproductive behavior of *Varroa jacobsoni* Oud.: Techniques and preliminary results. In *Present Status of Varroatosis in Europe and Progress in the Varroa Mite Control.* Proc. Meeting, Udine, Italy, 1988. Cavalloro, E. (Ed.). EC-Experts Group, Luxembourg, pp. 213–222.

Currah, R. S., L. Sigler, and S. Hambleton. 1987. New records and new taxa of fungi from the mycorrhizae of terrestrial orchids of Alberta. *Can. J. Bot.* 65: 2473–2482.

Fisher, F., and N. B. Cook. 1998. *Fundamentals of Diagnostic Mycology.* W. B. Saunders, Philadelphia, Pennsylvania, U.S.

Folk, D. G., C. Han, and T. J. Bradley. 2001. Water acquisition and partitioning in *Drosophilia melanogaster*: Effect of selection for desiccation-resistance. *J. Exp. Biol.* 204: 3323–3331.

Gilliam, M. 1979. Microbiology of pollen and bee bread: The yeasts. *Apidologie* 10: 43–53.

Gilliam, M. 1997. Identification and roles of non-pathogenic microflora associated with honey bees. *Federation of European Microbiological Societies (FEMS) Microbiol. Lett.* 155: 1–10.

Gilliam, M., and J. D. Vandenberg. 1997. Fungi. In *Honey Bee Pests, Predators and Diseases.* Morse, R. A., and K. Flottum (Eds.). A. I. Root Company, Medina, Ohio, U.S., pp. 79–112.

Gilliam M., B. J. Lorenz, A. M. Wenner, and R. W. Thorp. 1997. Occurrence and distribution of *Ascophaera apis* in North America: Chalkbrood in feral honey bee colonies that had been in isolation on Santa Cruz Island, California for over 110 years. *Apidologie* 28: 329–338.

Gilliam M., D. B. Prest, and B. J. Lorenz. 1989. Microbiology of pollen and bee bread: Taxonomy and enzymology of molds. *Apidologie* 20: 53–68.

Gilliam, M., S. Taber III, B. J. Lorenz, and D. B. Prest. 1988. Factors affecting development of chalkbrood disease in colonies of honey bees, *Apis mellifera*, fed pollen contaminated with *Ascosphaera apis*. *J. Invertebr. Pathol.* 52: 314–325.

Gliñski Z., and K. Buczek. 2003. Response of the Apoidea to fungal infections. *Apiacta* 38: 183–189.

Hitchcock J. D., J. O. Moffett, J. J. Lackett, and J. R. Elliot. 1970. Tylosin for control of American foulbrood disease in honey bees. *J. Econ. Entomol.* 63: 204–207.

Hua, L., and M.-G. Feng. 2006. New use of broomcorn millets for production of granular cultures of aphid-pathogenic fungus *Pandora neoaphidis* for high sporulation potential and infectivity to *Myzus persicae*. *Federation of European Microbiological Societies (FEMS) Microbiol. Lett.* 227: 311–317.

Jennings, D. H., and G. Lysek. 1999. *Fungal Biology: Understanding the Fungal Lifestyle.* Springer-Verlag, New York, NY, U.S.

Katznelson, H., and C. A. Jamieson. 1952. Control of Nosema disease of honeybees with fumagillin. *Science* 115: 70–71.

Klungness, L. M., and Y. Peng. 1983. A scanning electron microscopic study of pollen loads collected and stored by honeybees. *J. Apic. Res.* 22: 264–271.

Kubik, M., J. Nowacki, A. Pidek, Z. Warakomska, L. Michalczuk, W. Goszczyński, and B. Dwuznik. 2000. Residues of captan (contact) and difenoconazole (systemic) fungicides in bee products from an apple orchard. *Apidologie* 31: 531–541.

Kubik, M., J. Nowacki, A. Pidek, Z. Warakomska, L. Michalczuk, and W. Goszczyński. 1999. Pesticide residues in bee products collected from cherry trees protected during blooming period with contact and systemic fungicides. *Apidologie* 30: 521–532.

Ladurner, E., J. Bosch, W. P. Kemp, and S. Maini. 2008. Foraging and nesting behavior of *Osmia lignaria* (Hymenoptera: Megachilidae) in the presence of fungicides: Case studies. *J. Econ. Entomol.* 101: 647–653.

Ladurner, E., J. Bosch, W. P. Kemp, and S. Maini. 2005. Assessing delayed and acute toxicity of five formulated fungicides to *Osmia lignaria* Say and *Apis mellifera*. *Apidologie* 36: 449–460.

Mullin, C. A., M. Frazier, J. L. Frazier, S. Ashcraft, R. Simonds, D. vanEngelsdorp, and J. S. Pettis. 2010. High levels of miticides and agrochemicals in North American apiaries: Implications for honey bee health. *PLoS ONE* 5(3): e9754. doi: 10.1371/journal.pone.0009754

Mussen, E. C., J. E. Lopez, and C. Y. S. Peng. 2004. Effects of selected fungicides on growth and development of larval honey bees, *Apis mellifera* L. (Hymenoptera: Apidae). *Environ. Entomol.* 33: 1151–1154.

Osintseva, L. A., and G. P. Chekryga. 2008. Fungi of melliferous bees pollenload. *Mikologiya i Fitopatologiya* 42: 464–469.

Peng, C. Y. S., C. Mussen, A. Fong, P. Cheng, G. Wong, and M. A. Montague. 1996. Laboratory and field studies on the effects of the antibiotic tylosin on honey bee *Apis mellifera* L. (Hymenoptera: Apidae) development and prevention of America foulbrood disease. *J. Invertebr. Pathol.* 67, 65–71.

Sautour, M., P. Dantigny, C. Divies, and M. Bensoussan. 2001. A temperature-type model for describing the relationship between fungal growth and water activity. *Int. J. Food Microbiol.* 67: 63–69.

Smodiš-Škerl, M. I., S. V. Bolta, H. B. Česnik, and A. Gregorc. 2009. Residues of pesticides in honeybee (*Apis mellifera carnica*) bee bread and in pollen loads from treated apple orchards. *Bull. Environ. Contam. Toxicol.* 83: 374–377.

Sokal, R. R., and F. J. Rohlf. 1995. *Biometry: The Principles and Practice of Statistics in Biological Research*, W. H. Freeman, San Francisco, California, U.S.

Solomon, M. G., and K. J. M. Hooker. 1989. Chemical repellents for reducing pesticide hazard to honeybees in apple orchards. *J. Api. Res.* 28: 223–227.

Tomlin, C. 2006. *The Pesticide Manual*. British Crop Protection Council, Nottingham, UK.

United States Environmental Protection Agency. 2006. Registration Eligibility Decision (RED) for Propiconazole, EPA 738R-06-027, Washington, DC, U.S.

Williams, G. R., M. A. Sampson, D. Sutler, and R. E. L. Rogers. 2008. Does fumagillin control the recently detected invasive parasite *Nosema ceranae* in western honey bees (*Apis mellifera*)? *J. Invertebr. Pathol.* 99: 342–344.

Yoder, J. A., B. S. Christensen, T. J. Croxall, J. L. Tank, and D. Sammataro. 2008. Suppression of growth rate of colony-associated fungi by high fructose corn syrup feeding supplement, formic acid, and oxalic acid. *J. Apic. Res.* 47: 126–130.

Chapter 17

Fungicides Reduce Symbiotic Fungi in Bee Bread and the Beneficial Fungi in Colonies

Abdel-Fattah, G. M., Y. M. Shabana, A. E. Ismail, and Y. M. Rashad. 2007. *Trichoderma harzianum*: A biocontrol agent against *Bipolaris oryzae*. *Mycopathology* 164: 81–89.

Alarcón, R., G. DeGrandi-Hoffman, and G. Wardell. 2009. Fungicides can reduce, hinder pollination potential of honey bees. *Western Farm Press* 31: 17–21.

Al-Ghamdi, A., Y. Y. Molan, and S. El-Hussieni. 2004. Identification of associated fungi recovered from honey bee combs and biocontrol using *Trichoderma* spp (*in-vitro*). *Alexandria Sci. Exch.* 25: 671–677.

Baldrian, P., and J. Gabriel. 2002. Intraspecific variability in growth response to cadmium of the wood-rotting fungus *Piptoporus betulinus*. *Mycologia* 94: 428–436.

Barbosa, M. A. G., K. G. Rehn, M. Menezes, and R. L. R. Mariano. 2001. Antagonism of *Trichoderma* species on *Cladosporium herbarum* and their enzymatic characterization. *Braz. J. Microbiol.* 32: 98–104.

Barnett, H. L., and B. B. Hunter. 1998. *Illustrated Genera of Imperfect Fungi* (4th ed.). American Phytopathological Society Press, St. Paul, Minnesota, U.S.

Batra, L. R., S. W. T. Batra, and G. E. Bohart. 1973. The mycoflora of domesticated and wild bees (Apoidea). Mycopathol. *Mycol. Appl.* 49: 13–44.

Benoit J. B., J. A. Yoder, and L. W. Zettler. 2004. *Scopulariopsis brevicaulis* (Deuteromycota) affords protection from secondary fungus infection in the American dog tick, *Dermacentor variabilis* (Acari: Ixodidae): Inference from competitive fungal interactions *in vitro*. *Int. J. Acarol.* 30: 375–381.

Brown, A. E. 2007. *Benson's Microbiological Applications: Laboratory Manual in General Microbiology* (10th ed.). McGraw-Hill, New York, NY, U.S.

Chapman, P. 1993. *Caves and Cave Life.* Harper Collins, London, UK.

Charlton, A. J. A., and A. Jones. 2007 Determination of imidazole and triazole fungicide residues in honeybees using gas chromatography-mass spectrometry. *J. Chromatography A* 1141: 117–122.

Chiesa, F., N. Milani, and M. D'Agaro. 1989. Observations of the reproductive behavior of *Varroa jacobsoni* Oud.: Techniques and preliminary results. In *Present Status of Varroatosis in Europe and Progress in the Varroa Mite Control.* Proc. Meeting, Udine, Italy, 1988. Cavalloro, E. (Ed.). EC-Experts Group, Luxembourg, pp. 213–222.

Currah, R. S., L. Sigler, and S. Hambleton. 1987. New records and new taxa of fungi form the mycorrhizae of terrestrial orchids of Alberta. *Can. J. Bot.* 65: 2473–2482.

Cooper, B. 1980. Fluctuating broodnest temperature rhythm. *British Isles Bee Breeders News* 18: 12–16.

Elad, Y. 2000. *Trichoderma harzianum* T39 preparation for biocontrol of plant diseases—Control of *Botrytis cinerea, Sclerotinia sclerotiorum* and *Cladosporium fulvum. BioControl Sci. Technol.* 10: 499–507.

Folk, D. G., C. Han, and T. J. Bradley. 2001. Water acquisition and partitioning in *Drosophilia melanogaster*: Effect of selection for desiccation-resistance. *J. Exp. Biol.* 204: 3323–3331.

Gilliam, M. 1979. Microbiology of pollen and bee bread: The yeasts. *Apidologie* 10: 43–53.

Gilliam, M. 1997. Identification and roles of non-pathogenic microflora associated with honey bees. *Federation of European Microbiological Societies (FEMS) Microbiol. Lett.* 155: 1–10.

Gilliam, M., and J. D. Vandenberg. 1997. Fungi. In *Honey Bee Pests, Predators and Diseases.* Morse, R. A., and K. Flottum (Eds.). A. I. Root Company, Medina, Ohio, U.S., pp. 79–112.

Gilliam, M., B. J. Lorenz, A. M. Wenner, and R. W. Thorp. 1997. Occurrence and distribution of *Ascophaera apis* in North America: Chalkbrood in feral honey bee colonies that had been in isolation on Santa Cruz Island, California for over 110 years. *Apidologie* 28: 329–338.

Gilliam, M., D. B. Prest, and B. J. Lorenz. 1989. Microbiology of pollen and bee bread: Taxonomy and enzymology of molds. *Apidologie* 20: 53–68.

Gilliam, M., S. Taber III, B. J. Lorenz, and D. B. Prest. 1988. Factors affecting development of chalkbrood disease in colonies of honey bees, *Apis mellifera*, fed pollen contaminated with *Ascosphaera apis. J. Invertebr. Pathol.* 52: 314–325.

Gliñski, Z., and K. Buczek. 2003. Response of the Apoidea to fungal infections. *Apiacta* 38: 183–189.

Hua, L., and M.-G. Feng. 2006. New use of broomcorn millets for production of granular cultures of apid-pathogenic fungus *Pandora neoaphidis* for high sporulation potential and infectivity to *Myzus persicae. Federation of European Microbiological Societies (FEMS) Microbiol. Lett.* 227: 311–317.

Jennings, D. H., and G. Lysek. 1999. *Fungal Biology: Understanding the Fungal Lifestyle.* Sci. Pub. Ltd., Oxford, UK.

Klepzig, K. D., and R. T. Wilkens. 1997. Competitive interactions among symbiotic fungi of the southern pine beetle. *Appl. Environ. Microbiol.* 63: 621–627.

Klungness, L. M., and Y. Peng. 1983. A scanning electron microscopic study of pollen loads collected and stored by honeybees. *J. Apic. Res.* 22: 264–271.

Kubik, M., J. Nowacki, A. Pidek, Z. Warakomska, L. Michalczuk, and W. Goszczyñski. 1999. Pesticide residues in bee products collected from cherry trees protected during blooming period with contact and systemic fungicides. *Apidologie* 30: 521–532.

Kubik, M., J. Nowacki, A. Pidek, Z. Warakomska, L. Michalczuk, W. Goszczyñski, and B. Dwuznik. 2000. Residues of captan (contact) and difenoconazole (systemic) fungicides in bee products from an apple orchard. *Apidologie* 31: 531–541.

Mónaco, C., M. Sisterna, A. Perelló, and G. Dal Bello. 2004. Preliminary studies on biological control of the blackpoint complex of wheat in Argentina. *World J. Microbiol. Biotechnol.* 20: 285–290.

Novak, M .G., L. G. Highley, C. A. Christianssen, and W. A. Rowley. 1993. Evaluating larval competition between *Aedes albopictus* and *A. triseriatus* (Diptera: Culicidae) through replacement series experiments. *J. Environ. Entomol.* 22: 311–318.

Osintseva, L. A., and G. P. Chekryga. 2008. Fungi of melliferous bees pollenload. *Mikologiya i Fitopatologiya* 42: 464–469.

Phillips, D. J., B. Mackey, W. R. Ellis, and T. N. Hansen. 1979. Occurrence and interaction of *Aspergillus flavus* with other fungi on almonds. *Phytopathology* 69: 829–831.

Roco, A., and L. M. Pérez. 2001. *In vitro* biocontrol activity of *Trichoderma harzianum* on *Alternaria alternata* in the presence of growth regulators. *Electron. J. Biotechnol.* 4: 2.

Sammataro, D., and A. Avitabile, 1998. *The Beekeeper's Handbook*. Cornell University Press, Ithaca, New York, U.S.

Scardaci, S. C., and R. K. Webster. 1981. Antagonism between the cereal root rot pathogens *Fusarium graminearum* and *Bipolaris sorokiniana*. *Plant Disease* 65: 965–966.

Škerl, M. I. S., S. Velikonja Bolta, H. Baša Česnik, and A. Gregorc. 2009. Residues of pesticides in honeybee (*Apis mellifera carnica*) bee bread and in pollen loads from treated apple orchards. *Bull. Environ. Contam. Toxicol.* 83: 374–377.

Sokal, R. R., and F. J. Rohlf. 1995. *Biometry: The Principles and Practice of Statistics in Biological Research*. W. H. Freeman, San Francisco, California, U.S.

United States Environmental Protection Agency. 2006. Registration Eligibility Decision (RED) for Propiconazole, EPA 738R-06-027, Washington, DC, U.S.

Vásquez, A., and T. Olofsson. 2009. The lactic acid bacteria involved in the production of bee pollen and bee bread. *J. Apic. Res.* 48: 189–195.

West, S. A., S. P. Diggle, A. Buckling, A. Gardner, and A. S. Griffin. 2007. The social lives of microbes. *Ann. Rev. Ecol. Evol. System.* 38: 53–57.

Wilson-Rich, N., M. Spivak, N. H. Fefferman, and P. T. Starks. 2009. Genetic, individual, and group facilitation of disease resistance in insect societies. *Annu. Rev. Entomol.* 54: 405–423.

Yoder, J. A., B. S. Christensen, T. J. Croxall, J. L. Tank, and D. Sammataro. 2008. Suppression of growth rate of colony-associated fungi by high fructose corn syrup feeding supplement, formic acid, and oxalic acid. *J. Apic. Res.* 47: 126–130.

Chapter 18

Interactions Between Risk Factors in Honey Bees

Alaux, C., J. L. Brunet, C. Dussaubat, F. Mondet, S. Tchamitchan, M. Cousin, J. Brillard, A. Baldy, L. P. Belzunces, and Y. Le Conte. 2010a. Interactions between Nosema microspores and a neonicotinoid weaken honeybees (*Apis mellifera*). *Environ. Microbiol.* 12: 774–782.

Alaux, C., F. Ducloz, D. Crauser, and Y. Le Conte. 2010b. Diet effects on honeybee immunocompetence. *Biol. Lett.* 6: 562–565.

Anderson, D., and I. J. East. 2008. The latest buzz about Colony Collapse Disorder. *Science* 319: 724–725.

Bailey, L., B. V. Ball, and J. N. Perry. 1983. Association of viruses with two protozoal pathogens of the honey bee. *Ann. Appl. Biol.* 103: 13–20.

Belzunces, L. P., S. Garin, and M. E. Colin. 1993. Synergistic effects of pyrethroid insecticides and azole fungicides in honey bees, at sublethal doses. *Mésogée* 53: 13–16.

Benoit, J. B., J. A. Yoder, D. Sammataro, and L. W. Zettler. 2004. Mycoflora and fungal vector capacity of the parasitic mite *Varroa destructor* (Mesostigmata: Varroidae) in honey bee (Hymenoptera: Apidae) colonies. *Int. J. Acarol.* 30: 103–106.

Biesmeijer, J. C., S. P. Roberts, M. Reemer, R. Ohlemuller, M. Edwards, T. Peeters, A. P. Schaffers, S. G. Potts, R. Kleukers, C. D. Thomas, J. Settele, and W. E. Kunin. 2006. Parallel declines in pollinators and insect-pollinated plants in Britain and the Netherlands. *Science* 313: 351–354.

Bogdanov, S. 2006. Contaminants of bee products. *Apidologie* 37: 1–18.

Bonmatin, J. M., P. A. Marchand, R. Charvet, I. Moineau, E. R. Bengsch, and M. E. Colin. 2005. Quantification of imidacloprid uptake in maize crops. *J. Agric. Food. Chem.* 53: 5336–5341.

Bonmatin, J. M., I. Moineau, R. Charvet, C. Fleche, M. E. Colin, and E. R. Bengsch. 2003. A LC/APCI–MS/MS method for analysis of imidacloprid in soils, in plants, and in pollens. *Ann. Chem.* 75: 2027–2033.

Brattsten, L. B., D. A. Berger, and L. B. Dungan 1994. *In vitro* inhibition of midgut microsomal P450s from *Spodoptera eridania* caterpillars by demethylation inhibitor fungicides and plant growth regulators. *Pestic. Biochem. Physiol.* 49: 234–243.

Brodschneider, R., and K. Crailsheim. 2010. Nutrition and health in honey bees. *Apidologie* 41: 278–294.

Bromenshenk, J. J., C. B. Henderson, C. H. Wick, M. F. Stanford, A. W. Zulich, R. E. Jabbour, S. V. Deshpande, P. E. McCubbin, R. A. Seccomb, P. M. Welch, T. Williams, D. R. Firth, E. Skowronski, M. M. Lehmann, S. L. Bilimoria, J. Gress, K. W. Wanner, and R. A. Cramer, Jr. 2010. Iridovirus and microsporidian linked to honey bee colony decline. *PLoS ONE* 5(10): e13181. doi: 10.1371/journal.pone.0013181

Charvet, R., M. Katouzian-Safadi, M. E. Colin, P. A. Marchand, and J. M. Bonmatin. 2004. Systemic insecticides: New risk for pollinator insects. *Ann. Pharmac. Franc.* 62: 29–35.

Chauzat, M. P., P. Carpentier, A. C. Martel, S. Bougeard, N. Cougoule, P. Porta, J. Lachaize, F. Madec, M. Aubert, and J. P. Faucon. 2009. Influence of pesticide residues on honey bee (Hymenoptera: Apidae) colony health in France. *Environ. Entomol.* 38: 514–523.

Claudianos, C., H. Ranson, R. M. Johnson, S. Biswas, M. A. Schuler, M. R. Berenbaum, R. Feyereisen, and J. G. Oakeshott. 2006. A deficit of detoxification enzymes: Pesticide sensitivity and environmental response in the honeybee. *Insect Mol. Biol.* 15: 615–636.

Colin, M. E., and L. P. Belzunces. 1992. Evidence of synergy between prochloraz and deltamethrin: A convenient biological approach. *Pestic. Sc.* 36: 115–119.

Cox-Foster, D. L., S. Conlan, E. C. Holmes, G. Palacios, J. D. Evans, N. A. Moran, P.-L. Quan, T. Briese, M. Hornig, D. M. Geiser, V. Martinson, D. vanEngelsdorp, A. L. Kalkstein, A. Drysdale, J. Hui, J. Zhai, L. Cui, S. K. Hutchison, J. F. Simons, M. Egholm, J. S. Pettis, and W. I. Lipkin. 2007. A metagenomic survey of microbes in honey bee Colony Collapse Disorder. *Science* 318: 283–287. doi: 10.1126/science.1146498

Dainat, B., T. Ken, H. Berthoud, and P. Neumann. 2009. The ectoparasitic mite *Tropilaelaps mercedesae* (Acari, Laelapidae) as a vector of honeybee viruses. *Insect. Soc.* 56: 40–43.

Decourtye, A., E. Mader, and N. Desneux. 2010. Landscape enhancement of floral resources for honey bees in agro-ecosystems. *Apidologie* 41: 264–277.

Degrandi-Hoffman, G., Y. Chen, E. Huang, and M. H. Huang. 2010. The effect of diet on protein concentration, hypopharyngeal gland development and virus load in worker honey bees (*Apis mellifera* L.). *J. Insect Physiol.* 56: 1184–1191.

Delpuech, J. M., F. Frey, and Y. Carton. 1996. Action of insecticides on the cellular immune reaction of *Drosophila melanogaster* against the parasitoid *Leptopilina boulardi*. *Environ. Toxicol. Chem.* 15: 2267–2271.

Downey, D. L., T. T. Higo, and M. L. Winston. 2000. Single and dual parasitic mite infestations on the honey bee, *Apis mellifera*. *Insect. Soc.* 47: 171–176.

Downey, D. L., and M. L. Winston. 2001. Honey bee colony mortality and productivity with single and dual infestations of parasitic mite species. *Apidologie* 32: 567–575.

Elbert, A., M. Haas, B. Springer, W. Thielert, and R. Nauen. 2008. Applied aspects of neonicotinoid uses in crop protection. *Pest. Manag. Sci.* 64: 1099–1105.

Ellis, J. D., J. D. Evans, and J. S. Pettis. 2010. Colony losses, managed colony population decline and Colony Collapse Disorder in the United States. *J. Apic. Res.* 49: 134–136.

Evans, J. D. 2004. Transcriptional immune responses by honey bee larvae during invasion by the bacterial pathogen, *Paenibacillus larvae*. *J. Invertebr. Pathol.* 85: 105–111.

Field, C. J., I. R. Johnson, and P. D. Schley. 2002. Nutrients and their role in host resistance to infection. *J. Leukocyte Biol.* 71: 16–32.

Forsgren, E., J. R. de Miranda, M. Isaksson, S. Wei, and I. Fries. 2009. Deformed wing virus associated with *Tropilaelaps mercedesae* infesting European honey bees (*Apis mellifera*). *Exp. Appl. Acarol.* 47: 87–97.

Fries, I. 2010. Nosema ceranae in European honey bees (*Apis mellifera*). *J. Invertebr. Pathol.* 103: 73–79.

George, P. J. E., and D. P. Ambrose. 2004. Impact of insecticides on the haemogram of *Rhynocoris kumarii* Ambrose and Livingstone (Hem., Reduviidae). *J. Appl. Entomol.* 128: 600–604.

Haydak, M. H. 1970. Honey bee nutrition. *Ann. Rev. Entomol.* 15: 143–156.

Hardstone, M. C., and J. G. Scott. 2010. Is *Apis mellifera* more sensitive to insecticides than other insects? *Pest Manag. Sci.* 11: 1171–1180. doi: http://onlinelibrary.wiley.com/doi/10.1002/ps.2001/abstract

Iwasa, T., N. Motoyama, J. T. Ambrose, and R. M. Roe. 2004. Mechanism for the differential toxicity of neonicotinoid insecticides in the honey bee, *Apis mellifera*. *Crop Prot.* 23: 371–378.

Jaffe, R., V. Dietemann, M. H. Allsopp, C. Costa, R. M. Crewe, R. Dall'olio, P. De La Rua, M. A. A. El-Niweiri, I. Fries, N. Kezic, M. S. Meusel, R. J. Paxton, T. Shaibi, E. Stolle, and R. F. A. Moritz. 2010. Estimating the density of honeybee colonies across their natural range to fill the gap in pollinator decline censuses. *Conserv. Biol.* 24: 583–593.

Janmaat, A. F., and M. L. Winston. 2000. The influence of pollen storage area and *Varroa jacobsoni* Oudemans parasitism on temporal caste structure in honey bees (*Apis mellifera* L.). *Insect. Soc.* 47: 177–182.

Jaramillo, J., C. Borgemeister, L. Ebssa, A. Gaigl, R. Tobón, and G. Zimmermann. 2005. Effect of combined applications of *Metarhizium anisopliae* (Metsch.) Sorokin (Deuteromycotina: Hyphomycetes) strain CIAT 224 and different dosages of imidacloprid on the subterranean burrower bug *Cyrtomenus bergi* Froeschner (Hemiptera: Cydnidae). *Biol. Control* 34: 12–20.

Johnson, R. M., M. D. Ellis, C. A. Mullin, and M. Frazier. 2010. Pesticides and honey bee toxicity—USA. *Apidologie* 41: 312–331.

Johnson, R. M., J. D. Evans, G. E. Robinson, and M. R. Berenbaum. 2009a. Changes in transcript abundance relating to Colony Collapse Disorder in honey bees (*Apis mellifera*). *PNAS* 106: 14790–14795.

Johnson, R. M., H. S. Pollock, and M. R. Berenbaum. 2009b. Synergistic interactions between in-hive miticides in *Apis mellifera*. *J. Econ. Entomol.* 102: 474–479.

Johnson, R. M., Z. Wen, M. A. Schuler, and M. R. Berenbaum. 2006. Mediation of pyrethroid insecticide toxicity to honey bees (Hymenoptera: Apidae) by cytochrome P450 monooxygenases. *J. Econ. Entomol.* 99: 1046–1050.

Jones, J. C., M. R. Myerscough, S. Graham, and B. P. Oldroyd. 2004. Honey bee nest thermoregulation: Diversity promotes stability. *Science* 305: 402–404.

Kaakeh, W., B. Reid, T. J. Bohnert, and G. W. Bennet. 1997. Toxicity of imidacloprid in the German cockroach (Dictyoptera: Blattellidae), and the synergism between imidacloprid and *Matarhizium anisopliae* (Imperfect fungi: Hyphomycetes). *J. Econ. Entomol.* 90: 473–482.

Keller, I., P. Fluri, and A. Imdorf. 2005. Pollen nutrition and colony development in honey bees, Part II. *Bee World* 86: 27–34.

Kremen, C., N. M. Williams, and R. W. Thorp. 2002. Crop pollinators from native bees at risk from agricultural intensification. *PNAS* 99: 16812–16816.

Ladas, A. 1972. The influence of some internal and external factors upon the insecticide resistance of honeybee. *Apidologie* 3: 55–78.

Laurent, F. M., and E. Rathahao. 2003. Distribution of [^{14}C]-imidacloprid in sunflowers (*Helianthus annuus* L.) following seed treatment. *J. Agric. Food Chem.* 51: 8005–8010.

Le Conte, Y., M. Ellis, and W. Ritter. 2010. Varroa mites and honey bee health: Can Varroa explain part of the colony losses? *Apidologie* 41: 353–363.

Le Conte, Y., and M. Navajas. 2008. Climate change: Impact on honey bee populations and diseases. *Rev. Sci. Tech. Off. Int. Epiz.* 27: 499–510.

Li, P., Y. L. Yin, D. Li, S. W. Kim, and G. Wu. 2007. Amino acids and immune function. *Brit. J. Nutr.* 98: 237–252.

Lydy, M., J. Belden, C. Wheelock, B. Hammock, and D. Denton. 2004. Challenges in regulating pesticide mixtures. *Ecol. Soc.* 9: 1.

Magnus, R. M., and A. L. Szalanski. 2010. Genetic evidence of honey bees belonging to the Middle East lineage in the United States. *Sociobiology* 55: 285–296.

Maredia, K. M., D. Dakouo, and D. Mota-Sanchez. 2003. *Integrated Pest Management in the Global Arena*. CABI Publishing, Wallingford, Connecticut, U.S.

Mattila, H. R., K. M. Burke, and T. D. Seeley. 2008. Genetic diversity within honeybee colonies increases signal production by waggle-dancing foragers. *Proc. R. Soc. Biol. Sci.* 275: 809–816.

Mattila, H. R., and T. D. Seeley. 2007. Genetic diversity in honey bee colonies enhances productivity and fitness. *Science* 317: 362–364.

Mayack, C., and D. Naug. 2009. Energetic stress in the honeybee *Apis mellifera* from *Nosema ceranae* infection. *J. Invertebr. Pathol.* 100: 185–188.

Meixner, M. D., C. Costa, P. Kryger, F. Hatjina, M. Bouga, E. Ivanova, and R. Büchler. 2010. Conserving diversity and vitality for honey bee breeding. *J. Apic. Res.* 49: 85–92.

Meled, M., A. Thrasyvoulou, and L. P. Belzunces. 1998. Seasonal variations in susceptibility of *Apis mellifera* to the synergistic action of prochloraz and deltamethrin. *Environ. Toxicol. Chem.* 17: 2517–2520.

Monosson, E. 2005. Chemical mixtures: Considering the evolution of toxicology and chemical assessment. *Environ. Health Persp.* 113: 383–390.

Moritz, R. F. A., J. de Miranda, I. Fries, Y. Le Conte, P. Neumann, and R. J. Paxton. 2010. Research strategies to improve honeybee health in Europe. *Apidologie* 41: 227–242.

Moritz, R. F. A., B. Kraus, P. Kryger, and R. Crewe. 2007. The size of wild honey bee populations (*Apis mellifera*) and its implications for the conservation of honey bees. *J. Insect Conserv.* 1: 391–397.

Mullin, C. A., M. Frazier, J. L. Frazier, S. Ashcraft, R. Simonds, D. vanEngelsdorp, and J. S. Pettis. 2010. High levels of miticides and agrochemicals in North American apiaries: Implications for honey bee health. *PLoS ONE* 5(3): e9754. doi: 10.1371/journal.pone.0009754

Naug, D. 2009. Nutritional stress due to habitat loss may explain recent honeybee colony collapses. *Biol. Conserv.* 142: 2369–2372.

Naug, D., and A. Gibbs. 2009. Behavioral changes mediated by hunger in honeybees infected with *Nosema ceranae*. *Apidologie* 40: 595–599.

Neumann, P., and N. L. Carreck. 2010. Honey bee colony losses. *J. Apicul. Res.* 49: 1–6.

Oldroyd, B. P. 2007. What's killing American honey bees? *PLoS Biology* 5: 1195–1199.

Pilling, E. D., K. A. C. Bromley-Challenor, C. H. Walker, and P. C. Jepson. 1995. Mechanism of synergism between the pyrethroid insecticide—cyhalothrin and the imidazole fungicide orochloraz, in the honeybee (*Apis mellifera* L.). *Pestic. Biochem. Physiol.* 51: 1–11.

Pilling, E. D., and P. C. Jepson. 1993. Synergism between EBI fungicides and a pyrethroid insecticide in the honeybee (*Apis mellifera*). *Pestic. Sci.* 39: 293–297.

Potts, S. G., S. P. M. Roberts, R. Dean, G. Marris, M. A. Brown, H. R. Jones, P. Neumann, and J. Settele. 2010. Declines of managed honey bees and beekeepers in Europe. *J. Apic. Res.* 49: 15–22.

Purwar, J. P., and G. C. Sachan. 2006. Synergistic effect of entomogenous fungi on some insecticides against Bihar hairy caterpillar *Spilarctia obliqua* (Lepidoptera: Arctiidae). *Microbiol. Res.* 161: 38–42.

Ramakrishnan, R., D. R. Suiter, C. H. Nakatsu, R. A. Humber, and G. W. Bennett. 1999. Imidacloprid-enhanced *Reticulitermes flavipes* (Isoptera: Rhinotermitidae) susceptibility to the Entomopathogen *Metarhizium anisopliae*. *J. Econ. Entomol.* 92: 1125–1132.

Rinderer, T. E., and K. D. Elliott. 1977. Worker honey bee response to infection with *Nosema apis*. *J. Econ. Entomol.* 70: 431–433.

Rinderer, T. E., W. C. Rothenbuhler, and T. A. Gochnauer. 1974. The influence of pollen on the susceptibility of honey bee larvae to *Bacillus*. *J. Invertebr. Pathol.* 23: 347–350.

Rosenkranz, P., P. Aumeier, and B. Ziegelmann. 2010. Biology and control of *Varroa destructor*. *J. Invertebr. Pathol.* 103: 96–119.

Santos, A. V., B. L. de Oliveira, and R. I. Samuels. 2007. Selection of entomopathogenic fungi for use in combination with sub-lethal doses of imidacloprid: Perspectives for the control of the leaf-cutting ant *Atta sexdens rubropilosa* Forel (Hymenoptera: Formicidae). *Mycopathologia* 163: 233–240.

Schiff, N. M., and W. S. Sheppard. 1995. Genetic analysis of commercial honey bees (Hymenoptera: Apidae) from the southern United States. *J. Econ. Entomol.* 88: 1216–1220.

Schmuck, R., T. Stadler, and H.-W. Schmidt. 2003. Field relevance of a synergistic effect observed in the laboratory between an EBI fungicide and a chloronicotinyl insecticide in the honeybee (*Apis mellifera* L, Hymenoptera). *Pest. Manag. Sci.* 59: 279–286.

Seeley, T. D. 2007. Honey bees of the Arnot Forest: A population of feral colonies persisting with *Varroa destructor* in the northeastern United States. *Apidologie* 38: 19–29.

Seeley, T. D., and D. R. Tarpy. 2007. Queen promiscuity lowers disease within honeybee colonies. *Proc. R. Soc. Biol. Sci.* 274: 67–72.

Stokstad, E. 2007. The case of the empty hives. *Science* 316: 970–972.

Tarpy, D. R. 2003. Genetic diversity within honeybee colonies prevents severe infections and promotes colony growth. *Proc. R. Soc. Biol. Sci.* 270: 99–103.

Tasei, J. N., and P. Aupinel. 2008. Nutritive value of 15 single pollens and pollen mixes tested on larvae produced by bumblebee workers (*Bombus terrestris*, Hymenoptera: Apidae). *Apidologie* 39: 397–409.

Thompson, H. M. 2003. Behavioural effects of pesticides in bees—their potenital for use in risk assessment. *Ecotoxicology* 12: 317–330.

Vandame, R., and L. P. Belzunces. 1998. Joint actions of deltamethrin and azole fungicides on honey bee thermoregulation. *Neurosci. Lett.* 251: 57–60.

vanEngelsdorp, D., J. D. Evans, C. Saegerman, C. Mullin, E. Haubruge, B. K. Nguyen, M. Frazier, J. Frazier, D. L. Cox-Foster, Y. Chen, R. M. Underwood, D. R. Tarpy, and J. S. Pettis. 2009. Colony Collapse Disorder: A descriptive study. *PLoS ONE* 4(8): e6481. doi: 10.1371/journal.pone.0006481

vanEngelsdorp, D., J. Hayes, Jr., R. M. Underwood, and J. Pettis. 2008. A survey of honey bee colony losses in the U.S., fall 2007 to spring 2008. *PLoS ONE* 3(12): e4071. doi: 10.1371/journal.pone.0004071

Wahl, O., and K. Ulm. 1983. Influence of pollen feeding and physiological condition on pesticide sensitivity of the honey bee *Apis mellifera carnica*. *Oecologia* 59: 106–128.

Wallner, K. 1999. Varroacides and their residues in bee products. *Apidologie* 30: 235–248.

Chapter 19

Understanding the Impact of Honey Bee Disorders on Crop Pollination

Aizen, M. A., and L. D. Harder. 2009. The global stock of domesticated honey bees is growing slower than agricultural demand for pollination. *Curr. Biol.* 19: 915–918. doi: 10.1016/j.cub.2009.03.071.

Anderson, D. L., and H. Giacon. 1992. Reduced pollen collection by honey bee (Hymenoptera: Apidae) colonies infected with *Nosema apis* and sacbrood virus. *J. Econ. Entomol.* 85(1): 47–51.

Cane, J. H., and J. A. Payne. 1990. Native bee pollinates rabbiteye blueberry. *Alabama Agricultural Experiment Station* 37(4): 4.

Dedej, S., and K. S. Delaplane. 2003. Honey bee (Hymenoptera: Apidae) pollination of rabbiteye blueberry *Vaccinium ashei* var. "Climax" is pollinator density-dependent. *J. Econ. Entomol.* 96: 1215–1220.

Delaplane, K. S., and D. F. Mayer. 2000. *Crop Pollination by Bees*. CABI, Oxon, UK.

Ellis, A., and K. S. Delaplane. 2008. Effects of nest invaders on honey bee (*Apis mellifera*) pollination efficacy. *Agric. Ecosyst. Environ.* 127: 201–206. doi: 10.1016/j.agee.2008.04.001

Ellis, A., and K. S. Delaplane. 2009. Individual forager profits in *Apis mellifera* unaffected by a range of colony *Varroa destructor* densities. *Insect. Soc.* doi: 10.1007/s00040-009-0040-2

Ellis, J. D., Jr., R. Hepburn, K. S. Delaplane, P. Neumann, and P. J. Elzen. 2003. The effects of adult small hive beetles, *Aethina tumida* (Coleoptera: Nitidulidae), on nests and foraging activity of Cape and European honey bees (*Apis mellifera*). *Apidologie* 34: 399–408.

Gallai, N., J.-M. Salles, J. Settele, and B. E. Vaissière. 2009. Economic valuation of the vulnerability of world agriculture confronted with pollinator decline. *Ecol. Econ.* 68(3): 810–821. doi: 10.1016/j.ecolecon.2008.06.014

Janmaat, A. F., M. L. Winston, and R. C. Ydenberg. 2000. Condition-dependent response to changes in pollen stores by honey bee (*Apis mellifera*) colonies with different parasitic loads. *Behav. Ecol. Sociobiol.* 47: 171–179.

Klein, A.-M., B. E. Vaissière, J. H. Cane, I. Steffan-Dewenter, S. A. Cunningham, C. Kremen, and T. Tscharntke. 2007. Importance of pollinators in changing landscapes for world crops. *Proc. Royal Soc. B.* 274: 303–313. doi: 10.1098/rspb.2006.3721

Moritz, R. F. A., and E. E. Southwick. 1992. *Bees As Superorganisms: An Evolutionary Reality*. Springer, Heidelberg, Germany.

Murilhas A. M. 2002. *Varroa destructor* infestation impact on *Apis mellifera carnica* capped worker brood production, bee population and honey storage in a Mediterranean climate. *Apidologie* 33: 271–281.

Neumann, P., and N. L. Carreck. 2010. Honey bee colony losses. *J. Apicult. Res.* 49(1): 1–6. doi: 10.3896/IBRA.1.49.1.01s

Potts, S. G., S. P. M. Roberts, R. Dean, G. Marris, M. A. Brown, R. Jones, P. Neumann, and J. Settele. 2010. Declines of managed honey bees and beekeepers in Europe. *J. Apic. Res.* 49(1): 15–22. doi: 10.3896/IBRA.1.49.1.02

Sampson, B. J., and J. H. Cane. 2000. Pollination efficiencies of three bee (Hymenoptera: Apoidea) species visiting rabbiteye blueberry. *J. Econ. Entomol.* 93(6): 1726–1731.

Schmid-Hempel, P. 1998. *Parasites in Social Insects*. Princeton University Press, Princeton, New Jersey, U.S.

Schmid-Hempel P., M. L. Winston, and R. C. Ydenberg. 1993. Foraging of individual workers in relation to colony state in the social Hymenoptera. *Can. Entomol.* 125: 129–160.

vanEngelsdorp, D., J. Hayes, Jr., R. M. Underwood, and J. Pettis. 2008. A survey of honey bee colony losses in the U.S., fall 2007 to spring 2008. *PLoS ONE* 3(12): e4071. doi: 10.1371/journal.pone.0004071

vanEngelsdorp, D., J. Hayes Jr., R. M. Underwood, and J. Pettis. 2010. A survey of honey bee colony losses in the United States, fall 2008 to spring 2009. *J. Apic. Res.* 49 (1): 7–14. doi: 10.3896/IBRA.1.49.1.03

Chapter 20

Calculating and Reporting Managed Honey Bee Colony Losses

Abdi, H. 2007. The Bonferonni and Šidák Corrections for Multiple Comparisons. In *Encyclopedia of Measurement and Statistics*. Salkind, N. (Ed.), Sage Publications, Thousand Oaks, California, U.S.

Aizen, M. A., and L. D. Harder. 2009. The global stock of domesticated honey bees is growing slower than agricultural demand for pollination. *Current Biology* 19: 915–918.

Brodschneider, R., R. Moosbeckhofer, and K. Crailsheim. 2010. Surveys as a tool to record winter losses of honey bee colonies: A two-year case study in Austria and South Tyrol. *J. Apic. Res.* 49: 23–30.

Brown, L. D., T. T. Cai, and A. DasGupta. 2001. Interval estimation for a binomial proportion. *Statistical Science* 16: 101–117.

Brown, L. D., T. T. Cai, and A. DasGupta. 2002. Confidence intervals for a binomial proportion and asymptotic expansions. *The Annals of Statistics* 30: 160–201.

COLOSS. www.coloss.org.

Daberkow, S., P. Korb, and F. Hoff. 2009. Structure of the U.S. beekeeping industry: 1982–2002. *J. Econ. Entomol.* 103: 19.

FAO. 2009. FAOSTAT.

Koepsell, T. D., and N. S. Weiss. 2003. *Epidemiologic Methods: Studying the Occurrence of Illness*. Oxford University Press, New York, NY, U.S.

Lowry, R. 2010. *VassarStats*: Website for Statistical Computation.

McCullagh, P., and J. Nelder. 1989. *Generalized Linear Models* (2nd ed.). Chapman & Hall/CRC, Boca Raton, Florida, U.S.

Nguyen, B. K., R. Van der Zee, F. Vejsnæs, S. Wilkins, Y. Le Conte, and W. Ritter. 2010. COLOSS Working Group 1: Monitoring and diagnosis. *J. Apic. Res.* 49: 97–99.

Nguyen, B. K., J. Mignon, D. Laget, D.C. de Graaf, F. J. Jacobs, D. vanEngelsdorp, Y. Brostaux, C. Saegerman and E. Haubruge. Honey bee colony losses in Belgium during the 2008–2009 winter. *J. Apic. Research* 49: 337–339.

Paoli, B., L. Haggard, and G. Shah. 2002. Confidence intervals in public health, 8 pp. Office of Public Health Assessment, Utah Department of Health, Utah, U.S.

R Development Core Team, v.2.10.0. 2009. R: A language and environment for statistical computing. R Foundation for Statistical Computing. Vienna, Austria.

USDA–NASS. 2010. Honey, 6 pp. Department of Agriculture, Washington, DC, U.S.

vanEngelsdorp, D., and M. D. Meixner. 2010. A historical review of managed honey bee populations in Europe and the United States and the factors that may affect them. *J. Invert. Path.* 103: S80–S95.

vanEngelsdorp, D., J. Hayes, Jr., R. M. Underwood, and J. Pettis. 2008. A survey of honey bee colony losses in the U.S., fall 2007 to spring 2008. *PLoS ONE* 3(12): e4071. doi: 10.1371/journal.pone.0004071

vanEngelsdorp, D., J. Hayes Jr., R. M. Underwood, and J. Pettis. 2010. A survey of honey bee colony losses in the United States, fall 2008 to spring 2009. *J. Apic. Res.* 49 (1): 7–14. doi: 10.3896/IBRA.1.49.1.03

Zar, J. H. 1996. *Biostatistical Analysis*. Prentice Hall, Upper Saddle River, New Jersey, U.S.

Chapter 21

Conservation of Plant–Pollinator Mutualisms

Abu-Asab, M. S., P. M. Peterson, S. G. Shetler, and S. S. Orli. 2001. Earlier plant flowering in spring as a response to global warming in the Washington, DC area. *Biodiversity and Conservation* 10: 597–612.

Aldridge, G., D. W. Inouye, J. Forrest, W. A. Barr, and A. J. Miller-Rushing. 2011. Emergence of a mid-season period of low floral resources in a montane meadow ecosystem associated with climate change. *J. Ecology* 99: 905–913.

Batáry, P., A. Báldi, M. Sárospataki, F. Kohler, J. Verhulst, E. Knop, F. Herzog, and D. Kleijn. 2010. Effect of conservation management on bees and insect-pollinated grassland plant communities in three European countries. *Agriculture Ecosystems & Environment* 136: 35–39.

Buchmann, S. L., and G. P. Nabhan. 1996. *The Forgotten Pollinators*. Island Press, Washington, DC, U.S.

Cameron, S. A., J. D. Lozier, J. P. Strange, J. B. Koch, N. Cordes, L. F. Solter, and T. L. Griswold. 2011. Patterns of widespread decline in North American bumble bees. *Proc. Natl. Acad. Sci. USA* 108: 662–667.

Committee on the Status of Pollinators in North America, National Research Council. 2007. *Status of Pollinators in North America*. National Academy Press, Washington, DC, U.S. doi: www.nap.edu (entire report online)

Ekroos, J., M. Piha, and J. Tiainen. 2007. Role of organic and conventional field boundaries on boreal bumblebees and butterflies. *Agriculture Ecosystems & Environment* 124: 155–159.

Frankie, G. W., R. W. Thorp, M. Schindler, J. Hernandez, B. Ertter, and M. Rizzardi. 2005. Ecological patterns of bees and their host ornamental flowers in two northern California cities. *J. Kansas Entomol. Soc.* 78: 227–246.

Free, J. B. 1970. *Insect Pollination of Crops*. Academic Press, New York, NY, U.S.

Goulson, D., and K. Sparrow. 2009. Evidence for competition between honeybees and bumblebees; effects on bumblebee worker size. *J. Insect Conservation* 13: 177–181.

Grixti, J. C., L. T. Wong, S. A. Cameron, and C. Favret. 2008. Decline of bumble bees (*Bombus*) in the North American Midwest. *Biological Conservation* 142: 75–84.

Hernandez, J. L., G. W. Frankie, and R. W. Thorp. 2009. *Ecology* of urban bees: A review of current knowledge and directions for future study. *Cities and the Environment* 2: 3–15.

Inouye, D. W. 1977. Species structure of bumblebee communities in North America and Europe. In *The Role of Arthropods in Forest Ecosystems*. Mattson, W. J. (Ed.). Springer-Verlag, New York, NY, U.S., pp. 35–40.

Kearns, C. A., and D. W. Inouye. 1997. Pollinators, flowering plants, and conservation biology. *Bioscience* 47: 297–307.

Kearns, C. A., D. W. Inouye, and N. M. Waser. 1998. Endangered mutualisms: The conservation of plant–pollinator interactions. *Annu. Rev. Ecol. Systematics* 29: 83–112.

Lambert, A. M., A. J. Miller-Rushing, and D. W. Inouye. 2010. Changes in snowmelt date and summer precipitation affect the flowering phenology of *Erythronium grandiflorum* (Glacier Lily; Liliaceae). *Am. J. Bot.* 97: 1431–1437.

Memmott, J., P. G. Craze, N. M. Waser, and M. V. Price. 2007. Global warming and the disruption of plant–pollinator interactions. *Ecology Letters* 10: 710–717.

Miller-Rushing, A. J., and D. W. Inouye. 2009. Variation in the impact of climate change on flowering phenology and abundance: An examination of two pairs of closely related wildflower species. *Am. J. Botany* 96: 1821–1829.

Otterstatter, M. C., and J. D. Thomson. 2008. Does pathogen spillover from commercially reared bumble bees threaten wild pollinators? *PLoS ONE* 3(7): e2771. doi: 10.1371/journal.pone.0002771

Paini, D. R., and J. D. Roberts. 2005. Commercial honey bees (*Apis mellifera*) reduce the fecundity of an Australian native bee (*Hylaeus alcyoneus*). *Biol. Conserv.* 123: 103–112.

Roubik, D. W. (Ed.) 1995. *Pollination of Cultivated Plants in the Tropics*. Food and Agriculture Organization of the United Nations, Rome, Italy.

Rundlöf, M., H. Nilsson, and H. G. Smith. 2007. Interacting effects of farming practice and landscape context on bumble bees. *Biol. Conserv.* 141: 417–426.

Thomson, D. M. 2006. Detecting the effects of introduced species: A case study of competition between *Apis* and *Bombus*. *Oikos* 114: 407–418.

Turner, R. C., J. J. Midgley, and S. D. Johnson. 2011. Evidence for rodent pollination in *Erica hanekomii* (Ericaceae). *Botanical J. Linnean Society* 166: 163–170.

Walther-Hellwig, K., G. Fokul, R. Frankl, R. Büchler, K. Ekschmitt, and V. Wolters. 2006. Increased density of honeybee colonies affects foraging bumblebees. *Apidologie* 37: 517–532.

Werrell, P. A., G. A. Langellotto, S. U. Morath, and K. C. Matteson. 2009. The influence of garden size and floral cover on pollen deposition in urban community gardens. *Cities and the Environment 2*. Berkely Electronic Press, http://digitalcommons.lmu.edu/cate/vol2/iss1/6

Williams, P., S. Colla, and Z. Xie. 2009. Bumblebee vulnerability: Common correlates of winners and losers across three continents. *Conserv. Biol.* 23: 931–940.

Williams, P. H., and J. L. Osborne. 2009. Bumblebee vulnerability and conservation world-wide. *Apidologie* 40: 367–387.

Winfree, R., N. M. Williams, J. Dushoff, and C. Kremen. 2007. Native bees provide insurance against ongoing honey bee losses. *Ecology Letters* 10: 1105–1113.

Winfree, R., N. M. Williams, H. Gaines, J. S. Ascher, and C. Kremen. 2008. Wild bee pollinators provide the majority of crop visitation across land-use gradients in New Jersey and Pennsylvania, U.S. *J. Appl. Ecology* 45: 793–802.

Index

Page numbers followed by f *indicate figures; numbers followed by* t *indicate tables.*

— D —

— E —

— F —

— G —

— N —

— O —

— P —

— Q —

— R —

Credits

Cover *Top left*: honeycomb, Jennifer Finley, USDA-ARS; *top center*: honey bee on corn stalk, Maryann Frazier, Penn State University; honeydripper, Jack Martin; *center left*: Diana Sammataro, USDA-ARS; *center*: honey bee on raspberry blossom, Maryann Frazier, Penn State University; *bottom center*: Diana Sammataro, USDA-ARS; *bottom right*: Bradley C. Ellis, Fresno, CA. **i** Bradley C. Ellis, Fresno, CA. **iii** Gloria DeGrandi-Hoffman, USDA-ARS. **iv, v** Diana Sammataro, USDA-ARS. **vi, vii** Maryann Frazier, Penn State University. **2** *Top*: P. Greb, USDA-ARS; *remaining*: Gloria DeGrandi-Hoffman, USDA-ARS. **40** *Top*: William Styer; *bottom*: Diana Sammataro, USDA-ARS; *illus.*: Signe Nordin. **41** Gard W. Otis and J. Kralj, used with permission. **41** Signe Nordin. **43** *A–B*: William Styer; *C*: Chris Pooley, Electron and Confocal Microscope Unit, USDA-ARS. **46–48** Diana Sammataro, USDA-ARS. **56–58** Diana Sammataro, USDA-ARS. **59** Guy Mercadier, USDA-ARS. **60** Diana Sammataro, USDA-ARS. **61** James D. Ellis, courtesy of University of Florida. **65, 69** Jay Evans, USDA-ARS. **75** *A B*: © Scott Camazine, used with permission; *C*: Yan Ping Chen, USDA-ARS; *D*: M. Ribière, French Food Safety Agency. **76** © Magali Ribière and Elsevier; used with permission. **77** © Laurent Gauthier, IBRA and BioMed Central Ltd.; used with permission. **116–118** Thomas C. Webster, Kentucky State University. **119** Signe Nordin. **120–127** Katherine Aronstein, USDA-ARS. **137, 140** James D. Ellis, courtesy of University of Florida. **153–154** Diana Sammataro, USDA-ARS. **166** Aleš Gregorc, Agricultural Institute of Slovenia. **174** Elaine C. M. Silva-Zacarin, Federal University of São Carlos. **175** M. I. Smodiš-Škerl, Agricultural Institute of Slovenia. **176** Aleš Gregorc, Agricultural Institute of Slovenia. **200** Jay Yoder, Wittenburg University. **231** Dennis vanEngelsdorp, Penn State University. **238** David Inouye, University of Maryland. **239** *Top left, right*: David Inouye; *bottom*: Maryann Frazier, Penn State University. **240** *Left, right*: David Inouye, University of Maryland; *center*: Maryann Frazier, Penn State University. **Design, Layout:** Watch This Space, Inc. **Cover Design:** Jack Martin. **Index:** Cindy Coan.

For Product Safety Concerns and Information please contact our EU representative GPSR@taylorandfrancis.com Taylor & Francis Verlag GmbH, Kaufingerstraße 24, 80331 München, Germany

T - #0322 - 160425 - C320 - 280/216/17 [19] - CB - 9781439879405 - Gloss Lamination